Quantum Statistics of Charged Particle Systems

Quantum Statistics of Charged Particle Systems

by
Wolf-Dietrich Kraeft,
Dietrich Kremp,
Werner Ebeling,
and Gerd Röpke

With 63 Figures and 10 Tables

Akademie-Verlag Berlin
1986

Authors:

Prof. Dr. Wolf-Dietrich Kraeft
Ernst-Moritz-Arndt-Universität, Greifswald
Sektion Physik/Elektronik

Prof. Dr. Dietrich Kremp
Wilhelm-Pieck-Universität, Rostock
Sektion Physik

Prof. Dr. Werner Ebeling
Humboldt-Universität zu Berlin
Sektion Physik

Dr. Gerd Röpke
Wilhelm-Pieck-Universität, Rostock
Sektion Physik

Erschienen im Akademie-Verlag Berlin, DDR-1086 Berlin, Leipziger Straße 3—4

Lizenznummer: 202 · 100/426/85
Printed in the German Democratic Republic
Gesamtherstellung: VEB Druckerei „Thomas Müntzer", 5820 Bad Langensalza
Einband und Schutzumschlag: Annemarie Wagner
LSV 1175
Bestellnummer: 7634123 (6864)

06000

Preface

The year 1985 represents a special anniversary for people dealing with Coulomb systems. 200 years ago, in 1785, Charles Auguste de Coulomb (1736–1806) found „Coulomb's law" for the interaction force between charged particles. The authors want to dedicate this book to the honour of the great pioneer of electrophysics.

Recent statistical mechanics is mainly restricted to systems of neutral particles. Except for a few monographs and survey articles (see, e. g., ICHIMARU, 1973, 1982; KUDRIN, 1974; KLIMONTOVICH, 1975; EBELING, KRAEFT and KREMP, 1976, 1979; KALMAN and CARINI, 1978; BAUS and HANSEN, 1980; GILL, 1981, VELO and WIGHTMAN, 1981; MATSUBARA, 1982) the extended material on charged particle systems, which is now available thanks to the efforts of many workers in statistical mechanics, is widely dispersed in many original articles.

It is the aim of this monograph to represent at least some part of the known results on charged particle systems from a unified point of view. Here the method of Green's functions turns out to be a powerful method especially to overcome the difficulties connected with the statistical physics of charged particle systems; some of them are mentioned in the introduction. Here we can point, e.g., to the appearance of bound states in a medium and their role as new entities.

The presentation begins (in the second Chapter) with the basic physical ideas and a short survey of the exact results known for quantum Coulomb systems. The authors — being no experts in the field of mathematical physics — believe that the exact results obtained by DYSON and LENARD (1967, 1968), LEBOWITZ and LIEB (1969, 1972), LIEB and THIRRING (1975) and many other authors (see THIRRING, 1980; VELO and WIGHTMAN, 1981) are of fundamental importance and should be explained therefore — without proofs — at the very beginning. The third Chapter is devoted to the systematic representation of the main quantum statistical methods used in this book. Chapter 4. covers the entire density-temperature plane by using the apparatus of Green's functions. There especially the single- and two-particle properties as well as the dielectric behaviour of charged particle systems are dealt with. Chapter 5. gives some information about the classical case and is further devoted to the treatment of nearly-classical (non-degenerate) plasmas. The results obtained so far are used in the sixth Chapter in order to give thermodynamic functions in wide density-temperature regions. Chapters 7. and 8. are devoted to transport and optical properties, respectively.

Finally several restrictions have to be explained. Apart from a few remarks, the material presented here is restricted to the three-dimensional case. Recently, the interest in two-dimensional and in one-dimensional conductors is increasing rapidly due to their unusual properties and to possible technological applications (GINTSBURG, 1971, 1981; ALASTUEY and JANCOVICI, 1979, 1980, 1981; ALASTUEY, 1980, 1982;

Wilson et al., 1980, 1981; Aoki and Ando, 1981; Williams, 1982; Gallet et al., 1982).

A comprehensive treatment of this quickly developing field is outside the scope of this book, the reader is refered to the original work cited at several places throughout the volume. Such references to topics beyond the scope of this monograph are added in small print.

Other problems, also related to the Coulombic interaction, which we omitted from this discussion, are treated in the same manner. We may mention here inhomogeneous systems and band structure calculations (see, e.g., Brauer, 1972, Stolz, 1974), properties of matter at extreme conditions (see, e.g., Kirzhnits et al., 1975), order-disorder transitions and delocalization effects (see, e.g., Bonch-Bruevich et al., 1981), computer simulations (see, e.g., Zamalin, Norman, and Filinov, 1977), superconductivity (see, e.g., Tinkham, 1975; Lifshits and Pitaevskii, 1980), and far-from-equilibrium effects (see, e.g., Ebeling and Feistel, 1982).

The authors hope to contribute to the theory of Coulomb systems, a vastly growing field of statistical physics, which is of high interest for science and technology. In order to help graduate students and scientific workers to become familiar with this modern area of physics the authors decided to include a part where the fundamentals of the theory are presented in the manner of a textbook; this refers especially to parts of Chapters 2.—4.

In conclusion, the authors would like to express their sincere thanks to Klaus Kilimann, Hubertus Stolz, and Roland Zimmermann, who participated substantially in the development of the results presented in this book. They contributed especially to Chapters 4. and 6.—8. The authors would like to thank further J. Blümlein, F. E. Höhne, M. Luft, C.-V. Meister, T. Meyer, I. Orgzall, R. Redmer, W. Richert, T. Rother, M. Schlanges, T. Seifert, W. Stolzmann, and H. Wegener for invaluable help and cooperation. Moreover, we express our gratitude for discussions and for providing us with material to H. Böttger, V. E. Fortov, B. Jancovici, G. Kalman, G. Kelbg, Yu. L. Klimontovich, K. Suchy, I. T. Yakubov, and D. N. Zubarev.

H. Stolz and H. Böttger read the entire manuscript and made many useful comments and suggestions. Our thanks are moreover due to Lindsay Mann, Edinburgh, who checked the English version of the manuscript. For technical assistance in the preparation of the manuscript we are grateful to H. Bahlo, C. Berndt, R. Nareyka, D. Rosengarten, and E. Wendt. Finally, our cordial thanks are due to Renate Trautmann, Akademie-Verlag Berlin, for excellent editorial cooperation.

Contents

1. Introduction

Matter consists of charged particles, the negative electrons and the positive nuclei; this fundamental discovery, due to Rutherford, belongs to the mile-stones in the history of physics. The famous communication of Rutherford "Scattering of α- and β-rays and the structure of atoms" was presented on April 7, 1911 to the Philosophical Society in Manchester.

Using the simple picture that matter consists of nonrelativistic point charges, the founders of quantum mechanics and quantum statistics developed the idea that quantum physics is able now to calculate most properties of particle systems from first principles. Having in mind the properties of atoms and molecules DIRAC wrote in 1929: "The underlying physical laws necessary for the mathematical theory of a larger part of physics and the whole of chemistry are thus completely known, and the difficulty is only that the application of these laws leads to equations much too complicated to be soluble". The optimistic program of the pioneers of quantum physics was not easy to realize since enormous mathematical difficulties had to be overcome. As a result of their great efforts, we now have a well-developed quantum chemistry which is able to calculate even the properties of quite complicated molecules, although of course there are still many open problems (LUDWIG, 1980; PRIMAS, 1981). Furthermore, we have at least the fundamentals of a quantum statistics of macroscopic matter consisting of very many nuclei and electrons. Here we are concerned with the latter subject only, the theory of macroscopic charged particle systems. Of course, a theory of macroscopic matter based on first principles only is of fundamental interest for our understanding of the properties of matter. In order to underline the importance of the quantum statistics of Coulomb systems let us quote LEBOWITZ (1980): "In some sense, all of statistical mechanics, which is the microscopic theory of macroscopic matter, deals with Coulomb systems. The properties of the materials we see and touch are almost entirely determined by the nature of the Coulomb force as it manifests itself in the collective behaviour of interacting electrons and nuclei. In most applications of statistical mechanics however this fact is not explicit at all. One starts with an effective short range microscopic Hamiltonian appropriate to the problem at hand." For example, in order to discuss the properties of a real gas we describe the system as a collection of neutral molecules interacting via Lennard-Jones potentials. There are two reasons for considering explicitly systems with a starting microscopic Hamiltonian with Coulomb interactions. The first is primarily a theoretical one; we would like to understand how the Coulomb forces give rise to the effective interactions, which appear in the usual statistical mechanics, and we would like to have a theory which is based on first principles only. The second reason is more practical. There are many systems, e.g., plasmas, metals, semiconductors, electrolytes, molten salts, ionic crystals, etc., where bare Coulomb interactions play a very important role. In this way the quantum statistics of Coulomb systems is able to predict the thermodynamical, optical, and transport properties for systems of high practical importance, e.g., states of dense matter with very great energy density. Such states are of great importance for many modern technical devices such as nuclear fusion plants, gas-phase fission reactors, plasma-chemical devices, etc.

In this way it is obvious that the development of a consistent statistical mechanics of systems of point charges is of principal and practical interest, and it is of special importance to deal with the specific Coulomb difficulties, such as the instability (unboundedness of the energy) of a classical system of point charges; the long range character of the Coulomb forces; and the divergence of the atomic partition function.

One of the most elegant and powerful techniques to overcome such difficulties is the method of quantumstatistical thermodynamic Green's functions (MARTIN and SCHWINGER, 1959; BONCH-BRUEVICH and TYABLIKOW, 1961; KADANOFF and BAYM, 1962; ABRIKOSOV, GOR'KOV and DZYALOSHINSKII, 1962; FETTER and WALECKA, 1971). The main part of this book is based on this technique. The great advantage of such an approach is that it covers, at least in principle, the entire region of density and temperature. Other methods describe — as a rule — only a certain part of the density temperature plane.

There is for instance, on the one hand, the well developed theory of highly degenerate electron liquids (see, e.g., ABRIKOSOV, GOR'KOV and DZYALOSHINSKII, 1962; KADANOFF and BAYM, 1962; PINES, 1962; NOZIÈRES, 1966; GLICKSMAN, 1971; STOLZ, 1974; ZIMAN, 1974; LIFSHITS et al., 1975; ABRIKOSOV, 1976; RICE et al., 1977; HERRMANN and PREPPERNAU, 1979). On the other hand, there exist theories of non-degenerate plasmas (BALESCU, 1963; KLIMONTOVICH, 1964, 1975; SILIN, 1971; ECKER, 1972; KADOMTSEV, 1976; GRYAZNOV et al., 1980; KHRAPAK and YAKUBOV, 1981; ZHDANOV, 1982; EBELING et al., 1983, 1984; GÜNTHER and RADTKE, 1984).

A unified Green's functions approach covers, as already mentioned, the entire density-temperature plane; of course, one still needs different approximation schemes in different regions.

The program to be dealt with by the theory is the following: Starting from the elementary units (electrons and nuclei) and their equation of motion (quantum mechanics and Coulomb's interaction law), it is to obtain the properties of matter under different external conditions (temperature and pressure). This includes the derivation of plasma, gaseous, fluid or solid states under certain conditions and the transition between them.

At the present stage of the theory such a wide program seems to be unrealistic, of course; therefore we have to restrict ourselves here to certain aspects of the program.

Some of the material given in this book is new and is of interest for theoretical and practical reasons. Here we mention

(i) the quantum-statistical treatment of two-particle properties as, e.g., energy levels and the life time of two-particle states in dependence on density and temperature

(ii) the foundation of the chemical picture (i.e., bound states are treated as new species) and its application in thermodynamics and transport phenomena;

(iii) formulae and numerical results for thermodynamic functions in wide regions of density and temperature;

(iv) formulae and numerical results for electrical and thermoelectrical properties covering degenerate and non-degenerate plasmas;

(v) formulae and numerical results for optical properties especially for line shift and shape.

The contents of Chapters 6.—8. represent some applications using the general methods of quantum statistics. The material given may serve as a basis for further developments.

2. Physical Concepts and Exact Results

2.1. Basic Concepts for Coulomb Systems

One of the simplest Coulomb systems which already shows all the qualitative effects of interest is the two-component charge-symmetrical system consisting of N_e electrons with the charge $(-e)$ and N_i positive ions with the charge $(+e)$; the masses are m_e and m_i respectively. We note that in some respect the unsymmetrical pseudo-one-component plasma (OCP), where one of the charged species is uniformly distributed over the space, is in some respect even still more simple and represents a very useful model system (BAUS and HANSEN, 1980; ICHIMARU, 1982). However since we are also interested here in bound states which do not exist in the OCP we prefer to start here from a charge-symmetrical system, the particles of which obey Fermi statistics. The most simple representatives of this kind of systems are the electron-positron plasmas or special electron-hole plasmas where the masses of electrons and ions are equal, i.e. $m_i = m_e$. More important are of course plasmas with certain asymmetry with respect to the masses $m_i \gg m_e$. Of special importance are hydrogen plasmas (H-plasmas) with $m_i = 1836.3\, m_e$. We know that a large part of our metagalaxis consists of hydrogen in the plasma state. Furthermore we may think about charge-symmetrical alkali plasmas or noble gas plasmas. However, in this case the Coulomb potential has to be modified (pseudopotential). The charged particles are considered in the following as nonrelativistic points with the masses m_e and m_i, respectively, which are moving according to the laws of quantum mechanics. Short range interaction effects as well as radiation effects are not taken into account. Let the plasma be fully ionized, enclosed in a box of volume V and in thermal equilibrium with a heat bath of temperature T. Assuming that the free energy is given by

$$F(T, V, N_e, N_i)$$

we get for the internal energy U, the pressure p, the chemical potentials μ_k

$$U = F - T\left(\frac{\partial F}{\partial T}\right)_{V,N}, \qquad p = -\left(\frac{\partial F}{\partial V}\right)_{T,N}$$
$$\mu_e = \left(\frac{\partial F}{\partial N_e}\right)_{T,V,N_i}, \qquad \mu_i = \left(\frac{\partial F}{\partial N_i}\right)_{T,V,N_e} \tag{2.1}$$

Let us split the free energy into an ideal part and an interaction part:

$$F = F^{\mathrm{id}} + F^{\mathrm{int}}.$$

In the nondegenerate case

$$n_e\Lambda_e^3 \ll 1, \quad n_i\Lambda_i^3 \ll 1, \quad \Lambda_k = h[2\pi m_k k_B T]^{-1/2} \tag{2.2}$$

the ideal contribution is given by

$$F^{\mathrm{id}} = N_e k_B T[\ln (n_e \Lambda_e^3) - 1] + N_i k_B T[\ln (n_i \Lambda_i^3) - 1] \,. \tag{2.3}$$

We note that nondegeneracy of a species k means that the mean kinetic energy is far above the corresponding Fermi energy:

$$E_k^{\mathrm{kin}} \gg E_k^{\mathrm{F}} = \frac{\hbar^2}{2m_k}\left(\frac{6\pi^2}{2s_k + 1}\right)^{2/3} N_k n_k^{2/3} \,.$$

The contribution F^{int} is due to the pair interaction potentials

$$V_{ee} = V_{ii} = \frac{e^2}{4\pi\varepsilon_0 r}, \qquad V_{ei} = V_{ie} = -\frac{e^2}{4\pi\varepsilon_0 r} \,. \tag{2.4}$$

We note that the Coulomb potential is of infinite range; however we may define a length

$$l = (e^2/4\pi\varepsilon_0 k_B T) \tag{2.5}$$

which characterizes the distance where the Coulomb potential gets smaller than the thermal energy

$$|V_{ab}(l)| = k_B T \,.$$

Another characteristic length is the mean distance between two electrons or two ions (protons) respectively

$$d = d_e = d_i \,, \qquad d_e = \left(\frac{3}{4\pi n_e}\right)^{1/3}, \qquad d_i = \left(\frac{3}{4\pi n_i}\right)^{1/3} \,. \tag{2.6}$$

This length is usually referred to as the Wigner-Seitz radius. Further we define the thermal De Broglie wave lengths of relative motion

$$\lambda_{ii} = \frac{\hbar}{\sqrt{m_i k_B T}}, \qquad \lambda_{ee} = \frac{\hbar}{\sqrt{m_e k_B T}} = \frac{\Lambda_e}{\sqrt{2\pi}} \,.$$

By using these lengths we may define interaction parameters ξ of the electrons and ions and a nonideality parameter Γ by

$$\xi_e = \frac{l}{\lambda_{ee}}, \qquad \xi_i = \frac{l}{\lambda_{ii}}, \qquad \Gamma = \frac{l}{d} = \frac{e^2}{4\pi\varepsilon k_B T} \,. \tag{2.7}$$

We note that Γ is a pure classical parameter and ξ_e and ξ_i are quantum parameters which are connected with the ionization energy

$$I = \frac{m_{ei} e^4}{(4\pi\varepsilon)^2 2\hbar^2}, \qquad m_{ei} = \frac{m_e m_i}{m_e + m_i},$$

by the relations

$$I = \tfrac{1}{4} k_B T(\xi_e^{-2} + \xi_i^{-2})^{-1} \approx \tfrac{1}{4} k_B T \xi_e^2, \qquad \xi_e \approx 2[I/k_B T]^{1/2} \,. \tag{2.8}$$

The relations (2.5)—(2.8) may easily be generalized to the degenerate case if the mean kinetic energy is taken from the theory of ideal Fermi gases:

$$k_B T \to \theta_k = \frac{2k_B T}{n_k \Lambda_k^3} I_{3/2}(\alpha_k) \,. \tag{2.9}$$

Here θ_k denotes the kinetic energy per two degrees of freedom and further α_k is given by the chemical potential of the ideal electrons or ions, respectively, at the temperature and density in question.

$$n_k\Lambda_k^3 = 2I_{1/2}(\alpha_k)$$
$$\alpha_k = \beta\mu_k^{\text{id}} = I_{1/2}^{-1}(\tfrac{1}{2}\, n_k\Lambda_k^3) \,. \tag{2.10}$$

Further the $I_n(x)$ are the Fermi functions defined by

$$I_\nu(y) = \sum_{r=1}^{\infty} (-1)^{r+1} \frac{\exp(ry)}{r^{\nu+1}}, \qquad \nu > (-1) \,. \tag{2.11}$$

In this way we get, e.g.,

$$\xi_e = \frac{e^2\sqrt{m_e}}{4\pi\varepsilon\hbar\sqrt{\theta_e}}, \qquad \xi_i = \frac{e^2\sqrt{m_i}}{4\pi\varepsilon\hbar\sqrt{\theta_i}}, \qquad \Gamma_e = \frac{e^2}{8\pi\varepsilon d\theta_e}, \qquad \Gamma_i = \frac{e^2}{8\pi\varepsilon d\theta_i} \,. \tag{2.12}$$

In Figs. 2.1 and 2.2 the regions in the density-temperature plane are shown where the parameters ξ_e, Γ_e and Γ_p are larger than one. Physically $\Gamma_k \gtrapprox 1$ means that the mean potential energy is larger than the mean kinetic energy and $\xi_k \gtrapprox 1$ means that

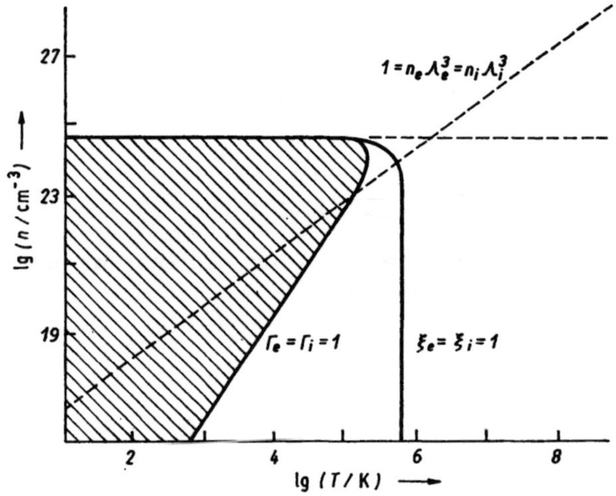

Fig. 2.1. The region of nonideality in the density-temperature plane for symmetrical plasmas (electron-positron plasmas)

the interactions are essential for the microscopic scattering processes, i.e. the perturbation theory (Born series) breaks down. It is interesting to note that the interaction region as well as the nonideality region are bounded to the left lower corner in the $n-T$-plane. In this respect high density and high temperatures have the same qualitative effect: interactions and nonideality break down, the kinetic energy only governs the behaviour. The reason for this somewhat surprising effect is of course the strong increase of the kinetic energy of Fermi gases at high temperatures and/or high densities; we shall come back to this important effect in the sections 2.3., 6.1.—6.4 Here let us further note that the upper border is approximately given by the so-called Brueckner parameter:

$$r_s = \frac{d}{a_B}, \qquad r_s \approx 0.9\xi_e \approx 0.7\,\Gamma_e$$

(a_B — Bohr radius). At very high densities where the mean distance between the electrons becomes smaller than the Bohr radius, interactions loose their significance, and they may be neglected. Interactions are restricted to a small island in our world, however just this very island is of high importance since interactions provide cooperations, and cooperations are responsible for all structure and organisation in the universe (ZELDOVICH and NOVIKOV, 1975; WEINBERG, 1977; NICOLIS and PRIGOGINE, 1977, HAKEN 1978; EBELING and FEISTEL, 1982).

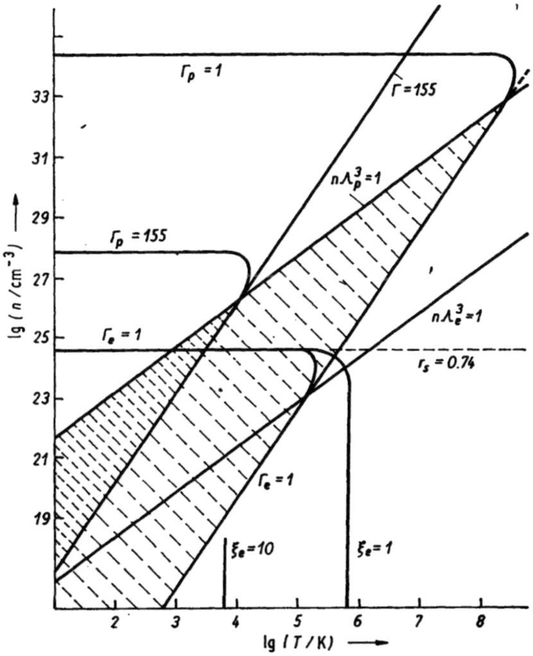

Fig. 2.2. The region of nonideality in the density-temperature plane for hydrogen plasmas

The influence of quantum effects strongly depends on the masses of the particles. Therefore the n—T-plane of mass-symmetrical systems (Fig. 2.1) has a much simpler structure than that of hydrogen plasmas (Fig. 2.2). Looking at Fig. 2.2 we see that the high-density behaviour of electrons and ions is completely different; this is due to their different masses $m_p = 1836.3\, m_e$. The lighter electrons reach degeneracy at a density which is by a factor

$$(m_e/m_p)^{3/2} = 1.27 \cdot 10^{-5}$$

lower than the density of proton degeneracy. There exists therefore an intermediate region

$$\Lambda_e^{-3} < n < \Lambda_p^{-3},$$

where the protons behave classically but the electrons form a degenerate quantum gas. That is the region where we may expect the formation of an ionic lattice or eventually of a molecular lattice (KRANENDONK, 1982). We note that the competition between both types of lattices depends on the binding energy of atoms and molecules. At lower densities binding may be stronger than degeneracy i.e. the electrons are fixed to protons; at higher densities however the degeneracy will be so strong that collectivization of the electrons occurs. In other words we expect the formation of hydrogen metal consisting of a proton lattice immersed in a uniform electron sea (MATSUBARA, 1982).

The crystallization of one-component charged particles immersed in a uniform neutralizing background is called Wigner-crystallization. The lattice formed by the protons is of the *bbc*-type. POLLOCK and HANSEN (1973) carried out extensive Monte Carlo calculations for *bbc* lattice configurations and found that Wigner crystallization takes place at

$$\Gamma_{\mathrm{m}} = 155 \pm 10 \, .$$

Subsequently SLATTERY, DOOLEN and DE WITT (1980) performed improved Monte Carlo calculations and found a somewhat higher transition point near to 170. For an estimate we have used the Pollock-Hansen value which yields the crystallization line shown in Fig. 2.2. We note that a *bcc* lattice corresponds to the potential energy

$$U^{\mathrm{pot}} = -\frac{1}{2} N_{\mathrm{i}} \cdot \alpha_{\mathrm{M}} \left(\frac{e^2}{4\pi\varepsilon d} \right) \tag{2.13}$$

where α_{M} is the Madelung constant. Another form of writing is

$$U^{\mathrm{pot}} = -(N_{\mathrm{i}} \cdot k_{\mathrm{B}} T)\, 0.895925\, \Gamma \, .$$

Wigner crystallization in a two-dimensional plasma has been studied experimentally (GRIMES and ADAMS, 1979) and theoretically (HOCKNEY and BROWN, 1975; KALIA et al., 1981). Experimentally the existence of an electron solid was observed on a liquid helium surface. FISHER, HALPERIN and PLATZMAN (1979) analyzed the experiment and found that the results can be explained by a Wigner lattice formed below a transition point $\Gamma_m = 137 \pm 15$ which is in good agreement with the Monto Carlo result $\Gamma_m = 125 \pm 15$ (GANN et al., 1979).

Let us now say a few words about screening effects. Considering nondegenerate electrons in the vicinity of an ion which is held fixed we may assume that the mean electron density in a distance r is given by a Boltzmann distribution

$$n_{\mathrm{ei}}(r) = n_{\mathrm{e}} \exp \left[+ e\psi(r)/k_{\mathrm{B}} T \right] \tag{2.14}$$

where $\psi(r)$ is the mean potential in distance r which may be related to the Poisson equation

$$\Delta\psi(r) = -\frac{1}{\varepsilon} \left(e n_{\mathrm{ii}}(r) - e n_{\mathrm{ei}}(r) \right) \tag{2.15}$$

where

$$n_{\mathrm{ii}}(r) = n_{\mathrm{i}} \exp \left(-e\psi(r)/k_{\mathrm{B}} T \right) .$$

is the mean ionic density around the central (fixed) ion. Linearizing with respect to $(e\psi/k_{\mathrm{B}}T)$ we get

$$\Delta\psi = +\frac{1}{\varepsilon} \left[n_{\mathrm{e}} e (1 + e\psi/k_{\mathrm{B}}T) - n_{\mathrm{i}} e (1 - e\psi/k_{\mathrm{B}}T) \right] = \varkappa^2 \psi \, ,$$

$$\varkappa^2 = \frac{1}{\varepsilon k_{\mathrm{B}} T} \left(n_{\mathrm{e}} e^2 + n_{\mathrm{i}} e^2 \right) .$$

Under appropriate boundary conditions the solution is given by

$$\psi(r) = \frac{e}{4\pi\varepsilon r} \exp(-\varkappa r) . \tag{2.16}$$

In this way we get the average electron density

$$n_{ei}(r) = n_e \exp\left(\frac{e^2}{4\pi\varepsilon k_B T r} \exp(-\varkappa r)\right). \tag{2.17}$$

For short distances this simple classical calculation is of course unrealistic. Qualitatively however it represents the effect of screening for nondegenerate plasmas in a correct way. Due to the influence of the other charges the interaction region is restricted to distances smaller than the Debye radius

$$r_D = \varkappa^{-1} = \frac{1}{e}\sqrt{\frac{\varepsilon k_B T}{2n_i}} .$$

The next effect we want to discuss briefly is the formation of bound states. The attractive interaction between electrons and ions

$$V_{ei}(r) = -\frac{e^2}{4\pi\varepsilon r}$$

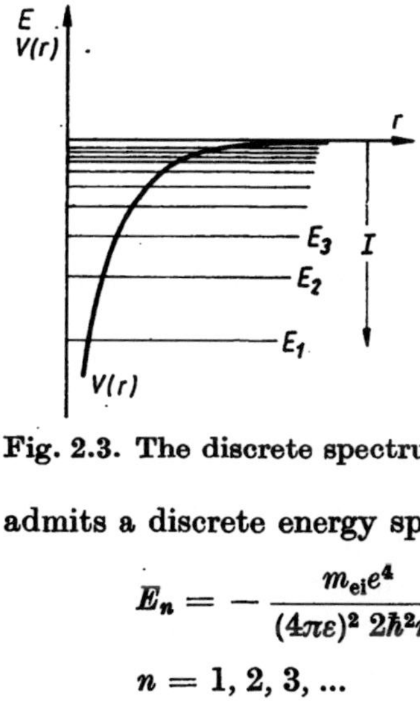

Fig. 2.3. The discrete spectrum of the relative energy of an electron-ion pair

admits a discrete energy spectrum (Fig. 2.3.)

$$E_n = -\frac{m_{ei}e^4}{(4\pi\varepsilon)^2\, 2\hbar^2 n^2}, \qquad m_{ei} = \frac{m_e m_i}{m_e + m_i}, \tag{2.18}$$

$$n = 1, 2, 3, \ldots$$

The corresponding wave functions are

$$\psi_{nlm}(r, \vartheta, \varphi) = R_{nl}(\varrho)\, Y_{em}(\vartheta, \varphi) ,$$

$$\varrho = r/a_B , \quad a_B = (4\pi\varepsilon\hbar^2)/\, m_{ei}e^2 ,$$

$$R_{nl}(\varrho) = C_{nl}\varrho^l \exp(-\varrho/n)\, {}_1F_1(l + 1 - n;\; 2l + 2;\; 2\varrho/n) \tag{2.19}$$

where ${}_1F_1$ are the confluent hypergeometric functions and Y_{lm} the spherical functions. The ground state is given by

$$E_{10} = -\frac{e^2}{(4\pi\varepsilon)\, 2a_B}, \qquad R_{10}(\varrho) = C_{10} \exp(-\varrho) .$$

The numerical value of the ground state energy of hydrogen is 13.6 eV. The binding of two atoms into an hydrogen molecule may give the additional binding energy of about 4.5 eV.

We note that there exist other bound states which correspond to charged complexes, e.g., H^- (binding energy: 0.92 eV) and H_2^+ (binding energy: 2.7 eV).

So far we have considered the bound states of pairs, triples, quadruples of charged particles imbedded into a vacuum. In a real plasma any bound complex has neighbours which have a disturbing influence. This effect will be considered in detail in Chapters 4.—7. Here we shall give some qualitative considerations only. In order to study the influence of screening on the ground state of the hydrogen atom we have to solve the eigenvalue problem for the screened potential

$$\frac{\hbar^2}{2m_e}\frac{d^2}{dr^2}(rR_{10}) + \left[\tilde{E}_{10}(r_D) + \frac{e^2}{4\pi\varepsilon r}\exp\left(-\frac{r}{r_D}\right)\right](rR_{10}) = 0 \,. \tag{2.20}$$

The result of numerical calculations (Rogers et al., 1970; Lam and Varshni, 1971; Bühring, 1977; Bessis et al., 1978; Green et al., 1981, 1982), is shown in Fig. 2.4. One sees that $\tilde{E}_{10}(r_D)$ converges to zero with increasing density (Mott effect).

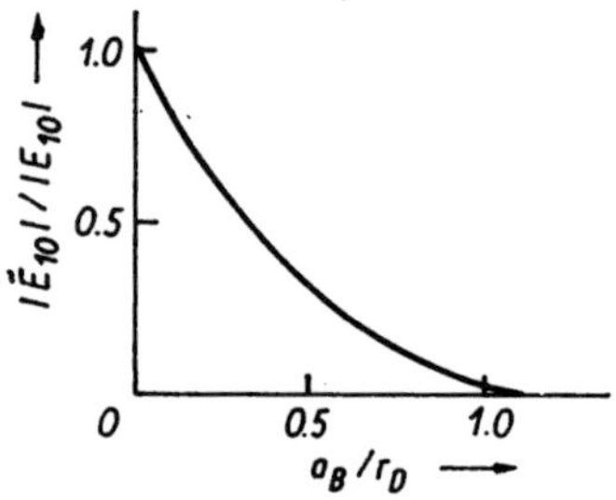

Fig. 2.4. The dependence of the ground state energy on the reciprocal Debye length

At densities with

$$r_D \lessapprox 0.84 a_B \tag{2.21}$$

the ground state merges into the continuum. Therefore atomic bound states may exist only at low or moderate densities. The border of atomic bound states is given by the Mott effect according to eq. (2.21). The border for the existence of molecular bound states is given by atomic scattering effects. If we denote the scattering length by a_S, which indicates the beginning of the repulsive region of the H—H-interaction, then we may expect the H_2 molecules to break up at mean distances

$$d \lessapprox a_S \,.$$

The scattering length for H atoms may be estimated to be $a_S \approx a_B$; therefore we get the border

$$d = a_B \,, \qquad r_S = 1 \,.$$

The upper limit of the formation of molecules is, as a rough estimate, given by the border of electron nonideality. Wigner and Huntington (1935) have shown already that the molecular hydrogen crystal at high density becomes unstable above $r_S \approx 1$ and changes into the metallic phase (Ashcroft, 1968, 1982; Schneider, 1969; Ginzburg, 1971, 1981; Matsubara, 1982).

The last effect we shall discuss briefly here, is the Anderson effect, the delocalization due to long-range order. As is well known the electrons moving in a perfect crystal are collectivized (see, e.g., MOTT and DAVIS, 1971, 1974; ZIMAN, 1974, 1979; ABRIKOSOV, 1976). Energy bands are formed which correspond to delocalized states of the electrons. With increasing disorder one observes the appearance of band tails corresponding to localized states of the electrons (Fig. 2.5). The one-particle energy corresponding to the transition between localized and delocalized states is called the mobility edge. The transition dielectric-metal which is observed if the Fermi energy crosses the mobility edge is called the Anderson transition (ANDERSON, 1958; THOULESS, 1974; EFROS 1978; ABRAHAMS et al., 1979; SADOYSKY, 1981; BONCH-BRUEVICH et al., 1981; ZIMAN, 1979, 1982; LIFSHITS et al. 1982; BÖTTGER 1983; NAGAOKA and FUKUYAMA, 1983).

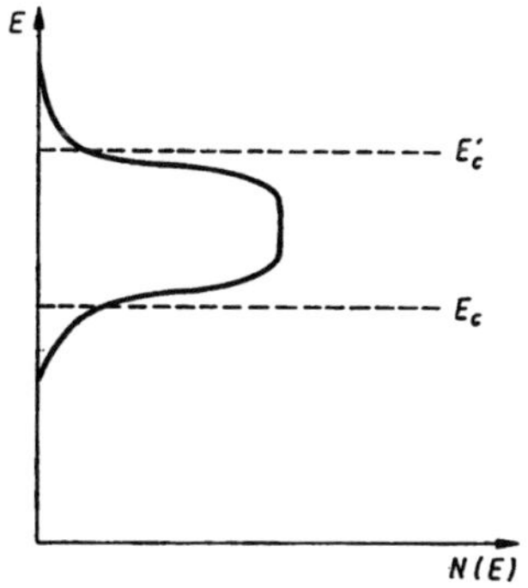

Fig. 2.5. Energy bands corresponding to extended states and band tails corresponding to localized states
E_c, E'_c — mobility edges

Summarizing we may state that the electrons may be trapped in bound states in the interaction region $\xi \gtrapprox 1$ only. However there are still further restrictions which are given by the Mott effect (delocalization due to screening) and by the Anderson effect (delocalization due to long-range order). Due to these effects the bound-state region may be much smaller than the interaction region.

Up to now we have assumed that we consider charges distributed in a three-dimensional space. However charges may be bound to a plane or to a line. Two-dimensional systems of particles interacting through a Coulomb potential ($e^2/4\pi\varepsilon r$) are realized in the laboratory, e.g., by trapping electrons at the surface of liquid helium (GRIMES, 1978; WILLIAMS, 1982; VALDESH, 1982; GALLET et al. 1983). These electron sheets on liquid helium are in the range of validity of classical stastistical mechanics. Quantum mechanical two-dimensional systems are formed, if electrons in a solid are trapped in z-direction within a narrow potential well but move freely within a perpendicular plane ($x-y$-plane). The potential must be so narrow that the resulting energy eigenvalues for the motion in z-direction are so well separated that the kinetic energy is smaller than the distance between the bands. Such an electron system can be formed at interfaces between two semiconductors (DINGLE, 1975; ESAKI and CHANG, 1976) or at the insulator-semiconductor interface of a metal-oxide-semiconductor (MOS) structure (FOWLER et al., 1966). Such quasi two-dimensional electron systems are in general in a strongly quantum regime (FUKUYAMA, 1976; WILSON et al., 1981; VALDESH, 1982). The classical system has a first-order solid-fluid phase transition at $\Gamma \cong 125$ as mentioned above (GANN et al., 1979; FISHER, HALPERIN and PLATZMAN, 1979). For the fluid phase extensive analytical work (ALASTUEY and JANCOVICI,

1979, 1980, 1981; CHACRAVARTY and DASGUPTA, 1980; SJÖGREN, 1980) and computer simulations (GANN et al. 1979; HANSEN et al., 1979; KALIA et al., 1981) have been done. The thermodynamical function, the pair correlations, the dynamic structure factor, the current correlation functions, the velocity autocorrelation function, the diffusion coefficient and many other properties have been studied in those papers. Taking into account the quantum effects the phase diagram may be studied in the whole density-temperature plane (CEPERLEY, 1978; STUDART and HIPOLITO, 1979, 1980; ALASTUEY, 1980, 1982; MEHROTRA et al. 1982). Further the effects of magnetic field have been investigated; a remarkable quantized Hall effect allows a new very accurate determination of the value of the fine-structure constant (KLITZING et al., 1980; KLITZING, 1981).

Another case which has been studied extensively are systems with $\ln r$-interactions which correspond to the forces between charged wires. Many results have been obtained for one-component plasmas (DEUTSCH and LAVAUD, 1974; DEUTSCH, DE WITT and FURUTANI, 1979; BAKSHI et al., 1979, 1981; JANCOVICI, 1981, 1982; SMITH, 1982) and for two-component plasmas (KOSTERLITZ and THOULESS, 1973; KOSTERLITZ, 1974; VILLAIN, 1975; FRÖHLICH and SPENCER, 1981). In the one-dimensional case also two variants for the interaction potential are of interest: the true Coulomb potential corresponding to charges located at polymers etc. and further the r-potential corresponding to the interaction of parallel charged plates (LIEB and MATTIS, 1966; EMERY, 1979; AIZENMAN and MARTIN, 1980; FISCHBECK, 1982; BERNASCONI and SCHNEIDER, 1981). For compouter experiments see FEIX (1978).

We would like to mention that the more general treatment of systems of charged particles should be based on a quantum-electrodynamical approach. Only in such a theory radiative processes are described in a correct manner. The quantum electrodynamical approach is usually applied only to the special case $T = 0$; see, e.g., AKHIEZER and BERESTETSKII (1962). In many cases, the radiation field is coupled to the many particle system in a semiclassical way, see, e.g., KLIMONTOVICH (1980, 1982). In the present book the interaction between the charged particles is assumed to be nonrelativistic (instantaneously); and only in Chapter 8. the coupling to the radiation field is taken into account in an approximative manner.

2.2. Survey of Exact Quantum-Mechanical Results for Coulomb Systems

Let us consider a system consisting of N_e electrons with the charge

$$e_0 = -e$$

(e — proton charge) and M nuclei with the charges

$$e_k = +z_k e$$

and the Hamiltonian

$$H = T + U ,$$

$$T = \sum_{l=1}^{N_e} \frac{p_l^2}{2m_e} + \sum_{n=1}^{M} \frac{p_k^2}{2m_k}, \quad p_k = \frac{\hbar}{i} \nabla_k , \quad p_l = \frac{\hbar}{i} \nabla_l .$$

There are two limiting cases where the quantum mechanical problem is exactly solvable. The first refers to the one-electron atom $N_0 = 1$, $M = 1$. Here one finds, e.g., for the ground state energy and for the excited levels,

$$E_0 = -\frac{m_{ek} e^4 Z^2}{2\hbar^2 (4\pi\varepsilon_0)^2}, \quad E_{nl} = -\frac{m_{ek} e^4 Z^2}{2\hbar^2 n^2 (4\pi\varepsilon_0)^2}, \quad m_{ek} = \frac{m_e m_k}{m_e + m_k}. \tag{2.22}$$

Furthermore we have a lot of exact results for the scattering states; e.g., phase shifts, Jost functions etc. (LANDAU and LIFSCHITZ, 1973; EBELING, KRAEFT and KREMP, 1976, 1979).

The second case where exact results are available is the limit of "large atoms", i.e., $M = 1$, $N_e \to \infty$. Here the Hamiltonian reduces to

$$H = \sum_{l=1}^{N_e} \left(\frac{p_l^2}{2m_e} - \frac{Ze^2}{4\pi\varepsilon_0 |r_l|} \right) + \frac{1}{2} \sum_{i,j} \frac{e^2}{(4\pi\varepsilon_0) |r_i - r_j|} . \tag{2.23}$$

It has been proved that the Thomas-Fermi theory yields exact results for the limit $N_e \to \infty$. In order to characterize these results we introduce the density in the phase space of the electrons

$$\varrho(\boldsymbol{r}, \boldsymbol{p}) = \left[\exp \beta \left\{ \frac{p^2}{2m_e} - \frac{Ze^2}{4\pi\varepsilon_0 |\boldsymbol{r}|} + \int \mathrm{d}r' \frac{e^2 n(\boldsymbol{r}')}{4\pi\varepsilon_0 |\boldsymbol{r} - \boldsymbol{r}'|} - \mu \right\} + 1 \right]^{-1} , \tag{2.24}$$

$\beta = 1/k_B T$, $n(r) = \int \varrho(\boldsymbol{r}, \boldsymbol{p}) \, \mathrm{d}\boldsymbol{p}/(2\pi\hbar)^3$.

In the limit of zero temperature $T \to 0$ follows

$$n(\boldsymbol{r}) = \frac{1}{3\pi^2} |\Phi(\boldsymbol{r}) + \mu|_+^{3/2}, \qquad |f|_+ = |f| \, \theta(f) ,$$

$$\Phi(\boldsymbol{r}) = \frac{Ze^2}{4\pi\varepsilon_0 |\boldsymbol{r}|} - \frac{e^2}{4\pi\varepsilon_0} \int \mathrm{d}\boldsymbol{r}' \frac{n(\boldsymbol{r}')}{|\boldsymbol{r} - \boldsymbol{r}'|} . \tag{2.25}$$

Applying the Laplace operator follows

$$\Delta\Phi(\boldsymbol{r}) = -\frac{1}{\varepsilon_0} Ze^2 \, \delta(\boldsymbol{r}) + \frac{4}{3\pi} |\Phi(\boldsymbol{r}) + \mu|_+^{3/2} . \tag{2.26}$$

In this way we obtained a closed equation for the potential, the so called Thomas-Fermi equation. The energy is a functional of the density (Thomas-Fermi functional)

$$\begin{aligned} E(\{n(\boldsymbol{r})\}) &= K(n) - A(n) + R(n) , \\ K(n) &= \tfrac{3}{5} (2\pi^2)^{2/3} \int \mathrm{d}\boldsymbol{r} \, n^{5/3}(\boldsymbol{r}) , \\ A(n) &= Z(e^2/4\pi\varepsilon_0) \int \mathrm{d}\boldsymbol{r} \, n(\boldsymbol{r})/|\boldsymbol{r}| , \\ R(n) &= \tfrac{1}{2} (e^2/4\pi\varepsilon_0) \int \mathrm{d}\boldsymbol{r} \, \mathrm{d}\boldsymbol{r}' \, n(\boldsymbol{r}) \, n(\boldsymbol{r}')/|\boldsymbol{r} - \boldsymbol{r}'| . \end{aligned} \tag{2.27}$$

It has been proved that the Thomas-Fermi energy gives a lower bound to the exact energy

$$E_{TF} \leqq E . \tag{2.28}$$

In the limit $N_e \to \infty$ equality holds and therefore in this limit the Thomas-Fermi theory is correct (THIRRING, 1980). We note that the Thomas-Fermi theory gives not only correct limiting results but also practical results of high importance for the modelling of matter at high temperatures and high density (KIRSHNITS et al., 1975).

Let us now consider the case of real matter, i.e.

$$M \gg 1, \qquad N_e \gg 1 .$$

Due to the large asymmetry of the masses

$$m_k \gg m_e$$

it is possible to apply the Born-Oppenheimer approximation (BOA), i.e. we split the Hamiltonian into two parts:

$$H = H_\infty + \sum_{l=1}^{M} \frac{p_k^2}{2m_k} . \tag{2.29}$$

The Born-Oppenheimer approximation means the following procedure: First we solve the quantum-mechanical problem for nuclei with infinite masses fixed at $\boldsymbol{R}_1, \dots, \boldsymbol{R}_M$:

$$H_\infty \psi = E(\boldsymbol{R}_1, \dots, \boldsymbol{R}_M)\, \psi \,. \tag{2.30}$$

Due to the assumption $m_k/m_e \to \infty$ the wavefunction ψ factorizes, and one gets the energy levels of the electrons, where the positions of the nuclei appear as parameters. Afterwards the electron energies are used as potentials for the motion of the nuclei itself. The ground state energy function $E(\boldsymbol{R}_1, \dots, \boldsymbol{R}_M)$ is decreasing with the distance $|\boldsymbol{R}_k - \boldsymbol{R}_l|$ of the nuclei, i.e. in the ground state the electrons would like to have the nuclei together (THIRRING, 1980). However the collapse of the nuclei is prevented by the Coulomb repulsion of the nuclei. In this way a stable configuration of the whole system is formed. We shall see that in infinite systems the Fermi character of the electrons is absolutely necessary to prevent the collapse of the system, as shown first by DYSON and LENARD (1967) in a fundamental paper.

Let us consider now the ground state energy in the BO-approximation. From dimensional reasoning we expect

$$E_0(\boldsymbol{R}_1, \dots, \boldsymbol{R}_M) = m_e e^2 C(\boldsymbol{R}_1, \dots, \boldsymbol{R}_M)/\hbar^2 \tag{2.31}$$

where the function C does not depend on the masses and the charges. By differentiating

$$E_0(\boldsymbol{R}_1, \dots, \boldsymbol{R}_M) = \langle \psi_0 | H_\infty | \psi_0 \rangle = m e^4 c/\hbar^2$$

with respect to e^2 we find

$$e^2 \frac{\partial}{\partial e^2} E_0(\boldsymbol{R}_1, \dots, \boldsymbol{R}_M) = \left\langle \psi_0 \left| \frac{1}{2} \sum_{ij} \frac{e^2}{4\pi\varepsilon_0 |\boldsymbol{r}_i - \boldsymbol{r}_j|} - \sum_{ik} \frac{Z_k e^2}{4\pi\varepsilon_0 |\boldsymbol{r}_i - \boldsymbol{R}_k|} \right| \psi_0 \right\rangle$$
$$= \langle U \rangle = 2 m_e e^4 C/\hbar^2 = 2E_0 \,. \tag{2.32}$$

In other words we have for Coulomb systems instead of the more general virial theorem $\langle U \rangle = -\langle p\dot{q} \rangle$

$$\begin{aligned} E_0 &= \langle T \rangle + \langle U \rangle = \tfrac{1}{2} \langle U \rangle \,, \\ \langle T \rangle &= -\tfrac{1}{2} \langle U \rangle \,, \\ E_0 &= -\langle T \rangle \,. \end{aligned} \tag{2.33}$$

This is the famous virial theorem valid for Coulomb systems. Since $\langle T \rangle > 0$ the ground state energy for Coulomb systems is always negative as well as all the other eigenvalues (THIRRING, 1980).

Now we come to the important question whether there exists a lower bound for the ground state energy proportional to N_e, i.e.,

$$E_0 \geqq -C N_e; \qquad 0 < C < \infty \,. \tag{2.34}$$

The existence of such a bound i.e. the extensivity of the ground state energy of the system was proven by DYSON and LENARD (1967) under the assumption that at least one type of charge obeys Fermi statistics — as indeed electrons do. The proof given by DYSON and LENARD (1967) is a mathematical "tour de force"; these authors obtained $C \sim 10^{14}$ Ry. A simple elegant derivation of (2.34) with a greatly improved constant ($C \approx 23$ Ry) was later given by Lieb and Thirring (LIEB, 1976). This derivation is based on the representation of the energy in the BOA as a functional of the electron density

$$E = E\{n(\mathbf{r})\}\,, \tag{2.35}$$

$$n(\mathbf{r}) = N_e \sum_{\sigma_i} \int \mathrm{d}\mathbf{r}_2 \ldots \mathrm{d}\mathbf{r}_{N_e} \, |\psi(\mathbf{r}_1, \ldots, \mathbf{r}_{N_e}, \sigma_1, \ldots, \sigma_{N_e})|^2$$

where

$$\psi(\mathbf{r}_1, \ldots, \mathbf{r}_{N_e}; \sigma_1, \ldots, \sigma_{N_e})$$

is the electron wave function. There is a theorem due to Kohn, Sham and Hohenberg (Hohenberg and Kohn, 1964; Kohn and Sham, 1965; Schneider, 1971; Singwi, 1976; Chihara, 1978; Bamzai and Deb, 1981; Dolgov et al., 1981) which states that the functional $E = E\{n(\mathbf{r})\}$ is unique if the exact energy is considered. This exact functional is of course not known. However here we are interested in a lower bound only and therefore it is enough to construct an approximate functional. Let us consider first the kinetic energy

$$T_\psi = N_e \sum_{\sigma_i} \int \mathrm{d}\mathbf{r}_1 \ldots \mathrm{d}\mathbf{r}_{N_e} \, |\nabla_1 \psi|^2. \tag{2.36}$$

A lower bound to T is given by the Lieb-Thirring theorem

$$T_\psi \geqq K \int \mathrm{d}\mathbf{r} \, n^{5/3}(\mathbf{r}), \qquad K = \tfrac{3}{5}\,(2\pi^2)^{2/3} \approx 4.382 \tag{2.37}$$

which is in fact a Thomas-Fermi estimate. Using the Hölder inequality we get further

$$\int \mathrm{d}\mathbf{r} \, n^{5/3}(\mathbf{r}) \geqq \left(\int \mathrm{d}\mathbf{r} \, n(\mathbf{r})\right)^{5/3} \cdot \left(\int \mathrm{d}\mathbf{r} \, 1\right)^{-2/3} = N_e^{5/3} V^{-2/3}\,,$$

$$T_\psi \geqq T_{\mathrm{id}} \tag{2.38}$$

where T_{id} is the kinetic energy of an ideal uniform Fermi gas. The increase of the kinetic energy with $N_e^{5/3}$ is essential for the stability of matter. Now we give the so-called Thomas-Fermi-Lenz functional.

$$E_{\mathrm{TF}}\{n(\mathbf{r})\} = K \int \mathrm{d}\mathbf{r} \, n^{5/3}(\mathbf{r}) - \sum_k \frac{Z_k e^2}{4\pi\varepsilon_0} \int \mathrm{d}\mathbf{r} \frac{n(\mathbf{r})}{|\mathbf{r} - \mathbf{R}_k|}$$

$$+ \frac{e^2}{8\pi\varepsilon_0} \int \mathrm{d}\mathbf{r} \, \mathrm{d}\mathbf{r}' \frac{n(\mathbf{r})\, n(\mathbf{r}')}{|\mathbf{r} - \mathbf{r}'|} + U(\mathbf{R}_1, \ldots, \mathbf{R}_M)\,. \tag{2.39}$$

The last three terms represent respectively, the electron-nuclear, the electron-electron, and nuclear-nuclear energy. The second and the fourth terms on the right hand side of (2.39) are exact but the first and third are not. The Thirring-Lieb-Simon theorem states that (Lieb, 1976; Lieb and Simon, 1977)

$$E_{\mathrm{TF}}\{n(\mathbf{r})\} = \inf \left\{E\{n(\mathbf{r})\};\ \int n(\mathbf{r})\, \mathrm{d}\mathbf{r} = N_e\right\}. \tag{2.40}$$

It is now understood that TF theory is really a large Z theory; to be precise it is exact in the limit $Z \to \infty$. An important point is that TF theory is well defined. In particular the density $n_{\mathrm{TF}}(\mathbf{r})$ is unique — a state of affairs in marked contrast to that of Hartree-Fock theory. It would be too much to try to reproduce the details of the proof of H-stability of matter. The idea is based on the theorem (2.40) and on the H-stability of TF-systems. The H-stability of TF-systems however follows from the no binding theorem due to Teller (1962). TF theory is not precise enough to give binding. Let us quote Lieb (1976): "Atomic binding is a fine quantum effect. Nevertheless, TF theory deserves to be well understood because it is exact in a limit;

the TF theory is to the many-electron system as the hydrogen atom is to the few-electron system."

The existence of a lower bound proportional to N_e which is called H-stability of the system is of fundamental importance. In this way the Fermi character of the electrons was proven to be both sufficient and necessary for the H-stability of Coulomb systems. For systems consisting of charged bosons only, DYSON and LENARD (1967) proved

$$-AN_e^{5/3} \leqq E_0 \leqq -BN_e^{7/3} \tag{2.41}$$

with $0 < A < \infty$, $0 < B < \infty$. Therefore charged bosons do not have H-stability. In order to demonstrate the role of the Pauli principle for H-stability let us consider again the kinetic energy and the pressure for ideal fermion systems (with spin 1/2)

$$T_{\rm id} = KV^{-2/3}N_e^{5/3}, \quad p_{\rm id} = \tfrac{2}{3}K(N_e/V)^{5/3}. \tag{2.42}$$

The increase of the kinetic energy with $N_e^{5/3}$ is the essential reason for the stability of matter, the Fermi pressure prevents the collapse. In order to demonstrate the importance of the Pauli principle for stability of matter let us quote DYSON (1967). He writes that without the Pauli principle "we (can) show that not only individual atoms but matter in bulk would collapse into a condensed high-density phase. The assembly of any two macroscopic objects would release energy comparable to that of an atomic bomb."

2.3. Survey of Exact Quantum-Statistical Results for Macroscopic Coulomb Systems

The existence of a lower bound for the energy, i.e., H-stability is not sufficient to guarantee the basic thermodynamic conditions, extensitivity and stability of the thermodynamic functions. In other words the free energy density

$$f = -\beta^{-1} \ln Z(N, V, \beta)/V, \qquad Z = \mathrm{Tr} \exp(-\beta H_N), \tag{2.43}$$

should exist in the limit

$$N \to \infty, \qquad V \to \infty, \qquad \varrho = N/V = \sum n_a = \text{const}. \tag{2.44}$$

Further thermodynamics requires convexity of $f(\varrho, \beta)$ as a function of ϱ and concavity of $\beta f(\varrho, \beta)$ as a function of β in order to guarantee the necessary positivity of the compressibility as well as of the heat capacity.

$$\frac{1}{K} = \frac{\partial p}{\partial \varrho} = \varrho \frac{\partial^2}{\partial \varrho^2} f(\varrho, \beta) \geqq 0, \qquad c_v = -\beta^2 \frac{\partial^2}{\partial \beta^2} f(\varrho, \beta) \geqq 0. \tag{2.45}$$

The problem is therefore: Does a thermodynamic limit and a thermodynamic description exist for matter consisting of point charges? In order to explain the specificity of the problem let us quote LIEB (1976): "The problem here centers around the long range r^{-1} nature of the Coulomb potential, not the short range singularity. Put another way, the question is, that if matter does not implode, how do we know that it does not explode? Normally systems with potentials that fall off less slowly than $r^{-3-\varepsilon}$ for some $\varepsilon > 0$ cannot be expected to have a thermodynamic limit. The crucial fact was discovered by Newton in 1687: Outside an isotropic distribution of charge, all the other charge appears to be concentrated at the center. This fact is the basis for screening." Screening is possible because the charges have different signs. However an additional hypothesis is necessary, namely neutrality of the matter in bulk. In order to provide extensitivity, different parts of the system far from each other must be approximately independent. The physical fact that makes this possible, despite the long range of Coulomb forces is screening. In other words the distribution of charges must be sufficiently neutral locally so that the electric potential far away will be zero. The following important result from Newton's Principia Matematica is the basis for this property: Let $\varrho(r)$ be an integrable charge density on the 3-dimensional space such that

$$\varrho(\boldsymbol{r}) = 0 \quad \text{if} \quad |r| > R\,,$$

i.e. outside some radius around the center of the distribution. Then Newton's theorem states that the potential generated by $\varrho(r)$ is

$$\Phi(\boldsymbol{r}) = \frac{1}{4\pi\varepsilon_0\,|\boldsymbol{r}|} \int \varrho(\boldsymbol{r}')\,\mathrm{d}\boldsymbol{r}' \quad \text{if} \quad |r| \gg R\,. \tag{2.46}$$

In other words a point far away feels only the total charge of the distribution irrespective of how the charge is distributed radially. Neutral charge distributions with

$$\int \varrho(\boldsymbol{r}')\,\mathrm{d}\boldsymbol{r} = 0$$

generate zero potentials outside its support. This is the physical basis for the screening property which is the most important peculiarity of Coulomb systems. Because the Coulomb forces are long range they are also very strong (their integral diverges at infinity) and cause the system to be locally neutral. The effective interaction between different regions is therefore greatly attenuated. This general idea was turned into a formal proof of extensitivity and stability of the thermodynamic potentials for Coulomb systems in fundamental papers by Lebowitz and Lieb (1969) (see also Lieb and Lebowitz, 1972, 1973; Lieb, 1976; Thirring, 1979; Lebowitz, 1980; Velo and Wightman, 1981).

We shall not discuss details of the proof here, let us mention just that it is based on the Bogolyubov-Peierls inequality

$$\begin{aligned} &\operatorname{Tr} \exp\,(\boldsymbol{A} + \boldsymbol{B}) \geqq \operatorname{Tr} \exp\,(\boldsymbol{A}) \exp\,(\langle \boldsymbol{B} \rangle)\,, \\ &\langle \boldsymbol{B} \rangle = \operatorname{Tr} \boldsymbol{B} \exp\,(\boldsymbol{A}) / \operatorname{Tr} \exp\,(\boldsymbol{A})\,. \end{aligned} \tag{2.47}$$

Let us discuss now the low density form of the free energy for Coulomb systems Low density means here the limit

$$n_e \Lambda_e^3 \to 0\,.$$

Lebowitz and Peña (1973) proved the theorem

$$K_1 \varrho \leqq \Big[\beta f(n_1, \ldots, n_s, \beta) - \sum_{a=1}^{s} n_a \ln (n_a \Lambda_a^3)\Big] \leqq K_2 \varrho \quad \text{if} \quad \varrho < \bar{\varrho}\,, \quad \varrho = \sum_{a=1}^{s} n_a \tag{2.48}$$

where K_1, K_2 are some constants $K_1 < K_2$ independent of the density ϱ. Therefore it is proved now rigorously that the thermodynamic free energy per unit volume of an overall-neutral system, composed of electrons and nuclei interacting via Coulomb forces, has the low density asymptotic form of an ideal Boltzmann gas consisting of non-interacting electrons and nuclei

$$f(n_1, \dots, n_s, \beta) = \beta^{-1} \sum_{a=1}^{s} n_a \ln n_a + \text{const.} \tag{2.49}$$

We may consider the expression (2.49) as the beginning of a series and may ask the question: What is the mathematical structure of the series? There is a theorem due to FRIEDMAN (1962) which we believe to be exact, whether the proof is rigorous or not. The Friedman theorem states

$$\begin{aligned}\beta f(n_1, \dots, n_s, \beta) &= \sum_{a=1}^{s} n_a \ln (n_a \Lambda_a^3/\mathrm{e}) - A_0(\beta)\, \varrho^{3/2} \\ &-A_1(\beta)\, \varrho^2 \ln \varrho - A_2(\beta)\, \varrho^2 - A_3(\beta)\, \varrho^{5/2} \ln \varrho - A_4(\beta)\, \varrho^{5/2} \\ &- A_5(\beta)\, \varrho^3 \ln \varrho - A_6(\beta)\, \varrho^3 - \dots\end{aligned} \tag{2.50}$$

where

$$\begin{aligned}\mathrm{e} &= 2.71, \\ A_0(\beta) &= (\varkappa^3/12\pi\varrho^{3/2}), \\ A_1(\beta) &= \frac{\pi}{6} \sum_{ab} n_a n_b (\beta e_a e_b/4\pi\varepsilon)^3\, \varrho^{-2}, \\ \varkappa^2 &= \beta \sum_a n_a e_a^2/\varepsilon\end{aligned} \tag{2.51}$$

denote the coefficients of the classical Debye-Hückel limiting law and of the extended limiting law. Some of the following coefficients which are influenced by quantum effects are also known. The derivation of these coefficients has been given by a group of physicists working in the sixties at the Rostock University including Kelbg, Ebeling, Kraeft, Kremp, Hoffmann, Schmitz and Czerwon (for a survey see KELBG, 1972; EBELING, KRAEFT and KREMP, 1976, 1979). The results which are believed to be exact, in spite of the fact that a rigorous proof is still missing, are in short

$$\begin{aligned}&A_2(\beta)\, \varrho^2 + A_3(\beta)\, \varrho^{5/2} \ln \varrho + A_4(\beta)\, \varrho^{5/2} \\ &= 2\pi \sum_{ab} n_a n_b \Bigg\{ (1 + \beta e_a e_b \varkappa/4\pi\varepsilon)\, \lambda_{ab}^3 \left[Q(\xi_{ab}) + \delta_{ab} \frac{(-1)^{2sq}}{2s_a + 1} E(\xi_{ab}) \right] \\ &+ \frac{1}{6} (\beta e_a e_b/4\pi\varepsilon)^3 \left[\ln (\varkappa\lambda_{ab}/\varrho^{1/2}) + (\beta e_a e_b \varkappa/4\pi\varepsilon) \ln (\varkappa\lambda_{ab})\right] \\ &- \frac{\pi}{3} (\beta e_a e_b/4\pi\varepsilon)^4\, \varkappa \left(1 - \ln \frac{4}{3}\right) \Bigg\} - 2\pi^2 a_1 \sum_{abc} n_a n_b n_c\, \varkappa^{-1} (\beta/4\pi\varepsilon)^5\, e_a^4 e_b^3 e_c^3 \\ &+ a_2 \sum_{abcd} n_a n_b n_c n_d \varkappa^{-3} (\beta/4\pi\varepsilon)^6\, (e_a e_b e_c e_d)^3\end{aligned} \tag{2.52}$$

where $a_1 \approx 0.543$, $a_2 \approx 10.13$ are numbers which are known only approximately. We note that the last two terms containing the factors a_1 and a_2 disappear in the case of symmetrical plasmas, i.e. especially for H-plasmas. Further the quantum virial function Q and E are given by converging series

$$\begin{aligned}Q(\xi) &= -\frac{1}{6} \xi - \frac{1}{8} \sqrt{\pi}\, \xi^2 - \frac{1}{6} \left(\frac{1}{2} C + \ln 3 - \frac{1}{2} \right) \xi^3 \\ &+ \sum_{n=4}^{\infty} \frac{\sqrt{\pi}\, \zeta(n-2)}{\Gamma(1 + n/2)} \left(\frac{\xi}{2} \right)^n,\end{aligned} \tag{2.53}$$

$$E(\xi) = \frac{1}{4}\sqrt{\pi} + \frac{1}{2}\,\xi + \frac{1}{4}\sqrt{\pi}\,(\ln 2)\,\xi^2 + \sum_{n=3}^{\infty} \frac{\sqrt{\pi}\,(1 - 2^{2-n})\,\zeta(n-1)}{\Gamma(1 + n/2)} \left(\frac{\xi}{2}\right)^n \tag{2.54}$$

and finally the interaction parameter ξ is given by

$$\xi_{ab} = -\frac{\beta e_a e_b}{4\pi\varepsilon\lambda_{ab}}, \qquad \lambda_{ab} = \frac{\hbar}{[2m_{ab}k_B T]^{1/2}}. \tag{2.55}$$

If the dimensionless density of the electrons $n_e\Lambda_e^3$ is increased we would expect from physical reasons that there exists a wide region of densities and temperatures where the system behaves like a gas consisting of neutral atoms and molecules interacting via effective forces of Lennard-Jones type. The *strong* derivation of such an effective Hamiltonian is a very difficult task. Let us quote again LEBOWITZ (1980): "I believe, that the problem of deriving effective (e.g., Lennard-Jones) interactions between neutral atoms from the Coulomb Hamiltonian while very difficult may not be entirely hopeless. I regard the work by FRÖHLICH and SPENCER (1981) on the two-dimensional charged lattice gas as having (despite of its being two-dimensional and entirely classical) the right flavor for this problem. They show in a precise way, that in a certain range of temperature and density the weight of the Gibbs measure is concentrated entirely on 'neutral molecules of final extent.' We don't expect this to happen in three dimensions where the Coulomb force is weaker and there should always be a finite density of loose charges (electrons and ions) around — we do expect that these play a negligible role at low temperatures and moderate densities. At very low densities the system will always be almost completely ionized". In recent time there have been many papers which deal with the question how to introduce composite particles into statistical mechanics (DASHEN, MA and BERNSTEIN, 1969; VOROB'EV and KHOMKIN, 1971, 1976; EBELING, 1974; GIRARDEAU, 1975, 1978; FLECKINGER, GOMES and SOULET, 1976, 1978; GILBERT, 1977; EBELING, KRAFT and KREMP, 1976, 1979).

The question is even more difficult in the kinetic theory where also certain progress has been reached (PELETMINSKY, 1971; KOLESNICHENKO, 1977; HOFFMANN, KOURI and TOP, 1979; KLIMONTOVICH, 1975, 1980, 1982; KLIMONTOVICH and KREMP, 1982; VASHUKOV and MARUSIN, 1982). The application of these new ideas to Coulomb systems is especially difficult due to the long range nature of the Coulomb forces. Let us discuss here only one idea we believe to be a fundamental requirement for all theories of composite particles. This is the idea of consistency of any two descriptions of real matter in the limit of low densities (EBELING, 1969; EBELING and SÄNDIG, 1973).

Let us explain here this requirement which leads to an infinite set of exact relations for the special case of H-plasmas. On the level of the electrons and nuclei the system is described by a series ($\varrho = n_e + n_p$)

$$\begin{aligned}\beta f(\varrho, \beta) &= 2\varrho \ln \varrho + A(\beta)\,\varrho - A_0(\beta)\,\varrho^{3/2}\\ &-A_2(\beta)\,\varrho^2 - A_3(\beta)\,\varrho^{5/2} \ln \varrho - A_4(\beta)\,\varrho^{5/2} - A_5(\beta)\,\varrho^3 \ln \varrho\\ &- A_6(\beta)\,\varrho^3 - A_7(\beta)\,\varrho^{7/2} \ln \varrho - A_8(\beta)\,\varrho^{7/2} - A_9(\beta)\,\varrho^4 \ln \varrho\\ &- A_{10}(\beta)\,\varrho^4 - \dots\end{aligned} \tag{2.56}$$

Here $A(\beta)$ describes the temperature coefficient of an ideal Boltzmann gas consisting of electrons and protons which depends on the masses but does not depend on the charges, and the coefficients $A_i(\beta)$ ($i = 0, 1, 2, \dots$) describe the interaction effects.

From eq. (2.56) follows the pressure

$$\beta p = \varrho \frac{\partial}{\partial \varrho} (\beta f) - (\beta f) = \varrho - \frac{1}{2} A_0(\beta) \varrho^{3/2} - A_2(\beta) \varrho^2$$
$$- \frac{3}{2} A_3(\beta) \varrho^{5/2} \ln \varrho - \left(A_3(\beta) + \frac{3}{2} A_4(\beta)\right) \varrho^{5/2}$$
$$-2 A_5(\beta) \varrho^3 \ln \varrho - (A_5(\beta) + 2 A_6(\beta)) \varrho^3$$
$$- \frac{5}{2} A_7(\beta) \varrho^{7/2} \ln \varrho - \left(A_7(\beta) + \frac{5}{2} A_8(\beta)\right) \varrho^{7/2}$$
$$- 3A_9(\beta) \varrho^4 \ln \varrho - (A_9(\beta) + 3 A_{10}(\beta)) \varrho^4 + \dots \tag{2.57}$$

On the other hand the same system may be described as gas consisting of free electrons and free protons with the density ϱ^* of atoms with the density n_A and of free molecules with the density n_M. The pressure is then expected to be

$$\beta p = \varrho^* + n_A + n_M - \tfrac{1}{2} A_0(\beta) (\varrho^*)^{3/2} - A_2^*(\beta) (\varrho^*)^2$$
$$- B_{AA}(\beta) n_A^2 - B_{MM}(\beta) n_M^2 - B_{CA}(\beta) \varrho^* n_A - B_{CM}(\beta) \varrho^* n_M - \dots \tag{2.58}$$

where B_{AA}, B_{MM}, B_{CA} and B_{CM} are the second virial coefficients for the interactions of atom-atom, molecule-molecule, charge-atom and charge-molecule type.

The corresponding free energy series is

$$\beta f(\varrho^*, n_A, n_M, \beta) = \varrho^* \ln \varrho + A(\beta) \varrho^* - A_0(\beta) (\varrho^*)^{3/2} - A_2^*(\beta) (\varrho^*)^2 - \dots$$
$$+ n_A \ln [n_A/K_A(\beta)] + B_A(\beta) n_A - B_{AA}(\beta) n_A^2 - B_{CA}(\beta) n_A \varrho^*$$
$$+ n_M \ln [n_M/K_M(\beta)] + B_M(\beta) n_M - B_{MM}(\beta) n_M^2 - B_{CM}(\beta) n_M \varrho^* - \dots \tag{2.59}$$

where $K_A(\beta)$ and $K_M(\beta)$ are the so-called mass-action constants. Further the conservation requires

$$\varrho = \varrho^* + 2n_A + 4n_M \,. \tag{2.60}$$

The second description is completely equivalent to the first one (2.57) if the correct effective interactions are chosen for the calculation of the effective virial coefficients B_{AA}, B_{MM}, B_{CA} and B_{CM}. We see that the description of the system in the composite-particle picture contains more temperature functions as we had in the elementary-particle description. However there are several exact relations which connect both pictures. These relations are simply obtained by expansion of eq. (2.58) with respect to the density and excluding the new densities ϱ^*, n_A and n_M by means of mass-action relations which follow from eq. (2.59) after minimization with respect to particle distributions conserving the balance eq. (2.60). We find e.g. (Ebeling and Sändig, 1973)

$$n_A = \tfrac{1}{4} K_A(\beta) \varrho^2 + \dots, \qquad n_M = \tfrac{1}{16} K_M(\beta) \varrho^4 + \dots \tag{2.61}$$

By introducing this into eq. (2.58) and ordering with respect to powers of ϱ we obtain after comparison with eq. (2.57) a whole series of exact relations between the coefficients in both pictures. Some examples are

$$\begin{aligned}
A_0(\beta) &= A_0^*(\beta)\,,\\
A_2(\beta) &= A_2^*(\beta) + \tfrac{1}{4}\,K_A(\beta)\,,\\
A_3(\beta) &= A_3^*(\beta)\,,\\
A_4(\beta) &= A_4^*(\beta) - \tfrac{1}{8}\,A_0^*(\beta)\,K_A(\beta)\,,\\
A_5(\beta) &= A_5^*(\beta)\,,\\
A_6(\beta) &= A_6^*(\beta) + \tfrac{1}{8}\,B_{CA}(\beta)\,K_A(\beta) - \tfrac{1}{4}\,A_2^*(\beta)\,K_A(\beta)\,.
\end{aligned} \tag{2.62}$$

These are balance relations which express the fact that in the picture of composite particles one part of the interactions determines the mass action constants which contain the bound states and the other part determines the new effective interactions between the quasiparticles. However, the origin of all these interactions is the same, the bare Coulomb potential which determines the coefficients in the picture of elementary particles $A_i(\beta)$.

In the last part of this chapter let us consider the region of very high densities where we expect again ideal behaviour of the system. This expectation is based on the Lieb-Thirring theorem which states that the mean kinetic energy per electron increases whith the density as $n_e^{2/3}$:

$$\frac{T_\psi}{N_e} \geqq K \frac{\int n(r)^{5/2}\,\mathrm{d}\boldsymbol{r}}{\int n(r)\,\mathrm{d}\boldsymbol{r}} \geqq \left(\frac{N_e}{V}\right)^{2/3}. \tag{2.63}$$

On the other hand we can hardly expect that the potential energy of a Coulomb system increases faster than

$$e^2/d \sim e^2\,n_e^{1/3}\,. \tag{2.64}$$

In this way the kinetic energy of the electrons will always dominate over the potential energy contributions. In this way one expects that the Coulomb systems at very high densities behave like ideal gases. Any rigorous proof of this is not known to the authors.

So far we have discussed only exact results for the thermodynamical systems. Many results have been obtained for the correlation function and screening properties of these systems (see Stillinger and Lovett, 1968; Fröhlich and Park, 1978, 1980; Brydges and Federbush, 1980; Gruber and Martin, 1980; Blum et al., 1981, 1982; Velo and Wightman, 1981; Gruber et al., 1981; Lebowitz, 1983). Finally let us mention exact results obtained for the case of two-dimensional systems (see Jancovici, 1982; Forrester et al., 1983) and for one-dimensional systems (Lieb and Mattis, 1966; Aizenman and Martin, 1980; Velo and Wightman, 1981; Bernasconi and Schneider, 1982).

3. Quantum Statistics of Many-Particle Systems

3.1. Elements of Quantum Statistics

3.1.1. Quantum Mechanics of Many-Particle Systems

The properties and the behaviour of systems composed of many elementary particles, atomic nuclei, atoms etc. are described by the quantum statistics. We assume that the reader is familiar with the most important features of quantum theory and statistical mechanics (see, e.g., DIRAC, 1958; MESSIAH, 1961; FICK, 1968; LANDAU and LIFSHITS, 1967). Therefore in this chapter we will give only a brief introduction to this field, in order to explain the main ideas, relations and notations.

Let $b_1, \dots, b_N$ be a set of observables (maximal set of simultaneously measurable physical quantities) and b_1 a complete observable of a one particle system, e.g., $b_1 = r_1, s_1$ where r_1 is the position vector of the particle and s_1 the spin projection or $b_1 = p_1, s_1$ where p_1 is the momentum of the particle 1. Then the microscopic state of the quantum many body system is given by a vector $|b_1 \dots b_N\rangle$ of the linear vector space $\mathscr{H}_N$.

The observable physical quantities, i.e., the sets $(b_1, \dots, b_N)$, are represented by commuting Hermitean operators acting in the state space $\mathscr{H}_N$. From the system of eigenvalue equations

$$\boldsymbol{b}_i \, |b_1 \dots b_N\rangle = b_i \, |b_1 \dots b_N\rangle \tag{3.1}$$

we get the possible measurable values of the physical quantities as the eigenvalues and the states as the eigenvectors.

The essential properties of the eigenstates are the orthogonality

$$\langle b_1 \dots b_N \mid b_{1'} \dots b_{N'}\rangle = \delta(b_1 - b_{1'}) \dots \delta(b_N - b_{N'}) \tag{3.2}$$

and the completeness given by the relation

$$1 = \sum\!\!\!\!\!\!\int |b_1 \dots b_N\rangle \langle b_N \dots b_1| \, \mathrm{d}b_1 \dots \mathrm{d}b_N \, . \tag{3.3}$$

Here we have to take δ-distributions and integrals in the case if the eigenvalues are continuous, and Kronecker symbols and sums if the eigenvalues are discrete, respectively.

If the state of the system is given by the vector $|\psi\rangle$ then the result of the measurement of any observable A is characterized by the expectation value given by

$$\langle A\rangle = \langle\psi| \, A \, |\psi\rangle \, . \tag{3.4}$$

Of course the system develops in time and therefore the average value will change, too. We can describe this change in the Heisenberg picture in which the time evolution of the system is determined by the Heisenberg dynamical equation for the operator

$$\frac{\mathrm{d}A}{\mathrm{d}t} = \frac{i}{\hbar}[H, A]\,. \tag{3.5}$$

H is the Hamiltonian of the system given by the operators of the kinetic energy and the potential energy of the interaction:

$$H = \sum_i \frac{p_i^2}{2m_i} + \sum_{i<j} V_{ij}(\boldsymbol{r}_i - \boldsymbol{r}_j)\,. \tag{3.6}$$

We have special problems in systems of identical particles. Here there is a further great difference in the behaviour of classical and quantum many particle systems. Instead of orbits, we have in the quantum case probabilistic statements on the particles, and therefore the identical particles loose there individuality. Identical microparticles are on principle indistinguishable. In the quantum mechanical formalism that means that the state vector is an eigenstate of the permutation operator P_N,

$$P_N\,|b_1 \dots b_N\rangle = \lambda\,|b_1 \dots b_N\rangle \qquad \text{with} \quad |\lambda|^2 = 1\,, \tag{3.7}$$

and further that for any observable A we have

$$[A, P_N] = 0\,.$$

Especially, of course, this relation is valid for the Hamiltonian. That means the actual state vector of a system of identical particles is symmetric or antisymmetric for any time. So we have to solve the question, whether we have to choose the symmetric or the antisymmetric state vector. The answer follows from the experiments and is given by the symmetry postulate:

The space of states for fermions is the antisymmetric subspace $\mathcal{H}_N^- \subset \mathcal{H}_N$, and the space of states for bosons is the symmetric subspace $\mathcal{H}_N^+ \subset \mathcal{H}_N$.

This postulate is important for the behaviour of many-particle systems. Because of the symmetry postulate one obtaines fundamentally different statistical mechanics for Bose and Fermi particles, respectively. The Fermi statistics is characterized by the Pauli exclusion principle, and the Bose statistics by the possible condensation of the particles in the ground state, the so-called Bose-Einstein condensation. Therefore this postulate determines essential properties of macroscopic systems.

Now let us consider the more mathematical question, how the symmetrical or antisymmetrical subspaces may be constructed. For this purpose we consider the basic vectors

$$|b_1 \dots b_N\rangle = |b_1\rangle \dots |b_N\rangle \in \mathcal{H}_N\,. \tag{3.8}$$

It is easy to show that we get a basic system in $\mathcal{H}_N^\pm$ with

$$\Lambda_N^\pm\,|b_1\rangle \dots |b_N\rangle = |b_1 \dots b_N\rangle^\pm \in \mathcal{H}_N^\pm\,. \tag{3.9}$$

Here $\Lambda_N^\pm$ are projection operators onto the subspaces $\mathcal{H}_N^\pm$ given by

$$\Lambda_N^\pm = \frac{1}{N!}\sum_{P_N} \lambda^\pm(P_N)\,\mathrm{P}_N \tag{3.10}$$

with

$$\lambda^+(P_N) = 1 \qquad \text{for Bose particles}$$

and

$$\lambda^-(P_N) = \begin{cases} 1 & \text{for even permutations } P_N\,, \\ -1 & \text{for odd permutations } P_N \end{cases} \tag{3.11}$$

in the case of Fermi particles.

3.1.2. The Method of Second Quantization

The properties of a quantum mechanical system composed of many identical particles are essentially determined by the symmetry postulate. A very useful and convenient method in order to fulfil this postulate is the method of second quantization (see, e.g., DE BOER, 1965), that means we describe the system in terms of so-called annihilation and creation operators. These operators act in the "grand space of states" (Fock space), defined as the direct sum of symmetrical or antisymmetrical spaces of states $\mathcal{H}_N^\pm$ of N particles, $|b_1 \ldots b_N\rangle$,

$$\mathcal{H}^\pm = \mathcal{H}_0 \oplus \mathcal{H}_1 \oplus \mathcal{H}_2^\pm \oplus \ldots \oplus \mathcal{H}_N^\pm \oplus \ldots$$

with the completeness relation

$$\sum_N \int \mathrm{d}b_1 \ldots \mathrm{d}b_N \, |b_1 \ldots b_N\rangle^\pm \, {}^\pm\langle b_N \ldots b_1| = 1\,. \tag{3.12}$$

In this space the vector $|b_1 \ldots b_N\rangle$ is an eigenstate of the particle number operator N and we have the relation

$$\langle b_1 \ldots b_N \mid b_{N'} \ldots b_1\rangle = \delta_{NN'}\,. \tag{3.13}$$

In order to construct the symmetrical or antisymmetrical basis states in $\mathcal{H}_N^\pm$ we introduce construction operators, so-called creation operators, $a^+(b_1)$, which construct the ket-vector of state by the following relation (MESSIAH, 1961):

$$|b_1 \ldots b_N\rangle^\pm = (n_\alpha!\, n_\beta! \ldots n_\varkappa!)^{-1/2}\, a^+(b_1) \ldots a^+(b_N)\, |0\rangle\,,$$

$$n_\alpha \text{ — occupation number, } n_\alpha + n_\beta + \ldots = N\,. \tag{3.14}$$

Here $|0\rangle$ is the state without particles, the vacuum state. By taking the adjoint relation we get the construction operators of bra-vectors of state

$$\langle 0|\, a(b_1) \ldots a(b_N)\, (n_\alpha!\, n_\beta! \ldots n_\varkappa!)^{-1/2} = {}^\pm\langle b_1 \ldots b_N|\,. \tag{3.15}$$

Further it is necessary to determine the effect of the operator a^+ on the bra-vector of state and that of a on the ket-vector of state.

Using the completeness relation (3.12) we find the following representation of a:

$$a(b) = |0\rangle \langle b| + \sqrt{2} \int \mathrm{d}b_1 \, |b_1\rangle \, {}^\pm\langle b_1 b| + \sqrt{3} \int \mathrm{d}b_1 \, \mathrm{d}b_2 \, |b_1 b_2\rangle \, {}^\pm\langle b_1 b_2 b| + \ldots \tag{3.16}$$

Taking into account the orthogonality of states with different numbers of N, it follows easily

$$a(b)\, |b_1' \ldots b_N'\rangle^\pm = \{\delta(b - b_1')\, |b_2' \ldots b_N'\rangle^\pm \pm \delta(b - b_2')\, |b_1' b_2' \ldots b_N'\rangle^\pm + \ldots\} \frac{1}{\sqrt{N}}\,, \tag{3.17}$$

that means the action of $a(b)$ on a ket-vector is that of an annihilation operator. The effect of $a^+(b)$ on a bra-vector we get by taking the adjoint relation

$${}^\pm\langle b_N' \ldots b_1'|\, a^+(b) = \{\delta(b - b_1')^\pm \langle b_2' \ldots b_N'| \pm \delta(b - b_2')\, {}^\pm\langle b_N' \ldots b_3' b_1'| + \ldots\} \frac{1}{\sqrt{N}}\,. \tag{3.18}$$

The condition, that the state which we have constructed must be symmetrical for Bose particles or antisymmetrical for Fermi particles, respectively, can be expressed by the commutation relations between the construction operators. Using the relations (3.14), (3.15) it is easy to show that the following set of commutation relations holds

$$\begin{aligned} &a(b)\, a(b') \mp a(b')\, a(b) = 0\,, \\ &a^+(b)\, a^+(b') \mp a^+(b')\, a^+(b) = 0\,, \\ &a(b)\, a^+(b') \mp a^+(b')\, a(b) = \delta(b - b') \end{aligned} \tag{3.19}$$

where the upper sign refers to Bose and the lower sign to Fermi particles, respectively. Consequently the construction operators for bosons commute and for fermions anti-commute. The operators $a(b)$ and $a^+(b)$ do not commute. In this way the requirements of the symmetry postulate are very simply represented by the commutation relations between the construction operators. That is one of the main advantages of the second-quantization formalism.

We consider the special case of the coordinate representation. In this case we will write

$$a(b_1) = a(\boldsymbol{r}_1, s_1) = \psi(\boldsymbol{r}_1, s_1) = \psi(1)\,. \tag{3.20}$$

These operators are usually called field operators. An important question in this formalism is how the operator A of any macroscopic observable A_N can be represented in terms of the construction operators $a(b)$, $a^+(b)$. Let us consider an N particle operator which does not change the number of particles N. Using the completeness relation (3.12) and relations (3.14) one gets

$$\begin{aligned} A_N = &\int db_1 \dots db_N\, db_{1'} \dots db_{N'}\, |b_N \dots b_1\rangle^\pm \\ &\cdot {}^\pm\langle b_1 \dots b_N|\, A_N\, |b_{N'} \dots b_{1'}\rangle^\pm \langle b_{1'} \dots b_{N'}| = \frac{1}{N!} \int db_1 \dots db_N \\ &\cdot db_{1'} \dots db_{N'}\, {}^\pm\langle b_1 \dots b_N|\, A_N\, |b_{N'} \dots b_{1'}\rangle^\pm a^+(b_N) \dots a^+(b_1)\, |0\rangle \langle 0|\, a(b_{1'}) \dots a(b_{N'})\,. \end{aligned} \tag{3.21}$$

Because of the orthogonality between states with different particle number N in the case that A_N operates on an N particle state vector, it is possible to extend the projector $|0\rangle \langle 0|$ to the identity operator 1 (3.12) in the grand space of states. The final result is (see e.g., SCHWEBER, 1961):

$$\begin{aligned} A_N = &\frac{1}{N!} \int db_1 \dots db_N\, db_{1'} \dots db_{N'} \\ &\cdot {}^\pm\langle b_1 \dots b_N|\, A_N\, |b_{N'} \dots b_{1'}\rangle^\pm\, a^+(b_N) \dots a^+(b_1)\, a(b_{1'}) \dots a(b_N)\,. \end{aligned} \tag{3.22}$$

So we have expressed the operator A_N by the construction operators a, a^+.

An essential simplification we get in the case that A_N may be decomposed, and is, e.g., of an additive or binary type, respectively, i.e.,

$$A_N = \sum_{i=1}^{N} A_i\,; \qquad A_N = \sum_{i<j}^{N} A_{ij}\,.$$

In such cases from (3.22) follows

$$A_N = \int db\, db'\, \langle b|\, A_1\, |b'\rangle\, a^+(b)\, a(b') \tag{3.23}$$

for additive operators and

$$A_N = \tfrac{1}{2} \int db_1\, db_2\, db_1'\, db_2'\; {}^{\pm}\langle b_1 b_2 | \, A_{12} \, | b_1' b_2' \rangle^{\pm} \, a^+(b_1)\, a^+(b_2)\, a(b_2')\, a(b_1') \tag{3.24}$$

for binary operators.

In the coordinate representation we get for example from (3.23) and (3.24)

$$H = -\frac{\hbar^2}{2m} \int d\boldsymbol{r}_1\, \psi^+(\boldsymbol{r}_1)\, \nabla^2_{\boldsymbol{r}_1} \psi(\boldsymbol{r}_1)$$

$$+ \frac{1}{2} \int d\boldsymbol{r}_1\, d\boldsymbol{r}_2\, \psi^+(\boldsymbol{r}_1)\, \psi^+(\boldsymbol{r}_2)\, V(\boldsymbol{r}_1 - \boldsymbol{r}_2)\, \psi(\boldsymbol{r}_2)\, \psi(\boldsymbol{r}_1)\,, \tag{3.25}$$

for the Hamiltonian of the system.

In many cases it is more convenient to use the momentum representation. Then we get

$$H = \sum_{s_1} \int d\boldsymbol{p}_1 \frac{p_1^2}{2m} a^+(p_1, s_1)\, a(p_1, s_1)$$

$$+ \frac{1}{2} \sum_{s_1 s_2} \sum_{\bar{s}_1 \bar{s}_2} \int d\boldsymbol{p}_1\, d\boldsymbol{p}_2\, d\bar{\boldsymbol{p}}_1\, d\bar{\boldsymbol{p}}_2 \, \langle p_1 s_1 p_2 s_2 | \, V_{12} \, | \bar{s}_2 \bar{p}_2 \bar{s}_1 \bar{p}_1 \rangle$$

$$\cdot\, a^+(p_1, s_1)\, a^+(p_2, s_2)\, a(\bar{p}_2, \bar{s}_2)\, a(\bar{p}_1, \bar{s}_1) \tag{3.26}$$

with

$$\langle p_1 s_1 p_2 s_2 | \, V_{12} \, | \bar{s}_2 \bar{p}_2 \bar{s}_1 \bar{p}_1 \rangle = V_{12}(\boldsymbol{p}_1 - \bar{\boldsymbol{p}}_2)\, \delta(\bar{\boldsymbol{p}}_1 + \bar{\boldsymbol{p}}_2 - \boldsymbol{p}_1 - \boldsymbol{p}_2)\, \delta_{s_1 \bar{s}_1}\, \delta_{s_2 \bar{s}_2}\,.$$

Now let us consider the time development of the many particle system in thelanguage of second quantization. In the Heisenberg picture the time development is determined by the equation of motion for the construction operators given for example by

$$\left(\frac{\hbar}{i}\right) \frac{\partial}{\partial t} a(b, t) = [a(b, t), H]\,. \tag{3.27}$$

Using this equation in the coordinate representation the equation of motion for the field operators $\psi(r, t) = \psi(1)$ is

$$\left(i\hbar \frac{\partial}{\partial t} + \frac{\hbar^2 \nabla^2}{2m}\right) \psi(r, t) = \int d\bar{\boldsymbol{r}}\, V(\boldsymbol{r} - \bar{\boldsymbol{r}})\, \psi^+(\bar{r}, t)\, \psi(\bar{r}, t)\, \psi(\boldsymbol{r}, t)\,. \tag{3.28}$$

It is interesting to remark that this equation is very similar to the one-particle Schrödinger equation.

3.1.3. Quantum Statistics. Density Operator

Up to now we assumed that we have the maximum information on the many-particle system. In that case the so-called micro-state is given by a vector of the "grand space of states" $|\psi_m\rangle$, and the average of a large number of measurements of the observable A_N is given by

$$\langle \psi_m | \, A_N \, | \psi_m \rangle\,.$$

Such a state is called according to von Neumann a pure state.

Now we will consider a large many-body system in the so-called thermodynamic limit:

$$N \to \infty\,, \qquad V \to \infty\,, \qquad n = \frac{N}{V} \text{ — finite}\,.$$

Such a system can be realized by a gas, a plasma, a liquid or a solid, respectively, that means, by macroscopic systems. For such system in this sense the maximum information, that means the micro-state (pure state), in general is not available. By the unavoidable interaction of each subsystem of the system in question with the infinite environment, an uncertainty about the micro-state is produced. In this way we know only an ensemble of possible states $|\psi_m\rangle$, which are considered to be normalizable, and the probabilities P_m of the system being in states $|\psi_m\rangle$, respectively. The information on the system is therefore given by the following scheme

$$\begin{array}{llll} |\psi_1\rangle, & |\psi_2\rangle, \ldots, & |\psi_m\rangle, \ldots & \text{— possible states,} \\ P_1, & P_2, \ldots, & P_m, \ldots & \text{— probabilities} \end{array}$$

with

$$\sum_m P_m = 1\,, \qquad 0 \leqq P_m \leqq 1\,.$$

It is useful to condense this information into the so-called density operator

$$\varrho = \sum_m P_m |\psi_m\rangle \langle\psi_m| \tag{3.29}$$

which was first introduced by J. von Neumann. The average value of any observable represented by the operator A_N is given by

$$\langle A_N\rangle = \sum_m P_m \langle m|\, A_N\, |m\rangle = \mathrm{Tr}(A_N\, \varrho_N)\,. \tag{3.30}$$

The state which is determined by ϱ was called by von Neumann a mixed state. We call this concept also the quantum statistical description.

The density operator has the following important properties:

1. From the assumption that $|\psi_m\rangle$ are normalized and because P_m must satisfy the condition $\sum P_m = 1$ we find the normalization condition for the density operator

$$\mathrm{Tr}\, \varrho = 1\,.$$

2. It is easy to see from (3.29), that the density operator is Hermitean

$$\varrho^+ = \varrho\,.$$

3. A pure state occurs if the density operator is given by

$$\varrho = |\psi_m\rangle \langle\psi_m|$$

This is the case, if $P_m = 1$, $P_{n \neq m} = 0$.

4. If $|b_1 \ldots b_N\rangle = |b^N\rangle$ is a system of basic vectors, normalization relations and relation (3.30) assume the form

$$\mathrm{Tr}\, \varrho = \sum_{\{b^N\}} \langle b^N|\, \varrho\, |b^N\rangle = 1\,,$$

$$\langle A\rangle = \sum_{\{b^N\}\{\bar{b}^N\}} \langle b^N|\, A\, |\bar{b}^N\rangle \langle\bar{b}^N|\, \varrho\, |b^N\rangle\,. \tag{3.31}$$

The matrix

$$\langle b^N | \varrho | \bar{b}^N \rangle = \langle b_N \dots b_1 | \varrho | \bar{b}_1 \dots \bar{b}_N \rangle$$

is called the density matrix.

The diagonal elements of the density matrix

$$\langle b^N | \varrho | b^N \rangle$$

give the probability for finding the values $b_1, \dots, b_N$ in case of a b-measurement.

5. Finally we remark that the density operator is because of the relation

$$\langle b^N | \varrho | b^N \rangle = \sum_m P_m |\langle b^N | \psi_m \rangle|^2 \geqq 0$$

a positive operator and

$$\mathrm{Tr}\, \varrho^2 = \sum_{\{b^N\}\{\bar{b}^N\}} \langle b^N | \varrho | \bar{b}^N \rangle \langle \bar{b}^N | \varrho | b^N \rangle \leqq 1 \, .$$

In order to complete the quantum statistical description it is necessary to introduce a measure for the information which is contained in the density operator. A quantity suitable for this purpose is the statistical entropy (Jaynes, 1957; Zubarev, 1971; Ingarden, 1973) given by

$$S = -k_\mathrm{B} \,\mathrm{Tr}\, (\varrho \ln \varrho) \, . \tag{3.32}$$

Therefore S is the average value of the entropy operator

$$S = -k_\mathrm{B} \ln \varrho \, .$$

Because of the properties of the statistical operator we have the following properties of S:

1. S is positively definite, that means

$$S = -k_\mathrm{B} \,\mathrm{Tr}\, \varrho \ln \varrho \geqq 0 \, .$$

2. The entropy is zero in the case of a pure state

$$\varrho = |\psi_m\rangle \langle \psi_m|$$

and has a maximum in the case of maximal uncertainty of the state.

3. If ϱ describes two statistically independent systems 1 and 2 with the density operators ϱ_1 and ϱ_2, i.e., if $\varrho = \varrho_1 \varrho_2$, the entropy is additive:

$$S = S_1 + S_2 \, .$$

Finally let us consider the dynamics in quantum statistics. In the Heisenberg picture ϱ does not depend on the time. In the Schrödinger picture the time dependence of the states $|\psi_m(t)\rangle$ is given by the Schrödinger equation

$$-\frac{\hbar}{i} \frac{\mathrm{d}}{\mathrm{d}t} |\psi_m(t)\rangle = H |\psi_m(t)\rangle \, . \tag{3.33}$$

Using the definition of ϱ (3.29) and the Schrödinger equation one can show, that the time development of ϱ is given by the so-called von Neumann equation

$$\frac{\mathrm{d}\varrho}{\mathrm{d}t} = \frac{i}{\hbar} [\varrho, H] \, . \tag{3.34}$$

This equation is the quantum version of the classical Liouville equation.

3.1.4. Reduced Density Operators. Bogolyubov Hierarchy

In many cases of interest the physical observables are represented by operators of additive-, binary- or s-particle type, respectively,

$$A_N = \sum_{\{s\}} A_{\{s\}} \,. \tag{3.35}$$

In such cases we can simplify the expression (3.30) for the average of observables by introduction of reduced density operators defined by (Bogolyubov, 1970)

$$F_s = V^s \operatorname*{Tr}_{s+1\ldots N} \varrho_N \tag{3.36}$$

with the normalization (because of (3.31))

$$\operatorname*{Tr}_{1\ldots s} F_s = V^s \,. \tag{3.37}$$

Then the average of the s-particle observable A_N is given by

$$\langle A_N\rangle = \frac{n^s}{s!} \operatorname*{Tr}_{1\ldots s} F_s A_s \,. \tag{3.38}$$

If one introduces a basis we get the reduced density matrices

$$V^{-1}\langle b_1|\, F_1\, |\bar{b}_1\rangle = \langle b_1|\, \operatorname*{Tr}_{2\ldots N} \varrho_N\, |\bar{b}_1\rangle \,,$$

$$V^{-2}\langle b_1 b_2|\, F_{12}\, |\bar{b}_2\bar{b}_1\rangle = \langle b_1 b_2|\, \operatorname*{Tr}_{3\ldots N} \varrho_N\, |\bar{b}_2\bar{b}_1\rangle \text{ etc } \ldots$$

These matrices we will call one-, two-, s-particle density matrices, respectively.

Let us consider in more detail the one-particle density matrix in the momentum p representation for homogeneous systems. In this case we have $[F_1, p_1] = 0$, and therefore the equation

$$F_1|p\rangle = F(p)\,|p\rangle \,,$$

where $|p\rangle$ are normalized eigenstates of p, and $F(p)$ is the momentum distribution function which is connected with the diagonal elements of the one-particle density matrix $\langle p|\, F_1\, |p\rangle$ by

$$\langle p|\, F_1\, |p\rangle = F(p)\,\langle p\mid p\rangle = F(p)\,\delta(0), \qquad \delta(0) = V/(2\pi\hbar)^3, \tag{3.39}$$

so we can write

$$F(p) = \frac{(2\pi\hbar)^3}{V} \langle p|\, F_1\, |p\rangle \,.$$

Because of (3.37) we may obtain the normalization

$$\int F(p) \frac{\mathrm{d}p}{(2\pi\hbar)^3} = 1 \,. \tag{3.40}$$

The average of an additive operator with $\langle p|\, A_1\, |p\rangle = A_1(p)\,\delta(0)$ is given by

$$\langle A\rangle = N \int \frac{\mathrm{d}p}{(2\pi\hbar)^3} F_1(p)\, A_1(p) \,. \tag{3.41}$$

In general the mean value of an operator of the form (3.35) is

$$\langle A_N\rangle = \frac{n^s}{s!} \int \mathrm{d}p^s\, \mathrm{d}\bar{p}^s\, \langle p^s|\, F_s\, |\bar{p}^s\rangle\, \langle \bar{p}^s|\, A_s\, |p^s\rangle \,. \tag{3.42}$$

On the other hand we can express the average value in the second quantized form

$$\langle A_N \rangle = \frac{1}{s!} \int \mathrm{d}\boldsymbol{p}^s \, \mathrm{d}\bar{\boldsymbol{p}}^s \, \langle b^s | \, A_s \, |\bar{b}^s \rangle \, , \; \mathrm{Tr} \, [\varrho a^+(b_s) \, \dots \, a^+(b_1) \, a(\bar{b}_1) \dots a(\bar{b}_s)] \, . \quad (3.43)$$

Comparing (3.42) and (3.43) we get the very important representation

$$n^s \langle p_1 \dots p_s \, |F_s| \, \bar{p}_s \dots \bar{p}_1 \rangle = \mathrm{Tr} \, \{ \varrho a^+(p_1) \dots a^+(p_s) \, a(\bar{p}_s) \dots a(\bar{p}_1) \} \quad (3.44)$$

for the reduced density matrices (Bogolyubov, 1970). Therefore the matrix elements of the density operator are given by the statistical average of the construction operators $a^+(b_1), \dots, a(\bar{b}_1)$. The introduction of the second quantization formalism in quantum statistics is very useful. It is especially possible on the basis of this formulation to use the very effective methods of quantum field theory in quantum statistics.

Now let us consider the equation of motion for the reduced density operators. We may obtain such equations from the von Neumann equation by taking the trace over the $(N-s)$-particle space of states. For this purpose it is useful to decompose the N-particle Hamiltonian in the following way:

$$H_s + H_{N-s} + V_{s,\,N-s} = H_N \quad (3.45)$$

with

$$H_s = \sum_{i=1}^{s} \frac{p_i^2}{2m} + \sum_{i<j=1}^{s} V_{ij} \, , \quad (3.46)$$

$$H_{N-s} = \sum_{i=s+1}^{N} \frac{p_i^2}{2m} + \sum_{i<j=s+1}^{N} V_{ij} \, , \quad (3.47)$$

$$V_{s,\,N-s} = \sum_{i=1}^{s} \sum_{j=s+1}^{N} V_{ij} \, . \quad (3.48)$$

Taking the trace we get

$$\frac{\hbar}{i} \frac{\partial}{\partial t} F_s = [F_s, H_s] + V^s \mathop{\mathrm{Tr}}_{s+1 \dots N} [\varrho_N, H_{N-s}] + V^s \mathop{\mathrm{Tr}}_{s+1 \dots N} [\varrho_N, V_{s,\,N-s}] \, . \quad (3.49)$$

The second term on the right hand side is zero because of the cyclic invariance of the trace. The last contribution can be transformed into

$$V^s \mathop{\mathrm{Tr}}_{s+1 \dots N} [\varrho_N, V_{s,N-s}] = \frac{N-s}{V} \mathop{\mathrm{Tr}}_{s+1} [F_{s+1}, \sum_{i=1}^{s} V_{i,\,s+1}] \, .$$

With this expression one finds the following equation of motion:

$$\frac{\hbar}{i} \frac{\partial}{\partial t} F_s = [F_s, H_s] + n \sum_{i=1}^{s} \mathop{\mathrm{Tr}}_{s+1} [F_{s+1}, V_{i,\,s+1}] \, , \quad (3.50)$$

$$s = 1, 2, \dots$$

This equation was first obtained by Bogolyubov and is fundamental for the quantum statistics of equilibrium and nonequilibrium systems. An essential feature of this equation is the coupling between F_s and F_{s+1} because of the interaction.

Is is easy to obtain a formal solution of this equation in the form

$$F_s(t) = e^{-\frac{i}{\hbar}H_s\tau} F_s(t-\tau)\, e^{\frac{i}{\hbar}H_s\tau}$$
$$-n\frac{i}{\hbar}\int_0^\tau d\tau'\, e^{-\frac{i}{\hbar}H_s\tau'} \operatorname{Tr}_{s+1} \sum_{j=1}^{s} [V_{j,s+1}, F_{s+1}]\, e^{\frac{i}{\hbar}H_s\tau'} . \tag{3.51}$$

This equation is very useful in several connections. In the case of nonequilibrium systems, one can derive, from the first equation of (3.50), quantum kinetic equations using Bogolyubovs condition of the weakening of initial correlations

$$\lim_{t\to-\infty} F_s(t) = \prod_{i=1}^{s} F_i \tag{3.52}$$

see, e.g., KLIMONTOVICH and KREMP, 1981; EBELING et al., 1983).

3.1.5. The Classical Limit, BBGKY Hierarchy

Now we will consider the classical limit of the reduced density matrices. Let us consider the coordinate representation of any operator F

$$\langle x_1 \dots x_N|\, F\, |x_{N'} \dots x_{1'}\rangle\,, \qquad x_i = \boldsymbol{r}_i s_i\,.$$

In order to get a representation that has many properties analogous to the classical phase space formalism we will introduce the Wigner representation. First we write the matrix elements of F in terms of the new variable (Wigner transposition)

$$\boldsymbol{r}_i \to \boldsymbol{r}_i + \frac{\boldsymbol{y}_i}{2}\,; \qquad \boldsymbol{r}_{i'} \to \boldsymbol{r}_i - \frac{\boldsymbol{y}_i}{2}\,. \tag{3.53}$$

Then we have (for the purpose of this consideration the spin variables are omitted)

$$\left\langle \boldsymbol{r}_1 + \frac{\boldsymbol{y}_1}{2}, \boldsymbol{r}_2 + \frac{\boldsymbol{y}_2}{2} \dots |F| \dots \boldsymbol{r}_1 - \frac{\boldsymbol{y}_1}{2}\right\rangle .$$

The Wigner representation is obtained by Fourier transformation with respect to y

$$F(r_1, \dots, r_N, p_1, \dots, p_N) = \int e^{\frac{i}{\hbar}\sum_{i=1}^{N} p_i y_i} \left\langle \boldsymbol{r}_1 + \frac{\boldsymbol{y}_1}{2} \dots |\, F\, | \dots \boldsymbol{r}_1 - \frac{\boldsymbol{y}_1}{2}\right\rangle d\boldsymbol{y}^N . \tag{3.54}$$

In the Wigner representation the average of any observable A is given by

$$\langle A\rangle = \frac{n^s}{s!}\int d\boldsymbol{r}^s\, d\boldsymbol{p}^s\, A_s(r_1, \dots, r_s, p_1, \dots, p_s)\, F_s(r_1, \dots, r_s, p_1, \dots, p_s)\,. \tag{3.55}$$

This mixed representation has many similarities to the classical phase space representation of average values. Especially for an operator $A(\boldsymbol{r}, \boldsymbol{p})$ with the matrix elements

$$\langle r_1 \dots |\, A\, | \dots r_{1'}\rangle = \prod_{i=1}^{s} \delta(\boldsymbol{r}_i - \boldsymbol{r}_{i'})\, A\left(\frac{\hbar}{i}\frac{\partial}{\partial r_1}, \dots, r_1, \dots\right)$$

it is possible, using (3.54), to show that the Wigner representation $A_s(r_1, \dots, r_s, p_1, \dots, p_s)$ is identical with the classical phase space function of the observable A. Taking the limit $\hbar \to 0$ of the reduced density matrix in Wigner representation we get the classical reduced distribution function

$$F_s^{\mathrm{cl}}(r_1, \dots, r_s, p_1, \dots, p_s) = \lim_{\hbar \to 0} F_s(r_1, \dots, r_s, p_1, \dots, p_s) \tag{3.56}$$

and the classical form of the average of the observable A

$$\langle A_{\mathrm{cl}} \rangle = \frac{n^s}{s!} \int \mathrm{d}r^s \, \mathrm{d}p^s \, A_s(r_1, \dots, r_s, p_1, \dots, p_s) \, F_s^{\mathrm{cl}}(r_1, \dots, r_s, p_1, \dots, p_s) \, . \tag{3.57}$$

From (3.36) follows that F_s^{cl} is connected with the N-particle distribution function ϱ^{cl} by

$$F_s^{\mathrm{cl}} = V^s \int \mathrm{d}p^{N-s} \, \mathrm{d}r^{N-s} \, \varrho^{\mathrm{cl}}(r_1, \dots, r_N, p_1, \dots, p_N) \, . \tag{3.58}$$

The normalization is given by

$$\int F_s^{\mathrm{cl}}(r^s, p^s) \, \mathrm{d}r^s \, \mathrm{d}p^s = V^s \, .$$

So in the limit $\hbar \to 0$ the quantum statistics contains the well known phase space formalism of the classical statistical mechanics.

In order to get the classical limit of the Bogolyubov hierarchy we must take the Wigner representation of the equations (3.50) and the limit $\hbar \to 0$. In this way, the BBGKY hierarchy may be obtained for the classical reduced distribution functions

$$\frac{\partial F_s^{\mathrm{cl}}}{\partial t} = [H_s; F_s]_{\mathrm{cl}} + n \int [\sum_{1 \leqq i \leqq s} V_{i,s+1}; \quad F_{s+1}]_{\mathrm{cl}} \, \mathrm{d}r_{s+1} \mathrm{d}p_{s+1} \; ;$$

here is

$$[H_s; F_s]_{\mathrm{cl}} = \sum_{i=1}^{s} \left(\frac{\partial H_s}{\partial r_i} \frac{\partial F_s}{\partial p_i} - \frac{\partial H_s}{\partial p_i} \frac{\partial F_s}{\partial r_i} \right)$$

the Poisson bracket and H_s the Hamiltonian of an interacting s-particle system.

This hierarchy was first obtained by Bogolyubov, Born, Green, Kirkwood and Yvon and plays an important role in equilibrium and nonequilibrium problems (see, e.g., Bogolyubov, 1946).

3.1.6. Systems in Thermodynamical Equilibrium

The existence of the thermodynamic equilibrium states for macroscopic systems is guaranteed from the experimental observation. In this connection we have the following questions: How can this state be defined from the point of view of the quantum statistics, and how can the density operator be obtained for systems in the thermodynamic equilibrium?

Obviously in the thermodynamic equilibrium state the density operator is independent of the time, that means

$$\frac{\partial \varrho}{\partial t} = \frac{i}{\hbar} [H, \varrho] = 0 \, . \tag{3.59}$$

From this condition follows that ϱ depends only on the integrals of the motion. Further, in generalization of the H-theorem (Huang, 1964), we make the assumption that in the state of thermodynamic equilibrium the statistical entropy has a maximum,

$$S = -k_{\mathrm{B}} \operatorname{Tr} \varrho \ln \varrho = \max \, , \tag{3.60}$$

and is identical with the thermodynamic entropy if we choose k_B in a suitable manner. It is clear that this condition has the character of a postulate, because this relation is proved only on the basis of the Boltzmann equation and for the one-particle operator. The Boltzmann equation is valid only for dilute gases.

In general it is necessary to show that the N-particles density operator gives the behaviour (3.60) for the entropy. Because of the invariance of the von Neumann equation for time reflections $t \to -t$ (reversible behaviour), that is a difficult and fundamental problem of quantum statistics. It is necessary, on the basis of additional conditions, such as weakening of the initial correlations and taking the thermodynamical limit, to break the symmetry of the solutions of the von Neumann equations under time reflections (ZUBAREV, 1971; ZUBAREV and NOVIKOV, 1973).

With (3.60) it is possible to determine the statistical operator (JAYNES 1957, ZUBAREV 1971). We consider a system in the thermodynamic equilibrium, which exchanges energy and particles with the surrounding. But the average of the energy, which we will identify with the thermodynamical internal energy, and the average of the particle number are given by

$$\langle H \rangle = \mathrm{Tr}\,(\varrho H) = U\,, \tag{3.61a}$$

$$\langle N \rangle = \mathrm{Tr}\,(\varrho N) = \overline{N}\,. \tag{3.61b}$$

Further we have to take into account the normalization condition

$$\mathrm{Tr}\,\varrho = 1\,. \tag{3.61c}$$

Therefore we have to solve the variational problem

$$\delta(k_B\,\mathrm{Tr}\,\varrho \ln \varrho) = 0; \tag{3.61d}$$

with the additional conditions (3.61a—c) it is easy to find the solution in the following form:

$$\varrho = Z_{\mathrm{gr}}^{-1}\,\mathrm{e}^{-\beta(H-\mu N)}\,. \tag{3.62}$$

This operator is called grand canonical density operator. Z_{gr} is the grand canonical partition function which is given by the condition (3.61c):

$$Z_{\mathrm{gr}}(V, \beta, \mu) = \mathop{\mathrm{Tr}}_{(\mathscr{H})}\,\mathrm{e}^{-\beta(H-\mu N)}\,. \tag{3.63}$$

The quantities β and μ are Lagrangian multipliers which are connected with the conditions (3.61a) and (3.61b).

Now we have to correlate the quantum statistical averages to their thermodynamical counterparts and to explain the physical meaning of the parameters β and μ. The starting point is the identification of

$$S = -k_B\,\mathrm{Tr}\,\varrho \ln \varrho$$

with the thermodynamical entropy and that of

$$\langle H \rangle = \mathrm{Tr}\,(\varrho H) = U$$

with the internal energy. Moreover we use the well known thermodynamical relation

$$S = \frac{U}{T} + P\frac{V}{T} - \frac{\mu N}{T} \tag{3.64}$$

where μ is the chemical potential and T the temperature. We compare this expression with the expression for the statistical entropy which we obtain using the density operator (3.62)

$$S = k_B \ln Z_{gr} + k_B\beta U - k_B\beta\mu\langle N\rangle . \tag{3.65}$$

So we get the physical meaning of β and μ

$$\beta = \frac{1}{k_B T},$$

$$\mu = \text{chemical potential.}$$

Further there may be obtained a statistical expression for the equation of states

$$pV = k_B T \ln Z_{gr}(V, \mu, T) . \tag{3.66}$$

This equation gives the connection between the Hamiltonian of the many-body system and the equation of states. This is therefore one of the most important results of quantum statistics. Finally the macroscopic number of particles is given by

$$\overline{N} = Z_{gr}^{-1} \operatorname{Tr} (N\, e^{-\beta(H-\mu N)}) = \langle N\rangle .$$

Using (3.63) it can be shown that

$$\langle N\rangle = k_B T \frac{\partial}{\partial\mu} \ln Z_{gr}(\mu, T, V) = \frac{\partial}{\partial\mu} pV ,$$

$$\overline{N}/V = n(\mu, T) = \frac{\partial}{\partial\mu} p . \tag{3.67}$$

Eqs. (3.66) and (3.67) determine the thermodynamical quantities of the many-body system as functions of T, V and μ. Eq. (3.67) can, in principle, be inverted to obtain the chemical potential as a function of the number of particles. Substitution into eq. (3.66) then yields the thermodynamical variables as a function of T, V and n. We get a very useful relation if we write the Hamiltonian in the form

$$H_\lambda = H_0 + \lambda V .$$

By differentiating (3.63) with respect to λ one finds

$$\frac{\partial}{\partial\lambda} \ln Z_{gr} = -\beta\langle V\rangle = -\beta\, Z_{gr}^{-1} \operatorname{Tr} (V\, e^{-\beta(H-\mu N)}) .$$

Integrating both sides of this equation with respect to λ from $\lambda = 0$ to $\lambda = 1$ and taking intó account (3.66), the equation of states is given by the average of the nteraction energy

$$pV - p_0 V = -\int_0^1 \frac{d\lambda}{\lambda} \langle\lambda V\rangle . \tag{3.68}$$

Using expression (3.38) for the average of a binary operator, one may determine the equation of states with the help of the binary density operator

$$\langle\lambda V\rangle = \frac{n^2}{2} \operatorname{Tr} (\lambda V_{12} F_{12}) . \tag{3.69}$$

In some cases it is more convenient to consider a system only under energy exchange with the surrounding. That means the number of particles is fixed. Then we have to solve the variational problem (3.61d) with the conditions (3.61a) and (3.61c) only.

The solution is the canonical density operator

$$\varrho = \frac{1}{Z} e^{-\beta H} . \tag{3.70}$$

For systems with a fixed number of particles eq. (3.64) is to be replaced by

$$S = \frac{U}{T} - \frac{F}{T} . \tag{3.71}$$

For the thermodynamical functions we get in this case similar as before the following statistical expression

$$\beta = \frac{1}{k_B T}, \qquad F = -k_B T \ln Z(T, V, N) .$$

It is possible to show, that for large systems (thermodynamic limit) the canonical and grand canonical density operators yield the same macroscopic results. In systems in which phase transitions may occur, transitions are indicated in a different way in both ensembles.

The relation between these two different approaches is established by the van Hove theorem (HUANG, 1964).

3.2. The Method of Green's Functions in Quantum Statistics

3.2.1. Definition of Green's Functions

The information about the behaviour and the properties of a many-body system is given by the reduced density matrices which may be represented with the construction operators; for example we have for the one- and two-particle density matrices

$$\begin{aligned} n\langle x_1 | F_1 | x_1' \rangle &= \langle \psi^+(x_1)\, \psi(x_1') \rangle , \\ n^2 \langle x_1 x_2 | F_{12} | x_2' x_1' \rangle &= \langle \psi^+(x_1)\, \psi^+(x_2)\, \psi(x_2')\, \psi(x_1') \rangle , \\ x_1 &= (\boldsymbol{r}_1, s_1) \text{ etc.} \end{aligned} \tag{3.72}$$

These expressions are special cases of the more general averages of field operators

$$\begin{aligned} G_1(11') &= \frac{1}{i\hbar} \operatorname{Tr} \{ \varrho T(\psi(1)\, \psi^+(1')) \} = \frac{1}{i\hbar} \langle T\{\psi(1)\, \psi^+(1')\} \rangle , \\ G_2(121'2') &= \frac{1}{(i\hbar)^2} \operatorname{Tr} \{ \varrho T(\psi(1)\, \psi(2)\, \psi^+(2')\, \psi^+(1')) \} \\ &= \frac{1}{(i\hbar)^2} \langle T\{\psi(1)\, \psi(2)\, \psi^+(2')\, \psi^+(1')\} \rangle \end{aligned} \tag{3.73}$$

which are called one- and two-particle (causal) Green's functions. Here ϱ is the density operator. We use the notation $1 = \boldsymbol{r}_1 s_1 t_1$ etc. T represents the time ordering operator which orders the operators ψ, ψ^+ in such a way that the operators with the earliest time appear on the right and the latest on the left, and include a factor $\lambda^{\pm}(P_N)$ defined by (3.10). P_N is the permutation of the field operators in order to obtain the T-order from the original order. For example in the one-particle Green's function we have

$$T\{\psi(1)\,\psi^+(1')\} = \psi(1)\,\psi^+(1')\,,\ t_1 > t_1'\,, \qquad (3.74)$$
$$= \pm\psi^+(1')\,\psi(1),\ t_1 < t_1'\,.$$

Therefore we can write the one-particle Green's function in the following form:

$$G_1(11') = \theta(t - t')\,G^>(11') + \theta(t' - t)\,G^<(11') \qquad (3.75)$$

where the functions

$$G^>(11') = \frac{1}{i\hbar}\langle\psi(1)\,\psi^+(1')\rangle,\quad t_1 > t_1'\,,$$

$$G^<(11') = \pm\frac{1}{i\hbar}\langle\psi^+(1')\,\psi(1)\rangle,\quad t_1 < t_1'\,, \qquad (3.76)$$

are called the two-time correlation functions. From this representation it is easy to show that the connection between the one-particle Green's function and the one-particle density matrix is

$$n\langle x|\,F_1\,|x'\rangle = i\hbar G^<(11')|_{t_1=t_1'} = i\hbar G_1(11')|_{t'=t^+}\,. \qquad (3.77)$$

Here the superscript + indicates an infinitesimally later time. There are several reasons for the generalization of (3.44) to the Green's function G_1.

The Green's functions introduced here are an appropriate combination of the formalism of the reduced density matrices of quantum statistics and the Green's functions of quantum field theory and allow to a large extent for a formal unification of the relativistic many-particle theory, of the quantum field theory and of the nonrelativistic many-particle theory.

The physical ideas of statistical physics and the powerful mathematical techniques of quantum field theory yields in this way solutions to problems of elementary particle physics, of nuclear physics, of plasma and solid state physics.

The Green's functions are determined by the equation of motion of the field operators, its commutation relations and by the manner of averaging. Obviously the Green's functions of relativistic physics are established by the relativistic field equations. In connection with the vacuum averaging we get the Green's function of quantum field theory

$$G_{\alpha\beta}(11') = \langle 0|\,T\{\psi_\alpha(1)\,\psi_\beta^+(1')\}\,|0\rangle\,.$$

If, however, a relativistic generalization of the grand canonical density operator of the shape

$$\varrho = \frac{1}{Z}\,e^{-\beta_\nu(p^\nu + \mu N)}$$

is applied (with $\beta_\nu = \dfrac{U\nu}{K_B T}$ as fourdimensional temperature and p^ν as the corresponding energy momentum vector) we may develop a relativistic many-body theory (relativistic quantum statistics).

The nonrelativistic many-particle theory considered in this monograph follows if we use the nonrelativistic equations of motion for the field operators and the averaging with the nonrelativistic density operator.

In practical applications it is sometimes sufficient to replace the density operator by the projector to the ground state

$$P_0 = |\Phi_0\rangle\langle\Phi_0|.$$

Then the Green's function has the following shape:

$$G_1(11') = \langle\Phi_0|\,T\{\psi(1)\,\psi^+(1')\}\,|\Phi_0\rangle\,.$$

The method of Green's functions is outlined in a series of monographs and review papers. Among them we mention the fundamental work by MARTIN and SCHWINGER (1959) and the monograph by KADANOFF and BAYM (1962). On the basis of the Kubo-Martin-Schwinger (KMS) condition the technique of the imaginary time Green's functions is developed.

Other fundamental developments are outlined in the books by ABRIKOSOV, GOR'KOV and DZYALOSHINSKII (1962), BONCH-BRUEVICH and TYABLIKOV (1961), and in the excellent review paper of ZUBAREV (1960). More recent representations are to be found in the book of FETTER and WALECKA (1971) and in the monograph of STOLZ (1974).

Finally, for systems in thermodynamic equilibrium the one-particle Green's function contains more information about the many-particle system than the one-particle density matrix does; for example:

1. The one-particle Green's function determines not only the average of additive operators, as the one-particle density matrix does, but also the average of the interaction and therefore the equation of state.
2. The one-particle Green's function describes for $t > t'$ the propagation of an additional particle inserted in a many-particle system. Therefore G_1 contains informations about the properties of a particle in a many-body system as the one-particle energy, one-particle states and the lifetime of these states.

We will show the points 1 and 2 later by consideration of the equation of motion for G and its analytical properties. In connection with the physical interpretation of the Green's function it is useful in many cases to introduce a notation by graphs. We represent the one-particle Green's function by a straight line with an arrow running from 1 to $\bar{1}$:

$G_1(1\bar{1}) =$ 

and for the two-particle Green's function which describes the correlated motion of two particles in a many-body system the graphical representation will be

$G_2(121'2') =$ 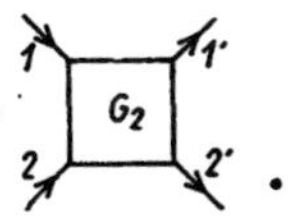

.

The graphs are known as Feynman diagrams because they were first introduced by Feynman in quantum electrodynamics.

For the two-particle Green's function such a simple decomposition in correlation functions as given in (3.75) is not possible for G_2. Because of the time ordering operator, G_2 describes, in dependence of the ordering of t_1, t_2, $t_{\bar{1}}$, $t_{\bar{2}}$, 4! different correlation functions, corresponding to 4! different scattering processes.

Of special interest is the time ordering $t_1 = t_2$; $\bar{t}_1 = \bar{t}_2$ which determines a function G_2 in the particle-particle channel (s-channel). In Feynman diagrams

$G_2(12\bar{1}\bar{2})_{t_1=t_2;\, t_{\bar{1}}=t_{\bar{2}}} =$ 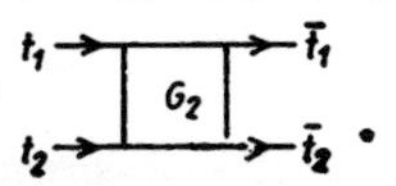

.

This function describes the propagation of two particles in the system. In this case we can write

$$G_2(t, \bar{t}) = \theta(t - \bar{t})\, G_2^{>}(x_1, x_2, \bar{x}_1, \bar{x}_2, t, \bar{t}) + \theta(\bar{t} - t)\, G_2^{<}(x_1, x_2, \bar{x}_1, \bar{x}_2, t, \bar{t}) \quad (3.78)$$

where the functions

$$G^{>}(t, \bar{t}) = \frac{1}{(i\hbar)^2} \langle \psi(x_1, t)\, \psi(x_2, t)\, \psi^+(\bar{x}_2, \bar{t})\, \psi^+(\bar{x}_1, \bar{t}) \rangle ,$$

$$G^{<}(t, \bar{t}) = \frac{1}{(i\hbar)^2} \langle \psi^+(\bar{x}_1, \bar{t})\, \psi^+(\bar{x}_2, \bar{t})\, \psi(x_2, t)\, \psi(x_1, t) \rangle \tag{3.78a}$$

are the two-time correlation functions corresponding to G_2. Another important time ordering is $t_1 = \bar{t}_1$; $t_2 = \bar{t}_2$ with G_2 in the t-channel:

$$G_2(12\bar{1}\bar{2})_{t_1=\bar{t}_1;\, t_2=\bar{t}_2} =$$ 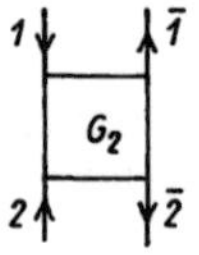

.

In Fermi systems this situation may be interpreted as the channel of particle-hole propagation.

So far we considered the Green's functions in the coordinate representation. In many cases, especially for homogeneous systems, it is more convenient to use the momentum representation. We find for example for the correlation functions

$$G^{<}(p_1, p_1', \tau) = \int d\boldsymbol{r}_1\, d\boldsymbol{r}_1'\, e^{-i(\boldsymbol{p}_1\boldsymbol{r}_1 - \boldsymbol{p}_1'\boldsymbol{r}_1')}\, G^{<}(r_1, r_1', \tau) ,$$

here

$$\tau = t_1 - \bar{t}_1, \text{ now } p\text{-wave-number};$$

and for homogeneous systems

$$\begin{aligned} G^{<}(p_1, p_1', \tau) &= G^{<}(p_1, \tau)\, \delta(\boldsymbol{p}_1 - \boldsymbol{p}_1') , \\ G^{<}(p_1, \tau) &= i\hbar \int d\boldsymbol{r}\, e^{-i\boldsymbol{p}_1\boldsymbol{r}}\, G^{<}(r, \tau) . \end{aligned} \tag{3.79}$$

The inverse transformation is given by

$$G^{<}(r_1 - r_1', \tau) = \frac{1}{i\hbar} \int \frac{d\boldsymbol{p}_1}{(2\pi)^3}\, e^{i\boldsymbol{p}_1(\boldsymbol{r}_1 - \boldsymbol{r}_1')} G^{<}(p_1) .$$

Taking into account that because of (3.77) and (3.72)

$$i\hbar G^{<}(p_1, p_1', \tau)|_{\tau=0} = n\langle p_1|\, F_1\, |p_1'\rangle = nF(p_1)\, \delta(\boldsymbol{p}_1 - \boldsymbol{p}_1')$$

$G^{<}(p)$ is to be interpreted as the momentum distribution

$$i\hbar G^{<}(p, \tau)|_{\tau=0} = n(p) \tag{3.79a}$$

with the normalization

$$\int \frac{d\boldsymbol{p}}{(2\pi)^3}\, n(p) = n .$$

3.2.2. General Properties of the Correlation Function and One-Particle Green's Function

In this section we will consider exact properties of correlation and Green's functions which are independent of the concrete many-particle system. That means we consider properties which follow from the definition and from general principles of quantum statistics.

First we consider $G^>(1\bar{1})$ and $G^<(1\bar{1})$ in the case of the thermodynamic equilibrium with the grand canonical density operator (3.62). Because of the cyclic invariance of the trace and the relation $[H, N] = 0$ (conservation of the number of particles) it is easy to show that $G^{\lessgtr}$ are functions only of the difference of times. That is valid also for the Green's function $G_1(1\bar{1})$:

$$G^{\gtrless}(t, \bar{t}) = G^{\gtrless}(t - \bar{t}) , \qquad G_1(t, \bar{t}) = G_1(t - \bar{t}) . \tag{3.80}$$

The exact correlation function $G^<$ is given by

$$G^<(1, \bar{1}) = \pm \frac{1}{i\hbar} Z^{-1} \operatorname*{Tr}_{\mathcal{H}} [\mathrm{e}^{-\beta(H-\mu N)} \psi^+(\bar{1}) \psi(1)] . \tag{3.81}$$

We consider the trace with respect to the complete set of common eigenstates of H and N:

$$\begin{aligned} &H \,|EN\rangle = E \,|EN\rangle, \ \mathrm{N} \,|EN\rangle = N \,|EN\rangle , \\ &1 = \sum_{EN} |EN\rangle \langle NE| = \sum_n |n\rangle \langle n|, \qquad |n\rangle = |EN\rangle. \end{aligned} \tag{3.82}$$

The field operators are given in the Heisenberg picture:

$$\psi(t) = \mathrm{e}^{\frac{i}{\hbar}Ht} \psi(0) \, \mathrm{e}^{-\frac{i}{\hbar}Ht} .$$

Applying this relations we get easily for $G^<(1\bar{1})$

$$G^<(1, \bar{1}) = \frac{\pm 1}{i\hbar} Z^{-1} \sum_{nm} \mathrm{e}^{-\beta(E_n-\mu N)} \langle n| \psi^+(\bar{x}_1) |m\rangle \cdot \langle m| \psi(x_1) |n\rangle \, \mathrm{e}^{+\frac{i}{\hbar}(E_m-E_n)(t-\bar{t})} . \tag{3.83}$$

In the same manner follows for $G^>(\bar{1}1)$

$$\begin{aligned} G^>(1, \bar{1}) = \frac{1}{i\hbar} Z^{-1} \sum_{nm} \mathrm{e}^{-\beta(E_n-\mu N)} \langle n| \psi(x_1) |m\rangle \\ \cdot \langle m| \psi^+(\bar{x}_1) |n\rangle \, \mathrm{e}^{\frac{i}{\hbar}(E_n-E_m)(t-\bar{t})} \end{aligned}$$

which can be rewritten in the form ($n \to m$)

$$\begin{aligned} G^>(1, \bar{1}) = \frac{1}{i\hbar} Z^{-1} \sum_{nm} \mathrm{e}^{-\beta(E_n-\mu N)} \langle n| \psi^+(\bar{x}_1) |m\rangle \\ \cdot \langle m| \psi(x_1) |n\rangle \, \mathrm{e}^{-\beta(E_m-E_n+\mu)} \, \mathrm{e}^{+\frac{i}{\hbar}(E_m-E_n)(t-\bar{t})} \end{aligned} \tag{3.84}$$

if we use the orthogonality of states with different N, i. e., $N_m = N_n - 1$.

Now we define the spectral function

$$I(\omega) = \frac{1}{Z} \sum_{nm} \langle n| \psi^+(\bar{x}_1) |m\rangle \langle m| \psi(x_1) |n\rangle \cdot \mathrm{e}^{-\beta(E_n-\mu N)} \, \delta(E_n - E_m - \hbar\omega) \, 2\pi . \tag{3.85}$$

This function is very important, because we may show that $I(\omega)$ contains the essential properties of the many-particle system. Using the spectral function, $G^>$ and $G^<$ may be written as

$$G^<(t-\bar{t}) = \pm \int I(\omega)\, e^{-i\omega(t-\bar{t})} \frac{d\omega}{2\pi i\hbar},$$
$$G^>(t-\bar{t}) = \int I(\omega)\, e^{\beta(\hbar\omega-\mu)}\, e^{-i\omega(t-\bar{t})} \frac{d\omega}{2\pi i\hbar}. \tag{3.86}$$

These fundamental relations are called Lehmann spectral representation for the correlation functions.

We can write these relations also in the form

$$(\mp) G^{\lessgtr}(t-\bar{t}) = \int \frac{d\omega}{2\pi i\hbar}\, e^{-i\omega(t-\bar{t})}\, G^{\gtrless}(\omega),$$
$$G^>(\omega) = I(\omega)\, e^{\beta(\hbar\omega-\mu)}, \quad G^<(\omega) = I(\omega).$$

An interesting question in connection with this representation is the question whether $G^{\gtrless}(\omega)$ can be interpreted as Fourier transforms of $G^{\gtrless}(\tau)$. This question will be discussed later.

We will consider now the essential properties of the correlation functions following from the spectral representation. The first important property is a connection between $G^>(\omega)$ and $G^<(\omega)$ following directly from (3.86):

$$G^<(\omega) = e^{-\beta(\hbar\omega-\mu)}\, G^>(\omega). \tag{3.87}$$

Introducing the more convenient spectral function $A(\omega)$ by

$$A(\omega) = G^>(\omega) \mp G^<(\omega)$$

we get because of (3.87) the useful relations

$$G^<(\omega) = f(\omega)\, A(\omega),$$
$$G^>(\omega) = \{1 \pm f(\omega)\}\, A(\omega). \tag{3.88}$$

Here $f(\omega)$ are the Bose (Fermi) like functions

$$f(\omega) = \frac{1}{e^{\beta(\hbar\omega-\mu)} \mp 1}. \tag{3.89}$$

The new function $A(\omega)$ is connected with $I(\omega)$ by

$$I(\omega) = f(\omega)\, A(\omega). \tag{3.90}$$

Using the integral representation for the step function $\theta(\tau)$

$$\theta(\tau) = \frac{1}{2\pi i} \int_{-\infty}^{\infty} \frac{e^{i\omega\tau}}{\omega + i\varepsilon}\, d\omega \tag{3.91}$$

we can express the Fourier transform of the one-particle Green's function also by the spectral function $A(p, \omega)$. We get easily from (3.75) and (3.88)

$$G_1(\omega) = \int_{-\infty}^{\infty} \frac{d\bar{\omega}}{2\pi} A(\bar{\omega}) \left\{ \frac{f(\bar{\omega})}{\omega - \bar{\omega} - i\varepsilon} \mp \frac{1 \pm f(\bar{\omega})}{\omega - \bar{\omega} + i\varepsilon} \right\}. \tag{3.92}$$

This is the Landau spectral representation (LANDAU and LIFSHITS, 1977) of the one-particle Green's function, which does, however, not provide for an analytic continuation to the complex energy plane. With the relations (3.92) and (3.88) it can be seen that the correlation functions and the one-particle Green's function are completely determined by $A(\omega)$.

A second group of properties of correlation and one-particle Green's function may be obtained by the consideration of the field operators for complex times t

$$\psi(x, t) = e^{\frac{i}{\hbar}Ht}\,\psi(x, 0)\,e^{-\frac{i}{\hbar}Ht}\,.$$

In this way we can define $G^{\gtrless}$ and G_1 in the complex $(t - t')$-plane (see KADANOFF and BAYM, 1962; MARTIN and SCHWINGER, 1959). If we assume that convergence of $G^<$ and $G^>$ is guaranteed by $e^{\beta(E_n-\mu N)}$ for real times, $G^<$ may be analytically continued in the region

$$\hbar\beta > \mathrm{Im}\,(t_1 - t_1') > 0 \tag{3.93}$$

and $G^>$ in the region

$$0 \geqq \mathrm{Im}\,(t_1 - t_1') > (-\hbar\beta)$$

of the complex $(t - t')$-plane. This is true because all exponentials in this regions are damped. Further it follows from (3.75) that the Green's function can also be extended into the complex $(t - t')$-plane as follows:

$$G_1(11') = \begin{cases} G^<(11') & \text{for } \hbar\beta > \mathrm{Im}\,(t_1 - t_1') > 0 \text{ and} \\ & \mathrm{Im}\,(t_1 - t_1') = 0\,,\ \mathrm{Re}\,(t_1 - t_1') < 0\,, \\ G^>(11') & \text{for } 0 > \mathrm{Im}\,(t_1 - t_1') > -\hbar\beta \\ & \text{and } \mathrm{Im}\,(t_1 - t_1') = 0\,,\ \mathrm{Re}\,(t_1 - t_1') > 0\,. \end{cases} \tag{3.94}$$

Between the correlation functions $G^>$ and $G^<$, in this extended region there exists a fundamental relation which we get directly from the spectral representation (3.86):

$$G^>(x_1, x_1', t_1 - t_1') = \pm\, e^{-\beta\mu}\, G^<(x_1, x_1', t_1 - t_1' + i\hbar\beta)\,. \tag{3.95}$$

This is the Kubo-Martin-Schwinger (KMS) condition.

From these considerations follows that the imaginary axis from $-i\beta$ to $i\beta$ plays a particularily important role. Along this line we can extend the definition of time ordering when the times are imaginary. We define in agreement with (3.94) t_1 earlier than t_1' if $\mathrm{Im}\,t_1 > \mathrm{Im}\,t_1'$. That means, the further down at the imaginary axis a time is, the later it is. Now we restrict t_1 and t_2 to imaginary values within $(0, -i\hbar\beta)$; then $t = 0$ is the earliest and $t = -i\beta\hbar$ the latest time possible, and we have

$$G_1(11')_{t_1=0} = G^<(11')_{t_1=0}\,, \qquad G_1(11')_{t_1=-i\hbar\beta} = G^>(11')_{t_1=-i\hbar\beta}\,; \tag{3.96}$$

therefore we obtain because of (3.45) the condition

$$G_1(11')_{t_1=0} = \pm\, e^{\beta\mu} G_1(11')_{t_1=-i\hbar\beta} \tag{3.97}$$

which is again the Kubo-Martin-Schwinger (KMS) condition (see MARTIN and SCHWINGER, 1959). This condition connects the Green's function at the boundary of the region $(0, -i\hbar\beta)$. The KMS condition is automatically fulfilled if we expand G as a Fourier series with respect to t and t' in the region $(0, -i\hbar\beta)$. For this purpose it is necessary to extend the Greens' function into stripes parallel to the real axis.

Then we get

$$G_1(11') = \frac{i}{\hbar\beta} \sum_\nu G_1(x_1, x_1', z_\nu)\, \mathrm{e}^{-iz_\nu(t_1 - t_1')} . \tag{3.98}$$

In order to reproduce the KMS condition correctly the so-called Matsubara frequencies z_ν are to be used; the latter are given by

$$z_\nu = \frac{\pi\nu}{-i\hbar\beta} + \frac{\mu}{\hbar}, \qquad \nu = \begin{cases} \pm 1, \pm 3, \dots \text{ for fermions,} \\ 0, \pm 2, \pm 4, \dots \text{ for bosons.} \end{cases}$$

The Fourier coefficients can be obtained in the well known manner:

$$G_1(z_\nu) = \int_0^{-i\hbar\beta} \mathrm{d}t_1\, \mathrm{e}^{-iz_\nu(t_1 - t_1')}\, G_1(11') . \tag{3.99}$$

For the evaluation we may take $t' = 0$ because $G_\nu(z_\nu)$ is independent of t'. Using (3.96), (3.87) and (3.88), it follows

$$G_1(z_\nu) = \int_{-\infty}^{\infty} \frac{\mathrm{d}\omega}{2\pi\hbar} \frac{A(x_1, x_1', \omega)}{z_\nu - \omega} . \tag{3.100}$$

This result is very interesting because it shows that the Fourier coefficient $G_1(z_\nu)$ is also determined by the spectral function $A(x, x', \omega)$. In the case of Bose particles, we observe a pole for $z_\nu = 0$, and expression (3.100) is not well defined. This question will be considered in more detail in Section 3.2.3.

Now the question is important, whether the function $A(x, x', \omega)$ can be obtained from $G_1(z_\nu)$, that means, whether eq. (3.100) may be inverted. In order to consider this problem we continue $G_1(z_\nu)$ from the discrete points z_ν along the straight line parallel to the imaginary axis into the full complex z-plane:

$$G_1(z) = \int_{-\infty}^{\infty} \frac{\mathrm{d}\omega}{2\pi\hbar} \frac{A(x_1, x_1', \omega)}{z - \omega} . \tag{3.101}$$

This is an integral of Cauchy type which has the following properties:

1. By (3.101) there are defined two analytical functions

$$G_{\mathrm{I}}(z) \quad \text{for} \quad \mathrm{Im}\, z > 0 ,$$

$$G_{\mathrm{II}}(z) \quad \text{for} \quad \mathrm{Im}\, z < 0 .$$

2. $G_1(z)$ has a branch cut with the following limiting behaviour (Plemlj formulae):

$$G_1(\omega \pm i\varepsilon) = \mp \frac{\pi i}{\hbar} A(x_1, x_1', \omega) + P \int \frac{A(x_1, x_1', \omega)}{\omega - \overline{\omega}} \frac{\mathrm{d}\overline{\omega}}{2\pi\hbar} \tag{3.102}$$

or more compact (Dirac identity)

$$\frac{1}{\omega \pm i\varepsilon} = \mp i\pi\, \delta(\omega) + P \frac{1}{\omega} . \tag{3.103}$$

3. From 2. follows the discontinuity at the cut:

$$A(\omega) = i\hbar[G_1(\omega + i\varepsilon) - G_1(\omega - i\varepsilon)] . \tag{3.104}$$

4. It is possible to continue $G_{\rm II}(z)$ and $G_{\rm I}(z)$ analytically beyond the cut. We get

$$G_{\rm I}(z) = G_{\rm II}(z) + A(z), \quad \operatorname{Im} z < 0;$$
$$G_{\rm II}(z) = G_{\rm I}(z) - A(z), \quad \operatorname{Im} z > 0\,.$$

These functions are no longer analytic; they have singularities which are those of $A(z)$.

These results are very important, since it is difficult to determine the real time Green's functions, for example by perturbation theory or from the equation of motion with suitable boundary conditions. In practice it is usually most simple to calculate the Green's function along the imaginary time axis. Especially by using the KMS condition and the equation of motion, $G_1(z_\nu)$ can be determined, and by analytical continuation, $G_1(z_\nu) \to G_1(z)$, the spectral function $A(\omega)$ follows with (3.104). With $A(\omega)$ we can determine the real time correlation and Green's functions by using the spectral representations (3.88), (3.92).

3.2.3. Long Time Behaviour of Correlation Functions

In order to determine the long time behaviour of the correlation function which is essential for the interpretation of $G^{\lessgtr}(\omega)$ as Fourier transform of $G^{\gtrless}(t - t')$, it is necessary to consider the spectral function $I(\omega)$ for $\omega \to \mu/\hbar$. Since $I(\omega)$ has in the case $E_n = E_m$, for $\omega = 0$, a δ-type singularity, it is useful to divide the spectral function into two parts (Kwok and Schultz, 1969; Ramos and Gomes, 1971):

$$I(\omega) = I^0(\omega) + I'(\omega) \tag{3.105}$$

with

$$\begin{aligned} I^0(\omega) &= \frac{1}{Z}\frac{1}{i\hbar}\sum_{nm}\langle n|\,\psi^+(x_1)\,|m\rangle\,\langle m|\,\psi(x_1')\,|n\rangle\,\mathrm{e}^{-\beta(E_n-\mu N)}\delta\left(\omega-\frac{\mu}{\hbar}\right)2\pi \\ &= C(x_1, x_1')\,\delta\left(\omega-\frac{\mu}{\hbar}\right), \quad E_n - E_m = \mu \end{aligned} \tag{3.106}$$

and

$$\begin{aligned} I'(\omega) &= \frac{1}{Z}\frac{1}{i\hbar}\sum_{nm}\langle n|\,\psi^+(x_1)\,|m\rangle\,\langle m|\,\psi(x_1')\,|n\rangle\,\mathrm{e}^{-\beta(E_n-\mu N)}\,\delta(\hbar\omega - E_n + E_m)\,2\pi, \\ E_n &- E_m \neq \mu\,, \end{aligned} \tag{3.107}$$

$I'(\omega)$ is the part of $I(\omega)$ which is regular for $\omega = 0$. We can write

$$I(\omega) = C(x_1, x_1')\,\delta\left(\omega - \frac{\mu}{\hbar}\right) + I'(\omega)\,. \tag{3.108}$$

Using eq. (3.86) we get therefore for the correlation function $G^<(t - t')$

$$G^<(t-t') = \pm \int_{-\infty}^{\infty} I'(\omega)\,\mathrm{e}^{i\omega(t'-t)}\frac{\mathrm{d}\omega}{2\pi i\hbar} + C(x_1, x_1')\,\mathrm{e}^{\frac{i}{\hbar}(t'-t)\mu} \tag{3.109}$$

with the long time behaviour

$$\lim_{t\to\infty} G^<(t - t') = C(x_1, x_1') \lim_{t\to\infty} \mathrm{e}^{\frac{i}{\hbar}(t-t')\mu}\,.$$

Such a behaviour is called "time long range order" (LRO). LRO for the one-particle correlation function can therefore be observed if the diagonal contribution is nonvanishing, i. e.,

$$\frac{1}{Z}\frac{1}{i\hbar}\sum_{\substack{nm\\E_n=E_m}} \langle n|\,\psi^+\,|m\rangle\,\langle m|\,\psi\,|n\rangle\,\mathrm{e}^{-\beta(E_n-\mu N)} \neq 0$$

in the thermodynamic limit $N \to \infty$ $V \to \infty$.

The concept of LRO was introduced by ONSAGER and PENROSE and YANG (1962) in connection with the problem of macroscopic quantum phenomena as Bose condensation and suprafluidity. In this consideration the so-called "spatial long range order" for the one- and two-particle density matrix was observed:

$$\lim_{r-r'\to\infty} \langle r|\,F\,|r'\rangle \neq 0\,.$$

The "time long range order" in Green's function first time was introduced in papers of CALLEN, SWENDSON and TAHIR KHELI (1967), KWOK and SCHULZ (1969). An excellent introduction to the problem of LRO and "macroscopic quantum phenomena" can be found in papers of STOLZ (1975, 1976) and PUFF (1976). See also KREMP, KRAEFT and KILIMANN (1978).

It is easy to show that for ideal Fermi systems $C(x_1x_1') = 0$ for $V \to \infty$. However, for ideal Bose systems we find

$$C = \frac{N_0}{V}\,\delta_{\mu,0}\,\delta_{p,0}\,. \tag{3.110}$$

Here N_0 is the macroscopic occupation of the ground state. Therefore, if LRO takes place, the spectral representations are to be modified. From the definition of $A(\omega)$ and using (3.85) it can be seen, that the contribution with $E_n = E_m$ is cancelled out in $A(\omega)$. Therefore, in the Bose case the correlation function is not completely determined by $A(\omega)$, because $A(\omega)$ is related only to $I'(\omega)$ by $I'(\omega) = f(\omega)\,A(\omega)$. Instead of (3.88), we have to write correctly for Bose particles (PUFF, 1979)

$$\begin{aligned} G^<(x_1, x_1', \omega) &= Pf(\omega)\,A(x_1, x_1', \omega) + 2\pi\,\delta(\omega - \mu/\hbar)\,C(x_1, x_1')\,,\\ G^>(x_1, x_1', \omega) &= [1 + Pf(\omega)]\,A(x_1, x_1', \omega) + 2\pi\,\delta(\omega - \mu/\hbar)\,C(x_1, x_1') \end{aligned} \tag{3.111}$$

with

$$f(\omega) = \frac{1}{\mathrm{e}^{\beta(\hbar\omega-\mu)} - 1}\,.$$

It is clear that for ideal Bose systems in this way the LRO is connected with the well known phenomena of Bose condensation. In the case of ideal Bose systems is

$$A(p, \omega) = 2\pi\hbar\delta\left(\hbar\omega - \frac{\hbar^2p^2}{2m}\right), \qquad C = \frac{N_0}{V}\,\delta_{\mu,0}\delta_{p,0}\,.$$

With (3.77) for homogeneous systems, the particle density is

$$n = (2s + 1)\left[\frac{N_0}{V}\,\delta_{\mu,0} + P\int\limits_{-\infty}^{\infty}\frac{dp}{(2\pi)^3}\,\frac{1}{\mathrm{e}^{\beta(\hbar^2p^2/2m-\mu)} - 1}\right]. \tag{3.112}$$

This result is to be interpreted in the well known manner. In the case $\mu = 0$ we have a macroscopic number of particles N_0 in the ground state, which form the Bose condensate. The remaining particles are still distributed according to the Bose function.

In the same manner the spectral representation for the Fourier coefficient of G_1 (3.100) is to be modified. In the expression (3.100), LRO is indicated for Bose particles (more general for Bose like Matsubara frequencies) by the pole for $\nu = 0$ and $\hbar\omega = \mu$ on the real axis. Because of this pole the expression (3.100) is not entirely defined. We get the correct spectral representation if we use expression (3.111). The result is

$$G_1(z_\nu) = -\beta C(x_1, x_1')\, \delta_{\nu,0} + P \int\limits_{-\infty}^{\infty} \frac{d\omega}{2\pi\hbar} \frac{A(x_1, x_1', \omega)}{z_\nu - \omega}. \tag{3.113}$$

The weight function $A(x_1, x_1', \omega)$ follows from the analytic continuation $G_1(z)$:

$$A(x_1, x_1', \omega) = \{G_1(\omega + i\varepsilon) - G_1(\omega - i\varepsilon)\}\, i\hbar\,.$$

A special problem is the determination of $C(x_1, x_1')$. In order to get $C(x_1, x_1')$ we consider the expression

$$\theta(t - t')\,\{G^>(11') + G^<(11')\} = G^{\mathrm{r}}_+(11') \tag{3.114}$$

which is called retarded anticommutator Green's function. In this combination the diagonal contribution ($E_n = E_m$) is not cancelled out as in the case of $A(x_1x_1'\omega)$. It is not difficult to show the following spectral representation

$$G^{\mathrm{r}}_+(\omega) = \frac{1}{2\pi}\, n_{\mathrm{F}}(0)^{-1}\, \frac{C(x, x')}{\omega} + \frac{1}{2\pi} \int\limits_{-\infty}^{\infty} (e^{\beta\hbar\omega} + 1)\, I'(\omega)\, \frac{d\bar{\omega}}{\bar{\omega} - \omega}. \tag{3.115}$$

That means

1. $G^{\mathrm{r}}_+(\omega)$ has a branch cut along the real axis with the discontinuity $I'(\omega)$.
2. $G^{\mathrm{r}}_+(\omega)$ has a simple pole at $\omega = 0$ with the residuum

$$\frac{C}{2\pi} = -\lim_{\omega \to 0} G^{\mathrm{r}}_+(\omega)\, \frac{\omega}{e^{\beta\hbar\omega} + 1}. \tag{3.116}$$

Therefore it is possible to determine the structure of $C(x_1x_1')$ with the help of the function $G^{\mathrm{r}}_+(\omega)$ (Ramos and Gomes, 1971).

3.2.4. Equation of Motion for the One-Particle Green's Function. Self Energy

We have now to consider the problem of determining the Green's function. In order to solve this problem two methods were developed. The method of perturbation theory starts from the definition of the Green's function in the interaction picture using the adiabatic theorem and the well known perturbation expansion for the evolution operator. This method, which is represented in several text books (Abrikosov, Gor'kov and Dzyloshinskii, 1962; Landau and Lifshits, 1982), works in a simple way only at $T = 0$ and for $T \neq 0$ along the imaginary axis. For real times and $T \neq 0$ a more complicated technique of double real-time Green's function was developed (see Landau and Lifshits, 1982).

More general is a method which is based on the equation of motion for the field operators $\psi(x_1, t_1) = \psi(1)$:

$$\left(i\hbar \frac{\partial}{\partial t} + \frac{\hbar^2\nabla^2}{2m}\right)\psi(x_1, t_1) = \int d\bar{x}\; V(x_1 - \bar{x})\,\psi^+(\bar{x}, t_1)\,\psi(\bar{x}, t_1)\,\psi(x_1, t_1)\,.$$

If this equation is used, we find by differentiation of the Green's function the following equation of motion:

$$\left(i\hbar \frac{\partial}{\partial t_1} + \frac{\hbar^2}{2m}\nabla_1^2\right) G_1(1,1') = \delta(1 - 1') \pm i\hbar \int d\bar{1}\; V(1-\bar{1})\; G_2(1\bar{1}, 1'\bar{1}^+),$$

$$V(1 - \bar{1}) = \delta(t_1 - \bar{t}_1)\, V(x_1 - \bar{x}_1)\,,\; t_1^+ = t_1 + 0. \tag{3.117}$$

This equation is not a closed one, but the first member of a chain of equations for the many-particle Green's functions. Therefore we get, as in the case of density operators, a hierarchy of equations, coupled by the interaction. Furthermore we remark that in deriving these equations we do not use the concrete form of the averaging. The equations are therefore valid for any statistical operator. In order to find the Green's function for a certain averaging, boundary conditions are necessary. For example, we obtain thermodynamic Green's function defined by the grand canonical density operator with the KMS condition.

From (3.117) we find a formally closed equation of motion, if we introduce the so-called self-energy operator Σ by

$$\int d\bar{1}\; \Sigma(1\bar{1})\, G_1(\bar{1}1') = \pm i\hbar \int d2\; V(1 - 2)\, G_2(121'2^+)\,. \tag{3.118}$$

With this definition we get from (3.117) the Dyson equation

$$\left(i\hbar \frac{\partial}{\partial t_1} + \frac{\hbar^2}{2m}\nabla_1^2\right) G_1(11') = \delta(1 - 1') + \int d\bar{1}\; \Sigma(1\bar{1})\, G_1(\bar{1}1')\,. \tag{3.119}$$

It is possible to convert the Dyson equation into an integral equation if we define the inverse Green's function by

$$\int d\bar{1}\; G_1(1\bar{1})\, G_1^{-1}(\bar{1}1') = \delta(1 - 1')\,.$$

Then follows from the Dyson equation for $\Sigma = 0$

$$(G_1^0)^{-1}\,(11') = \left\{i\hbar \frac{\partial}{\partial t_1} + \frac{\hbar^2}{2m}\nabla^2\right\}\delta(1 - 1')$$

for the inverse of the single particle function of noninteracting systems. Using $(G_1^0)^{-1}$ we obtain the Dyson equation in the form of an integral equation

$$G_1(11') = G_1^0(11') + \int_0^{-i\beta\hbar} d\bar{1}\; d2\; G_1^0(1\bar{1})\, \Sigma(\bar{1}2)\, G_1(21') \tag{3.120}$$

where the region of the time integration follows from the KMS condition.

With Feynman diagrams we get

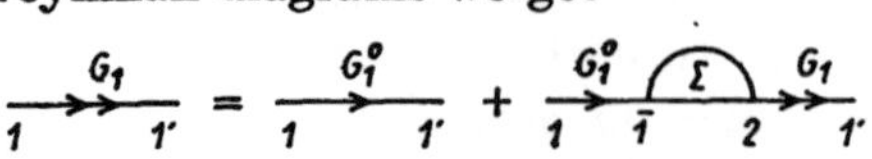

with

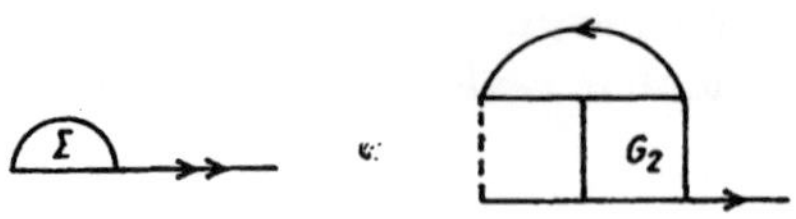

for the self energy and

 $= i\hbar\, V(1\bar{1})$,  $= G_1^0(1\bar{1})$

for the interaction and for the free particle Green's function. In the language of Feynman diagrams Σ consists of all diagrams which cannot be separated into two parts by cutting a one-particle line G_1^0.

Let us now consider in more detail the properties of the self-energy operator. The self energy is determined by the two-particle Green's function. The simplest approximation is the two-particle Green's function for ideal systems, given by

$$G_2^0(121'2') = G_1(11')\, G_1(22') \pm G_1(12')\, G_1(21')$$

= 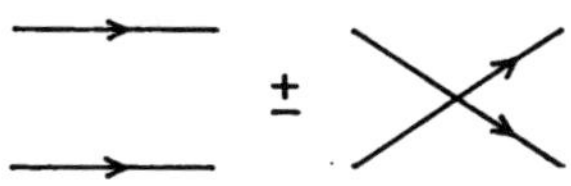 . (3.121)

The second term is a consequence of the symmetry postulate. We get the Hartree-Fock self energy to

$$\Sigma^{\mathrm{HF}}(11') = \pm i\hbar\{\delta(1-1') \int \mathrm{d}\bar{1}\; V(1-\bar{1})\, G_1(\bar{1}\bar{1}^+) \pm V(11')\, G_1(11'^+)\}$$

= 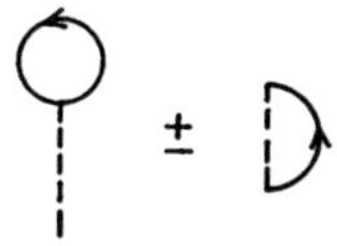 . (3.122)

With the representation $G_2 = G_2^0 + G_2^C$, Σ may be subdivided into

$$\Sigma = \Sigma^{\mathrm{HF}} + \Sigma^{\mathrm{C}} .$$

Because of eq. (3.120), Σ^{C} has a similar spectral representation as $G_1(11')$. We notice, that Σ^{C} is composed of the two functions $\Sigma^{\gtrless}$ by

$$\Sigma^{\mathrm{C}}(1\bar{1}) = \begin{cases} \Sigma^{>}(1\bar{1}) & \text{for} \quad it > i\bar{t}\,, \\ \Sigma^{<}(1\bar{1}) & \text{for} \quad it < i\bar{t} \end{cases} \tag{3.123}$$

which obey the same KMS conditions as $G^{>}$, $G^{<}$ and G_1 do. Therefore we get

$$\begin{aligned} \Sigma^{>}(x, x', \omega) &= \Gamma(x, x', \omega)\,[1 \pm f(\omega)]\,, \\ \Sigma^{<}(x, x', \omega) &= \Gamma(x, x', \omega)\, f(\omega)\,, \\ \Gamma(x, x', \omega) &= \Sigma^{>}(x, x', \omega) \mp \Sigma^{<}(x, x', \omega)\,. \end{aligned} \tag{3.124}$$

Further $\Sigma^{\mathrm{C}}(1\bar{1})$ may be expanded into a Fourier series with the spectral representation

$$\Sigma^{\mathrm{C}}(x_1, x_1', z_\nu) = \int_{-\infty}^{\infty} \frac{\mathrm{d}\omega}{2\pi}\, \frac{\Gamma(x_1, x_1', \omega)}{z_\nu - \omega} \tag{3.125}$$

for the Fourier coefficients.

If the system is translational invariant, the momentum representation of G_1 is given by

$$G_1(p_1, p_1', z_\nu) = G_1(p_1, z_\nu)\, \delta(p_1 - p_1') .$$

This will be used for the discussion of the spectral properties of Σ and G_1.

In order to solve Dyson's equation it is convenient to use the Matsubara momentum representation

$$G_1(11') = \int \frac{dp}{(2\pi)^3} e^{ip_1(r_1 - r_1')} \frac{1}{-i\beta\hbar} \sum_\nu e^{-iz_\nu(t-t')} G_1(p_1, z_\nu) .$$

In this representation Dyson's equation is an algebraic equation for $G_1(p, z_\nu)$ with the solution

$$G_1(p, z_\nu) = \frac{1}{\hbar z_\nu - E(p) - \Sigma^C(p, z_\nu)}, \qquad E(p) = \hbar^2 p^2/2m + \Sigma^{HF}(p) . \quad (3.126)$$

Because of (3.101), $A(p, \omega)$ is the discontinuity across the cut of the analytic continuation $G_1(z)$ of $G_1(z_\nu)$. We get for the spectral function

$$A(p, \omega) = \{G_1(\omega + i\varepsilon) - G_1(\omega - i\varepsilon)\} i\hbar$$
$$= \frac{\hbar\Gamma(p\omega)}{[\hbar\omega - E(p) - \operatorname{Re} \Sigma^C(p, \omega)]^2 + \left[\frac{\Gamma(p,\omega)}{2}\right]^2} . \quad (3.127)$$

As will be seen, $A(p, \omega)$ contains all information about the dynamical behaviour of a particle in an interacting many-body system and all possible information about the thermodynamic properties.

Let us now consider some special cases of the exact expression (3.127) for $A(p, \omega)$. In the case of noninteracting particles we get $\Gamma = 0$, $E(p) = \hbar^2 p^2/2m$; therefore may be written

$$A(p, \omega) = 2\pi\hbar\delta\left(\hbar\omega - \frac{\hbar^2 p^2}{2m}\right). \quad (3.128)$$

Using the Hartree-Fock approximation we find also $\Gamma = 0$ and therefore

$$A(p, \omega) = 2\pi\hbar\delta(\hbar\omega - E(p)) . \quad (3.129)$$

Only in this two cases we have $\Gamma = 0$. In other approximations we find a finite Γ. For simplification we consider first the approximation that Γ and $\operatorname{Re} \Sigma$ are independent of ω:

$$A(p, \omega) = \frac{\Gamma(p)}{[\hbar\omega - E(p) - \operatorname{Re} \Sigma^C(p)]^2 + \left[\frac{\Gamma(p)}{2}\right]^2} . \quad (3.130)$$

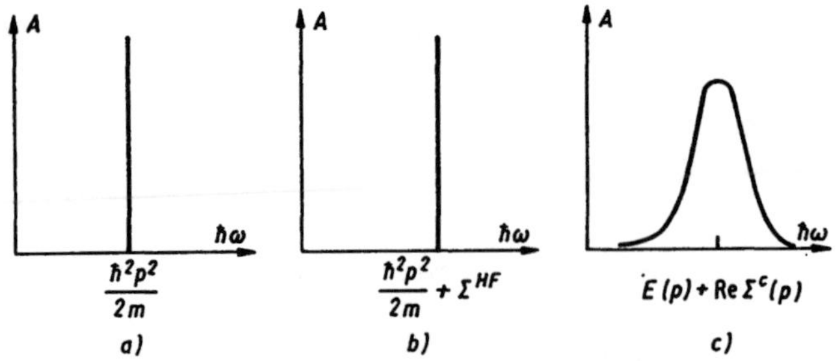

Fig. 3.1. Shape of the spectral function in different approximations
a) ideal gas, b) Hartree-Fock approximation, c) Lorentz form obtained for frequency-independent correlations

$A(p\omega)$ now has not the simple δ-shape but a Lorentz form with a finite width. In Fig. 3.1. we compare the different $A(p, \omega)$.
If $\Gamma \ll E(p) + \text{Re}\,\Sigma\,(p)$, $A(p, \omega)$ is strongly peaked for $\hbar\omega = E + \text{Re}\,\Sigma$.

In general $A(p, \omega)$ is characterized by poles in the complex z-plane $\hbar z = \varepsilon + i\,\frac{\Gamma}{2}$ which are determined by the dispersion relation (as to be seen from (3.130)):

$$\hbar\omega - [E(p) + \text{Re}\,\Sigma^{\text{C}}\,(p, \omega) \pm \frac{i}{2}\,\Gamma(p, \omega)] = 0\,.$$

If $\varepsilon \gg \frac{\Gamma}{2}$ we can calculate ε and Γ form the simpler relations

$$\varepsilon(p) = E(p) + \text{Re}\,\Sigma^{\text{C}}(p, \varepsilon(p))\,, \qquad \Gamma = 2\,\text{Im}\,\Sigma^{\text{C}}(p, \varepsilon(p))\,. \tag{3.131}$$

See also Section 4.3. As a consequence of this, the analytic continuations of $G_{\text{I}}(z_\nu)$ to $\text{Im}\,z < 0$, and $G_{\text{II}}(z_\nu)$ to $\text{Im}\,z > 0$ have poles at $\hbar z = \varepsilon \pm i\,\frac{\Gamma}{2}$.

3.2.5. Dynamical and Thermodynamical Information Contained in the Spectral Function $A\,(p, \omega)$

The spectral function contains all possible information about the behaviour of one particle in an interacting many-body system. Moreover $A(p, \omega)$ determines completely the thermodynamical properties of the system. The propagation of one particle in an interacting many-particle system is determined by the correlation function $G^{>}(p, t - t')$:

$$G^{>}(p, t - t') = \int\limits_{-\infty}^{\infty} \frac{d\omega}{2\pi}\,A(p, \omega)\,[1 \pm f(\omega)]\,e^{-i\omega(t-t')}\,. \tag{3.132}$$

The correlation function determines the averaged probability amplitude that the momentum of an extra particle is unchanged during the time $t - t'$, that means, $G^{>}(p, t - t')$ is the probability to remain in the momentum state p. It is intuitively clear, that in an interacting many particle system these states should decay and therefore the probability should be damped.

In the case of ideal particles we get

$$G^{>}(p, t - t') = e^{-\frac{i\hbar p^2}{2m}(t-t')}\,f\left(\frac{p^2}{2m}\right) \tag{3.133}$$

that means $|G^{>}|^2$ is independent of t and therefore the one-particle state is stable.

The situation is more complicated if $\Gamma \neq 0$. In order to study this case we consider for simplification a Lorentz like $A(p, \omega)$ (3.130). Using the residue theory we get easily (see, e.g., Stolz, 1974) for times $t - t' > (\hbar\Gamma)^{-1}$

$$G^{>}(p, t - t') = [1 \pm f(E(p) + \text{Re}\,\Sigma(p))]\,e^{-\left\{i(E(p)+\text{Re}\,\Sigma(p)) + \frac{\Gamma}{2}\right\}|t-t'|/\hbar} \tag{3.134}$$

if we assume that $\beta\Gamma \ll 1$ and $E(p) + \mathrm{Re}\,\Sigma(p) \gg \Gamma$. With respect to the physical meaning of $G^>$ we can give therefore the following interpretation:

$$\left.\begin{array}{l} \dfrac{\hbar}{\Gamma} = \tau - \text{life-time} \\ E(p) - \text{energy} \end{array}\right\} \text{of the one-particle states.}$$

In the case that $E(p) \gg \Gamma$, that means if the line width of $A(p, \omega)$ is very small, the Lorentzian is nearly the δ-like spectral function of ideal systems. This long lived one-particle states are usually interpreted as states of free quasiparticles.

Because the poles of the solution of the inhomogeneous Dyson equation are eigenvalues of the homogeneous Dyson equation

$$[\hbar z - E(p) - \Sigma^c(p, z)]\,\varphi(p) = 0 \tag{3.135}$$

this equation can be interpreted as an effective Schrödinger equation for the quasiparticles states $\varphi(p)$. In general the effective one-particle Hamiltonian H defined by this equation is non hermitean and so it is difficult to find the connection between the spectral function $A(p, \omega)$ and the solutions $\varphi_n(p)$. Only in the hermitean case we can write G_1 in terms of the solutions of (3.135), ω_n and φ_n:

$$G_1(p, z) = \sum_n \frac{\varphi_n(p)\,\varphi_n^*(p)}{z - \omega_n}. \tag{3.136}$$

This is the so-called bilinear expansion of G_1. From this we get $A(p, \omega)$ in terms of quasi particle energies and states:

$$A(p, \omega) = \sum_n \varphi_n(p)\,\varphi_n^*(p)\,\delta(\omega - \omega_n). \tag{3.137}$$

Now we will consider the thermodynamic behaviour. The thermodynamic properties can be obtained in the following way. From (3.79a) follows

$$n = i\hbar \int \frac{dp}{(2\pi)^3}\, G^<(p, t - t')_{t=t'}\,. \tag{3.138}$$

Using (3.87) and (3.88) we may write the following relation:

$$n(\mu, T) = \int \frac{dp\, d\omega}{(2\pi)^4}\, A(p, \omega)\, f(\omega)\,. \tag{3.139}$$

From the relation (3.67) we get the equation of state

$$p(\mu, T) = \int_{-\infty}^{\mu} n(T, \mu')\, d\mu'\,. \tag{3.139a}$$

The thermodynamic properties of ideal systems are completely determined by the spectral function

$$A(p, \omega) = 2\pi\hbar\, \delta(\hbar\omega - \hbar^2 p^2/2m)\,.$$

Using the relations (3.88) and (3.139) we find

$$n(\mu, T) = \int \frac{dp}{(2\pi)^3}\, [e^{\beta(\hbar^2 p^2/2m - \mu)} \mp 1]^{-1} \tag{3.140}$$

and for the equation of state we get

$$p(\mu, T) = k_B T \int \frac{dp}{(2\pi)^3}\, \ln[1 \mp e^{-\beta(\hbar^2 p^2/2m - \mu)}]\,. \tag{3.140a}$$

Let us now consider a system of interacting particles. Then, in many cases it is more convenient to start with the equation (3.68). This equation is simply written in terms of Green's functions

$$p(T, V, \mu) - p_0 = -\frac{1}{2}\int_0^1 \frac{d\lambda}{\lambda}\int d\bar{1}\; V(1\bar{1})\, G_2(1\bar{1}1^{++}\bar{1}^{+})|_{\bar{t}_1 = t_{\bar{1}}^{+}} . \tag{3.141}$$

Using the definition of the self energy this equation may be rewritten:

$$p(\mu, T, V) = p_0 - \frac{1}{2}\int_0^1 \frac{d\lambda}{\lambda}\int d\bar{1}\; \Sigma(1\bar{1})\, G_1(\bar{1}1^{+}) . \tag{3.142}$$

This equation is a useful starting point in order to obtain approximate expressions for the equation of state. Further, we find with the help of the equation of motion (3.19)

$$p(\mu, T) = p_0 + \frac{1}{V}\int_0^1 \frac{d\lambda}{\lambda}\int \left[i\hbar\frac{\partial}{\partial t} - \frac{\hbar^2}{2m}\Delta\right] G^{<}(x, t, x', t')|_{\substack{x=x' \\ t=t'}}\, dx\, dt$$

and finally by Fourier transform with respect to $r - r'$ and $t - t'$ and the relation (3.88)

$$p - p_0 = -\frac{1}{2}\int_0^1 \frac{d\lambda}{\lambda}\int \frac{dp}{(2\pi)^3}\,(\hbar\omega - \hbar^2p^2/2m)\, A(p, \omega)\, f(\omega)\, \frac{d\omega}{2\pi} . \tag{3.143}$$

In addition to this equation we use the equation

$$n(\mu, T) = \frac{\partial}{\partial\mu}\, p(\mu, T) \tag{3.143a}$$

in order to obtain the equation of state as a function of the density.

The essential quantity to derive the thermodynamic properties is the spectral weight function $A(p, \omega)$ given by (3.127). By inserting expression (3.127) in the thermodynamic relations (3.139) or (3.143) rather involved expressions are obtained. Therefore the following approximation for $A(p, \omega)$ seems to be very useful. We expand (3.127) near $\Gamma = 0$, i.e. near the undamped case:

$$A(p, \omega) = A(p, \omega)\Big|_{\Gamma=0} + \frac{1}{\hbar}\frac{dA(p, \omega)}{d\omega}\Big|_{\Gamma=0} \Gamma(p, \omega) . \tag{3.144}$$

In lowest order we then obtain

$$A(p, \omega)|_{\Gamma=0} = 2\pi\hbar\delta\left(\hbar\omega - \frac{\hbar^2p^2}{2m} - \mathrm{Re}\, \Sigma(p, \omega)\right).$$

Using the well known δ-function properties we find

$$A(p, \omega)|_{\Gamma=0} = \frac{2\pi\hbar}{1 - \frac{1}{\hbar}\frac{\partial}{\partial\omega}\mathrm{Re}\, \Sigma(p, \omega)|_{\hbar\omega=\varepsilon(p)}}\, \delta(\hbar\omega - \varepsilon(p))$$

where $\varepsilon(p)$ is the solution of the dispersion relation for the quasiparticle energy

$$\varepsilon(p) = \frac{\hbar^2 p^2}{2m} + \mathrm{Re}\, \Sigma\, (p, \omega)|_{\hbar\omega=\varepsilon(p)} \,.$$

The second term in equation (3.144) can be written in the form

$$\frac{\mathrm{d}A(p, \omega)}{\mathrm{d}\omega}\Big|_{\Gamma=0} \Gamma(p, \omega) = -\frac{1}{\pi} \frac{\partial}{\partial\omega} \frac{P}{\omega - \varepsilon(p)/\hbar} \Gamma(p, \omega) \,.$$

Now we expand the denominator and find the following expression for $A(p, \omega)$:

$$A(p, \omega) = 2\pi\hbar\delta(\hbar\omega - \varepsilon(p)) \left\{1 + \frac{1}{\hbar} \frac{\partial}{\partial\omega} \mathrm{Re}\, \Sigma(p, \omega)|_{\hbar\omega=\varepsilon(p)}\right\}$$
$$- \frac{1}{\pi} \left\{\frac{\partial}{\partial\omega} \frac{P}{\hbar\omega - \varepsilon(p)}\right\} \Gamma(p, \omega) \,. \qquad (3.145)$$

Because of (3.125), $\mathrm{Re}\, \Sigma(p, \omega)$ may be expressed by Γ:

$$\mathrm{Re}\, \Sigma(p, \omega)|_{\hbar\omega=\varepsilon(p)} = P \frac{1}{\pi} \int \mathrm{d}\bar{\omega}\, \frac{\Gamma(p, \bar{\omega})}{\bar{\omega} - \varepsilon(p)/\hbar} \,.$$

Then we can rewrite $A(p, \omega)$ as

$$A(p, \omega) = 2\pi\hbar\delta(\hbar\omega - \varepsilon(p)) \left\{1 + \frac{\partial}{\partial\varepsilon(p)} \frac{P}{\pi} \int \mathrm{d}\bar{\omega}\, \frac{\Gamma(p, \bar{\omega})}{\bar{\omega} - \varepsilon(p)/\hbar}\right\}$$
$$- \frac{1}{\pi} \frac{\partial}{\partial\omega} \frac{P}{\hbar\omega - \varepsilon(p)} \Gamma(p, \omega) \,.$$

This approximation has very convenient properties. For example it is easy to show that the exact relation

$$\int A(p, \omega) \frac{\mathrm{d}\omega}{2\pi} = 1$$

is fulfilled.

Inserting the expression (3.145) in the relation (3.139) we find the expression

$$n(T, \mu) = \int \frac{\mathrm{d}\boldsymbol{p}}{(2\pi)^3} f\{\varepsilon(p)\} - \frac{1}{\pi} \int \frac{\mathrm{d}\boldsymbol{p}}{(2\pi)^3} \frac{\mathrm{d}\bar{\omega}}{2\pi}$$
$$\cdot \left[\frac{\partial}{\partial\omega'} P \frac{\Gamma(p, \bar{\omega})}{\hbar\omega' - \varepsilon(p)}\right]\Bigg|_{\omega'=\bar{\omega}} \{f(\bar{\omega}) - f\{\varepsilon(p)\}\} \,. \qquad (3.146)$$

This expression can be interpreted in the way that the quasiparticle contribution is separated in closed form, the remaining contribution describes the correlations between the quasiparticles. For instance, in binary collision approximation for $\Gamma(p, \omega)$ bound states may be described as will be discussed in more detail in Section 3.2.9. and in Chapter 6.

3.2.6. The Two-Particle Green's Function

In order to obtain the self energy it is necessary to know the two-particle Green's function. Knowing this function in connection with Σ, the dynamical behaviour of one particle and the thermodynamic properties of the Systems are determined. Moreover, G_2 determines the dynamical behaviour of a pair of particles in an interacting many-body system. Therefore, it is important to study the properties of G_2 in more detail. Especially it is of interest to find the equation of motion for G_2 in order to construct suitable approximations for G_2.

The two-particle Green's function was introduced by

$$G_2(121'2') = \frac{1}{(i\hbar)^2} \langle T\{\psi(1)\,\psi(2)\,\psi^+(2')\,\psi^+(1')\}\rangle. \tag{3.147}$$

These functions contain in dependence of the four times $t_1, \dots, t_{2'}$ in general 24 different correlation functions, and therefore the spectral properties are very complicated (cf. PUFF, 1976). But it is possible to show, that the KMS condition is valid for any of the times $t_1, \dots, t_{2'}$ along the imaginary axis:

$$G_2(121'2')_{t_j=0} = \pm e^{\beta\mu}\, G_2(121'2')|_{t_j=-i\hbar\beta}\,. \tag{3.148}$$

Because of this quasi periodicity condition it is possible to expand G_2 in the well known manner in a Fourier series. Taking into account the homogeneity of the time we may write

$$G_2(t_1, \dots, t_4) = \frac{1}{(-i\hbar\beta)^3} \sum_{z_{\nu_1} z_{\nu_2} z_{\nu_3}} G_2(z_{\nu_1}, z_{\nu_2}, z_{\nu_3})$$

$$\cdot \exp\left[iz_{\nu_1}(t_1 - t_4) + iz_{\nu_2}(t_2 - t_4) + iz_{\nu_3}(t_3 - t_4)\right], \tag{3.149}$$

$$z_\nu = \frac{\pi}{-i\hbar\beta}\nu + \frac{\mu}{\hbar} = \text{Matsubara frequency},$$

$$\nu = \begin{cases} 0, \pm 2, \pm 4, \dots & \text{Bose case,} \\ \pm 1, \pm 3, \pm 5, \dots & \text{Fermi case}\,. \end{cases}$$

If some time arguments are equal, Green's functions may be obtained with less then three frequencies, e.g.,

$$G_2(t_1, \dots, t_4)_{t_1=t_2=t} = \frac{1}{(-i\hbar\beta)^2} \sum_{z_{12} z_3} G_2(z_{12}, z_3) \exp\left[iz_{12}(t - t_4) - i\,z_3(t_3 - t_4)\right] \tag{3.150}$$

=

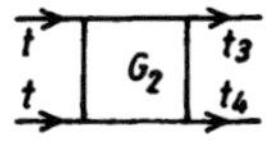

where the two-frequency function $G_2(z_{12}, z_3)$ is connected with $G_2(z_1, z_2, z_3)$ by

$$G_2(z_{12}, z_3) = \frac{1}{-i\hbar\beta} \sum_{z_2} G_2(z_{12} - z_2, z_2, z_3)\,,$$

$$z_{12} = z_1 + z_2 = \frac{\pi}{-i\hbar\beta}\nu_{12} + \frac{\mu_1}{\hbar} + \frac{\mu_2}{\hbar}. \tag{3.151}$$

z_{12} is a Bose like Matsubara frequency. Of special interest is the Green's function

$$G_2(t_1, \dots, t_4)_{\substack{t_1=t_2=t \\ t_3=t_4=t'}} = \frac{1}{-i\hbar\beta} \sum_{z_{12}} G_2(z_{12}) \exp\left[-iz_{12}(t - t')\right] \tag{3.152}$$

=

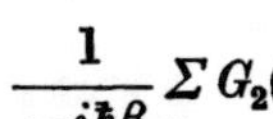

with

$$G_2(z_{12}) = \frac{1}{-i\hbar\beta} \sum_{z_3} G(z_{12}, z_3) . \tag{3.153}$$

This function describes the propagation of two particles in an interacting many-body system and determines the two-particle states. To show this fact we study the analytic properties of $G_2(t - t')$, which are similar to that of $G_1(t - t')$ because we have the decomposition

$$G_2(\tau) = \theta(\tau)\, G_2^{>}(\tau) + \theta(-\tau)\, G_2^{<}(\tau), \tau = t - t' , \tag{3.154}$$

with

$$\begin{aligned} G_2^{>}(t - t') &= \frac{1}{(i\hbar)^2} \langle \psi(x_1, t)\, \psi(x_2, t)\, \psi^+(x_2', t')\, \psi^+(x_1', t')\rangle , \\ G_2^{<}(t - t') &= \frac{1}{(i\hbar)^2} \langle \psi^+(x_1', t')\, \psi^+(x_2', t')\, \psi(x_2, t)\, \psi(x_1, t)\rangle . \end{aligned} \tag{3.155}$$

In the same manner as in the case of the one-particle two-time correlation functions, $G_2^{>}$ and $G_2^{<}$ are completely determined by the spectral function

$$I_2(\omega) = \frac{2\pi\hbar}{Z} \sum_{nm} \langle n|\, \psi^+\psi^+\, |m\rangle \langle m|\, \psi\psi\, |n\rangle\, \mathrm{e}^{-\beta(E_n - \mu N)}\, \delta(\hbar\omega - E_n - E_m) . \tag{3.156}$$

With help of this function we get the spectral representation

$$G_2^{>}(t - t') = \int \frac{\mathrm{d}\omega}{2\pi}\, \mathrm{e}^{-i\omega(t-t')} G_2^{>}(\omega) , \qquad G_2^{<}(t - t') = \int \frac{\mathrm{d}\omega}{2\pi}\, \mathrm{e}^{-i\omega(t-t')} G_2^{<}(\omega) , \tag{3.156a}$$

$$G_2^{>}(\omega) = I(\omega)\, \mathrm{e}^{\beta(\hbar\omega - 2\mu)} , \qquad G^{<}(\omega) = I(\omega)$$

from which follows the relation

$$G^{>}(\omega) = G^{<}(\omega)\, \mathrm{e}^{\beta(\hbar\omega - 2\mu)} .$$

So it is useful to introduce the spectral function $A_2(\omega)$ by

$$A_2(\omega) = G_2^{>}(\omega) - G_2^{<}(\omega) . \tag{3.157}$$

Then we get the spectral representations

$$\begin{aligned} G_2^{<}(\omega) &= I_2(\omega) = n_{\mathrm{B}}(\omega)\, A_2(\omega) , \\ G_2^{>}(\omega) &= \{1 + n_{\mathrm{B}}(\omega)\}\, A_2(\omega) , \end{aligned} \qquad n_{\mathrm{B}}(\omega) = \frac{1}{\mathrm{e}^{\beta(\hbar\omega - 2\mu)} - 1} , \tag{3.158}$$

and for $G_2(\omega)$ we have the Landau spectral representation

$$G_2(\omega) = \int_{-\infty}^{+\infty} \frac{\mathrm{d}\bar{\omega}}{2\pi i}\, A_2(\bar{\omega}) \left\{ \frac{n_{\mathrm{B}}(\bar{\omega})}{\omega - \bar{\omega} - i\varepsilon} - \frac{1 + n_{\mathrm{B}}(\bar{\omega})}{\omega - \bar{\omega} + i\varepsilon} \right\} . \tag{3.159}$$

It is easy to show that the KMS condition can now be written

$$G_2(t - t')|_{t=0} = +\mathrm{e}^{2\beta\mu}\, G_2(t - t')|_{t=-i\hbar\beta} . \tag{3.160}$$

To satisfy this condition we express $G_2(t - t')$ as a Fourier series. Taking into account the Bose like behaviour, the Fourier coefficients are given by

$$G_2(z_{12}) = \frac{1}{2\pi\hbar} P \int \frac{A_2(\omega)}{z_{12} - \omega}\, \mathrm{d}\omega + \frac{1}{-i\hbar\beta}\, \delta_{z_{12}, 0} C . \tag{3.161}$$

We can continue $G_2(z_{12})$ to the z-plane and have (excluding the real axis)

$$G(z) = \int \frac{\mathrm{d}\omega}{2\pi\hbar} \frac{A_2(\omega)}{z - \omega}. \tag{3.162}$$

$G_2(z)$ has the shape of a Cauchy type integral and therefore it has the following analytic properties:

i) The integral defines two distinct analytic functions of $G_2^{\mathrm{I}}(z)$ for $\mathrm{Im}\, z > 0$ and $G_2^{\mathrm{II}}(z)$ for $\mathrm{Im}\, z < 0$.
ii) $G_2(z)$ has a branch cut at the positive real axis with the discontinuity $\hbar i(G_2(\omega + i\varepsilon) - G_2(\omega - i\varepsilon)) = A_2(\omega)$.
iii) It is possible to continue $G_2^{\mathrm{I}}(z)$ and $G_2^{\mathrm{II}}(z)$ analytically beyond the cut. We get
$G_2^{\mathrm{I}}(z) = G_2^{\mathrm{II}}(z) + A_2(z),\ \mathrm{Im}\, z < 0,$
$G_2^{\mathrm{II}}(z) = G_2^{\mathrm{I}}(z) - A_2(z),\ \mathrm{Im}\, z > 0.$

These functions are no longer analytic, they have the same singularities as $A_2(z)$.

Now it is easy to get the spectral function for a non interacting system

$$A_2(\omega) = 2\pi\hbar\delta(\hbar\omega - E_2^0(p_1, p_2)). \tag{3.163}$$

$E_2^0(p_1, p_2) = \dfrac{\hbar^2 p_1^2}{2m} + \dfrac{\hbar^2 p_2^2}{2m}$ is the two-particle energy of the noninteracting system.

Similar as in the case of the one-particle Green's function it is possible to show that in interacting systems a two-particle state is characterized by a peak with a finite width Γ for the spectral function. We assume that, in this case in the neighbourhood of the peak, $A_2(\omega)$ has the form

$$A_2(\omega) = \frac{\Gamma}{(\hbar\omega - E_2^0(p_1, p_2))^2 + \dfrac{\Gamma^2}{4}}. \tag{3.164}$$

In the complex plane we have then

$$A_2(z) = \frac{i}{\hbar z - \left(E_2^0 + i\dfrac{\Gamma}{2}\right)} - \frac{i}{\hbar z - \left(E_2^0 - i\dfrac{\Gamma}{2}\right)}. \tag{3.165}$$

Therefore $A_2(\omega)$ has poles in the complex z-plane. The poles at E_2^0 correspond to the two-particle states with

$E_2^0 =$ two-particle energy,

$\dfrac{\hbar}{\Gamma} = \tau =$ life-time of the states.

3.2.7. Equation of Motion for Higher Order Green's Functions

In order to determine G_2 it is necessary to write down the equation of motion for this function. For this purpose we remark that G_2 is a member of a chain of many-point functions which begins with the one particle Green's function

$$G_1(12,\mathrm{U}) = \frac{1}{i\hbar} \frac{\mathrm{Tr}\,\{\varrho T[S\psi(1)\,\psi^+(2)]\}}{\langle TS \rangle} \tag{3.166}$$

with

$$S = \exp\left[-\int_0^{-i\hbar\beta} \mathrm{d}\bar{1}\, \mathrm{d}\bar{2}\, U(\bar{1}\bar{2})\, \psi^+(\bar{1})\, \psi(\bar{2})\right]. \tag{3.167}$$

Here we have (following the paper of BAYM and KADANOFF, 1961) introduced the nonlocal external potential U, in order to obtain the higher order Green's functions by functional derivation with respect to $U(12)$. So we have e.g.

$$\left.\frac{\delta G_1(11', U)}{\delta U(2'2)}\right|_{U=0} = \pm [G_2(121'2') - G_1(11')\, G_1(22')]_{U=0} = \pm L(121'2'). \tag{3.168}$$

L is refered to as the density fluctuation function. Therefore it is in principle sufficient to know the equation of motion for $G_1(11', U)$. Such equation is the well known Dyson equation which we write in the from

$$G_1^{-1}(11', U) = G_1^{0-1}(11') - U(11') - \Sigma(11'). \tag{3.169}$$

Here Σ is the self energy given by (3.118) and $G_1^0(11')$ is the Green's function for free particles. In order to obtain an equation for the self energy, we express G_2 by a functional derivative of G_1. To do this we start with the relation $G_1 G_1^{-1} = 1$. From this we get

$$\begin{aligned} &\pm\{G_2(121'2') - G_1(11')\, G_1(22')\} = \pm L(121'2'), \\ &\frac{\delta G_1(11')}{\delta U(22')} = -\int \mathrm{d}\bar{3}\, \mathrm{d}\bar{4}\, G_1(1\bar{3})\, \frac{\delta G_1^{-1}(\bar{3}\bar{4})}{\delta U(22')}\, G(\bar{4}1'), \end{aligned} \tag{3.170}$$

and using the Dyson equation it follows

$$\begin{aligned} G_2(121'2') &= G_1(11')\, G_1(22') \pm G_1(12')\, G_1(21') \\ &\pm \int G_1(13)\, \frac{\delta\Sigma(34)}{\delta U(22')}\, G_1(41')\, \mathrm{d}3\, \mathrm{d}4 = G_1(11')\, G_1(22') \\ &\pm G_1(12')\, G_1(21') \pm \int G_1(13)\, G_1(41')\, \frac{\delta\Sigma(34)}{\delta G_1(65)}\, \frac{\delta G_1(65)}{\delta U(22')}\, \mathrm{d}3\, \mathrm{d}4\, \mathrm{d}5\, \mathrm{d}6. \end{aligned} \tag{3.171}$$

With (3.171) we get the following functional equation for Σ:

$$\begin{aligned} \Sigma(11', U) &= \delta(1,1')\left[\pm \int \mathrm{d}\bar{2}\, V(1\bar{2})\, G_1(\bar{2}\bar{2}^+ U) + V(1, 1')\, G_1(11', U)\right] i\hbar \\ &+ i\hbar \int \mathrm{d}\bar{1}\, \mathrm{d}\bar{2}\, V(1\bar{2})\, G_1(1\bar{1}\, U)\, \frac{\delta\Sigma(\bar{1}1' U)}{\delta U(\bar{2}\bar{2})}. \end{aligned} \tag{3.172}$$

With the help of the eqs. (3.169) and (3.172) it is possible to determine Green's functions to arbitrary order. However, in practice it is often more convenient to use an equation of motion for G_2.

Equation (3.171) may be rewritten to give an integral equation for G_2:

$$\begin{aligned} G_2(121'2') &= G_1(11')\, G_1(22') \pm G_1(12')\, G_1(21') \\ &+ \int G_1(1\bar{3})\, G_1(\bar{4}1')\, \Xi(\bar{3}\bar{5}\bar{4}\bar{6})\, L(\bar{6}2\bar{5}2')\, \mathrm{d}\bar{3}\mathrm{d}\bar{4}\mathrm{d}\bar{5}\mathrm{d}\bar{6}, \\ \Xi(\bar{3}\bar{5}\bar{4}\bar{6}) &= \frac{\delta\Sigma(\bar{3}\bar{4})}{\delta G_1(\bar{6}\bar{5})}. \end{aligned} \tag{3.173}$$

This equation is called Bethe-Salpeter equation with respect to the u-channel. With Feynman diagrams we get

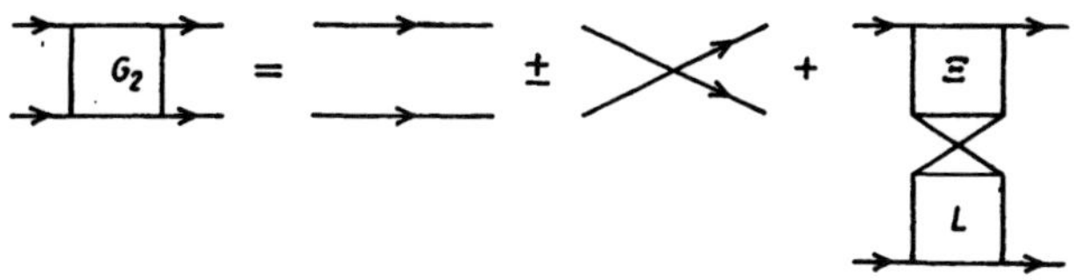

(3.173a)

Ξ can be interpreted as an effective particle-hole interaction (in Fermi systems). From the point of view of the diagram technique, Ξ is the sum of irreducible two-particle diagrams in the u-channel.

For many problems it is more useful to write the Bethe-Salpeter equation in the particle-particle channel (s-channel), namely

$$G_2(121'2') = G_2^0(121'2') + \int G_1(13)\, G_1(24)\, K(34\bar{3}\bar{4})\, G_2(\bar{3}\bar{4}1'2')\; d3\, d\bar{3}\, d4\, d\bar{4}\,, \tag{3.174}$$

and in Feynman diagrams

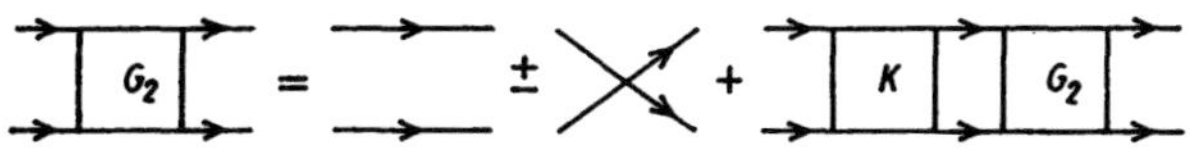

(3.175)

$K(34\bar{3}\bar{4})$ is an effective two-particle potential. In the language of diagrams, K is the sum of irreducible diagrams in the particle-particle channel. K is defined analytically by the equation

$$\pm \int G_1(13)\, \frac{\delta \Sigma(34)}{\delta U(2'2)}\, G_1(41')\; d3\, d4 = \int G_1(13)\, G_1(24)\, K(34\bar{3}\bar{4})\, G_2(\bar{3}\bar{4}1'2')\; d3\, d4\, d\bar{3}\, d\bar{4}\,. \tag{3.176}$$

In many cases it is more convenient to introduce the many particle T-matrix instead of the two-particle Green's function. The T-matrix is defined by

$$\frac{\delta \Sigma(11')}{\delta U(43)} = \pm \int G_1(32)\, T(121'2')\, G_1(2'4)\; d2\, d2'\,. \tag{3.177}$$

The connection with G_2 we get from the equation (3.172):

$$G_2(121'2') = G_2^0(121'2') + \int d\bar{1}\, d\bar{2}\, d\bar{3}\, d\bar{4}\, G_1(1\bar{1})\, G_1(2\bar{2})\, T(\bar{1}\bar{2}\bar{3}\bar{4})\, G_1(\bar{3}1')\, G_1(\bar{4}2')\,. \tag{3.178}$$

This equation shows that the correlation part of the two-particle Green's function is determined by T. Therefore, all the properties of the system may be determined with help of T. For example, the self energy is given by

$$\Sigma(11') = \Sigma^{\mathrm{HF}}(11') \pm i \int d\bar{1}\, d\bar{2}\, d\bar{3}\, d\bar{4}\, V(1\bar{1})\, G_1(1\bar{2})\, G_1(\bar{1}\bar{3})\, T(\bar{2}\bar{3}1'\bar{4})\, G_1(\bar{4}\bar{1}^{+})\,. \tag{3.179}$$

In correspondence to G_2 it is possible to determine T from two equivalent equations of motion in the s-channel and in the u-channel. We get in the s-channel the equation of motion in the form (if we use eq. (3.175))

$$T(121'2') = K(121'2') + \int K(12\bar{3}\bar{4})\, G_1(\bar{3}\bar{5})\, G_1(\bar{4}\bar{6})\, T(\bar{5}\bar{6}1'2')\; d\bar{3}\, d\bar{4}\, d\bar{5}\, d\bar{6}\,, \tag{3.180}$$

with Feynman diagrams

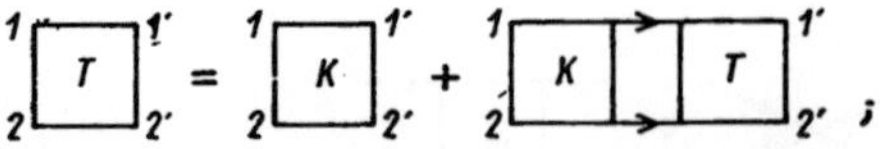

(3.180a)

and in the u-channel using equation (3.173)

$$T(121'2') = \Xi(121'2') \pm \int d\bar{5}\, d\bar{6}\, d\bar{7}\, d\bar{8}\, \Xi\,(1\bar{5}1'\bar{6})\, G_1(\bar{5}\bar{7})\, G_1(\bar{6}\bar{8})\, T(\bar{7}2\bar{8}2')\,, \tag{3.181}$$

with Feynman diagrams

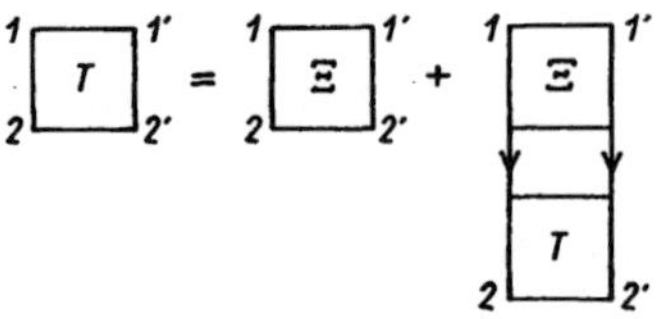

(3.181a)

The effective two-particle interation can also be expressed by the three-particle Green's function

$$G_3(1231'2'3') = \frac{1}{(i\hbar)^3} \langle T\{\psi(1)\psi(2)\psi(3)\psi^+(3')\psi^+(2')\psi^+(1')\}\rangle \tag{3.182}$$

or in terms of a functional derivation

$$G_3(1231'2'3') = \pm \frac{\delta G_2(121'2')}{\delta U(3'3)} + G_2(121'2')\, G_1(33')\,. \tag{3.183}$$

Using the definition (3.118) of the self energy Σ by the two-particle Green's function we can rearrange the left-hand side of eq. (3.176) in the following way:

$$i\hbar \int G_1(13) \frac{\delta \Sigma(34)}{\delta U(2'2)} G_1(41')\, d3\, d4 = i\hbar \int G_1(13) \left\{ \frac{\delta \Sigma(34)\, G(41')}{\delta U(2'2)} \right.$$
$$\left. - \Sigma\,(34) \frac{\delta G_1(41')}{\delta U(2'2)} \right\} d3\, d4 = (i\hbar)^2 \int G_1(13) \left\{ V(34) \frac{\delta G_2(341'4^+)}{\delta U(2'2)} \right.$$
$$\left. - \Sigma\,(34) \frac{\delta G_1(41')}{\delta U(2'2)} \right\} d3 d4\,. \tag{3.184}$$

Because of eq. (3.173) we obtain then

$$\int G_1(13)\, G_1(24)\, K(34\bar{3}\bar{4})\, G_2(\bar{3}\bar{4}1'2')\, d3\, d4\, d\bar{3}\, d\bar{4}$$
$$= \pm i\hbar \int G_1(13)\, \{\, V(34)\, G_3(3241'2'4^+) - \Sigma\,(34)\, G_2(421'2')\}\, d3\, d4\,. \tag{3.185}$$

Therefore, we can write the Bethe-Salpeter equation as a member of a hierarchy of equations for G_1, G_2, G_3, ...:

$$G_2(121'2') = G_1(11')\, G_1(22') \pm G_1(12')\, G_1(21')$$
$$\pm\, i\hbar \int d\bar{1}\, d3\, G_1(1\bar{1})\, \{\, V(\bar{1}3)\, G_3(\bar{1}231'2'3^+) - \Sigma\,(\bar{1}3) G_2(321'2')\}\,. \tag{3.186}$$

The general form of this hierarchy can be found in the paper of Martin and Schwinger (1959). In this paper the hierarchy is given by integrodifferential equations of the form

$$\left\{\frac{\hbar^2}{2m}\Delta_1 + i\hbar\frac{\partial}{\partial t_1}\right\} G_n(1, \dots, n, 1', \dots, n') = \sum_{\nu=1}^{n} (\pm 1)^{\nu-1}\delta(1-\nu') , \tag{3.187}$$

$$G_{n-1}(2, \dots, n, 1', \dots, \nu'-1, \nu'+1, \dots, n')$$

$$\pm i\hbar \int d\bar{1}\; V(1\bar{1})\, G_{n+1}(1, \dots, n, \bar{1}1'2', \dots, n', \bar{1}^+).$$

As can be seen G_n is coupled with the functions G_{n-1} and G_{n+1}. For instance we have explicitely

$$\left\{i\hbar\frac{\partial}{\partial t_1} + \frac{\hbar^2}{2m}\Delta_1\right\} G_2(121'2')$$

$$= G_1(22')\,\delta(11') \pm G_1(21')\,\delta(12') \pm i \int d\bar{3}\; V(1\bar{3})\, G_3(123 1'2'\bar{3}^+) .$$

The formal solution of the hierarchy for the Green's function is obtained by iteration starting from the zeroth order corresponding to the ideal system

$$G_1^{(0)} = G_1^0(11') ,$$

$$G_n^{(0)}(1, \dots, n; 1', \dots, n') = \begin{vmatrix} G_1^0(11') & \dots & G_1^0(1\,n') \\ \vdots & & \vdots \\ G_1^0(n\,1') & \dots & G_1^0(n\,n') \end{vmatrix}_{\pm} . \tag{3.188}$$

Let us represent now the n-particle Green's function by a perturbation series:

$$G_n = G_n^{(0)} + G_n^{(1)} + G_n^{(2)} + \dots$$

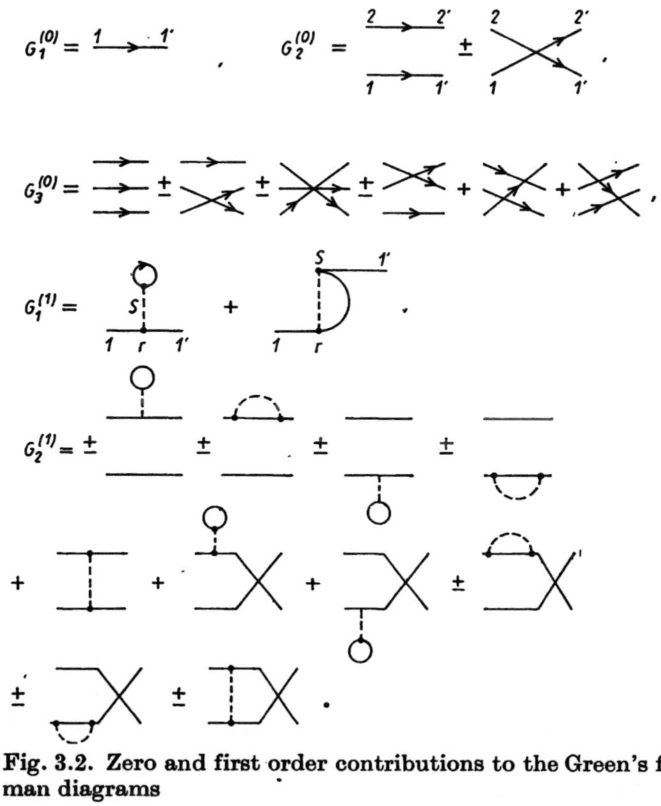

Fig. 3.2. Zero and first order contributions to the Green's functions represented by Feynman diagrams

where the upper index gives the order in the interaction parameter e^2. In this way one gets the hierarchy.

$$\left\{i\hbar \frac{\partial}{\partial t_1} + \frac{\hbar^2}{2m} \Delta_1\right\} G_n^{(m)}(1, \dots, n; 1', \dots, n') = \{G_{n-1}^{(m)}(2, \dots, n; 2', \dots, n')$$

$$\delta(11')\}_s \pm i\hbar \int d\overline{m}\ V(1\overline{m})\ G_{n+1}^{(m-1)}(1, \dots, n, \overline{m}; 1', \dots, n', \overline{m}^+) \qquad (3.189)$$

where $\{\dots\}_s$ means the symmetrized sum with respect to $1', \dots, n'$.

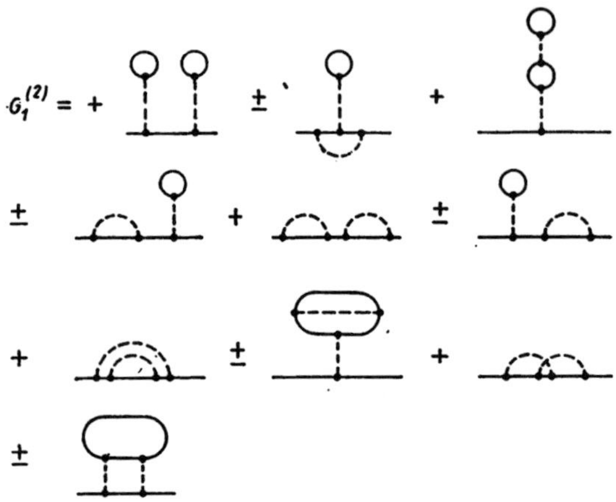

Fig. 3.3. Second order contributions to the one-particle Green's functions

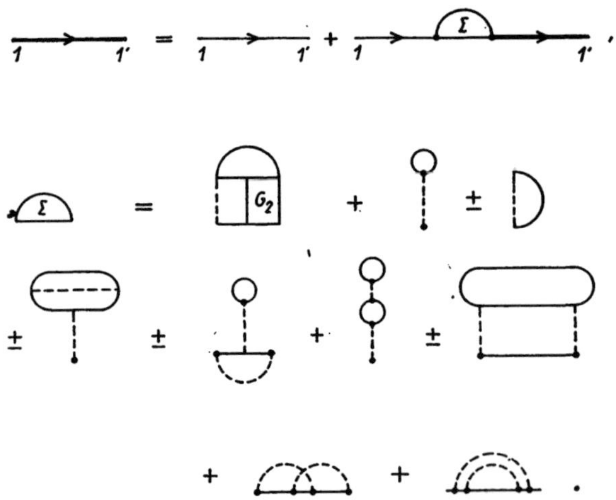

Fig. 3.4. Feynman diagrams contributing to the self energy

This decoupled form of the hierarchy is easily solved by iteration starting from the zeroth order. The result for the lower orders is shown in Figs. 3.2-4 graphically. Here as usual the thin lines denote free one-particle Green's functions, the dashed lines denote interactions and the nodes correspond to integrations. Following the scheme given below, contributions to any order of the Green's functions may be calculated

$$\begin{array}{llllll}
G_1^{(0)} & & & & & \\
\downarrow & & & & & \\
G_2^{(0)} \rightarrow & G_1^{(1)} & & & & \\
\downarrow & \downarrow & & & & \\
G_3^{(0)} \rightarrow & G_2^{(1)} \rightarrow & G_1^{(2)} & & & \\
\downarrow & \downarrow & \downarrow & & & \\
G_4^{(0)} \rightarrow & G_3^{(1)} \rightarrow & G_2^{(2)} \rightarrow & G_1^{(3)} & & \\
\downarrow & \downarrow & \downarrow & \downarrow & & \\
G_5^{(0)} \rightarrow & G_4^{(1)} \rightarrow & G_3^{(2)} \rightarrow & G_2^{(3)} \rightarrow & G_1^{(4)} & \\
\vdots & \vdots & \vdots & \vdots & \vdots & \\
G_n^{(0)} \rightarrow & G_{n-1}^{(1)} \rightarrow & G_{n-2}^{(2)} \rightarrow & G_{n-3}^{(3)} \rightarrow & \dots & G_1^{(n-1)}
\end{array} \tag{3.190}$$

Representing all contributions by Feynman diagrams one can state:

The n-particle Green's function may be represented as the sum of all topologically non-equivalent connected Feynman diagrams starting from the open ends 1, 2, ..., n and leading to the open ends 1, 2, ..., n'. The sign of a diagram contributing to G_n is given by

$$(\pm 1)^{F+K}$$

where F is the number of loops and K is the number of crossing points in the diagram.

The proof of this theorem follows by induction. One assumes that the theorem is correct for the m-th line of the scheme given above and shows that in the m-th line the non-equivalent connected Feynman diagrams are added. (For details of the proof see MIGDAL (1965) für $T = 0$ and RICHERT (1979) for $T \neq 0$.)

3.2.8. The Binary Collision Approximation (Ladder Approximation)

Let us now try to find approximative versions of the effective two-particle potential. A simple, but very useful approximation is the binary collision approximation. The basic idea is that only two particles of the system can interact simultaneously. Since $V(\bar{1}3)$ in (3.186) is different from zero only if the particles $\bar{1}$ and 3 are at a finite distance, in the binary collision approximation for G_3 we find the following expression:

$$G_3(1231'2'3') = G_2(131'3')\,G_1(22') + G_2(133'2')\,G_1(21') + G_2(132'1')\,G_1(23')\,. \tag{3.191}$$

Introducing this approximation into (3.176) and using the relation (3.118) of Σ and G_2, the Bethe-Salpeter equation reduces to the more simple equation

$$G_2(121'2') = G_1(11')\,G_1(22') \pm G_1(12')\,G_1(21')$$
$$+\, i\hbar \int \mathrm{d}\bar{1}\;\mathrm{d}3\; G(1\bar{1})\,G_1(23)\;V(3\bar{1})\,G_2(\bar{1}31'2')\,. \tag{3.192}$$

In terms of Feynman diagrams it may be written

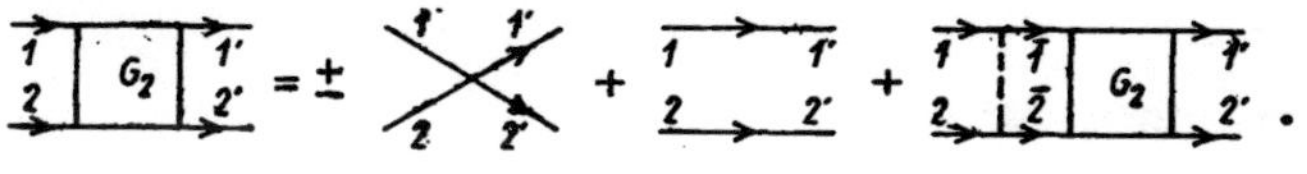

(3.192a)

This integral equation sums up the infinite subset of so called ladder diagrams as can be seen if we solve the equation by iteration

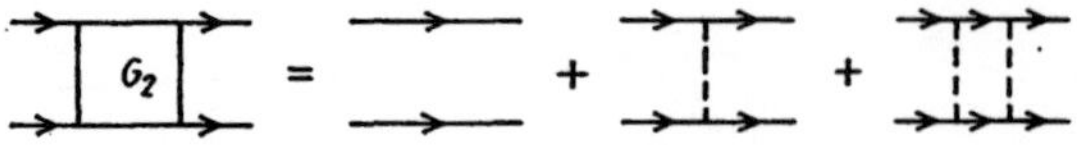

+ ... + exchange terms . (3.192b)

Because of the structure of this subset of diagrams, the binary collision approximation is often called ladder approximation. In order to describe the propagation of two particles in a many-particle system it is sufficient to take into account only the special Green's functions

$$G_2(121'2')\Big|_{\substack{t_1=t_2=t \\ t_1'=t_2'=t'}} = G_2(x_1, x_2, x_1', x_2', t-t') .$$

In the following we consider Fermi particles. In this special case, in eq. (3.192) the Fourier transformation with respect to the space variables and the Fourier series with respect to the time is performed. For this purpose we write

$$G_1(12) = \frac{1}{-i\beta\hbar} \sum_{\nu_1} \int \frac{\mathrm{d}\boldsymbol{p}}{(2\pi)^3} \mathrm{e}^{-iz_1(t-t')} \mathrm{e}^{-i\boldsymbol{p}(\boldsymbol{r}_1-\boldsymbol{r}_2)} G(z_1, \boldsymbol{p}) , \tag{3.193}$$

$$z_1 = \frac{\pi\nu_1}{-i\beta\hbar} + \frac{\mu}{\hbar}, \qquad \nu_1 = \pm 1, \pm 3, \pm 5 ,$$

and

$$G_2(r_1, r_2, r_1', r_2', t-t') = \frac{1}{-i\beta\hbar} \sum_{\nu_{12}} \int \frac{\mathrm{d}\boldsymbol{p}_1 \mathrm{d}\boldsymbol{p}_2 \mathrm{d}\boldsymbol{p}_1' \mathrm{d}\boldsymbol{p}_2'}{(2\pi)^{12}} \mathrm{e}^{-iz_{12}(t-t')}$$

$$\cdot\, \mathrm{e}^{-i(\boldsymbol{p}_1\boldsymbol{r}_1 + \boldsymbol{p}_2\boldsymbol{r}_2 - \boldsymbol{p}_2'\boldsymbol{r}_2' - \boldsymbol{p}_1'\boldsymbol{r}_1')} G(p_1, p_2, p_1', p_2', z_{12}),\ z_{12} = \frac{\pi\nu_{12}}{-i\hbar\beta},\ \nu_{12} = 0, \pm 2, \pm 4, \ldots \tag{3.194}$$

Using these relation we find from (3.192)

$$G_2(p_1, p_2, p_1', p_2', z_{12}) = \frac{1}{-i\hbar\beta} \sum_{\nu_2} \Big\{ G_1(p_1, z_{12} - z_2)\, G_1(p_2, z_2)$$

$$\cdot \big[(2\pi)^6 \big(\delta(\boldsymbol{p}_1 - \boldsymbol{p}_1')\, \delta(\boldsymbol{p}_2 - \boldsymbol{p}_2') - \delta(\boldsymbol{p}_1 - \boldsymbol{p}_2')\, \delta(\boldsymbol{p}_2 - \boldsymbol{p}_1')\big)$$

$$+ i \int V(\bar{\boldsymbol{p}}_2 - \boldsymbol{p}_2)\, (2\pi)^3\, \delta(\bar{\boldsymbol{p}}_1 + \bar{\boldsymbol{p}}_2 - \boldsymbol{p}_1 - \boldsymbol{p}_2)\, G_2(\bar{p}_1, \bar{p}_2, p_1', p_2', z_{12}) \frac{\mathrm{d}\bar{\boldsymbol{p}}_1, \mathrm{d}\bar{\boldsymbol{p}}_2}{(2\pi)^6} \big]\Big\}. \tag{3.195}$$

The summation over ν_2 can be replaced by an appropriate contour integral. For this purpose we consider the Fermi function

$$f(z) = \frac{1}{\mathrm{e}^{\beta(\hbar z - \mu)} + 1} .$$

$f(z)$ has poles at $z_1 = \pi\nu_1/(i\hbar\beta) + \mu/\hbar$ with the corresponding residues $(\hbar\beta)^{-1}$. With $h(z)$ being an analytic function having poles at $z_1', z_2', \ldots$ not located at the Matsubara frequencies z_1.

Under the condition

$$z f(z)\, h(z) \to 0 \quad \text{for} \quad |z| \to \infty ,$$

we may transform summations into contour integrals (see Fig. 3.5):

$$\frac{1}{-i\hbar\beta}\sum_{\nu_1} h(z_1) = \int_{C'} \frac{dz}{2\pi} f(z)\, h(z) = \int_{C} \frac{dz}{2\pi} f(z)\, h(z) \tag{3.196}$$

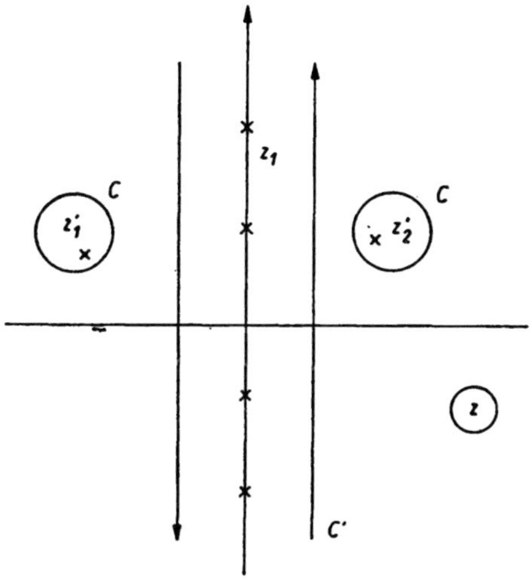

Fig. 3.5. Path of integration

C' is a contour encircling the imaginary axis in the positive sense which can be deformed into C which encloses the poles $z_1, z_2, \dots$ This technique is applied to the expression

$$I = \frac{1}{-i\beta}\sum_{\nu_1} G_1(p_1, z_1)\, G_1(p_2, z_{12} - z_1) = \mathcal{G}_2^0(z_{12})\,.$$

For the Green's function we use the Hartree-Fock approximation of (3.126) with the quasi particle energy $E\,(p)$. Then we have

$$I = \int_{C} \frac{d\hbar z}{2\pi} f(z) \frac{1}{\hbar z - E(p_1)} \frac{1}{\hbar z_{12} - \hbar z - E(p_2)}\,. \tag{3.197}$$

The complex integral yields

$$\mathcal{G}_2^0(z_\nu^{12}) = i\,\frac{f(E(p_1)) - f(\hbar z_{12} - E(p_2))}{\hbar z_{12} - E(p_1) - E(p_2)} = \frac{1 - f(E(p_1)) - f(E(p_2))}{\hbar z_{12} - E(p_1) - E(p_2)}\,. \tag{3.198}$$

Using this expression for I the equation (3.195) gets the form

$$\{E(\boldsymbol{p}_1) + E(\boldsymbol{p}_2) - \hbar z\}\, G_2(\boldsymbol{p}_1, \boldsymbol{p}_2, \boldsymbol{p}_1', \boldsymbol{p}_2', z) + [1 - f(E(\boldsymbol{p}_1)) - f(E(\boldsymbol{p}_2))]$$
$$\cdot\, (2\pi)^3 \int V(\bar{\boldsymbol{p}}_2 - \boldsymbol{p}_2)\, \delta(\bar{\boldsymbol{p}}_1 + \bar{\boldsymbol{p}}_2 - \boldsymbol{p}_1 - \boldsymbol{p}_2)\, G_2(\bar{\boldsymbol{p}}_1, \bar{\boldsymbol{p}}_2, \boldsymbol{p}_1', \boldsymbol{p}_2', z) \frac{d\bar{\boldsymbol{p}}_1\, d\bar{\boldsymbol{p}}_2}{(2\pi)^6}$$
$$= (2\pi)^6 [\delta(\boldsymbol{p}_1' - \boldsymbol{p}_1)\, \delta(\boldsymbol{p}_2' - \boldsymbol{p}_2) \pm \delta(\boldsymbol{p}_2' - \boldsymbol{p}_1)\, \delta(\boldsymbol{p}_1' - \boldsymbol{p}_2)][1 - f(E(\boldsymbol{p}_1)) - f(E(\boldsymbol{p}_2))]\,. \tag{3.199}$$

In order to discuss the physical meaning of this equation it is useful to consider the homogeneous Bethe-Salpeter equation

$$\{E(\mathbf{p}_1) + E(\mathbf{p}_2) - \hbar z\}\, \psi(\mathbf{p}_1, \mathbf{p}_2) + [1 - f(E(\mathbf{p}_1)) - f(E(\mathbf{p}_2))]$$
$$\cdot (2\pi)^3 \int V(\bar{\mathbf{p}}_2 - \mathbf{p}_2)\, \delta(\bar{\mathbf{p}}_1 + \bar{\mathbf{p}}_2 - \mathbf{p}_1 - \mathbf{p}_2)\, \psi(\bar{\mathbf{p}}_1, \bar{\mathbf{p}}_2) \frac{\mathrm{d}\bar{\mathbf{p}}_1\, \mathrm{d}\bar{\mathbf{p}}_2}{(2\pi)^6} = 0\,. \qquad (3.200)$$

Using $E(p) = \dfrac{\hbar^2 p^2}{2m} + \Sigma^{\mathrm{HF}}(p)$ and the substitution

$$\mathbf{p}_2 - \bar{\mathbf{p}}_2 = \mathbf{q}$$

we can write this equation in the form

$$\left(\frac{\hbar^2 p_1^2}{2m} + \frac{\hbar^2 p_2^2}{2m} - \hbar z\right) \psi(\mathbf{p}_1, \mathbf{p}_2) + \int V(q)\, \psi(\mathbf{p}_1 + \mathbf{q}, \mathbf{p}_2 - \mathbf{q}) \frac{\mathrm{d}\mathbf{q}}{(2\pi)^3}$$
$$= +\{f(E(\mathbf{p}_1)) + f(E(\mathbf{p}_2))\} \int \frac{\mathrm{d}\mathbf{q}}{(2\pi)^3} V(q)\, \psi(\mathbf{p}_1 + \mathbf{q}, \mathbf{p}_2 - \mathbf{q})$$
$$+ \int \frac{\mathrm{d}\mathbf{q}}{(2\pi)^3} V(q)\, \{f(E(\mathbf{p}_1 + \mathbf{q})) + f(E(\mathbf{p}_2 - \mathbf{q}))\}\, \psi(\mathbf{p}_1\, \mathbf{p}_2)$$
$$\equiv \int \frac{\mathrm{d}\mathbf{q}}{(2\pi)^3} \Delta H_{12}\, \psi(\mathbf{p}_1 + \mathbf{q}, \mathbf{p}_2 - \mathbf{q})\,. \qquad (3.201)$$

As can be seen easily the left hand side of this equation is the Schrödinger equation of the isolated two-particle problem in momentum representation. The many-particle effects are condensed in ΔH_{12}. These effects are

i) the so-called Pauli blocking or phase space occupation, represented by the Fermi functions $f(E(p))$. By this term the two-particle scattering in a many-particle system of finite density is restricted as a consequence of the Pauli principle.
ii) the exchange self energy Σ^{HF}.

Obviously the homogeneous Bethe-Salpeter equation may be interpreted therefore as the many particle version of the two particle Schrödinger equation. It is interesting to remark, that ΔH_{12} vanishes at $q = 0$; that means we can observe a compensation between the Pauli blocking and the self-energy effects. Thus, both of these terms must be taken into account.

Let us first review briefly the results for the isolated two particle problem ($\Delta H_{12}=0$). In this case we have two kinds of solutions:

a) scattering states with the energy

$$E_{p_1p_2} = \frac{\hbar^2 p_1^2}{2m} + \frac{\hbar^2 p_2^2}{2m}$$

and eigenfunctions with the normalization

$$\int \mathrm{d}\bar{\mathbf{p}}_1\, \mathrm{d}\bar{\mathbf{p}}_2\, \psi^0_{pP}(\bar{p}_1, \bar{p}_2)\, \psi^{0*}_{p'P'}(\bar{p}_1, \bar{p}_2) = \delta(\mathbf{p} - \mathbf{p}')\, \delta(P - P')$$

b) bound states with the energy

$$E_{nlmP} = \frac{\hbar^2 P^2}{2(m_1 + m_2)} - E_{nlm}$$

and eigenfunctions with the normalization

$$\int d\bar{p}_1\, d\bar{p}_2\, \psi^0_{nlmP}(\bar{p}_1, \bar{p}_2)\, \psi^{0*}_{n'l'm'P'}(\bar{p}_1, \bar{p}_2) = \delta_{nlm\, n'l'm'}\, \delta(\boldsymbol{P} - \boldsymbol{P}')\,. \tag{3.202}$$

Let us now assume that ΔH_{12} is small. The first-order perturbation theory gives the corrected eigenvalue

$$E = E_{\alpha P} + \langle \alpha P|\, \Delta H_{12}\, |\alpha P\rangle$$

where the energy shift is given in terms of the unperturbed wave functions by

$$\langle \alpha P|\, \Delta H_{12}\, |P\alpha\rangle = \int \frac{d\boldsymbol{q}}{(2\pi)^3} \int \frac{d\boldsymbol{p}_1\, d\boldsymbol{p}_2}{(2\pi)}\, V(\boldsymbol{q})\, \{f(\boldsymbol{p}_1) + f(\boldsymbol{p}_2)\}$$
$$\cdot\, \psi^0_{\alpha P}(\boldsymbol{p}_1 + \boldsymbol{q}, \boldsymbol{p}_2 - \boldsymbol{q})\, \{\psi^0_{\alpha P}(\boldsymbol{p}_1, \boldsymbol{p}_2) - \psi^0_{\alpha P}(\boldsymbol{p}_1 + \boldsymbol{q}, \boldsymbol{p}_2 - \boldsymbol{q})\}\,. \tag{3.203}$$

From this expression it is seen, that there is an essential difference between the bound state energy shift and that of the scattering energies. For bound states especially the low lying states are localized and thus extended in the momentum space. Therefore, the wave functions vary only weakly with the momenta, and thus the terms in the curly brackets compensate each other to a large extent. The bound state energy shifts down only slightly. However for scattering states the wave functions are sharply peaked in the momentum space. Especially the position of the continuum edge can be obtained by inserting the free state wave function for $p = 0$, $P = 0$ into (3.197):

$$\psi_{00}(p_1, p_2) = \delta_{p_1 0}\, \delta_{p_2 0}\,.$$

Then only the self-energy terms (quadratic terms in ψ) contribute to the shift of the continuum edge. So we can expect that the continuum edge shifts more rapidly than the bound state energy. This behaviour is shown in Fig. 3.6.

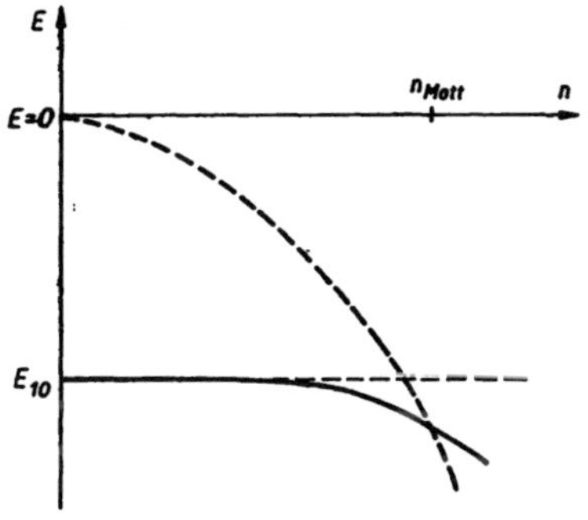

Fig. 3.6. Continuum edge and ground state energy as a function of the density. The Mott density is defined by the disappearance of the ground state.

The physical interpretation of this result is very interesting. At the so-called Mott density the bound state energy merges into the continuum and the bound state vanishes. The atom breaks up. Such an effect can be observed in some physical systems — in nuclear matter and, in a more complicated form, in electron-hole-plasmas and dense gaseous plasmas.

With the help of the antisymmetrized solutions ψ^{0-} of the homogeneous Bethe-Salpeter equation, $G_2(z_{12})$ may be represented as a bilinear expansion

$$G_2(p_1, p_2, p_1', p_2', z_{12}) = \sum_{\alpha P} \frac{\psi^{0-}_{\alpha P}(p_1, p_2)\, \psi^{0*-}_{\alpha P}(p_1', p_2')}{\hbar z_{12} - E_{\alpha P} - \langle \alpha P|\Delta H_{12}|\alpha P\rangle}\,. \tag{3.204}$$

The analytical properties of this function are well known from the relations (3.162). Especially the analytical continuation $G_2(z)$, of $G_2(z_{12})$ has a branch cut at the real axis which determines the spectral function. Because $\langle \alpha P| \Delta H_{12} |P\alpha\rangle$ is real, the two-particle states are stationary, and the spectral function has the simple form

$$A(p_1, p_2, p_1', p_2') = \sum_{\alpha P} \psi^{0}_{\alpha P}(p_1, p_2)\, \psi^{0*}_{\alpha P}(p_1', p_2')\, 2\pi\delta(\hbar\omega - E_{\alpha P} - \langle \alpha P| \Delta H_{12} |P\alpha\rangle)\,. \tag{3.205}$$

With this function all properties of two particles in an interacting many-body system are determined in a well-defined manner.

A special consideration is necessary in the case $\nu_{12} = 0$. In this case $z_{12} = 2\mu/\hbar$ is real, and $G_2(z_{12})$ has a pole located at $E_n + \langle n| \Delta H_{12} |n\rangle = 2\mu$. In order to obtain an unambiguous expression we must take into account (3.161). Then it follows for $G_2(z_{12})$

$$G_2(z_{12}) = \sum_{\alpha P} \frac{\psi^{0}_{\alpha P}(p_1, p_2)\, \psi^{0*}_{\alpha P}(p_1', p_2')}{\hbar z_{12} - E_{\alpha P} - \langle \alpha P| \Delta H_{12} |P\alpha\rangle} + \frac{1}{i\hbar\beta}\, \delta_{\nu_{12}0} C(p_1, p_2, p_1', p_2')\,. \tag{3.206}$$

That means, that similar to the considerations in 3.2.3.

i) the two particle Green's function can show long range order;

i) if $E_{\alpha P} + \langle \alpha P| \Delta H_{12} |P\alpha\rangle = 2\mu$, a Bose-Einstein condensation of bound states may take place.

3.2.9. T-Matrix and Thermodynamic Properties in Binary Collision Approximation

It is very convenient to express the binary collision approximation in terms of the T-matrix which was introduced by (3.178). Here we will consider the special T-matrix

$$T(121'2')\big|_{\substack{t_1=t_2=t\\ t_1'=t_2'=t'}} = \langle x_1x_2 | T(t-t') | x_2'x_1'\rangle\,.$$

The binary collision approximation follows from eq. (3.180) if we use

$$K(121'2') = V(1-2)\,\delta(1-1')\,\delta(2-2').$$

Then we obtain the equation

$$\begin{aligned}\langle 12| T(t-t') |2'1'\rangle &= V(1-2)\,\delta(1-1')\,\delta(2-2')\\ &+ i \int \mathrm{d}\bar{1}\, \mathrm{d}\bar{2}\, \langle 12| T(t-t') |\bar{2}\bar{1}\rangle\, G_1(\bar{1}1')\, G_1(\bar{2}2')\, V(1'-2')\,.\end{aligned} \tag{3.207}$$

The properties of this T-matrix are very similar to those of $G_2(x_1, x_2, x_1', x_2', t-t')$. Especially we have from (3.207)

$$\langle x_1x_2| T(t-t') |x_2'x_1'\rangle = \begin{cases} \langle x_1x_2| T^{>}(t-t') |x_2'x_1'\rangle\,, & it > it'\,,\\ \langle x_1x_2| V(1-2) |x_2'x_1'\rangle\,, & it = it'\,,\\ \langle x_1x_2| T^{<}(t-t') |x_2'x_1'\rangle, & it < it'\,.\end{cases}$$

Further the KMS condition (3.160) is valid also for T, that means

$$\langle x_1x_2| T(t-t')| x_2'x_1'\rangle_{t=0} = \mathrm{e}^{+2\beta\mu}\, \langle x_1x_2| T(t-t') |x_2'x_1'\rangle|_{t=-i\hbar\beta}\,. \tag{3.208}$$

This condition is automatically fulfilled if we express the T-matrix as a Fourier series

$$T(t-t') = \frac{1}{-i\hbar\beta} \sum_{z_{12}} e^{-iz_{12}(t-t')} T(z_{12}),$$

$$z_{12} = \frac{\pi\nu_{12}}{-i\hbar\beta} + \frac{\mu_1}{\hbar} + \frac{\mu_2}{\hbar}, \qquad \nu_{12} = 0, \pm 2, \pm 4, \ldots$$

The Fourier coefficient $T(z_{12})$ follows in well known manner from the spectral representation

$$T(z) - V = \int \frac{d\omega}{2\pi} \frac{T^>(\omega) - T^<(\omega)}{z-\omega};$$

$$T^>(\omega) - T^<(\omega) = T(\omega + i\varepsilon) - T(\omega - i\varepsilon) = \tfrac{1}{2} \operatorname{Im} T(\omega + i\varepsilon) \tag{3.209}$$

Let us consider now the Fourier transforms $T^<(\omega)$ and $T^>(\omega)$. From the KMS condition we find

$$T^>(\omega) = e^{\beta(\hbar\omega - 2\mu)} T^<(\omega) \tag{3.210}$$

and because of (3.209)

$$T^<(\omega) = \tfrac{1}{2} \operatorname{Im} T(\omega - i\varepsilon)\, n_B(\omega) \tag{3.210a}$$

with

$$n_B(\omega) = \frac{1}{e^{\beta(\hbar\omega - 2\mu)} - 1}.$$

For the following considerations is it useful to take the frequency-momentum representation. The eq. (3.178) takes the form

$$G_2(p_1, \ldots, p_4, z_{12}) = G_2^0(p_1, p_2, p_3, p_4, z_{12})$$
$$+ \int \frac{d\bar{p}_1\, d\bar{p}_2\, dp_3\, dp_4}{(2\pi)^{12}} \mathcal{G}_2^0(p_1, p_2, \bar{p}_1, \bar{p}_2, z_{12}) \langle \bar{p}_1\bar{p}_2 | T(z_{12}) | p_3 p_4 \rangle^{\pm} \mathcal{G}_2^0(p_3, p_4, p_1', p_2', z_{12}) \tag{3.211}$$

and the T-matrix equation

$$\langle p_1 p_2 | T(z_{12}) | \bar{p}_2 \bar{p}_1 \rangle^{\pm} = \langle p_1 p_2 | V_{12} | \bar{p}_2 \bar{p}_1 \rangle + \int \langle p_1 p_2 | V_{12} | p_1' p_2' \rangle$$
$$\cdot \mathcal{G}_2^0(p_1', p_2', p_3, p_4, z_{12}) \langle p_3 p_4 | T(z_{12}) | \bar{p}_2 \bar{p}_1 \rangle^{\pm} dp_3\, dp_4\, dp_1'\, dp_2'\, (2\pi)^{-12}. \tag{3.212}$$

Here the function $\mathcal{G}_2^0$ is given by (3.198):

$$\langle p_1 p_2 | \mathcal{G}_2^0(z_{12}) | p_2' p_1' \rangle = \langle p_1 p_2 | \frac{N_{12}}{H_{12}^0 - \hbar z_{12}} | p_1' p_2' \rangle$$
$$= \frac{1 - f(p_1) - f(p_2)}{\hbar z_{12} - E(p_1) - E(p_2)} \delta_{p_1 p_1'} \delta_{p_2 p_2'}.$$

Therefore eq. (3.212) is the momentum representation of the operator equation

$$T_{12}(z_{12}) = V_{12} + V_{12} \mathcal{G}_{12}^0(z_{12}) T_{12}(z_{12}). \tag{3.212a}$$

Similarly, eq. (3.211) is the momentum representation of

$$G_2(z_{12}) = \mathcal{G}_2^0(z_{12}) + \mathcal{G}_2^0(z_{12}) T(z_{12}) \mathcal{G}_2^0(z_{12}). \tag{3.211a}$$

In the case that N_{12} is replaced by 1 and after analytical continuation $z_{12} \to z$, these relations are well known from the two-particle scattering theory. Especially T

is the two-particle scattering operator, which determines, for $z = E + i\varepsilon$, the conventional scattering amplitude.

In general $T(E + i\varepsilon)$ describes the scattering of two particles in a many-particle system. The influence of the surrounding is given by N_{12}. In the Fermi case, the physical meaning of N_{12} was explained above.

As was shown by Kadanoff and Baym (1962), the many-particle T-operator satisfies an optical theorem. We consider

$$\begin{aligned}\text{Im } T(\omega + i\varepsilon) &= \tfrac{1}{2}\left(T(\omega + i\varepsilon) - T^+(\omega + i\varepsilon)\right) \\ &= \tfrac{1}{2}\left(T(\omega + i\varepsilon) - T(\omega - i\varepsilon)\right).\end{aligned}$$

The optical theorem follows from the two evident relations

$$\frac{\text{Im } T(\omega + i\varepsilon)}{|T(\omega - i\varepsilon)|^2} = \text{Im } T^{-1}(\omega - i\varepsilon)\,, \qquad \text{Im } T^{-1}(\omega + i\varepsilon) = \text{Im } \mathcal{G}_2^0(\omega + i\varepsilon)\,,$$

where the second relation follows from the T-matrix equation (3.212a). Using this relation we get

$$\text{Im } T(\omega + i\varepsilon) = T(\omega + i\varepsilon) \text{ Im } \mathcal{G}_2^0(\omega + i\varepsilon)\, T^+(\omega + i\varepsilon)\,. \tag{3.213}$$

That is the operator form of the generalized optical theorem. Here we have to take into account that the limit $\varepsilon \to 0$ is different for bound and scattering states (see KREMP et al., 1984a). Explicitly we have for

$$\text{Im } \mathcal{G}_2^0(\omega + i\varepsilon) = \frac{1}{2}\left\{\mathcal{G}_2^0(\omega + i\varepsilon) - \mathcal{G}_2^0(\omega - i\varepsilon)\right\} = \frac{\varepsilon N_{12}}{(\hbar\omega - H_{12}^0)^2 + \varepsilon^2}\,.$$

Next we consider the behaviour of the T-Matrix by using the bilinear expansion

$$\langle p_1 p_2|\, T(z)\, |\bar{p}_2 \bar{p}_1\rangle = \sum_{\alpha P} (E(p_1) + E(p_2) - \hbar z)^2\, \frac{\psi^0_{\alpha P}(p_1, p_2)\, \psi^{0*}_{\alpha P}(\bar{p}_1, \bar{p}_2)}{E_{\alpha P} - \hbar z} \tag{3.214}$$

which follows from (3.204) and (3.211a). In the case of the scattering spectrum, $T(z)$ remains finite, and therefore the limit $\varepsilon \to 0$ in (3.213) may be carried out immediately as

$$\text{Im } \mathcal{G}_2^0(\omega + i\varepsilon) = -\hbar\pi\, \delta(\hbar\omega - H_{12}^0)\, N_{12}\,.$$

Inserting this expression in (3.213) and taking the momentum representation we find the usual form of the optical theorem for scattering states

$$\begin{aligned}\text{Im } \langle p_1 p_2|\, T(E + i\varepsilon)\, |\bar{p}_2 \bar{p}_1\rangle^{\pm} &= -2\pi\hbar \int d\bar{p}_1\, d\bar{p}_2 \left\{1 - f(p_1) - f(p_2)\right\} \\ &\cdot \delta(E - E(p_1) - E(p_2))\, |\langle p_1 p_2|\, T(E + i\varepsilon)\, |\bar{p}_2 \bar{p}_1\rangle^{\pm}|^2\,.\end{aligned} \tag{3.213a}$$

For bound states, however, the T-operator has isolated poles at the bound states at the energies E_{nP}. Because $G_2^0(\omega)$ has no singularities in the region of the bound state spectrum, we find in this case

$$\text{Im } \mathcal{G}_2^0(\omega + i\varepsilon) = \varepsilon\, \frac{d\mathcal{G}_2^0(\omega)}{d\omega}\,. \tag{3.215}$$

In this way the limit $\varepsilon \to 0$ is well defined (KREMP et al., 1984a). Inserting (3.215) into the optical theorem and using (3.214) in the limit $\varepsilon \to 0$, it follows

$$\frac{|\psi^0_{n\bar{P}}(p_1p_2)|^2\,\varepsilon}{(\hbar\omega - E_{nP})^2 + \varepsilon^2}\,(E(p_1) + E(p_2) - \hbar\omega)^2 = \psi^0_{n\bar{P}}(p_1p_2)\,\psi^{0*}_{n\bar{P}}(p_1p_2)$$

$$\cdot\,(E(p_1) + E(p_2) - \hbar\omega)^2 \sum_{\bar{p}_1\bar{p}_2} \frac{|\psi^0_{n\bar{P}}(\bar{p}_1\bar{p}_2)|^2}{(\hbar\omega - E_{nP})^2 + \varepsilon^2}\,(E(p_1) + E(\bar{p}_2) - \hbar\omega)^2$$

$$\cdot\,\varepsilon\,\frac{\mathrm{d}\mathcal{G}^0_{12}}{\mathrm{d}\omega}\bigg|_{\omega=E_{nP}/\hbar},$$

and therefore the relation follows

$$1 = \sum_{\bar{p}_1\bar{p}_2} |\psi^0_n P(\bar{p}_1, \bar{p}_2)|^2\,.$$

For bound states the optical theorem is equivalent to the normalization of ψ_{nP}.

Another useful representation of the optical theorem may be obtained by using (3.209):

$$\langle p_1p_2|\,(T^>(\omega) - T^<(\omega))\,|p_2'p_1'\rangle = \int \frac{\mathrm{d}\bar{p}_1\,\mathrm{d}\bar{p}_2\,\mathrm{d}p_3\,\mathrm{d}p_4}{(2\pi)^{12}}$$

$$\cdot\langle p_1p_2|\,T(\omega + i\varepsilon)\,|\bar{p}_2\bar{p}_1\rangle\,\langle\bar{p}_1\bar{p}_2\,|(G^>_{12}(\omega) - G^<_{12}(\omega))\,|p_3p_4\rangle\,\langle p_4p_3|\,T(\omega - i\varepsilon)\,|p_2'p_1'\rangle\,. \tag{3.216}$$

Finally let us consider Σ in ladder approximation. Starting from the eq. (3.179) and using the relations (3.212) we find

$$\Sigma(11') = \pm i\hbar \int \mathrm{d}2\,\mathrm{d}\bar{2}\,\langle 12|\,T\,|\bar{2}1'\rangle\,G(\bar{2}2^+) = \tag{3.217}$$

Let us consider Im Σ and Re Σ given by

$$\begin{aligned}\mathrm{Im}\,\Sigma\,(p,\omega) &= \tfrac{1}{2}\,(\Sigma^>(p,\omega) - \Sigma^<(p,\omega))\,,\\ \mathrm{Re}\,\Sigma(p,\omega) &= \Sigma^{\mathrm{HF}}(p) + P\int\frac{\mathrm{d}\bar{\omega}}{2\pi}\,\frac{\mathrm{Im}\,\Sigma(p,\bar{\omega})}{\omega - \bar{\omega}}\,.\end{aligned} \tag{3.218}$$

In order to determine the self energy in binary collision approximation we must therefore consider $\Gamma(\omega) = \Sigma^> - \Sigma^<$. For this purpose we write

$$\Sigma^{\gtrless}(p_1,\omega_1) = \int \frac{\mathrm{d}p_2\,\mathrm{d}\omega_2}{(2\pi)^4}\,\langle p_1p_2|\,T^{\gtrless}\,(\omega_1 + \omega_2)\,|p_2p_1\rangle\,G^{\gtrless}(p_2,\omega_2)\,. \tag{3.217a}$$

Using (3.210a), (3.88), and (3.217a) we find

$$\Gamma(p_1,\omega_1) = \int \frac{\mathrm{d}p_2}{(2\pi)^3}\,2\,\mathrm{Im}\,\langle p_1p_2|\,T(\omega\hbar + E(p_2))\,|p_2p_1\rangle$$

$$\cdot\,\{n_{\mathrm{B}}(\hbar\omega_1 + E(p_2)) + f(E(p_2))\}\,, \tag{3.219}$$

where we have replaced

$$A(p,\omega) = 2\pi\hbar\delta(\hbar\omega - E(p))\,,\qquad E(p) = \frac{\hbar^2p^2}{2m} + \mathrm{Re}\,\Sigma(p,\omega)\,|_{\omega=E(p)/\hbar}\,.$$

On the basis of the eqs. (3.218) and (3.219) we can consider the properties of a Fermi system with strong short range potentials at finite densities and temperatures. Such a system can be realized by nuclear matter and He^3. For the ground state this was done in papers of Galitskii (1958). A systematic consideration at finite temperature including bound states can be found in Röpke et al. (1982,1983); see also Chapter 6.

In conclusion of this Section we give an explicit expression for the determination of thermodynamic quantities. We start from (3.143) and apply an approximation for the spectral function according to (3.145). The real and the imaginary parts of the self energy, $\mathrm{Re}\,\Sigma$ and $2\,\mathrm{Im}\,\Sigma = \Gamma$, respectively, are used in the binary collision approximation, by which the self energy is connected to the T-matrix via equations (3.219) and (3.218). Moreover, we expand the Fermi distribution function near the free particle energy in the manner

$$f_a(\varepsilon_a(p)) = f_a\left(\frac{\hbar^2p^2}{2m_a}\right) + \left[\varepsilon_a^{\mathrm{HF}} - \frac{\hbar^2p^2}{2m_a} + \mathrm{Re}\;\Sigma_a^{\mathrm{corr}}\left(p, \frac{\hbar^2p^2}{2m_a}\right)\right]$$
$$\cdot \frac{1}{k_{\mathrm{B}}T} f_a\left(\frac{\hbar^2p^2}{2m_a}\right)\left(f_a\left(\frac{\hbar^2p^2}{2m_a}\right) - 1\right) + \ldots$$

$\varepsilon_a(p)$ — quasiparticle energy; see also Section 4.3. and 6.2. For the pressure we get then

$$p - p_0 = p_{\mathrm{HF}} - \frac{1}{2}\sum_{ab}(2s_a+1)(2s_b+1)\int_0^1 \frac{\mathrm{d}\lambda}{\lambda}\int\frac{\mathrm{d}\boldsymbol{p}\,\mathrm{d}\boldsymbol{p}'}{(2\pi)^6}\left(\varepsilon_a(p) - \frac{\hbar^2p^2}{2m_a}\right)$$
$$\cdot \frac{\partial}{\partial\hbar\omega}\mathrm{Re}\langle pp'|\,\{T_{ab}(\omega) - V_{ab}\}\,|p'p\rangle \cdot g_{\mathrm{B}}^{ab}(\omega)\,\big(1 - f_b(\varepsilon_b(p'))\big)|_{\hbar\omega=\varepsilon_a(p)+\varepsilon_b(p')}$$
$$+ \frac{1}{2}\sum_{ab}(2s_a+1)(2s_b+1)\int_0^1\frac{\mathrm{d}\lambda}{\lambda}\int_{-\infty}^{+\infty}\frac{\mathrm{d}\omega}{2\pi}\int\frac{\mathrm{d}\boldsymbol{p}'\,\mathrm{d}\boldsymbol{p}}{(2\pi)^6}\left[\frac{\partial}{\partial\hbar\omega}\frac{P}{\omega - \frac{1}{\hbar}\varepsilon_b(p') - \frac{1}{\hbar}\varepsilon_a(p)}\right]$$
$$\cdot\, 2\,\mathrm{Im}\,\langle pp'|\,T_{ab}(\omega + i\varepsilon)\,|p'p\rangle\, g_{\mathrm{B}}^{ab}(\omega)\,\big(1 - f_b(\varepsilon_b(p'))\big)\left(\hbar\omega - \frac{\hbar^2p^2}{2m_a}\right). \qquad (3.220)$$

An expression of such type was derived for the density as a function of the chemical potential by Kremp, Kraeft and Lambert (1984) according to (3.146). See also Stolz and Zimmermann (1979), and Section 6.2. The pressure was given in terms of the T-matrix in Section 6.1. Here also the Hartree-Fock contribution to the pressure, p_{HF}, was evaluated (see eq. (6.34) ff.). The results of Sections 6.1. and 6.2. are, in the nondegenerate limit, in agreement with virial and fugacity expansions as discussed, e.g., in Ebeling (1976, 1979).

In terms of the second virial coefficients (second cluster coefficient) the pressure reads

$$p - p_0 = k_{\mathrm{B}}T\sum_{a,b} z_a z_b\, B_{ab}\,.$$

Here z_a are the fugacities, $z_a = \exp\,(\mu_a/k_{\mathrm{B}}T)$, and B_{ab} are the virial coefficients, which may include bound state and scattering state contributions.

The right hand side of (3.220) is essentially the second virial coefficient.

3.3. Quantum Statistics of Charged Many-Particle Systems

3.3.1. Basic Equations. Screening

Now we will develop a rigorous quantum statisticial theory of charged many-body systems. The Hamiltonian of such a system consisting of particles of species a, b, ... has the form

$$H = \sum_a \int \mathrm{d}x\, \psi_a^+(x) \left(-\frac{\hbar^2\nabla^2}{2m_a}\right) \psi_a(x) + \frac{1}{2} \sum_{ab} \int \mathrm{d}x\, \mathrm{d}x'\, V_{ab}(x - x')\, \psi_a^+(x)\, \psi_b^+(x')\, \psi_b(x')\, \psi_a(x)$$

with

$$V_{ab}(x - x') = \frac{e_a e_b}{4\pi\varepsilon_0\, |\boldsymbol{r} - \boldsymbol{r}'|}\, \delta_{s_a s_b}$$

beeing the interaction between the particles of the species a and b with the charges e_a and e_b, respectively and spin s_a, s_b. A characteristic feature of this interaction is that its range is very large. Therefore systems with Coulomb interaction show a collective behaviour. So we can observe for instance

i) the dynamical screening of the Coulomb potential,
ii) plasma oscillations.

Obviously we may expect, that the simple binary collision approximation, discussed in the previous section is not appropriate, because always a large number of particles interacts simultaneously.

Formally the binary collision approximation leads to many difficulties, for instance:

i) The second virial coefficient in the equation of state diverges for the Coulomb interaction.
ii) There is an infinite number of discrete energy values $E_n \sim n^{-2}$ at the negative energy axis converging at the point $E = 0$. Thus there is no energy gap between the discrete and the continuous part of the spectrum. If the second virial coefficient is divided into the two parts B_{ab}^{bound} and $\mathrm{B}_{ab}^{\mathrm{scatt}}$, it follows that both contributions are divergent especially the sum of bound states diverges,

$$B_{ab}^{\mathrm{bound}} = \sum_l (2l + 1) \sum_n \mathrm{e}^{-\beta E_{nl}} \to \infty\,.$$

iii) The transport coefficients determined on the basis of the binary collision approximation are divergent.

In order to solve these problems the screening of the Coulomb potential is essential; that means because of the collective behaviour the charged particles do not interact via a bare Coulomb potential but with a screened one. Thus a consequent many-body theory must take into account a proper screening. This may be achieved by reformulating the basic equations of quantum statistics in such a way that the Coulomb potential is replaced by a screened one.

The programme may be carried out using the method according to KADANOFF and BAYM (1962). (See also STOLZ, 1974; EBELING, KRAEFT and KREMP, 1976; 1979).

The starting point in the considerations is the fact that an external potential $U_a(1\bar{1})$ produces an inhomogeneity and thus a mean field of interaction (effective external field). This field is given by

$$U_a^{\text{eff}}(1\bar{1}) = U_a(1\bar{1}) + \Sigma_a^{\text{H}}(1\bar{1})$$
$$= U_a(1\bar{1}) + \sum_b i\hbar \int \mathrm{d}2\; V_{ab}(12)\, G_b(22^+)\, \delta(1\bar{1})\,. \tag{3.221}$$

Now we consider the equation of motion for G_a (3.117) and take into account the relation (3.168) for the two-particle Green's function. Then (3.117) assumes the form

$$\left(i\hbar \frac{\partial}{\partial t_1} + \frac{\hbar^2 \nabla_1^2}{2m_a}\right) G_a(11') - \int\limits_0^{-i\beta\hbar} U_a(1\bar{1})\, G_a(\bar{1}1')\, \mathrm{d}\bar{1}$$

$$\pm\, i\hbar \sum_b \int\limits_0^{-i\beta\hbar} \mathrm{d}2\; V_{ab}(12) \left\{ G_a(11')\, G_b(22^+) - \frac{\delta G_a(11')}{\delta U_b(22)} \right\} = \delta(11')\,. \tag{3.222}$$

Now we define the "screened self energy" $\bar{\Sigma}$ by

$$\int\limits_0^{-i\beta\hbar} \bar{\Sigma}_a(1\bar{1})\, G_a(\bar{1}1')\, \mathrm{d}\bar{1} = \pm\, i\hbar \sum_b \int\limits_0^{-i\beta\hbar} V_{ab}(12)\, \frac{\delta G_a(11')}{\delta U_b(22)}\, \mathrm{d}2 \tag{3.223}$$

and taking into account the definition of U_{eff} we find

$$\left\{ i\hbar \frac{\partial}{\partial t_1} + \frac{\hbar^2}{2m_a} \nabla_1^2 \right\} G_a(11') - \int\limits_0^{-i\beta\hbar} \{U_a^{\text{eff}}(1\bar{1}) + \bar{\Sigma}_a(1\bar{1})\}\, G_a(\bar{1}1')\, \mathrm{d}\bar{1} = \delta(11')\,.$$

From this equation may be obtained easily the integral equation

$$G_a(11') = G_a^0(11') + \int\limits_0^{-i\beta\hbar} G_a^0(1\bar{1})\, \{\bar{\Sigma}_a(\bar{1}\bar{2}) + U_a^{\text{eff}}(\bar{1}\bar{2})\}\, G_a(\bar{2}1')\, \mathrm{d}\bar{1}\mathrm{d}\bar{2}\,. \tag{3.224}$$

From this equation it can be seen that G_a depends on U only via U^{eff}. For this reason it is useful to replace derivatives according to U by those according to U^{eff}, i.e.,

$$\int\limits_0^{-i\beta\hbar} \bar{\Sigma}_a(1\bar{1})\, G_a(\bar{1}1')\, \mathrm{d}\bar{1} = \pm\, i\hbar \sum_{bc} \int\limits_0^{-i\beta\hbar} V_{ac}(12)\, \frac{\delta G_a(11')}{\delta U_b^{\text{eff}}(33')}$$
$$\cdot \frac{\delta U_b^{\text{eff}}(33')}{\delta U_c(22)}\, \mathrm{d}2\, \mathrm{d}3\, \mathrm{d}3'\,. \tag{3.225}$$

Further we introduce two quantities which are of fundamental meaning for a system with Coulomb interaction; namely the polarization function

$$\Pi_{ab}(13'1'3) = -\frac{\delta G_a(11')}{\delta U_b^{\text{eff}}(33')} \tag{3.226}$$

and the screened potential

$$V_{ab}^{\text{s}}(12)\; \delta(32)\; \delta(3'2') = V_{ab}(12)\, \frac{\delta U_b^{\text{eff}}(33')}{\delta U_b(22')} \tag{3.227}$$

(cf., e.g., Martin and Schwinger, 1959).

Then the relation (3.225) assumes the compact form

$$\int_0^{-i\beta\hbar} \overline{\Sigma}_a(1\bar{1})\, G_a(\bar{1}1')\, \mathrm{d}\bar{1} = i\hbar \sum_b \int_0^{-i\beta\hbar} V^s_{ab}(12)\, \Pi_{ab}(121'2)\, \mathrm{d}2\,.$$

First we discuss the polarization function Π in more detail. By functional derivation of (3.224) we get

$$\Pi_{ac}(12'1'2) = -\frac{\delta G_a(11')}{\delta U_c^{\mathrm{eff}}(22')}$$

$$= -\delta_{ac} G_a(12)\, G_a(2'1') - \int_0^{-i\beta\hbar} G_a(1\bar{1}) \frac{\delta\, \overline{\Sigma}_a(\bar{1}\bar{2})}{\delta U_c^{\mathrm{eff}}(22')} G_a(\bar{2}1')\, \mathrm{d}\bar{1}\, \mathrm{d}\bar{2}\,. \tag{3.228}$$

Inserting this equation into (3.225), we get for the self energy the functional equation

$$\overline{\Sigma}_a(11') = i\hbar\, V^s_{aa}(11')\, G_a(11') + i\hbar \sum_b \int \mathrm{d}\bar{1}\mathrm{d}\bar{2}\; V^s_{ab}(1\bar{2}) G_a(1\bar{1})\, \frac{\delta \overline{\Sigma}_a(\bar{1}1')}{\delta U_b^{\mathrm{eff}}(\bar{2}\bar{2})}\,. \tag{3.229}$$

In this way the Coulomb potential in the Dyson equation is replaced by a screened one. Further progress with this equation depends on the explicit form and the meaning of V^s. In order to deal with this problem we calculate $\delta U^{\mathrm{eff}}/\delta U$ from equation (3.221) and we find easily

$$V^s_{ab}(12) = V_{ab}(12) + i\hbar \sum_{cd} \int_0^{-i\beta\hbar} \mathrm{d}\bar{1}\, \mathrm{d}\bar{2}\; V^s_{ac}(1\bar{1})\, \Pi_{cd}(\bar{2}\bar{1}\bar{2}\bar{1})\; V_{db}(\bar{2}2)\,. \tag{3.230}$$

(3.230) represents a potential, which is modified on account of the many-particle effects, i.e., V^s is a screened potential. Now we are able to write down the basic equations for a many-particle system with Coulomb interaction (Kadanoff and Baym, 1962). The Dyson equation in integral form and the self energy are given by

$$\begin{aligned} G_a(11') &= G_a^0(11') - \int_0^{-i\beta\hbar} \mathrm{d}\bar{1}\, \mathrm{d}\bar{2}\, G_a^0(1\bar{1})\, \{U_a^{\mathrm{eff}}(\bar{1}\bar{2}) + \overline{\Sigma}_a(\bar{1}\bar{2})\}\, G_a(\bar{2}1')\,; \\ \overline{\Sigma}_a(\bar{1}1) &= i\hbar\, V^s_{aa}(1\bar{1})\, G_a(1\bar{1}) + i\hbar \sum_b \int_0^{-i\beta\hbar} \mathrm{d}\bar{\bar{1}}\, \mathrm{d}\bar{2}\; V^s_{ab}(1\bar{2}) \\ &\quad \cdot G_a(1\bar{\bar{1}})\, \delta\overline{\Sigma}_a(\bar{\bar{1}}\bar{1})/\delta U_b^{\mathrm{eff}}(\bar{2}\bar{2})\,. \end{aligned} \tag{3.231}$$

By the second group of equations the screened potential V^s is determined:

$$\begin{aligned} V^s_{ab}(12) &= V_{ab}(12) + i\hbar \sum_{cd} \int_0^{-i\beta\hbar} \mathrm{d}3\, \mathrm{d}4\; V^s_{ac}(13)\, \Pi_{cd}(4343)\; V_{db}(42)\,; \\ \Pi_{ac}(1234) &= -\delta_{ac} G_a(14)\, G_a(23) - \int_0^{-i\beta\hbar} G_a(11') \frac{\delta\overline{\Sigma}_a(1'2')}{\delta U_c^{\mathrm{eff}}(42)} G_a(2'3)\, \mathrm{d}1'\mathrm{d}2'\,. \end{aligned} \tag{3.232}$$

If we introduce the diagrams

$i\hbar\, V^{s}_{ab} =$  — screened potential ,

$\Pi_{ab}(121'2') =$ 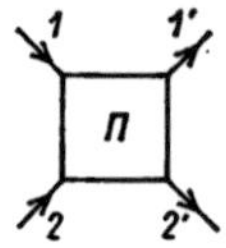

— polarization function, (3.233)

$\dfrac{\delta \overline{\Sigma}_a(1'2')}{\delta U^{\text{eff}}_c(12)} =$ 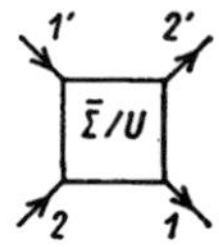

— vertex function

these equations may be expressed by Feynman diagrams (in the limit $U \to 0$)

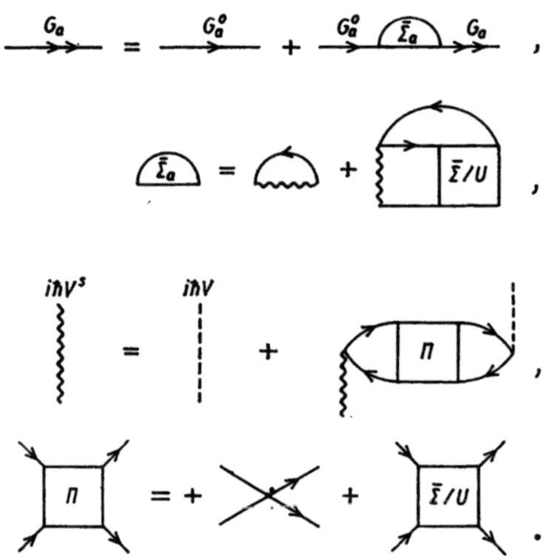

(3.234)

By means of eqs. (3.233) all functions which occur in a many particle theory may be expressed by the screened potential V^s. By this procedure the collective behaviour is automatically taken into account and the Coulomb divergencies are removed.

It is interesting to remark, that these equations are formally similar to the basic equations of quantum electrodynamics. Indeed these equations represent the non-relativistic limit of those equations, but the Green's functions here are defined for a system with finite densities and finite temperatures. Using this analogy the eqs. (3.233) were also obtained by many authors (e.g., Bonch-Bruevich and Tyablikov, 1961).

Finally let us show, that the thermodynamic functions may also be calculated by these quantities. The pressure is given by (cf. Section 3.2.)

$$V(p - p_0) = \frac{1}{2} \sum_{ab} \int_0^1 \frac{d\lambda}{\lambda} \int d2\, \lambda V_{ab}(12)\, G_{ab}(121^{++}2^{+})\,.$$

Using the eq. (3.225), (3.226) and (3.227) we get the pressure in terms of the screened potential:

$$V(p - p_0) = \frac{1}{2} \sum_{ab} \int_0^1 \frac{d\lambda}{\lambda} d2 \, \{ V^s_{ab}(12, \lambda) \, \Pi_{ab} \, (121^{++}2^+)$$

$$+ \lambda V_{ab}(12) \, G_a(11^{++}) \, G_b(22^+) \} \, .$$

In terms of Feynman diagrams this equation takes the form

$$V(p - p_0) = \frac{1}{2} \sum_{ab} \int_0^1 \frac{d\lambda}{\lambda} \left[\quad \right] \qquad (3.235)$$

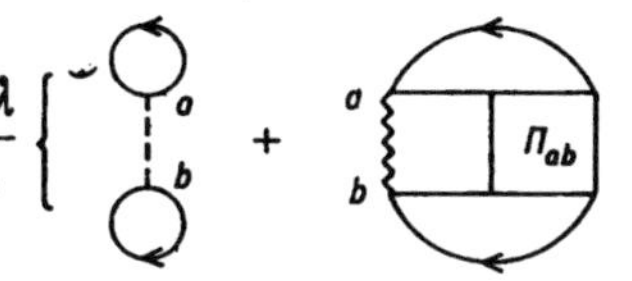

Starting from this expression for the equation of state the divergence problems are overcome.

3.3.2. Analytic Properties of V^s and Π

In the previous Section we have introduced the polarization function by a functional derivation. This function is therefore not directly defined by the field operators, and we cannot immediately apply the consideration of Section 3.2. in order to obtain the analytical properties. But it is easy to show, that Π_{ab} is connected with the function $L_{ab}(121'2')$ by the integral equation

$$\Pi_{ab}(121'2') = L_{ab}(121'2') + i\hbar \int d\bar{1} \, d\bar{2} \sum_{cd} L_{ad}(12\bar{1}\bar{1}^+)$$

$$\cdot \, V^s_{dc}(\bar{1}\bar{2}) \, \Pi_{cb}(\bar{2}\bar{2}^+1'2') \, . \qquad (3.236)$$

In this way the analytical properties of $\Pi(121'2')$ are determined by the properties of L given by

$$L_{ab}(121^+2^+) = \frac{1}{(i\hbar)^2} \langle T(\psi_a^+(1) \, \psi_a(1) \, \psi_b^+(2) \, \psi_b(2)) \rangle$$

$$- \frac{1}{(i\hbar)^2} \langle \psi_a^+(1) \, \psi_a(1) \rangle \, \langle \psi_b^+(2) \psi_b(2) \rangle \, . \qquad (3.237)$$

With $\hat{n}_a(1) = \psi_a^+(1) \, \psi_a(1) - \langle n(1) \rangle$ as density fluctuation operator, $L_{ab}(121^+2^+)$ can be interpreted as density correlation function

$$L_{ab}(121^+2^+) = \langle T\{\hat{n}_a(1) \, \hat{n}_b(2)\} \rangle$$

$$= \theta(t_1 - t_2) \, L^>_{ab}(t_1 - t_2) + \theta(t_2 - t_1) \, L^<_{ab}(t_1 - t_2) \, . \qquad (3.238)$$

The analytical properties are with slight modification in agreement with those of G_1. We obtain a Lehmann representation and especially the KMS condition in the form

$$L^>_{ab}(t_1 - t_2) = L^<_{ab}(t_1 - t_2 + i\beta) \, ,$$
$$L_{ab}(t_1 - t_2)|_{t_1=0} = L_{ab}(t_1 - t_2)|_{t_1=-i\beta} \, . \qquad (3.239)$$

In the same manner as in Section 3.2.2. we get then omitting the spin variables

$$\begin{aligned} L^{>}_{ab}(p,\omega) &= \{1 \pm f(\omega)\}\, \hat{A}_{ab}(p,\omega)\,, \\ L^{<}_{ab}(p,\omega) &= f(\omega)\, \hat{A}_{ab}(p,\omega)\,, \quad f(\omega) = [\mathrm{e}^{\beta\omega} - 1]^{-1}\,; \\ \hat{A}_{ab}(p,\omega) &= L^{>}_{ab}(p,\omega) \pm L^{<}_{ab}(p,\omega) \end{aligned} \tag{3.240}$$

with

$$L^{\gtrless}_{ab}(p,\omega) = \int \mathrm{d}\boldsymbol{r}\, \mathrm{d}\tau\, \mathrm{e}^{-ipr+i\omega\tau}\, L^{\gtrless}_{ab}(r,\tau)\,.$$

Further, because of the KMS condition, L_{ab} may be expanded in a Fourier series

$$L_{ab}(t_1 - t_2) = \frac{i}{\hbar\beta} \sum_n L_{ab}(\Omega_n)\, \mathrm{e}^{-i\Omega_n(t_1-t_2)} \tag{3.241}$$

with the Bose-like Matsubara frequencies

$$\Omega_n = \frac{\pi n}{-i\hbar\beta}\,, \qquad n = 0, \pm 2, \pm 4, \ldots$$

For the Fourier coefficients we find

$$L_{ab}(\Omega_n, p) = \frac{1}{2\pi} \int\limits_{-\infty}^{+\infty} \mathrm{d}\omega\, \frac{\hat{A}_{ab}(p,\omega)}{\Omega_n - \omega}\,. \tag{3.242}$$

$L_{ab}(\Omega_n, p)$ may be continued into the complex z-plane by

$$L_{ab}(p, z) = \int\limits_{-\infty}^{+\infty} \frac{\mathrm{d}\omega}{2\pi}\, \frac{\hat{A}_{ab}(p,\omega)}{z - \omega}\,. \tag{3.243}$$

From this Cauchy type integral we get in usual manner

$$i\hat{A}_{ab}(p,\omega) = L_{ab}(p, \omega + i\varepsilon) - L_{ab}(p, \omega - i\,\varepsilon)\,. \tag{3.244}$$

Since Π is connected with L by (3.236) the properties of Π are the same as that of L for a detailed discussion see e.g. STOLZ, 1974):

$$\begin{aligned} \Pi^{>}(p,\omega) &= \big(1 \pm f(\omega)\big)\, \hat{\Pi}(p,\omega)\,, \\ \Pi^{<}(p,\omega) &= f(\omega)\, \hat{\Pi}(p,\omega)\,. \\ \hat{\Pi}(p,\omega) &= \Pi^{>}(p,\omega) \mp \Pi^{<}(p,\omega)\,. \end{aligned} \tag{3.245}$$

The indication of species was omitted.

Analogous to L the analytic continuation may be written in the form of a Cauchy type integral

$$\Pi(p, z) = \int \frac{\mathrm{d}\omega}{2\pi}\, \frac{\hat{\Pi}(p,\omega)}{z - \omega}\,. \tag{3.246}$$

In many cases it is useful to separate this equation into its real and imaginary part

$$\begin{aligned} 2i\, \mathrm{Im}\, \Pi(p, \omega + i\varepsilon) &= i\, \hat{\Pi}(p,\omega) = \Pi(p, \omega + i\varepsilon) - \Pi(p, \omega - i\varepsilon)\,; \\ \mathrm{Re}\, \Pi(p,\omega) &= P \int \frac{\mathrm{d}\omega'}{2\pi}\, \frac{\hat{\Pi}(p,\omega')}{\omega - \omega'}\,. \end{aligned} \tag{3.247}$$

The analytic properties of V^{s} are given by the consideration of eq. (3.230).

In this case we get a spectral representation for the difference $V^s - V$

$$V^s(p, z) - V(p) = \int \frac{d\omega}{2\pi} \frac{V_s^>(p, \omega) - V_s^<(p, \omega)}{z - \omega}. \tag{3.248}$$

Separation into real and imaginary part gives the dispersion relations,

$$\begin{aligned} 2i \operatorname{Im} V^s(p, \omega + i\varepsilon) &= i(V_s^>(p, \omega) - V_s^<(p, \omega)) \\ &= V^s(p, \omega + i\varepsilon) - V^s(p, \omega - i\varepsilon); \\ \operatorname{Re} \{V^s(p, \omega) - V(p)\} &= P \int \frac{d\omega'}{2\pi} \frac{2 \operatorname{Im} V^s(p, \omega' + i\varepsilon)}{\omega - \omega'}. \end{aligned} \tag{3.249}$$

Using these properties of V^s and Π we can write the eq. (3.230) for the screened potential in Fourier-Matsubara representation, taking into account the spin dependence

$$V^s(p, \Omega_n) = [1 + (2s + 1)\, \Pi(p, \Omega_n)\, V^s(p, \Omega_n]\, V(p);$$

$$V(p) = \frac{e^2}{\varepsilon_0 p^2} \text{ — Fourier transform of the Coulomb potential.} \tag{3.250}$$

From this equation we obtain easily a formal expression for V^s. After analytic continuation we find

$$V^s(p, z) = \frac{V(p)}{1 - V(p)\,(2s+1)\, \Pi(p, z)}. \tag{3.251}$$

Because V^s is the interaction including all polarization effects this relation suggests that we define by

$$\varepsilon(p, z) = 1 - V(p)\,(2s + 1)\, \Pi(p, z) \tag{3.252}$$

the dielectric function of the charged particle system. This function plays an important role in the many-body theory of charged particle systems, because $\varepsilon(p, z)$ describes completely the collective behaviour.

Using the expressions (3.247) — (3.249) the following properties are simply to be verified:

$$\begin{aligned} 2 \operatorname{Im} V^s(p, \omega) &= |V^s(p, \omega)|^2 \{-\Pi^>(p, \omega) - \Pi^<(p, \omega)\}, \\ V_s^{\gtrless}(p, \omega) &= |V^s(p, \omega)|^2\, \Pi^{\gtrless}(p, \omega). \end{aligned} \tag{3.253}$$

3.3.3. The "Random Phase Approximation" RPA

Further progress in the consideration of systems with Coulomb interaction depends on the possibility to find approximations in order to handle the complicated self consistent system of basic equations (3.234).

It is useful to begin the discussion with the second group of equations. Let us consider especially the equation for the polarization function $\Pi(1234)$. Obviously we obtain the simplest approximation for Π by neglecting the integral contribution that means we neglect the so called vertex corrections, i.e., $\delta G^{-1}/\delta U^{\text{eff}} = 0$. Then Π reduces to

$$\Pi_{ab}(1234) = \pm\, \delta_{ab}\, G_a(14)\, G_a(23) \tag{3.254}$$

$= \pm\, \delta_{ab}$ 

and, in the special case which is necessary in order to determine the screened potential

$$\Pi_{ab}(1414) = \pm\,\delta_{ab}\,G_a(14)\,G_a(41) \tag{3.255}$$

$= \pm\,\delta_{ab}$ 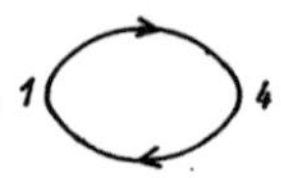 .

This simple approximation is called "random phase approximation" (RPA) (see also Section 5.2.). For the further considerations it is usefull to consider first the functions $\Pi^{>}$ and $\Pi^{<}$. Using (3.255) we find

$$\begin{aligned} \Pi^{>}(12) &= \pm\,G^{>}(12)\,G^{<}(21)\,, \\ \Pi^{<}(21) &= \pm G^{<}(12)\,G^{>}(21)\,. \end{aligned} \tag{3.256}$$

These expressions are most simply considered in momentum space. The Fourier transform is given by

$$\Pi^{\gtrless}(\boldsymbol{p},\omega) = \int \frac{d\boldsymbol{p}'}{(2\pi)^3}\frac{d\omega'}{(2\pi)}\,G^{\gtrless}(\boldsymbol{p}'+\boldsymbol{p},\omega'+\omega)\,G^{\lessgtr}(\boldsymbol{p}',\omega')\,. \tag{3.257}$$

In the following we use for $G^{\gtrless}$ the simple approximation

$$\begin{aligned} G^{>}(\boldsymbol{p},\omega) &= [1-f(\omega)]\,2\pi\hbar\,\delta\big(\hbar\omega-\varepsilon(\boldsymbol{p})\big)\,, \\ G^{<}(\boldsymbol{p},\omega) &= f(\omega)\,2\pi\hbar\,\delta\big(\hbar\omega-\varepsilon(\boldsymbol{p})\big)\,. \end{aligned} \tag{3.258}$$

Then $\Pi^{\gtrless}(\boldsymbol{p},\omega)$ takes the form

$$\begin{aligned} \Pi^{>}(\boldsymbol{p},\omega) &= \int \frac{d\boldsymbol{p}'}{(2\pi)^3}\{1-f[\varepsilon(\boldsymbol{p}')+\hbar\omega]\}\,f(\varepsilon(\boldsymbol{p}'))\,\hbar\delta\big(\hbar\omega+\varepsilon(\boldsymbol{p}')-\varepsilon(\boldsymbol{p}'+\boldsymbol{p})\big)\,, \\ \Pi^{<}(\boldsymbol{p},\omega) &= \int \frac{d\boldsymbol{p}'}{(2\pi)^3}\{1-f(\varepsilon(\boldsymbol{p}'))\}\,f[\varepsilon(\boldsymbol{p}')+\hbar\omega]\,\hbar\delta\big(\hbar\omega+\varepsilon(\boldsymbol{p}')-\varepsilon(\boldsymbol{p}'+\boldsymbol{p})\big). \end{aligned} \tag{3.259}$$

Inserting these expressions into equation (3.246) we get the spectral representation of Π

$$\Pi(\boldsymbol{p},z) = \int \frac{d\boldsymbol{p}'}{(2\pi)^3}\frac{f(\boldsymbol{p}'-\boldsymbol{p})-f(\boldsymbol{p}')}{\hbar z-\varepsilon(\boldsymbol{p}')+\varepsilon(\boldsymbol{p}'-\boldsymbol{p})}\,. \tag{3.260}$$

This equation leads to the following expressions for the real and imaginary parts of Π:

$$\begin{aligned} \operatorname{Im}\Pi(\boldsymbol{p},\omega) &= \tfrac{1}{2}\,[\Pi^{>}(\boldsymbol{p},\omega)-\Pi^{<}(\boldsymbol{p},\omega]\,, \\ \operatorname{Re}\Pi(\boldsymbol{p},\omega) &= P\int \frac{d\boldsymbol{p}'}{(2\pi)^3}\frac{f(\boldsymbol{p}'-\boldsymbol{p})-f(\boldsymbol{p}')}{\hbar\omega-\varepsilon(\boldsymbol{p}')+\varepsilon(\boldsymbol{p}'-\boldsymbol{p})}\,. \end{aligned} \tag{3.261}$$

Finally we write down the dielectric function in RPA

$$\varepsilon(\boldsymbol{p},z) = 1-\frac{e^2}{\varepsilon_0 p^2}\int \frac{d\boldsymbol{p}'}{(2\pi)^3}\frac{f(\boldsymbol{p}'-\boldsymbol{p})-f(\boldsymbol{p}')}{\hbar z-\varepsilon(\boldsymbol{p}')+\varepsilon(\boldsymbol{p}'-\boldsymbol{p})}\,. \tag{3.262}$$

By this expressions the collective behaviour of the charged many-particle system is explicitly determined. Especially the screened potential may be obtained in the form

$$V^{s}(q,z) = \frac{V(q)}{\varepsilon(q,z)}\,. \tag{3.263}$$

We shall consider this approximation in more detail in Section 4.2.

Let us now consider the first group of basic equations. In order to determine the single particle properties it is necessary to find approximations for the self energy. We neglect again the vertex corrections to $\bar{\Sigma}$ that means $\delta\bar{\Sigma}/\delta U^{\text{eff}} = 0$. Then we find the so-called V^s-approximation. In connection with the RPA for V^s we obtain in this way a well defined approximation

$$\bar{\Sigma}_a(1\bar{1}) = i\hbar\, V^s_{aa}(1\bar{1})\, G_a(\bar{1}1) = \tag{3.264}$$

In this approximation $\Sigma^{\gtrless}$ may be determined from the equation

$$\Sigma^{\gtrless}_a(1\bar{1}) = i\hbar\, V^{s\gtrless}_{aa}(1\bar{1})\, G^{\gtrless}_a(1\bar{1}) \tag{3.265}$$

and in Fourier momentum representation

$$\Sigma^{\gtrless}_a(\omega, p) = \int \frac{\mathrm{d}\boldsymbol{p}'}{(2\pi)^3} \frac{\mathrm{d}\omega'}{(2\pi)} V^{s\gtrless}_{aa}\,(\boldsymbol{p} - \boldsymbol{p}', \omega - \omega')\, G^{\gtrless}_a(p', \omega')\,. \tag{3.266}$$

In order to determine the single-particle properties we must determine explicitly the real and imaginary parts.

Using the spectral representation (3.125) we get

$$\begin{aligned} \operatorname{Re} \bar{\Sigma}(p, \omega) &= \Sigma_{\mathrm{HF}}(p) + P \int \frac{\mathrm{d}\omega'}{(2\pi)^3} \frac{\Gamma(p, \omega')}{\omega - \omega'}\,, \\ \Gamma(p, \omega) &= 2 \operatorname{Im} \bar{\Sigma}(p, \omega) = \Sigma^{>}(p, \omega) \mp \Sigma^{<}(p, \omega)\,. \end{aligned} \tag{3.267}$$

Let us now consider $\Gamma(p, \omega)$ in more detail. We get

$$\begin{aligned} \Sigma^{>} \mp \Sigma^{<} = \Gamma(p, \omega) &= \int \frac{\mathrm{d}\boldsymbol{p}'}{(2\pi)^3} \frac{\mathrm{d}\omega'}{2\pi} \{ V^{s>}(p', \omega') \\ &\cdot G^{>}(\boldsymbol{p} + \boldsymbol{p}', \omega + \omega') \mp V^{s<}(p', \omega')\, G^{>}(\boldsymbol{p} + \boldsymbol{p}', \omega + \omega')\}\,. \end{aligned} \tag{3.268}$$

Using the relation (3.88) and formula (3.248) we find

$$\begin{aligned} \Gamma_a(p, \omega) &= 4\pi\hbar \int \frac{\mathrm{d}\boldsymbol{p}'}{(2\pi)^3} \int \frac{\mathrm{d}\omega'}{2\pi} \delta\left(\hbar\omega - \hbar\omega' - \frac{(\boldsymbol{p} + \boldsymbol{p}')^2\, \hbar^2}{2m_a}\right) \\ &\cdot \{1 - f_a(\boldsymbol{p} + \boldsymbol{p}') + n_{\mathrm{B}}(\omega')\} \operatorname{Im} V^s_{aa}(p', \omega' + i\varepsilon)\,. \end{aligned} \tag{3.269}$$

The problem of single-particle properties is further dealt with in Chapter 5. and is applied in Section 6.2.

4. Application of the Green's Function Technique to Coulomb Systems

4.1. Types of Different Approximations

4.1.1. Diagram Representation of Σ and Π

It was outlined in the previous Sections that in Coulomb systems special effects arise which are due to the long range character of the Coulomb potential and which must be treated by many-body theory.

The typical effects to be described by the many-body theory are the formation of bound states, self-energy effects, and the dynamical screening of the interaction potential. A consistent approach to the statistical mechanics of a many-body system is given by the Green's function technique as described in Chapter 3. Now, in this Chapter, more detailed approaches to the one- and two- particle properties as well as the dielectric function will be given.

In the framework of the Green's function technique, a rather clear and systematic description of different approximations is given by the diagram representation of the corresponding analytical expressions to be derived from the equations given in Chapter 3.

Before discussing special types of Green's functions (G_1, G_2, L) we will give some general aspects how to specify the different approximations. Especially, we work out the case of the partially ionized plasma, where free particles as well as bound states are present. In particular, we will treat the free particles and the bound states on the same level; this description is referred to as the "chemical picture".

As shown in Chapter 3, the different Green's functions, such as G_1, G_2, L, are given by integral equations as the Dyson equation (3.224), the Bethe-Salpeter equation (3.192), or eq. (3.236), respectively. The iterative solution of those equations yields a perturbative expansion for the corresponding quantities, and each contribution of these series expansions with respect to the interaction potential can be represented by connected, topologically inequivalent Feynman diagrams. The diagram technique is well known in quantum field theory, see the textbooks of Schweber (1960) and of Akhiezer and Berestetski (1962), for instance. It is one of the main advantages of the thermodynamic Green's functions that perturbative expansions, diagram representations and partial summations can be performed in an analogous way. We outline shortly the diagram technique for the thermodynamic Green's function according to Kadanoff and Baym (1962), in the momentum-Matsubara frequency representation.

Elements of the diagrams are: The free particle propagator

$$G_1^0(1, z) \equiv G_{1,0}(1, z) \equiv \underset{1,\,z}{\longrightarrow} = \frac{1}{\hbar z - E(1)}, \qquad E(1) = \hbar^2 p_1^2/2m_1 \quad (4.1)$$

(in this Section, 1 denotes species, spin and momentum of the one-particle state); the Coulomb interaction

$i\hbar V_0(q, \omega) \equiv$ 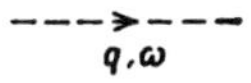 $= \frac{i\hbar}{\varepsilon_0 q^2} \equiv i\hbar\, V_0(q)$ (4.2)

and the vertex function

$\Gamma_{1,0}(1, z_1, 2, z_2, q, \omega) \equiv$ 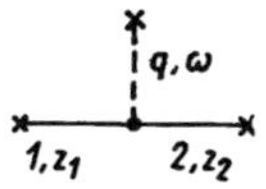 $= e_{a_1}\delta_{a_1 a_2}\delta_{\sigma_1\sigma_2}\delta_{p_1+q,\, p_2}\delta_{z_1+\omega,\, z_2}$ (4.3)

(crosses denote amputation with $G_{1,0}^{-1}$, V_0^{-1}.

These elements can be "dressed" to describe the single-particle propagator

$G_1(1, z) \equiv$ 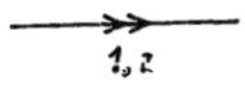 $= G_{1,0}(1, z) + G_{1,0}(1, z)\, \Sigma_1(1, z)\, G_1(1, z)$ (4.4)

(Dyson equation (2.322)), where $\Sigma_1(1, z)$ denotes the self energy, the screened interaction potential (3.232)

$V^s(q, \omega) \equiv$  $= V_0(q) + V_0(q)\, \hat{\Pi}(q, \omega)\, V^s(q, \omega)\,,$ (4.5)

where $\tilde{\Pi}(q, \omega) = \sum_{ab} e_a\, e_b\, \Pi_{ab}(q, \omega)$ denotes the polarization function, and the dressed vertex function (crosses denote amputation with G_1^{-1}, $(V^s)^{-1}$)

$\Gamma_1(1, z_1, 2, z_2, q, \omega) \equiv$ 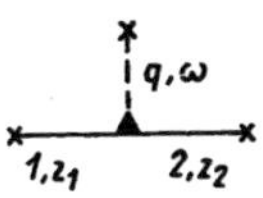 $= \Gamma_{1,0} + \hat{K} G_1 G_1 \Gamma_1\,,$ (4.6)

where $\hat{K}(1, z_1, 2, z_2, 1', z_1', 2', z_2')$ denotes the irreducible part of the T-matrix in the t-channel.

All the dressed elements can be represented by series of diagrams with given external lines or endpoints which are constructed from the primitive elements.

Due to the structure of the iterative solution of the equation of motion, only connected diagrams which do not decay into disconnected subgroups, must be taken into account (linked cluster theorem, cf., Akhiezer and Berestetskii, 1962), and diagrams which can be transformed into the same structure by deformation are topologically equivalent and should be dropped.

With these restrictions, G_1, V^s, and Γ_1 are represented by the sum over all possible diagrams, which may be constructed from G_0 and V_0 lines. The proof of this theorem may be found in Section 3.2.4.

In a similar way the quantities Σ_1, $\hat{\Pi}$, and $\hat{K}$ may be represented by irreducible diagrams. The self energy $\Sigma_1(1, z)$ is the sum of all connected, topologically nonequivalent irreducible diagrams with two external endpoints, which do not decay into two separate parts, if one free particle propagator is cut. Similarly, the polarization function $\hat{\Pi}(q, \omega)$ is represented by the sum of all connected, topologically inequivalent irreducible diagrams with two external endpoints (amputation of V_0) which do not decay into two separate parts if one Coulomb interaction line is cut. The irreducible part of the dressed vertex function will not be discussed in this Chapter more in detail; it is given by higher order (four point) diagrams.

We give the rules how to relate the diagrams to analytical expressions (see KADANOFF and BAYM, 1962). Each internal line is specified by asserting it with a momentum q and the bosonic Matsubara frequency $\omega_\lambda = 2\pi\lambda/(-i\beta\hbar)$, $\lambda = 0, \pm 1, \pm 2, \ldots$, in the case of a Coulomb interaction line and the variable 1 (species, spin and momentum) and the fermionic Matsubara frequency $z_{1\nu} = \pi\nu/(-i\hbar\beta) + \mu_1$, $\nu = \pm 1, \pm 3, \ldots$, in the case of the single-particle propagator $G_{1,0}(1, z_1)$. Summations have to be performed over all internal variables. Especially, summations over momenta are transformed into integrals according to

$$\sum_p \to \int \frac{\mathrm{d}p\,\Omega}{(2\pi)^3} \tag{4.7}$$

and the summations over Matsubara frequencies are represented by contour integrals according to eq. (3.196):

$$\frac{i}{\hbar\beta} \sum_{z_{1\nu}} h(z_{1\nu}) = \int\limits_C \frac{\mathrm{d}z}{2\pi} f_1(z)\, h(z)\,. \tag{4.8}$$

A factor (-1) has to be introduced for each closed fermion line.

4.1.2. The RPA and the V^s-Approximation for the Self Energy

The equations for G_1, V^s (4.4), (4.5) are solved in an approximative way if approximations for Σ_1 and $\tilde{\Pi}$ are introduced. If special diagrams are singled out to construct Σ_1 or $\tilde{\Pi}$, respectively, partial summations are performed over special classes of diagrams for G_1 and V^s.

The concept of partial summations is of importance because a series expansion with respect to the interaction do not converge near the poles of $G_{1,0}$ and V_0 (at $q = 0$ in the latter case). After performing a partial summation, these poles are shifted, as formally can be seen by the solution of (4.4), (4.5)

$$G_1(1, z) = \frac{1}{\hbar z - E(1) - \Sigma_1(1, z)}, \tag{4.9}$$

$$V^s(q, \omega) = \frac{V_0(q)}{1 - V_0(q)\,\hat{\Pi}(q, \omega)} = \frac{V_0(q)}{\varepsilon(q, \omega)}, \tag{4.10}$$

see (3.126) and (3.251).

The most simple approximation for $\hat{\Pi}$ is the random phase approximation (RPA):

$\tilde{\Pi}^{\mathrm{RPA}}(q, \omega_\lambda) = q, \omega_\lambda$ 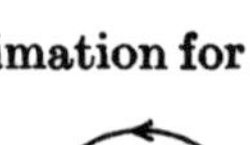 $q, \omega_\lambda = \sum_{a, s_a} \int \frac{\mathrm{d}p}{(2\pi)^3} e_a^2 \frac{f_a(p) - f_a(p + q)}{E_a(p) - E_a(p + q) + \hbar\omega_\lambda}.$ (4.11)

which leads to the corresponding expressions for $V_s^{\mathrm{RPA}}(q, \omega)$ and $\varepsilon^{\mathrm{RPA}}(q, \omega)$. This approximation is described in Section 4.2. The main idea taking only the ideal part to $\tilde{\Pi}$ (i.e., $\tilde{\Pi}$ is considered for non-interacting particles) consists in considering the constituents of the Coulomb system as free particles which are polarized by the influence of the internal screened field.

The most simple approximation for Σ_1 is the Hartree-Fock approximation (contributions to Σ_1 which are of first order with respect to e_a^2, see Section 3.2.4.):

$\Sigma_1^{\mathrm{HF}}(1, z) =$ 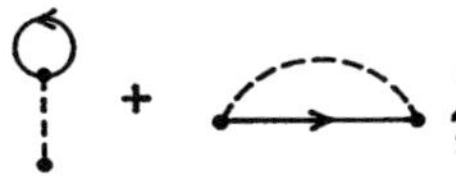 $\sum_2 e_{a_1} e_{a_2} \left(V_0(0) - \delta_{12} V_0(p_1 - p_2)\right) f_2(p_2)\,. \quad (4.12)$

Because of charge neutrality, the first contribution (Hartree term) vanishes for a homogeneous system. The second contribution is connected with statistical correlations (exchange interaction) due to the Pauli exclusion principle. See also (3.122) and (4.157).

In a next step, the V^s-approximation for Σ_1 is obtained replacing the bare potential $V_0(q)$ in Σ_1^{HF} by the dynamically screened potential,

$\Sigma_1(1, z) =$ 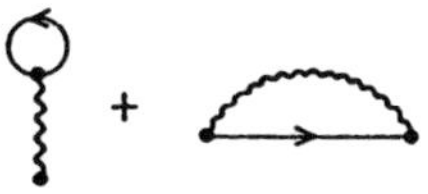 $= \sum_{p_2, \omega\lambda} e_{a_1}^2 V^s(p_1 - p_2, \omega_\lambda)\, G_{1,0}(p_2, z_1 - \omega_\lambda) \quad (4.13)$

for a neutral, homogeneous plasma. This approximation which also contains the Coulomb correlations in the formation of a charge density distribution around a particle is elaborated in Section 4.3. Of course, the V^s approximation has to be specified further by taking V^s in RPA, for instance.

A further important approximation is the self-consistent Hartree-Fock approximation

$\Sigma_1^{\mathrm{HF}}(1, z) =$ 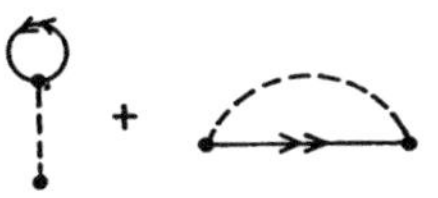 (4.14)

which contains the frequency independent part of the interaction with the environment. This approximation is the starting point for band structure calculations in solids and calculations of orbitals in atoms and molecules.

Similarly, a self-constistent formulation of the V^s approximation for the self-energy and the RPA would include the following diagrams:

$\tilde{\Pi}(q, \omega) =$ 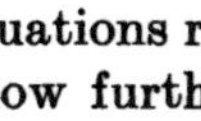 ,

$\Sigma_1(1, z) =$ 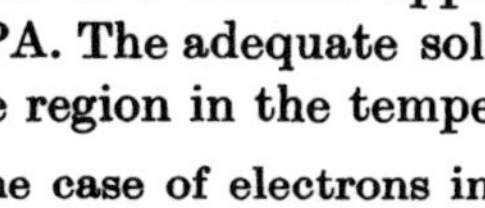 . (4.15)

In this approximations, $\tilde{\Pi}$ describes not the polarization of free particles but of quasiparticles under the influence of an internal, screened field. The consistent solution of this system of equations represents a rather involved problem.

The question arises how further approximations should be constructed in order to go beyond the RPA. The adequate solution of this problem depends on the special situation, i.e. on the region in the temperature-density plane.

For instance, in the case of electrons in a metal exchange terms are considered in $\tilde{\Pi}$ which lead after some approximations to the Hubbard corrections or the expression of Singwi, Tosi, Land, and Sjölander (STLS) (1968) for the dielectric function, see Hedin and Lundquist (1971), Ichimaru (1982), Bishop and Lührmann (1978), (1982), Golden and Kalman (1976, 1982). We will mainly concentrate to the case of a partially ionized plasma where bound states may be formed. Therefore we are looking for diagrams which represent the process of cluster formation.

4.1.3. Many-Particle Complexes and T-Matrices

In order to obtain two particle bound states (f.i. hydrogen atoms in the electron-proton plasma) we have at least to consider the two particle Green function $G_{2,0}$ in the ladder approximation. For the single frequency function (3.192a) we have

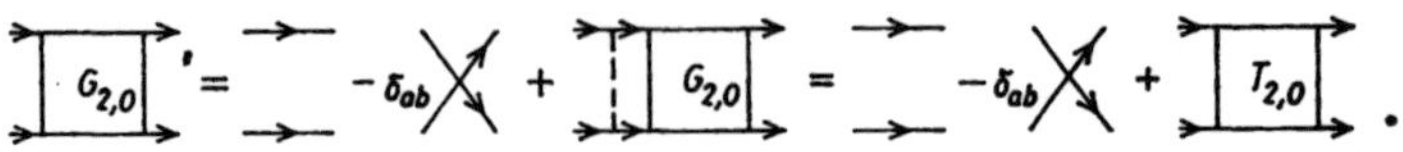

(4.16)

This Green function $G_{2,0}$ describes the propagation of isolated cluster of two particles at arbitrary order of mutual interaction. This equation is solved with the help of a bilinear expansion

$$G_{2,0}(121'2',z) = \sum_{nP} \frac{\psi^{0-}_{nP}(12)\, \psi^{0*-}_{nP}(1'2')}{\hbar z - E^0_{nP}}, \tag{4.17}$$

where ψ^{0-}_{nP} and E^0_{nP} denote the two particle wave function and the two-particle energy eigenvalue, respectively; n — internal quantum state, P — total momentum. The wave function $\psi^0_{nP}(1,2)$ which must be antisymmetrized for identical particles and the energy E^0_{nP} are determined by the corresponding wave equation (Schrödinger equation, see 3.2.6.)

$$\left(E(1) + E(2) - E_{nP}\right) \psi^0_{nP}(12) + \sum_{1'2'} V_{12}(121'2')\, \psi^0_{nP}(1'2') = 0\,. \tag{4.18}$$

The matrix elements of $G_{2,0}$ with respect to the two particle states $|nP\rangle$ read

$$G_{2,0}(n, P, z) = \frac{1}{\hbar z - E^0_{nP}}, \tag{4.19}$$

and are formally similar to $G_{1,0}(1, z)$ (4.1), and we introduce a corresponding new diagram element

$G_{2,0}(n, P, z) =$ 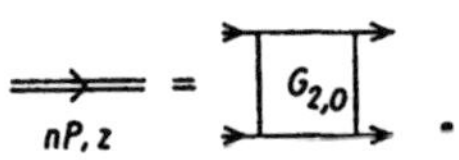

(4.20)

Notice that the spectrum of E^0_{nP} may contain a discrete part (bound states) and a continuous part (scattering states). As pointed out in Chapter 2., especially for bound states, $G_{2,0}\,(nP, z)$ takes over the role of the single-particle propagator $G_{1,0}(1, z)$, and each bound state with internal quantum number n can be considered as a special new species (chemical picture). This concept will be elaborated more in detail in the next Section.

For the T-matrix the bilinear expansion reads

$$T_{2,0}(121'2', z) = \sum_{nP} \psi^{0*}_{nP}(12)\left(z\hbar - E(1) - E(2)\right) \frac{1}{\hbar z - E^0_{nP}} \left(E^0_{nP} - E(1') - E(2')\right) \cdot \psi^0_{nP}(1'2')\,. \tag{4.21}$$

Of course, the formation of higher clusters can be described in the same way. The isolated n-particle cluster will be described by the n-particle propagator $G_{n,0}$. The corresponding n-particle T-matrix, T_n, obeys the Faddeev equation. Different channels must be considered, for instance, as three particle bound states, scattering states between a two-particle bound state and a free state, and scattering states between three free particles in the case of the three-particle problem. In this case, the three-particle bound states can be interpreted as new species within a chemical picture.

4.1.4. Cluster Formation and the Chemical Picture

The form of the n-particle cluster $G_{n,0}$ in the bilinear expansion, see (4.19), is very similar to the one particle propagator (4.1) and, in particular, it describes the propagation of a n-particle bound state which can be considered as a new species. On the other hand, we know from a chemical picture that free particles and composite particles (bound states) should be treated on the same footing (EBELING, 1974). From this we can formulate a heuristic principle how to construct approximations in a many-particle system if the formation of bound states is of interest: Expressions which are represented by diagrams with a single-particle propagator $G_{1,0}$ are extended to a class of contributions by replacing $G_{1,0}$ by $G_{n,0}$. Care must be taken in this procedure to avoid double counting of diagrams and to avoid the construction of disconnected diagrams so that an analysis must be performed which diagrams have to be subtracted from the extended class of contributions. This prescription guarantees the well founded introduction of a correct chemical description without the problems of statistical correlations which arise, e.g., within an empirical introduction of the chemical picture and in the Bose formalism as given by USUI (1960).

To realize the prescription for constructing a class of diagrams which treat free particles and bound states on the same level we must introduce also the vertex function for the interaction of the n-particle complex with the Coulomb field. For the two-particle complexes we find (RÖPKE and DER, 1979)

$$\Gamma_{2,0}(n, n', P, q, \Omega, \omega) =$$

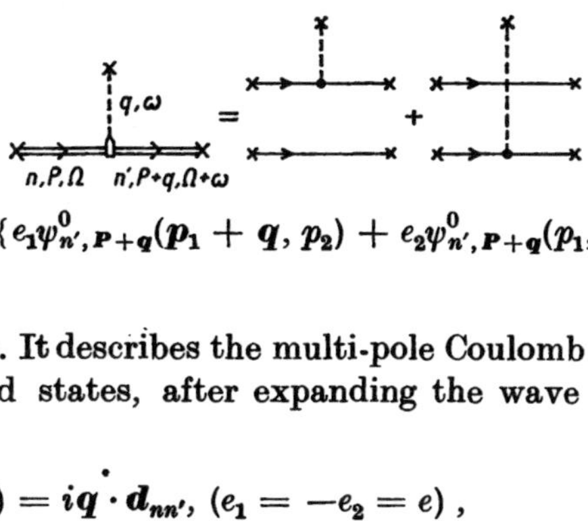

$$= \int \mathrm{d}\boldsymbol{p}_1 \int \mathrm{d}\boldsymbol{p}_2\, \psi^{0*}_{nP}(p_1, p_2)\, \{e_1\psi^0_{n',P+q}(\boldsymbol{p}_1 + \boldsymbol{q}, p_2) + e_2\psi^0_{n',P+q}(p_1, \boldsymbol{p}_2 + \boldsymbol{q})\}\,. \tag{4.22}$$

The physical meaning is as follows. It describes the multi-pole Coulomb interaction of the two-particle state. For bound states, after expanding the wave functions, we obtain in the limit $q \to 0$

$$\lim_{q \to 0} \Gamma_{2,0}(n, n', P, q, \Omega, \omega) = i\boldsymbol{q} \cdot \boldsymbol{d}_{nn'}, \; (e_1 = -e_2 = e)\,, \tag{4.23}$$

where $d_{nn'} = e\langle n|\, \boldsymbol{r}\, |n'\rangle$ is the dipole matrix element, $\boldsymbol{r}$ being the relative coordinate.

In the same manner, a vertex function can be introduced also for higher clusters:

$$\begin{aligned}\Gamma_{m,0}(n, n', P, q, \Omega, \omega) &= \int \mathrm{d}\boldsymbol{p}_1\, \mathrm{d}\boldsymbol{p}_2 \ldots \mathrm{d}\boldsymbol{p}_m\, \psi^{0*}_{nP}(p_1, \ldots, p_m)\\ &\cdot \{e_1\psi^0_{n',P+q}(\boldsymbol{p}_1 + \boldsymbol{q}, p_2, \ldots, p_m) + \ldots + e_m\psi^0_{n',P+q}(p_1, \ldots, \boldsymbol{p}_m + \boldsymbol{q})\}\end{aligned} \tag{4.24}$$

Using the generalized propagators $G_{m,0}$ and vertex functions $\Gamma_{m,0}$, we can formulate the extensions of the approximations described in 4.1.2. to a class of contributions which include the "chemical pictures."

The extended RPA for the polarization function can be given in the form of a cluster decomposition

$$\Pi^{\mathrm{RPA}}(q, \omega) = \sum_n \Pi^{\mathrm{RPA}}_n(q, \omega)|_{\mathrm{connected}}\,. \tag{4.25}$$

The m-particle cluster contribution Π_m^{RPA} contains the term

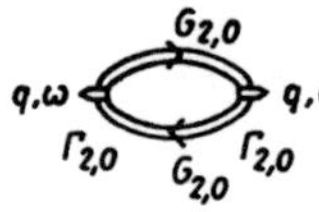

$$+ \ldots = \sum_m \sum_{nn'P\Omega} \Gamma_{m,0}(n, n', P, q, \Omega, \omega)\, \Gamma_{m,0}(n, n', \boldsymbol{P} + \boldsymbol{q}, -q, \Omega + \omega, -\omega) \cdot \frac{f_m(E_{nP}) - f_m(E_{n',\boldsymbol{P}+\boldsymbol{q}})}{E_{nP} - E_{n',\boldsymbol{P}+\boldsymbol{q}} - \hbar\omega}, \tag{4.26}$$

where

$$f_m(E) = (\mathrm{e}^{\beta(E - \sum_i \mu_i)} - (-1)^m)^{-1} \tag{4.27}$$

is the corresponding Bose or Fermi function. The summation over the internal quantum numbers n, n' runs over bound states and scattering states. If the scattering states are taken into account in (4.26) we have to subtract disconnected diagrams as shown in detail by Röpke and Der (1979), see also Section 4.4.

Similarly, the extension of the V^s-approximation for the self energy including two-particle complexes is given by

$$\Sigma_2(n, n', P, \Omega) =$$

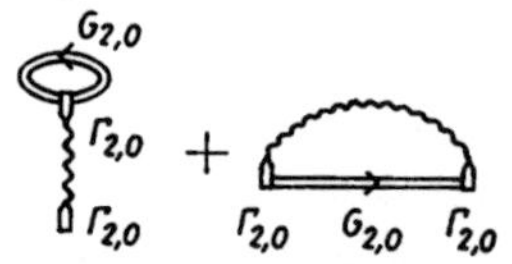

$$= \sum_{qn''\omega} \Gamma_{2,0}(n, n'', \boldsymbol{P} - \boldsymbol{q}, \Omega, -\omega) \cdot \Gamma_{2,0}(n'', n, \boldsymbol{P} - \boldsymbol{q}, q, \Omega - \omega, \omega)\, V^s(q, \omega)\, G_{2,0}(n'', \boldsymbol{P} - \boldsymbol{q}, \Omega - \omega)\,. \tag{4.28}$$

In this way, the Dyson equation for the one-particle propagator can be generalized to a Dyson equation for the two-particle propagator, and we can define the self energy for a two-particle complex. This approximation is discussed by Zimmermann et al. (1978), see Section 4.5. The generalization to a m-particle cluster is straightforward.

It should be mentioned that a self-consistent formulation can also be given if the polarization function Π_m^{RPA} as well as the self energy Σ_m are determined by dressed m-cluster states ("cluster-quasiparticle states").

4.1.5. Cluster Decomposition of the Self Energy

The chemical pictures can also be deduced in detail from the elementary physical picture via the decomposition of the T-matrix into connected parts. So, for the single particle self energy

$$\Sigma_1(1, z) = \tag{4.29}$$

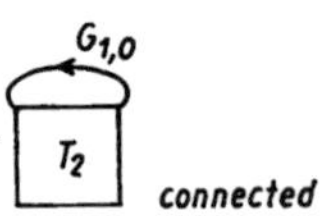

(see 3.2.4.), we have

$$\Sigma_1(1, z) = \tag{4.30}$$

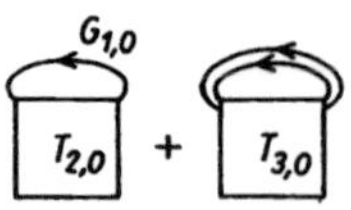

$+ \ldots |_{\text{conneted}}$

with

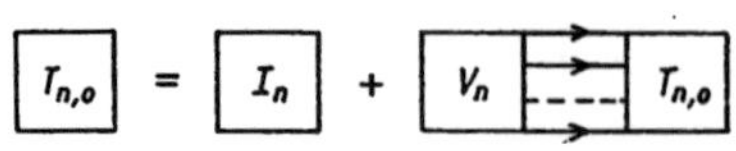

, (4.31)

$$I_n = \sum_i^n V_{1i} + \text{exchange terms.} \quad V_n = \sum_{i<j}^n V_{ij}\,.$$

This cluster decomposition for the self energy can be further developed replacing the bare one-particle Green's function $G_{1,0}$ by the full one-particle Green's function G_1 and taking into account for $T_{n,0}$ only the irreducible part (Röpke et al., 1983). Then we obtain the following system of equations which must be solved self-consistently:

Cluster decomposition of the self energy

$\Sigma_1(1, z) =$ 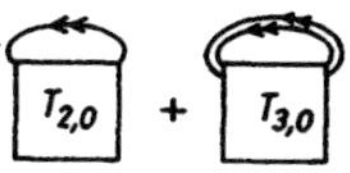

$+ \dots |_{\text{irred.}},$ (4.32)

Dyson equation for the one-particle Green's function G_1

$G_1(1, z) =$ 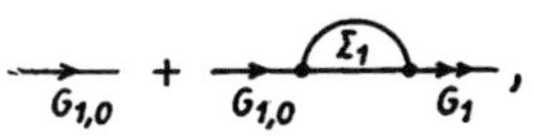

, (4.33)

Bethe-Salpeter equation for the T-matrix $T_{n,0}$

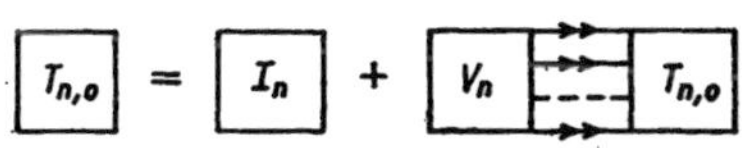

. (4.34)

Representing $T_{n,0}$ by a bilinear expansion and taking into account only the discrete part of the energy spectrum, we obtain immediately the description of an ideal mixture in the chemical picture. For instance we find the equation of state

$$n(\beta, \mu) = \sum_{nPm} g_{n,m} f_m(E^0_{nP}) \tag{4.35}$$

in the low density limit, (see Stolz and Zimmermann, 1979), $g_{n,m}$ denotes the degeneracy factor. The total density is decomposed into the free particle density and the density of bound states. Improvements of this elementary form of the equation of state (4.35) are obtained within the systematic quantumstatistical approach if the contribution of scattering states and self-energy corrections of the cluster energy eigenvalues E^0_{nP} of the clusters are taken into account, for instance.

4.2. Dielectric Properties of Charged Particle Systems. Random Phase Approximation

4.2.1. Linear Response to External Perturbations. General Remarks

If an external field E is applied to a system of charged particles a charge density $e\delta\langle n\rangle$ is induced, and we get, on account of the causality, for the vector of the dielectric displacement (α denotes a Cartesian component)

$$D_\alpha(r, t) = E_\alpha(r, t) + \sum_\beta \int_{-\infty}^{t} dt'\, d\boldsymbol{r}'\, K_{\alpha\beta}(\boldsymbol{r} - \boldsymbol{r}', t - t')\, E_\beta(r', t') \,. \tag{4.36}$$

For homogeneous media we may take the Fourier transform of eq. (4.36), and as a result we have

$$D_\alpha(p, \omega) = \sum_\beta \varepsilon_{\alpha\beta}(p, \omega)\, E_\beta(p, \omega) \tag{4.37}$$

where the dielectric tensor $\varepsilon_{\alpha\beta}$ is given in terms of the response function $K_{\alpha\beta}$ (the microscopic foundation of which is given below, (4.57) ff., see also Chapter 3.):

$$\varepsilon_{\alpha\beta}(p, \omega) = \delta_{\alpha\beta} + \int_0^\infty d\tau \int d\boldsymbol{q}\ K_{\alpha\beta}(q, \tau)\ e^{-i(\boldsymbol{pq} - \omega\tau)} . \tag{4.38}$$

The dielectric tensor may be split up into a transverse (t) and a longitudinal (l) part, respectively,

$$\varepsilon_{\alpha\beta}(p, \omega) = \left(\delta_{\alpha\beta} - \frac{p_\alpha p_\beta}{p^2}\right) \varepsilon_t(p, \omega) + \frac{p_\alpha p_\beta}{p^2} \varepsilon_l(p, \omega) . \tag{4.39}$$

While the transverse dielectric function plays a role if an electromagnetic field is applied to the system, we want to consider only the case of electric fields and thus only the longitudinal function ε_l, which we shall simply refer to as the dielectric function, $\varepsilon_l \to \varepsilon$. For many applications it is necessary to split up ε into real and imaginary parts,

$$\varepsilon(p, \omega) = \varepsilon'(p, \omega) + i\varepsilon''(p, \omega) . \tag{4.40}$$

The dielectric function is related to the complex electric conductivity σ by

$$\varepsilon(p, \omega) = 1 + \frac{i}{\varepsilon_0 \omega} \sigma(p, \omega) . \tag{4.41}$$

Using the equations

$$\begin{aligned} \operatorname{div} \boldsymbol{D} &= \frac{1}{\varepsilon_0} e n^{\text{ext}} , \\ \operatorname{div} \boldsymbol{E} &= \frac{1}{\varepsilon_0} e(\delta\langle n \rangle + n^{\text{ext}}) , \end{aligned} \tag{4.42}$$

the quantities $\varepsilon(p, \omega)$ and $\sigma(p, \omega)$ may be expressed in terms of the external charge density (en^{ext}) (external source) and the induced charge density ($e\delta\langle n \rangle$); n is the number density.

After Fourier transformation we get from (4.36), (4.42) for the dielectric function (DF)

$$\varepsilon(p, \omega) = \frac{n^{\text{ext}}(p, \omega)}{\delta\langle n(p, \omega) \rangle + n_{\text{ext}}(p, \omega)}$$

or

$$\frac{1}{\varepsilon(p, \omega)} - 1 = \frac{\delta\langle n(p, \omega) \rangle}{n_{\text{ext}}(p, \omega)} . \tag{4.43}$$

Thus the total charge density

$$en^{\text{tot}}(p, \omega) = e\delta\langle n(p, \omega) \rangle + en^{\text{ext}}(p, \omega)$$

is related to the density of the external source by

$$en^{\text{tot}}(p, \omega) = \frac{en^{\text{ext}}(p, \omega)}{\varepsilon(p, \omega)} . \tag{4.44}$$

The relevant Poisson equation reads

$$\Delta U^{\text{ext}} = -\frac{1}{\varepsilon_0} e^2 n^{\text{ext}}$$

with the Fourier transform (in the case of longitudinal displacement only)

$$U^{\text{ext}}(p, \omega) = \frac{e^2}{\varepsilon_0 p^2} n^{\text{ext}}(p, \omega) .$$

An effective external potential may be introduced by

$$U^{\text{eff}}(p, \omega) = \frac{U^{\text{ext}}(p, \omega)}{\varepsilon(p, \omega)} . \tag{4.45}$$

An interesting example of collective plasma behaviour are plasma oscillations, which we consider now in an electron gas over a positively charged background. We determine the reaction of the system on a pulse shaped external perturbation of the density

$$n^{\text{ext}}(r, t) = n_0^{\text{ext}} \, e^{iqr} \, \delta(t) \tag{4.46}$$

the Fourier transform of which reads

$$n^{\text{ext}}(p, \omega) = n_0^{\text{ext}} (2\pi)^3 \, \delta(p - q) . \tag{4.47}$$

The total density is determined from relation (5.44),

$$n^{\text{tot}}(p, \omega) = \frac{n^{\text{ext}}(p, \omega)}{\varepsilon(p, \omega)} = \frac{n_0^{\text{ext}} (2\pi)^3 \, \delta(p - q)}{\varepsilon(p, \omega)} . \tag{4.48}$$

The space time behaviour of n^{tot} we get from the Fourier transform

$$n^{\text{tot}}(r, t) = n_0^{\text{ext}} \, e^{iqr} \int \frac{d\omega}{2\pi} \frac{e^{i\omega t}}{\varepsilon(q, \omega)} . \tag{4.49}$$

In general, eq. (4.49) describes a damped oscillatory behaviour of the total charge density. For this reason we must investigate the poles of the analytic continuation of the function $1/\varepsilon(p, \omega)$ in the lower energy half plane (see below). The poles are, in general, complex and may be represented as

$$\hbar\tilde{\omega}(p) = \hbar\omega(p) - i\hbar\gamma(p) . \tag{4.50}$$

The evaluation of (4.49) is then possible with the help of the residuum theorem and, as a result, the plasma performs density or charge density oscillations, respectively, as a consequence of an external porturbation. These plasma oscillations are described by the complex plasmon energy (4.50). The frequency $\omega(p)$ is in simple approximations the plasma frequency (see below). The damping $\gamma(p)$ leads in the most situations to a vanishing of the spontaneous plasma oscillations. However, in plasma physics it is also possible that there is an increase in the oscillations, and that there are self-exciting plasma waves (see, e.g., Cap, 1972; Mikhailovskii, 1977).

The question of the determination of the complex plasmon energies is dealt with in Section 4.2.2. For practical and numerical tasks it is sometimes useful to apply sum rules for the DF; among them we have, e.g.,

$$\int_{-\infty}^{+\infty} d\omega \, \omega \, \text{Im} \frac{1}{\varepsilon(p, \omega)} = -\pi\omega_{\text{pl}}^2 , \tag{4.51}$$

where $\omega_{\text{pl}} = (ne^2/\varepsilon_0 m_e)^{1/2}$ is the plasma frequency in the case of an electron gas.

So far our considerations are based on phenomenological grounds. Let us now see, how the dielectric function ε and the response function $K_{\alpha\beta}$ are related to those functions which were defined in Chapter 3. on the basis of the microscopic theory. In Chapter 3. we introduced the dielectric function, the polarization function and the density correlation function. These functions are connected by the relations

$$\varepsilon(p, \Omega_n) = 1 - \sum_{ab} V_{ab}(p)\, \Pi_{ab}(p, \Omega_n)\,, \tag{4.52}$$

$$L_{ab}(p, \Omega_n) = \Pi_{ab}(p, \Omega_n) - \sum_{cd} L_{ac}(p, \Omega_n)\, V_{cd}(p)\, \Pi_{db}(p, \Omega_n)\,. \tag{4.53}$$

The Fourier coefficients $\varepsilon(p, \Omega_n)$, $L(p, \Omega_n)$, $\Pi(p, \Omega_n)$ are given by the spectral representations (3.243):

$$\begin{aligned} L(p, z) &= \int \frac{d\omega}{2\pi} \frac{\hat{A}(p, \omega)}{z - \omega}\,, \\ \varepsilon(p, z) - 1 &= \int \frac{d\omega}{2\pi} \frac{\hat{\varepsilon}(p, \omega)}{z - \omega}\,, \\ \varepsilon^{-1}(p, z) - 1 &= \int \frac{d\omega}{2\pi} \frac{\Phi(p, \omega)}{z - \omega}\,, \\ \Pi(p, z) &= \int \frac{d\omega}{2\pi} \frac{\hat{\Pi}(p, \omega)}{z - \omega} \end{aligned} \tag{4.54}$$

which must be taken at the Matsubara frequencies

$$z = \Omega_\nu = \frac{\pi\nu}{-i\beta\hbar}\,, \qquad \nu \text{ even}\,.$$

Let us now explain the physical meaning of these quantities. In particular we shall show that the collective behaviour which is due to the long range Coulomb interaction (such as screening, plasma oscillations and polarization) is completely determined by the functions L, ε, Π, which describe the properties of the medium on a microscopic basis. In order to demonstrate this fact let us consider the reaction of a system under the influence of an external potential with the Hamiltonian

$$H(t) = \int dx\, U^{\text{ext}}(x)\, \psi^+(x)\, \psi(x) = \int dx\, U^{\text{ext}}(x)\, n(x)\,. \tag{4.55}$$

A convenient basis for the description of such perturbations is the linear response theory, which shall be discussed now briefly. See also Chapter 7.

We consider a system with Coulomb interaction in the thermodynamic equilibrium with the density operator ϱ^{eq}. This system is perturbed at $t = t_0 = -\infty$ by the external potential $U^{\text{ext}}(x)$. Then the density operator will be changed, $\varrho(t) = \varrho^{\text{eq}} + \Delta\varrho$, and the response of any observable is given by

$$\delta\langle A\rangle = \text{Tr}\{(\varrho^{\text{eq}} + \Delta\varrho)\, A(t)\} - \text{Tr}\{\varrho^{\text{eq}} A(t)\}\,.$$

In order to determine this response it is necessary to determine $\Delta\varrho$.

For this purpose we consider the von Neumann equation

$$i\hbar \frac{\partial}{\partial t} \varrho + [(H + H^{\text{ext}}), \varrho] = 0$$

with the initial condition

$$\lim_{t=t_0\to-\infty} \varrho = \varrho^{\mathrm{eq}} .$$

Taking into account only linear terms in H^{ext}, i. e., we consider linear response, we obtain for $\Delta\varrho$

$$\frac{\partial}{\partial t}\Delta\varrho + \frac{1}{i\hbar}[H, \Delta\varrho] = -\frac{1}{i\hbar}[H^{\mathrm{ext}}, \varrho^{\mathrm{eq}}]$$

with the formal solution

$$\Delta\varrho = \frac{1}{\hbar}\int\limits_{t_0}^{t} \mathrm{e}^{\frac{i}{\hbar}H(t'-t)}\,[H^{\mathrm{ext}}(t'), \varrho^{\mathrm{eq}}]\,\mathrm{e}^{-\frac{i}{\hbar}H(t'-t)}\mathrm{d}t', \quad t > t_0 .$$

Using this solution we find for the linear response of the expectation value of A

$$\delta\langle A\rangle = \frac{1}{\hbar}\int\limits_{t_0}^{t} \mathrm{Tr}\,\{\varrho^{\mathrm{eq}}[H^{\mathrm{ext}}(t'), A(t)]\}\,\mathrm{d}t' .$$

As a specific example we consider a plasma with the particle density operator

$$n(x) = \psi^+(x)\,\psi(x) .$$

The external perturbation (4.55) produces a linear response for the average particle density given by

$$\delta\langle n(x, t)\rangle = \frac{1}{\hbar}\int\limits_{t_0}^{t} \mathrm{d}x'\langle[n(x', t'), n(x, t)]\rangle\, U^{\mathrm{ext}}(x', t')\,\mathrm{d}t' . \tag{4.56}$$

By this essential result the linear response of the density or charge density is connected with the retarded density fluctuation correlation function

$$\begin{aligned} L^{\mathrm{r}}_{ab}(1212) &= \frac{1}{i^2}\langle[n_a(1), n_b(2)]\rangle\,\theta(t_1 - t_2) \\ &= L^{\mathrm{r}}_{ab}(t_1 - t_2, x_1 - x_2) \\ &= \theta(t_1 - t_2)\,\{L^{>}_{ab}(t_1 - t_2) - L^{<}_{ab}(t_1 - t_2)\} . \end{aligned} \tag{4.57}$$

Since the system shall be assumed to be spatially homogeneous, it is convenient to take the Fourier transform of eq. (4.56). Then we get the more simple relation

$$\delta\langle n(p, \omega)\rangle = -L^{\mathrm{r}}(p, \omega)\, U^{\mathrm{ext}}(p, \omega) . \tag{4.58}$$

From this relation it can be seen that a uniform system reacts at the same momentum and the same energy as the perturbation.

The next problem is how to find $L^{\mathrm{r}}(p, \omega)$. It is not convenient to determine this function directly because its equation of motion is more complicated than that for the time ordered Green's functions. Moreover, there does not exist a Feynman perturbation technique for $L^{\mathrm{r}}(p, \omega)$. Therefore we consider the connection between the function $L^{\mathrm{r}}(p, \omega)$ and the associated time ordered density correlation function $L^{\mathrm{c}}(p, \omega) \equiv L(p, \omega)$ (c — causal). It is easy to find the spectral representation of L^{r}

using (3.242) and (3.243):

$$L^r(p, \omega) = \int \frac{d\bar{\omega}}{2\pi} \frac{\hat{A}(p, \bar{\omega})}{\omega + i\varepsilon - \bar{\omega}},$$

where the weight function $\hat{A}(p, \bar{\omega})$ is the same as that of $L(p, \Omega_n)$. As can be seen from (3.243), $L^r(p, \omega)$ is an analytic function in the upper half z-plane:

$$L^r(p, z) = \int \frac{d\bar{\omega}}{2\pi} \frac{\hat{A}(p, \bar{\omega})}{z - \bar{\omega}},$$

i.e., $L^r(p, z)$ is the analytic continuation $L(pz)$ of $L(p, \Omega_n)$ into the upper half plane.

The same result we find for the connection between $\Pi^r(p, \Omega_n)$ and $\Pi(p, \omega)$.

Using eq. (4.53) we can express finally $L^r(p, z)$ by the polarization function:

$$L^r(p, z) = \frac{\Pi^r(p, z)}{1 - V(p)\, \Pi^r(p, z)} \tag{4.59}$$

where

$$\Pi^r(p, z) = \int \frac{d\bar{\omega}}{2\pi} \frac{\hat{\Pi}(p, \bar{\omega})}{z - \bar{\omega}}. \tag{4.60}$$

The density response becomes in this way

$$\delta\langle n(p, \omega)\rangle = \frac{\Pi^r(p, \omega)}{1 - V(p)\, \Pi^r(p, \omega)}\, U^{ext}(p, \omega)\,. \tag{4.61}$$

The collective properties of the plasma are condensed in the denominator, by means of which the effective external potential is defined:

$$U^{eff}(p, \omega) = \frac{U^{ext}(p, \omega)}{\varepsilon^r(p, \omega)} = \frac{U^{ext}(p, \omega)}{1 - V(p)\, \Pi^r(p, \omega)}, \tag{4.62}$$

where ε^r is defined by (4.52) with $\Pi \to \Pi^r$. The "retarded" dielectric function $\varepsilon^r(p, \omega)$ is connected to $\varepsilon(p, \omega)$ (eq. (4.54)) by

$$\varepsilon^r(p, \omega) \to \varepsilon(p, \omega + i0)\,. \tag{4.63}$$

In all phenomenological considerations before one should understand the dielectric function as $\varepsilon(p, \omega) \to \varepsilon(p, \omega + i0)$. Let us now discuss some further properties of the (inverse) dielectric function. Stolz (1974) showed, e.g., that the quantity $\varepsilon^{-1}(q, z) - 1$ vanishes, for $z \to \infty$, as z^{-2}. Therefore we get from Cauchy's residuum theorem

$$\int_{-\infty}^{+\infty} d\omega' \frac{\varepsilon^{-1}(q, \omega') - 1}{\omega' - \omega + i\varepsilon} = 0$$

and with Dirac's identity

$$P \int_{-\infty}^{+\infty} d\omega' \frac{\varepsilon^{-1}(q, \omega') - 1}{\omega' - \omega} - i\pi\, [\varepsilon^{-1}(q, \omega) - 1] = 0\,.$$

Splitting up this relation into real and imaginary parts, we get the well known Kramers-Kronig dispersion relations for $\mathrm{Re}\, \varepsilon^{-1}$ and $\mathrm{Im}\, \varepsilon^{-1}$, namely

$$\mathrm{Re}\, \varepsilon^{-1}(q, \omega) - 1 = P \int_{-\infty}^{+\infty} \frac{2\, \mathrm{Im}\, \varepsilon^{-1}(q, \omega')}{\omega' - \omega} \frac{d\omega'}{2\pi}$$

and

$$\operatorname{Im} \varepsilon^{-1}(q, \omega) = -P \int_{-\infty}^{+\infty} \frac{d\omega'}{2\pi} \frac{2(\operatorname{Re} \varepsilon^{-1}(q, \omega') - 1)}{\omega' - \omega}.$$

Comparing with the third equation of (4.54) we confirm

$$\Phi(q, \omega) = 2 \operatorname{Im} \varepsilon^{-1}(q, \omega + i\varepsilon).$$

The spectral function of $\varepsilon^{-1}(q, z)$, $\Phi(q, \omega)$ is connected with the spectral function of $L(q, z)$, $\hat{A}(q, \omega)$, in agreement with (4.53), by the relation

$$2 \operatorname{Im} \varepsilon^{-1}(q, \omega) = -V(q)\, \hat{A}(q, \omega)$$

(cf. eqs. (3.243) — (3.245)).

According to eq. (3.238) and STOLZ (1974) we may write

$$\hat{A}(q, \omega) = \int_{-\infty}^{+\infty} e^{i\omega(t_1 - t_2)} \langle [\hat{n}(q, t_1), \hat{n}(q, t_2)] \rangle \, dt_1 \equiv S(q, \omega)\, \Omega,$$

where the quantity $S(q, \omega)$ is refered to as the dynamical form factor; Ω — volume.

For the commutators of the density operators $\hat{n}$ we get

$$\langle [\hat{n}(q, t_1), \hat{n}(q, t_2)] \rangle = \Omega \int_{-\infty}^{+\infty} \frac{d\omega}{2\pi} S(q, \omega)\, e^{-i\omega(t_1 - t_2)}$$

and

$$\left\langle \left[\frac{\partial \hat{n}(q, t_1)}{\partial t_1}, \hat{n}(q, t_2) \right] \right\rangle = \Omega \int \frac{d\omega}{2\pi i} \omega S(q, \omega)\, e^{-i\omega(t_1 - t_2)}.$$

The evaluation of the commutator, applying (3.26), gives for $t_2 = t_1$.

$$i\hbar \left[\frac{\partial \hat{n}(q, t_1)}{\partial t_1}, \hat{n}(q', t_1) \right] = \frac{\hbar^2 q^2}{2m} [(q + q')^2 - (q - q')^2]\, \hat{n}(q - q', t_1).$$

For the mean value we get thus

$$\Omega \int_{-\infty}^{+\infty} \frac{d\omega}{2\pi} \omega S(q\omega) = \frac{n}{m} q^2$$

and furthermore

$$2 \int_{-\infty}^{+\infty} \frac{d\omega}{2\pi} \omega \operatorname{Im} \varepsilon^{-1}(q\omega) = -V(q)\, nq^2/m.$$

With $\omega_p^2 = ne^2/\varepsilon_0 m$, this is the sum rule (4.51).

In most applications in Chapter 3. and in the considerations of this Section so far it was assumed that the quantities ε, ε^{-1}, Π and L depend only on differences of space coordinates, and consequently on one wave number only. For macroscopically inhomogeneous systems, however, all quantities really depend on the space coordinates and thus on two wave numbers. Taking into account this fact, we may write, e.g., for the screened potential (see, e.g., STOLZ, 1974)

$$V^s(r_1, r_2, t_1 - t_2) = \int dr_1' \int dt_1'\, \varepsilon^{-1}(r_1, r_1', t_1 - t_1')\, V(r_1' - r_2)\, \delta(t_1' - t_2)$$

and for the effective external potential (4.45)

$$U^{\text{eff}}(r, t) = \int \mathrm{d}r' \int \mathrm{d}t' \, \varepsilon^{-1}(r, r', t - t') \, U(r', t') \, .$$

For ε^{-1} we get, e.g., the spectral representation

$$\varepsilon^{-1}(q + l, q + l'; z) = \delta_{ll'} + \frac{e^2}{\varepsilon_0 |q + l|^2} \int\limits_{-\infty}^{+\infty} \frac{\mathrm{d}\omega}{2\pi} \frac{\hat{A}(q + l, q + l'; \omega)}{z - \omega}$$

with $\hat{A}$ being again the spectral function of L.

In order to apply the results, the polarization function, or its spectral function, $\hat{\Pi}$ must be approximated. Using the random phase approximation (RPA) for $\hat{\Pi}$,

$$\hat{\Pi} = \Pi^{>} - \Pi^{<} \, ,$$

we find (see Section 3.3.3.)

$$\Pi^{\mathrm{r}}(p, \omega) = \int \frac{\mathrm{d}\boldsymbol{q}}{(2\pi)^3} \frac{f(\boldsymbol{q}) - f(\boldsymbol{p} + \boldsymbol{q})}{\hbar(\omega + i\varepsilon) - \varepsilon(\boldsymbol{p} + \boldsymbol{q}) + \varepsilon(\boldsymbol{q})} \, . \tag{4.64}$$

This expression is due to LINDHARD (1954).

EHRENREICH and COHEN (1959) gave a derivation of eq. (4.64), see also KLIMONTOVICH and SILIN (1960). Ehrenreich and Cohen derived the density fluctuation as a result of an applied electrical field, and they got an approximation, in which in the Hamiltonian the Coulomb interaction between the charge carriers was totally neglected. This approximation we call Hartree-Fock approximation, $\varepsilon_{\mathrm{HF}}$:

$$\frac{1}{\varepsilon^{\mathrm{r}}_{\mathrm{HF}}(p, \omega)} = 1 + \frac{e^2}{\varepsilon_0 p^2} \Pi^{\mathrm{r}}(p, \omega) \tag{4.65}$$

with Π^{r} as in (4.64), while in the so called random phase approximation (RPA) the aforementioned Coulomb interaction was taken into account approximately, i.e., in summations of the type

$$\sum_{k} V(k) \, n_{k} \, c^{+}_{p} c_{q+p-k}$$

(c^+, c — creation and annihilation operators, respectively) only terms with $q = k$ are selected. The result is

$$\varepsilon^{\mathrm{r}}_{\mathrm{RPA}}(p, \omega) = 1 - \frac{e^2}{\varepsilon_0 p^2} \Pi^{\mathrm{r}}(p, \omega) \, . \tag{4.66}$$

4.2.2. Properties of the RPA Dielectric Function

In the preceding Subsection and in Chapter 3. it was mentioned that the dielectric function plays an essential role in the framework of the screening problem of charged particles' systems. In more detail, the dielectric function (DF) is responsible for the collective behaviour of the system and determines the plasma oscillations. Furthermore it describes the modification of single-particle and two-particle excitations with respect to the influence of the surrounding plasma. Among the single-particle properties the quasiparticle energy and its damping and the energy loss (stopping power) of a particle penetrating a plasma should be mentioned.

Further as shown in Section 5.1 the static dielectric function

$$\varepsilon(k, \omega \to 0) \equiv \varepsilon(k)$$

is a key quantity in the quasiclassical theory of Coulomb systems. In such theory (Born-Oppenheimer approximation) the ions are considered as heavy classical particles moving in a sea of electrons which give rise to (instantaneous) polarisation effects and screening of the ion-ion interaction. In this way the bare ion-ion potential is to be replaced by the screened one:

$$\frac{z^2e^2}{\varepsilon_0 k^2} \to \frac{z^2e^2}{\varepsilon_0 k^2 \varepsilon(k)}$$

where $\varepsilon(k)$ is the static dielectric function of an electron gas (HUBBARD and SLATTERY, 1971; DE WITT and HUBBARD, 1976; GALAM and HANSEN, 1976; BROVMAN and KAGAN, 1967, 1969; BROVMAN, KAGAN and KHOLAS, 1972; KRASNY and KOWALENKO, 1972; IYETOMI et al., 1981; IYETOMI and ICHIMARU, 1982; GOLDEN and KALMAN, 1976; GOLDEN and KALMAN, 1982; ICHIMARU, 1982). The modification of two-particle properties will be outlined in Section 4.4., while the role of the DF in thermodynamics, transport and optical properties becomes obvious from the details of Chapters 6.—8.

Let us now consider in detail the information contained in the RPA DF, by studying the real and imaginary part of the RPA DF (Fig. 4.1). In Chapter 3. and in Section 4.2.1. the complex DF was given, which reads

$$\varepsilon(p, z) = 1 - \sum_{ab} V_{ab}(p)\, \Pi_{ab}(p, z) \,. \tag{4.67}$$

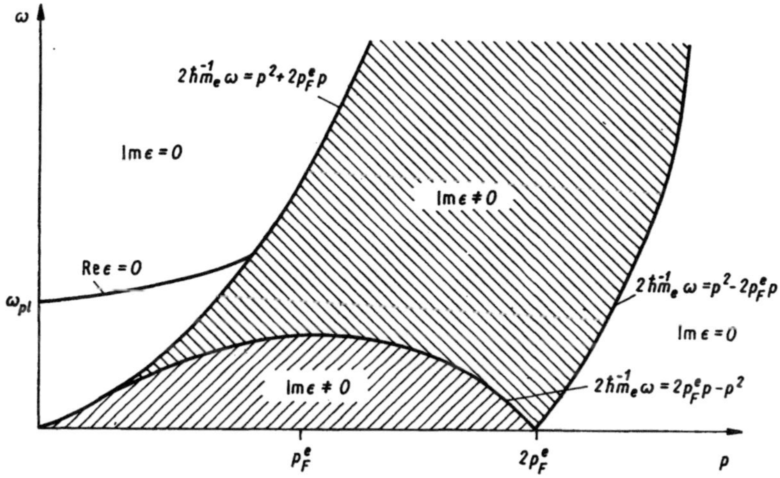

Fig. 4.1. Regions of nonvanishing imaginary part of the dielectric function of an electron gas at $T = 0$

In this Subsection we want to consider only retarded functions as discussed in Section 4.2.1., i.e., we consider eq. (4.67) for

$$z \to \omega + i\varepsilon\,, \quad \varepsilon \to +0\,.$$

In RPA we have then with $\Pi_{ab}^{\mathrm{RPA}} = \delta_{ab}\Pi_{ab}^{\mathrm{RPA}}$

$$\Pi_a^{\mathrm{r,RPA}}(p, \omega) \equiv \Pi_a^{\mathrm{RPA}}(p, \omega + i\varepsilon)$$

$$= (2s_a + 1) \int \frac{\mathrm{d}\boldsymbol{p}'}{(2\pi)^3} \frac{f_a\left(\boldsymbol{p}' + \frac{\boldsymbol{p}}{2}\right) - f_a\left(\boldsymbol{p}' - \frac{\boldsymbol{p}}{2}\right)}{\hbar(\omega + i\varepsilon) + \varepsilon_a\left(\boldsymbol{p}' - \frac{\boldsymbol{p}}{2}\right) - \varepsilon_a\left(\boldsymbol{p}' + \frac{\boldsymbol{p}}{2}\right)}. \tag{4.68}$$

Here f_a is the Fermi function, s_a is the spin and $\varepsilon_a(p)$ is the single particle energy.

Subsequently we shall drop the superscript RPA for brevity. For the real and imaginary parts, respectively, we get:

$$\mathrm{Re}\,\Pi_a(p, \omega) = (2s_a + 1)\, P \int \frac{\mathrm{d}\boldsymbol{p}'}{(2\pi)^3} \frac{f_a\left(\boldsymbol{p}' - \frac{\boldsymbol{p}}{2}\right) - f_a\left(\boldsymbol{p}' + \frac{\boldsymbol{p}}{2}\right)}{\hbar\omega + \varepsilon_a\left(\boldsymbol{p}' - \frac{\boldsymbol{p}}{2}\right) - \varepsilon_a\left(\boldsymbol{p}' + \frac{\boldsymbol{p}}{2}\right)}, \tag{4.69}$$

P — principle value integral. Without further restrictions, this integral may be evaluated only numerically.

If we take the single particle energy to be $\varepsilon_a(p) = \dfrac{\hbar^2 p^2}{2m_a}$ the imaginary part of the RPA DF may be evaluated in a closed form, as a consequence of the δ-distribution arising from (4.68); the result is (GLUCK, 1971)

$$\mathrm{Im}\,\Pi_a(p, \omega + i\varepsilon) = \frac{(2s_a + 1)\, k_B T m_a^2}{4\pi\hbar^4 p}$$

$$\cdot \ln \left| \frac{1 + \exp\left[-\dfrac{\beta}{2m_a}\left\{\left(\dfrac{m_a\omega}{p} + \dfrac{\hbar p}{2}\right)^2 - 2\mu_a m_a\right\}\right]}{1 + \exp\left[-\dfrac{\beta}{2m_a}\left\{\left(\dfrac{m_a\omega}{p} - \dfrac{\hbar p}{2}\right)^2 - 2\mu_a m_a\right\}\right]} \right|; \tag{4.70}$$

μ_a — chemical potential.

In the *nondegenerate case* ($n_a \Lambda_a^3 \ll 1$ or $T \gg T_F^a$; T_F^a — Fermi temperature) we may evaluate the integral in (4.69); the result is (KLIMONTOVICH and KRAEFT, 1974) ($s_a = \frac{1}{2}$)

$$\mathrm{Re}\,\Pi_a(p, \omega) = \frac{n}{\hbar p k_B T}\left\{\left(\frac{m_a\omega}{p} + \frac{\hbar p}{2}\right) {}_1F_1\left(1, \frac{3}{2}; -\left(\frac{m_a\omega}{p} + \frac{\hbar p}{2}\right)^2 \frac{\beta}{2m_a}\right)\right.$$

$$\left. - \left(\frac{m_a\omega}{p} - \frac{\hbar p}{2}\right) {}_1F_1\left(1, \frac{3}{2}; -\left(\frac{m_a\omega}{p} - \frac{\hbar p}{2}\right)^2 \frac{\beta}{2m_a}\right)\right\} \tag{4.71}$$

(${}_1F_1$ — confluent hypergeometric function), and for the imaginary part we get ($s_a = \frac{1}{2}$)

$$\mathrm{Im}\,\Pi_a(p, \omega + i\varepsilon) = \frac{n_a m_a \Lambda_a}{2\hbar^2 p}\left[\exp\left(-\left[\frac{m_a\omega}{p} + \frac{\hbar p}{2}\right]^2 \frac{\beta}{2m_a}\right)\right.$$

$$\left. - \exp\left(-\left[\frac{m_a\omega}{p} - \frac{\hbar p}{2}\right]^2 \frac{\beta}{2m_a}\right)\right]; \qquad \Lambda_a = h/(2\pi m_a k_B T)^{1/2}. \tag{4.72}$$

In particular, also in the nondegenerate limit, more simplified expressions than (4.71), (4.72) are sometimes of practical interest.

The approximation which follows from (4.69) for small momenta p reads for any degeneracy

$$\operatorname{Re} \Pi_a(p, \omega) = (2s_a + 1)\, P \int \frac{dp'}{(2\pi)^3} \frac{\hbar^2 pp'/m_a}{\hbar\omega - \frac{\hbar^2 pp'}{m_a}} \frac{\partial f_a(p')}{\partial \varepsilon_a(p')} . \tag{4.73}$$

From (4.73) we get with

$$f_a(p) = \frac{n_a \Lambda_a^3}{2s_a + 1} e^{-\beta\hbar^2 p^2/2m_a} \tag{4.74}$$

the classical limit which reads in the static (Debye) case

$$\operatorname{Re} \Pi_a^{\text{class}}(p, 0) = \frac{n_a}{k_B T} . \tag{4.75}$$

This expression is, indeed, independent on p. In this approximation the DF is real and reads according to (4.67)

$$\varepsilon^{\text{class}}(p, 0) = 1 + \sum_a V_{aa}(p) \frac{n_a}{k_B T} . \tag{4.76}$$

and for the Coulomb potential

$$\varepsilon^{\text{class}}(p, 0) = 1 + \frac{\varkappa_D^2}{p^2} , \tag{4.77}$$

$$r_D = \left(\sum_a \frac{e_a^2 n_a}{\varepsilon_0 k_B T} \right)^{-1/2} = \varkappa_D^{-1} \quad - \text{ Debye radius.}$$

This quantity serves for the screening to the simplest approximation.

For later purposes we give a series expansion with respect to $\hbar$-powers, yet

$$\begin{aligned} \operatorname{Re} \Pi_a(p, \omega) = \frac{n_a}{k_B T} \Bigg\{ 1 &- \left(\frac{\omega}{p}\right)^2 \frac{m_a}{k_B T} {}_1F_1\left(1, \frac{3}{2} ; -\frac{\omega^2 m_a}{2p^2 k_B T}\right) - \frac{2}{3} \frac{\hbar^2 p^2}{8 m_a k_B T} \\ &\cdot {}_1F_1\left(2, \frac{5}{2} ; -\frac{\omega^2 m_a}{2p^2 k_B T}\right) + \left(\frac{\hbar\omega}{2k_B T}\right)^2 \cdot \frac{8}{15} {}_1F_1\left(3, \frac{7}{2} ; -\frac{\omega^2 m_a}{2p^2 k_B T}\right) \\ &+ \left[{}_1F_1\left(1, \frac{3}{2} ; -\frac{\hbar^2 p^2}{8 m_a k_B T}\right) - 1 + \frac{2}{3} \frac{\hbar^2 p^2}{8 m_a k_B T} \right] + \dots \Bigg\} , \end{aligned} \tag{4.78}$$

$$\begin{aligned} \operatorname{Im} \Pi_a(p, \omega + i\varepsilon) &= \frac{\pi n_a m_a}{\hbar^2 p \Lambda_a} \exp\left(-\left(\frac{\omega m_a}{p}\right)^2 \frac{1}{2 m_a k_B T} \right) \\ &\cdot \left[\frac{\hbar\omega}{k_B T} + \frac{1}{24} \left(\frac{\hbar\omega}{k_B T}\right)^3 - \frac{\hbar^3 p^2 \omega}{8 m_a (k_B T)^2} + \dots \right] . \end{aligned} \tag{4.79}$$

For *arbitrary degeneracy* it is possible to give an expansion with respect to p/ω in the case $\hbar\omega \gg \hbar^2 pp'/m_a$. For the real part of the DF we get (Bonch-Bruevich and Tyablikov, 1961)

$$\operatorname{Re} \Pi_a(p, \omega) = -\frac{n_a}{m_a} \left(\frac{p}{\omega}\right)^2 \left(1 + \frac{\hbar^2}{m^2} \left(\frac{p}{\omega}\right)^2 \langle p^2 \rangle_a \right) . \tag{4.80}$$

Here we used the abbreviation

$$\langle p^k \rangle_a = \frac{2s_a + 1}{n_a} \int \frac{\mathrm{d}\boldsymbol{p}}{(2\pi)^3} p^k f_a(p) \,. \tag{4.81}$$

(4.81) represent Fermi integrals which are defined as

$$I_\nu(\alpha) = \frac{1}{\Gamma(\nu+1)} \int\limits_0^\infty \mathrm{d}x \, \frac{x^\nu}{\mathrm{e}^{x-\alpha} + 1} \,, \tag{4.82}$$

$$\alpha_a = \beta \mu_a \,.$$

Numerical evaluations gave SÄNDIG and MEISTER (1979).

For the case $p \to 0$ we get from (4.80) for the dielectric function

$$\operatorname{Re} \varepsilon(\omega) = 1 - \sum_a \frac{(\omega_{\mathrm{pl}}^a)^2}{\omega^2} \tag{4.83}$$

where $\omega_{\mathrm{pl}}^a = \left(\frac{n_a e_a^2}{\varepsilon_0 m_a}\right)^{1/2}$ is the plasma frequency of species a.

In order to describe the screening in the static situation ($\omega \to 0$) and for small p, we may again start from (4.73) and may write especially for $p \to 0$

$$\operatorname{Re} \Pi_a(0, 0) = \frac{\partial}{\partial \mu_a} n_a(T, \mu_a) \tag{4.84}$$

where the density n_a is a function of the chemical potential

$$n_a(T, \mu_a) = (2s_a + 1) \int \frac{\mathrm{d}\boldsymbol{p}}{(2\pi)^3} f_a(p) \,. \tag{4.85}$$

Ep. (4.85) is again a Fermi integral and describes the connection between the density and the chemical potential for a noninteracting fermion system.

In order to get explicit expressions for (4.80) we must, of course, again consider special situations. Applying (4.74), we get from (4.80), (4.81) for large frequencies and small momenta

$$\operatorname{Re} \Pi_a(p, \omega) = -\frac{n_a}{m_a} \left(\frac{p}{\omega}\right)^2 \left(1 + 3 \frac{k_B T}{m_a} \left(\frac{p}{\omega}\right)^2\right), \quad T \gg T_{\mathrm{F}}^a \,. \tag{4.86}$$

At low temperatures we may apply the Sommerfeld expansion, which reads for $f_a(p)$ (FENNEL et al., 1975)

$$f_a(p) = \theta(p_{\mathrm{F}}^a - p) - \frac{\pi^2}{24} \left(\frac{T}{T_{\mathrm{F}}^a}\right)^2 \left(p + \frac{(p_{\mathrm{F}}^a)^3}{p} \frac{\mathrm{d}}{\mathrm{d}p}\right) \delta(p_{\mathrm{F}}^a - p) \,, \quad T \ll T_{\mathrm{F}}^a, \tag{4.87}$$

(p_{F}^a — Fermi momentum). With (4.87) follows from (4.80), (4.81) (FENNEL et al., 1975)

$$\operatorname{Re} \Pi_a(p, \omega) = -\frac{n_a}{m_a} \left(\frac{p}{\omega}\right)^2 \left[1 + \left(\frac{\hbar p p_{\mathrm{F}}^a}{2 m_a \omega}\right)^2 \left(\frac{12}{5} + \pi^2 \left(\frac{T}{T_{\mathrm{F}}^a}\right)^2\right)\right], \quad T \ll T_{\mathrm{F}}^a \,. \tag{4.88}$$

For the static case and small momenta we get for $T \gg T_{\mathrm{F}}^a$ the result (4.75), (4.77).

For $T \ll T_F^a$ we get from (4.84), (4.87)

$$\operatorname{Re} \Pi_a(0, 0) = \frac{m_a p_F^a}{\pi^2 \hbar^2} \left(1 - \frac{\pi^2}{12} \left(\frac{T}{T_F^a}\right)^2\right) \tag{4.89}$$

and the dielectric function reads

$$\operatorname{Re} \varepsilon(p, 0) = 1 + \frac{k_{\mathrm{TF}}^2}{p^2}.$$

Here the Thomas-Fermi screening wave number is defined by

$$k_{\mathrm{TF}}^2 = \sum_a \frac{e_a^2 m_a p_F^a}{\varepsilon_0 \pi^2 \hbar^2} \left(1 - \frac{\pi^2}{12} \left(\frac{T}{T_F^a}\right)^2\right). \tag{4.90}$$

For any frequency and any momentum we get in the *highly degenerate case* (Fennel et al., 1975)

$$\begin{aligned} \operatorname{Re} \Pi_a (p, \omega) &= \frac{m_a p_F^{a2}}{2\pi^2 \hbar^2 p} \left\{ 2P - \frac{1}{2} [1 - P^2 - \Omega^2] \right. \\ &\cdot \ln \left[\frac{(1 - P)^2 - \Omega^2}{(1 + P)^2 - \Omega^2}\right] + P\Omega \ln \left[\frac{(1 - \Omega)^2 - P^2}{(1 + \Omega)^2 - P^2}\right] \\ &\left. + \frac{\pi^2}{12} \left(\frac{T}{T_F^a}\right)^2 \left(\frac{1}{2} \ln \left[\frac{(1 - P)^2 - \Omega^2}{(1 + P)^2 - \Omega^2}\right] + \frac{1 + P}{(1 + P)^2 - \Omega^2} - \frac{1 - P}{(1 - P)^2 - \Omega^2}\right)\right\}. \end{aligned} \tag{4.91}$$

Here $P = p/2p_F$ and $\Omega = \dfrac{m_a \omega}{\hbar p p_F}$.

For the imaginary part we get (Fennel et al., 1975)

$$\begin{aligned} \operatorname{Im} \Pi_a(p, \omega + i\varepsilon) &= \frac{m_a^2 k_B T}{2\pi \hbar^4 p} \exp(-\beta A) [e^{-\beta B^+} - e^{-\beta B^-}], \\ &\text{for } |2\hbar^{-1} m_a \omega| > 2p_F^a p + p^2; \\ &= -\frac{m_a^2 \omega}{2\pi \hbar^4 p} + \frac{m_a^2 k_B T}{2\pi \hbar^4 p} \exp \beta A\, [e^{\beta B^+} - e^{-\beta B^-}], \quad \text{for } |2\hbar^{-1} m_a \omega| < 2p_F^a p - p^2; \\ &= \frac{m_a^2}{2\pi \hbar^4 p} (B^- + A) + \frac{m_a^2 k_B T}{2\pi \hbar^4 p} (e^{-\beta(A + B^+)} - e^{\beta(A + B^-)}), \\ &\text{for } |2p_F^a p - p^2| < |2\hbar^{-1} m_a \omega| < 2p_F^2 p + p^2. \end{aligned} \tag{4.92}$$

Here we used the following abbreviations:

$$A = k_B T \frac{\pi^2}{12} (T/T_F^a); \qquad B^\pm = \frac{\hbar^2}{2m_a} \left[\left(\frac{m_a \omega}{\hbar p} \pm \frac{p}{2}\right)^2 - p_F^{a2}\right].$$

In the case $T = 0$ the real part of the polarization function reads ($\hbar = 1$) (SCHRIEFFER, 1970) (cf. Fig. 4.1)

$$\operatorname{Re} \Pi_a(p, \omega) = \frac{m_a p_F^a}{2\pi^2} \left\{ 1 + \frac{p_F^a}{2p} \left[1 - \left(\frac{m_a \omega}{p p_F^a} + \frac{p}{2p_F^a} \right)^2 \right] \right.$$

$$\cdot \ln \left| \frac{1 + \left(\dfrac{m_a \omega}{p p_F^a} + \dfrac{p}{2p_F^a} \right)}{1 - \left(\dfrac{m_a \omega}{p p_F^a} + \dfrac{p}{2p_F^a} \right)} \right| - \frac{p_F^a}{2p} \left[1 - \left(\frac{\omega m_a}{p p_F^a} - \frac{p}{2p_F^a} \right)^2 \right]$$

$$\left. \cdot \ln \left| \left[1 + \left(\frac{m_a}{p p_F^a} - \frac{p}{2p_F^a} \right) \right] \Big/ \left[1 - \left(\frac{m_a}{p p_F^a} - \frac{p}{2p_F^a} \right) \right] \right| \right\}.$$

The imaginary part reads (cf. Fig. 4.1)

$$\operatorname{Im} \Pi(p, \omega) =$$

$$= \begin{cases} 0, & 2m_a|\omega| > p^2 + 2pp_F^a ; \\ 0, & p > 2p_F^a \text{ and } 2m_a|\omega| < p^2 - 2pp_F^a ; \\ \dfrac{m_a^2 \omega}{2\pi p}, & p < 2p_F^2 \text{ and } 2m_a|\omega| < |p^2 - 2pp_F^a| ; \\ \dfrac{m_a (p_F^a)^2}{4\pi p} \left\{ 1 - \left(\dfrac{m_a \omega}{p p_F^a} - \dfrac{p}{2p_F^a} \right)^2 \right\}, & |p^2 - 2pp_F^a| < 2m_a |\omega| < |p^2 + 2pp_F^a| . \end{cases}$$

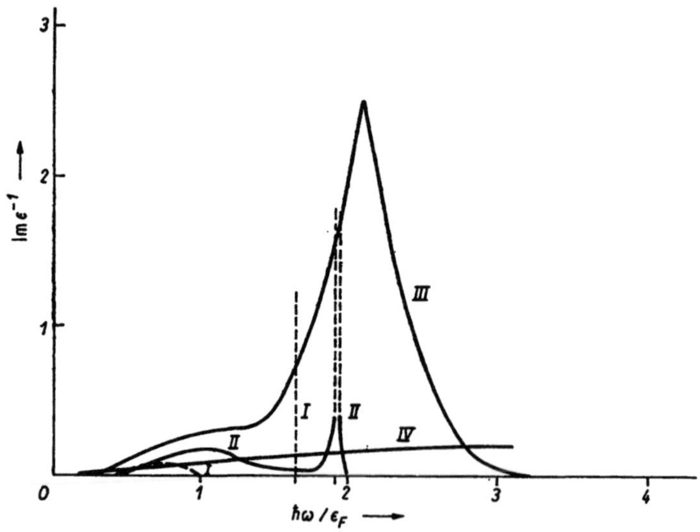

Fig. 4.2. Im ε^{-1} for an electron gas as a function of the frequency $r^s = 3$, $p/p^F = 0.5$; I — 0 K, II — 10^4 K, III — 10^5 K, IV — 10^6 K

In conclusion of this Subsection we want to deal with the (photon-) spectral function of the screened potential, V^s. According to Chapter 3. and eq. (4.52) this spectral function may be written as

$$\frac{1}{2} D_a(p, \omega) = \operatorname{Im} V^s_{aa}(p, \omega + i\varepsilon) \equiv V_{aa}(p) \operatorname{Im} \varepsilon^{-1}(p, \omega + i\varepsilon)$$

$$= \frac{V_{aa}(p) \sum_c V_{cc}(p) \operatorname{Im} \Pi_c(p, \omega + i\varepsilon)}{(1 + \sum_c V_{cc}(p) \operatorname{Re} \Pi_c(p, \omega))^2 + (\sum_c V_{cc}(p) \operatorname{Im} \Pi_c(p, \omega + i\varepsilon))^2} . \tag{4.93}$$

This quantity is used in many connections. Especially it may be sharply peaked if plasma oscillations are possible. If moreover the imaginary part of the DF tends to zero, (4.93) has the structure of a δ-distribution.
$\operatorname{Im} \varepsilon^{-1}$ is shown in Fig. 4.2.

4.2.3. Plasma Oscillations (Plasmons)

Plasma oscillations play an important role and may serve as a tool for the analysis of the structure in gaseous solid state plasma physics (Platzmann and Wolf, 1973). One of the tasks of many-particle physics is the determination of the complex plasmon energies

$$\hbar\tilde{\omega}(p) = \hbar\omega(p) - i\hbar\gamma(p) \tag{4.94}$$

(see, e.g., Mikhailovskii, 1977). In order to see the physical origin of plasmon excitations we refer to our considerations of Section 4.2.1.

According to the structure of eq. (4.49) we have to consider the analytic properties of ε^{-1} (p, z) Similar to the considerations given in Section 3.3.2. for the properties of Green's functions, we note that $1/\varepsilon^r(p, z)$ is analytic and does not have poles in the upper half z-plane including the real axis $z \to \omega + i0$ (except in the case $T = 0$, where the imaginary part of the polarization function may be zero under certain conditions; see eq. (4.92) for $T = 0$.) This is seen from eq. (4.54) which may be written in the shape

$$\varepsilon^r(p, \omega + i\varepsilon) = \operatorname{Re} \varepsilon^r(p, \omega + i\varepsilon) + i \operatorname{Im} \varepsilon^r(p, \omega + i\varepsilon) . \tag{4.95}$$

Let $\operatorname{Im} \varepsilon^r \ll 1$ and $\varepsilon \to 0$. Then one could try to find a solution of (4.95) near $\omega(p)$, what is defined by

$$\operatorname{Re} \varepsilon^r(p, \omega(p)) = 0 , \tag{4.96}$$

and instead of (4.95) we write

$$\varepsilon^r(p, \omega + i\varepsilon) = (\omega + i\varepsilon) \frac{d}{dz} \operatorname{Re} \varepsilon^r(p, z)|_{z=\omega(p)} + \dots + i \operatorname{Im} \varepsilon^r(p, \omega + i\varepsilon) . \tag{4.97}$$

The imaginary part of eq. (4.97) cannot be zero. The same is true for the advanced dielectric function. $\varepsilon^a(p, \omega - i\varepsilon)$, which is analytic and has no zeros in the lower half z-plane.

Instead of $\varepsilon(p, z)$, we have to consider the complex function

$$\frac{1}{\varepsilon(p, z)} = 1 + \int \frac{d\omega}{2\pi} \frac{\Phi(p, \omega)}{z - \omega} \tag{4.98}$$

where the spectral function is given by

$$\Phi(p,\omega) = \operatorname{Im} \varepsilon^{-1}(p, \omega + i\varepsilon)$$
$$= \frac{\sum\limits_{ab} V_{ab}(p) \operatorname{Im} \Pi^{r}_{ab}(p,\omega)}{\left(1 + \sum\limits_{ab} V_{ab}(p) \operatorname{Re} \Pi^{r}_{ab}(p,\omega)\right)^2 + \left(\sum\limits_{ab} V_{ab}(p) \operatorname{Im} \Pi^{r}(p,\omega)\right)^2}. \tag{4.99}$$

The function $1/\varepsilon^{r}(pz)$ is analytic in the upper half plane and does not have poles there, while the same statement holds for $1/\varepsilon^{a}(p, z)$ in the lower half plane. However, $1/\varepsilon^{r}(pz)$ may be extended to the lower half plane, the results is

$$1/\varepsilon^{r}(p,z) \to 1/\varepsilon^{a}(p,z) + \Phi(p,z), \quad \operatorname{Im} z < 0\,. \tag{4.100}$$

The continuation of $1/\varepsilon^{r}(p, z)$ to the lower half plane has poles which are that of the analytic continuation of the spectral function, $\Phi(p, \omega) \to \Phi(p, z)$. Thus we have to look for the poles of the function

$$\Phi(p,z) = \frac{-\sum\limits_{ab} V_{ab}(p) \operatorname{Im} \Pi_{ab}(p,z)}{\left(1 + \sum\limits_{ab} V_{ab}(p) \operatorname{Re} \Pi_{ab}(p,z)\right)^2 + \left(\sum\limits_{ab} V_{ab}(p) \operatorname{Im} \Pi_{ab}(p,z)\right)^2}. \tag{4.101}$$

$\Phi(p, z)$ may be decomposed as

$$\Phi(p,z) = \frac{1}{2i}\left[\frac{1}{\operatorname{Re} \varepsilon(p,z) + i \operatorname{Im} \varepsilon(p,z)} - \frac{1}{\operatorname{Re} \varepsilon(p,z) - i \operatorname{Im} \varepsilon(p,z)}\right]. \tag{4.102}$$

In contrast to (4.95) the second denominator of (4.102) has a zero in the upper half plane (while the first one has a zero in the lower half plane). The poles of $\Phi(p, z)$ in the upper half plane are given by

$$\hbar z \to \hbar\tilde{\omega}(p) = \hbar\omega(p) - \hbar i\gamma(p)$$

and are referred to as plasmon energies.

Approximately the plasmons are defined by eq. (4.96), especially if the imaginary part of ε is sufficiently small. Also for small imaginary part, the plasmons are described approximately by the position and the width of the peak of the function $\Phi(p, \omega)$ (eq. (4.99)).

The more rigorous discussion deals with the investigation of the equation

$$\operatorname{Re} \varepsilon(p,z) - i \operatorname{Im} \varepsilon(p,z) = 0 \tag{4.103}$$

and its approximate solution is (4.96) for the real part of the solution $\tilde{\omega}(p)$ and

$$\gamma(p) = \left.\frac{\operatorname{Im} \varepsilon(p, \omega + i\varepsilon)}{\dfrac{\partial}{\partial\omega} \operatorname{Re} \varepsilon(p,\omega)}\right|_{\omega=\omega(p)} \tag{4.104}$$

for the imaginary part, which gives the damping of the plasmons.

The general discussion of the plasmon properties becomes especially complicated by the occurrence of the summations over the species (see (4.93) and (4.52)). For this reason it is more easy to consider first the situation of the *electron gas* with a neutralizing background. First we will deal with some analytical evaluations, while lateron a numerical evaluation of (4.101) or (4.103) is discussed.

Of special interest is the region of small momenta, and it is of some interest, that eq. (4.103) may be satisfied only up to certain critical value p_c of the momentum, beyond which oscillations are not possible (PINES and NOZIÈRES, 1966).

Let us consider the situation $\hbar\omega \gg \frac{\hbar^2 pp'}{m_e}$. Then we get from an iterative solution of (4.103) for small α and using an expansion according to (4.80) (BONCH-BRUEVICH and TYABLIKOV, 1961; FENNEL et al., 1975)

$$\omega(p) = \omega_{pl}\left[1 + \frac{\hbar^2}{m_e^2}\frac{p^2}{\omega^2(p)}\langle p^2\rangle + \frac{\hbar^2}{4m^2}\frac{p^4}{\omega^2(p)} + \frac{\hbar^4}{m_e^4}\frac{p^4}{\omega^4(p)}\langle p^4\rangle\right]^{1/2},$$
$$\omega_{pl}^2 = \frac{n_e e^2}{\varepsilon_0 m_e}. \tag{4.105}$$

Again, the $\langle p^k\rangle$ are defined by (4.81).

An iterative solution of (4.105) yields

$$\omega(p) = \omega_{pl}\left\{1 + \frac{\hbar^2p^2}{2m_e^2\omega_{pl}^2}\langle p^2\rangle + \frac{\hbar^2p^4}{4m_e^2\omega_{pl}^2}\left[\frac{1}{2} + \frac{\hbar^2}{4m_e^2\omega_{pl}^2}\left(\langle p^4\rangle - \frac{17}{2}\langle p^2\rangle^2\right)\right]\right\}. \tag{4.106}$$

In the case $p = 0$ we have the well known result

$$\omega(0) = \omega_{pl} = \left(\frac{n_e e^2}{\varepsilon_0 m_e}\right)^{1/2}. \tag{4.107}$$

As before, the averaged momenta $\langle p^k\rangle$ may be evaluated analytically only in limiting situations. Evaluating (4.81) with (4.74) or (4.87), respectively, we get from (4.106)

$$\omega(p) = \omega_{p\_}\left[1 + \frac{3k_BT}{2m_e}\frac{p^2}{\omega_{pl}} + \frac{\hbar^2}{4m_e^2}\frac{p^4}{\omega_{pl}}\left(\frac{1}{2} + \frac{3}{8}\left(\frac{k_BT}{\hbar\omega_{pl}}\right)^2\right)\right], \quad T \gg T_F, \tag{4.108}$$

and

$$\omega(p) = \omega_{pl}\left[1 + \frac{3\hbar^2}{10m_e^2}\left(\frac{pp_F}{\omega_{pl}}\right)^2 + \frac{\hbar^2}{4m_e^2}\frac{p^4}{\omega_{pl}^2}\right.$$
$$\left.\cdot\left(\frac{1}{2} + \frac{0.105\,\hbar^2p_F^4}{m_e^2\,\omega_{pl}^2}\right) + \pi^2\left(\frac{T}{T_F}\right)^2\left(\frac{\hbar^2}{8m_e^2}\left(\frac{pp_F}{\omega_{pl}}\right)^2 + \frac{1{,}445}{16}\frac{\hbar^4}{m_e^4}\left(\frac{pp_F}{\omega_{pl}}\right)^4\right)\right], \quad T \ll T_F. \tag{4.109}$$

The damping of the plasmon, $\gamma(p)$, follows from the solution of the dispersion relation (4.103) and gives to the lowest approximation (4.104) what gives with (4.86)

$$\gamma(p) = \frac{\omega_{pl}}{2} V_{ee}(p)\,\mathrm{Im}\,\Pi_e(p, \omega_{pl} + i0). \tag{4.110}$$

With (4.110) we have for $T \gg T_F$

$$\gamma(p) = \frac{\omega_{pl}}{2} V_{ee}(p)\frac{n_e m_e \Lambda_e}{2\hbar^2 p}\left[\exp\left(-\left[\frac{m_e\omega}{p} + \frac{\hbar p}{2}\right]^2\frac{\beta}{2m_e}\right)\right.$$
$$\left.- \exp\left(-\left[\frac{m_e\omega}{p} - \frac{\hbar p}{2}\right]^2\frac{\beta}{2m_e}\right)\right]. \tag{4.111}$$

For $\hbar \to 0$, $\hbar\gamma(p)$ is the Landau damping. It represents the classical reversible damping and corresponds to the decay of a plasmon into an electron-hole pair.

For $T \ll T_F$ we have to take the case

$$2\hbar^{-1}m_e\omega \gtrsim 2p_F^e p + p^2$$

of (4.92), and we may write, neglecting T-corrections,

$$\gamma(p) = \frac{\omega_{pl}}{2} V_{ee}(p) \frac{m_e^2 k_B T}{\pi \hbar^4 p} \cdot \exp\left\{-\frac{\beta\hbar^2}{2m_e}\left[\left(\frac{m_e\omega_{pl}}{\hbar p}\right)^2 + \frac{p^2}{4} - (p_F^e)^2\right]\right\} \sinh\left(\frac{\beta\hbar\omega_{pl}}{2}\right). \tag{4.112}$$

It is to be seen that in this region the plasmon damping vanishes at $T = 0$.

Now let us discuss the question of the *critical momentum* mentioned above. Roughly speaking, we get this momentum from the condition that the real part of the DF

$$\mathrm{Re}\,\varepsilon(p, \omega) = 1 - V_{ee}(p)\,\mathrm{Re}\,\Pi_e(p, \omega)$$

has no longer zeros (see Fig. 4.3).

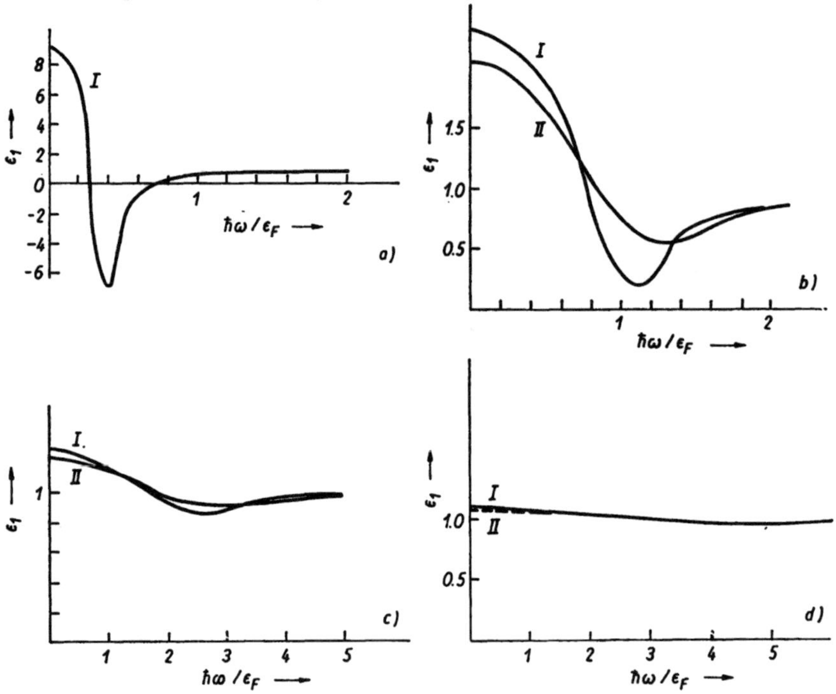

Fig. 4.3. Re $\varepsilon = \varepsilon_1$ as a function of the frequency. Electron gas with $r_s = 0.5$
I — 10 K, II — 10^6 K; $p/p_F = 0.2$ (a), $= 0.5$ (b), $= 1$ (c), $= 1.5$ (d)

We should mention here, that the zero at the smaller ω-value corresponds to a single particle excitation while the larger one corresponds to the plasmon frequency. However, we must take into account that there exists, in general, a nonvanishing imaginary part of the DF, and thus the plasmons are only approximately given by the zeros of the real part of the dielectric function.

In order to get a (numerical) determination of the plasmon frequency $\omega(p)$, its damping $\gamma(p)$ and the critical momentum p_c, we have to solve eq. (4.103) and take into account (4.69), (4.70). As we are interested in plasma oscillations, throughout our calculations we make use of the condition $\omega(p) \gg \gamma(p)$. Splitting up into real and imaginary parts, we get the set of equations

$$\frac{8\pi\hbar^2}{m_e e^2} p + \int_0^\infty \mathrm{d}x \, (\mathrm{e}^{x-\beta\mu_e} + 1)^{-1}$$

$$\cdot \ln \left| \frac{(\omega^2(p) + \gamma^2(p) - E)^{-2} + 4\omega^2(p)\,\gamma^2(p)}{(\omega^2(p) + \gamma^2(p) - E)^{+2} + 4\omega^2(p)\,\gamma^2(p)} \right| - \arctan \left(\frac{D_- \sin(a+b)}{1 + D_- \cos(a+b)} \right)$$

$$+ \arctan \left(\frac{D_+ \sin(b-a)}{1 + D_+ \cos(b-a)} \right) = 0 \tag{4.113}$$

and

$$2 \int_0^\infty \mathrm{d}x \, (\mathrm{e}^{x-\beta\mu_e} + 1)^{-1} \left\{ \arctan \left(\frac{2\omega(p)\,\gamma(p)}{E_-^2 + \gamma^2(p) - \omega^2(p)} \right) \right.$$

$$\left. - \arctan \left(\frac{2\omega(p)\,\gamma(p)}{E_+^2 + \gamma^2(p) - \omega^2(p)} \right) \right\} + \ln \left| \frac{1 + 2D_- \cos(a+b) + D_-^2}{1 + 2D_+ \cos(a-b) + D_+^2} \right| = 0\,. \tag{4.114}$$

Here the following abbreviations were applied:

$$E_\pm = \frac{1}{2m_e} \left(p^2 \pm 2p \left(x 2m_e \frac{k_B T}{\hbar^2} \right)^{1/2} \right),$$

$$D_\pm = \pm \frac{\hbar\omega(p)}{2k_B T} + \exp \left(- \frac{\hbar^2 p^2}{8m_e k_B T} + \beta\mu_e + \frac{m_e}{2p^2} \frac{\gamma^2(p) - \omega^2(p)}{k_B T} \right),$$

$$a = \frac{\hbar\gamma(p)}{2k_B T}, \qquad b = \frac{2\omega(p)\, m_e}{\hbar p^2} a\,.$$

In order to get the results as a function of the density, we have moreover to take into account

$$n_e = (2s_e + 1) \int \frac{\mathrm{d}p}{(2\pi)^3} \left[\mathrm{e}^{\frac{\beta\hbar^2 p^2}{2m_e} - \beta\mu_e} + 1 \right]^{-1}.$$

From eqs. (4.113) and (4.114) we may determine the frequency $\omega(p)$ and the damping $\gamma(p)$ of the plasmons, and we get solutions up to the critical momentum p_c. The results are shown in Figs. 4.4 and 4.5.

Analytical values for the critical momentum may be determined only in simple approximations. For $T = 0$ we take into account, that $\mathrm{Im}\,\Pi = 0$ for

$$2\,\hbar^{-1} m_e \omega \geqq 2p_F^e p + p^2\,,$$

and with the critical frequency ω_c

$$2\hbar^{-1} m_e \omega_c = 2p_F^e p_c + p_c^2$$

we get from the real part of the DF with the polarization (4.91) the following equation for the determination of the critical momentum:

$$\left(\frac{p_c}{p_F^e} \right)^2 = \frac{4e^2 m_e}{\pi\hbar^2 p_F^e} \left\{ \left(1 + \frac{p_c}{2p_F^e} \right) \ln \left(1 + \frac{2p_F^e}{p} \right) - 2 \right\}. \tag{4.115}$$

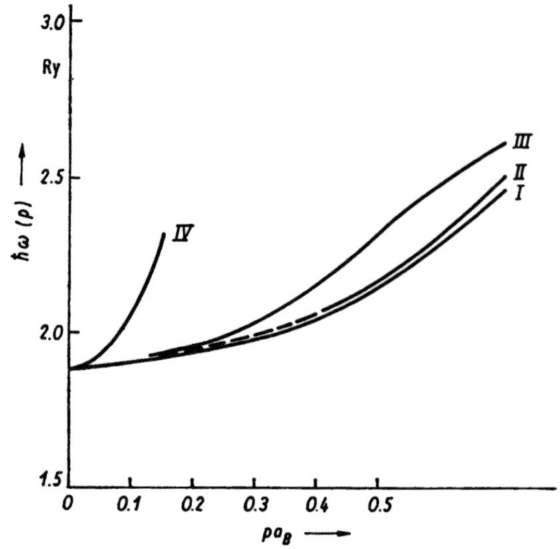

Fig. 4.4. Plasmon energy of an electron gas as a function of the momentum $r_s = 1.5$; I — 0 and 10^4 K, II — $2.56 \cdot 10^4$ K, III — 10^5 K, IV — $1.6 \cdot 10^6$ K

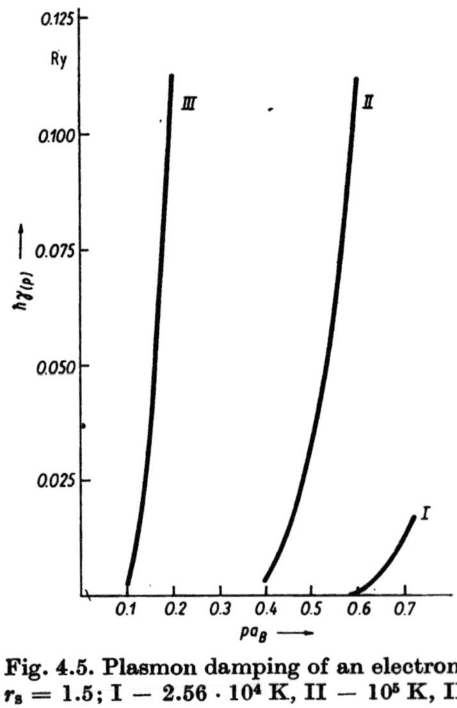

Fig. 4.5. Plasmon damping of an electron gas as a function of the momentum. $r_s = 1.5$; I — $2.56 \cdot 10^4$ K, II — 10^5 K, III — $1.6 \cdot 10^6$ K ($\hbar = 2m_e = e^2/2 = 1$)

The situation becomes more complicated if we consider a *multicomponent system*. Now any of the components may exhibit separate plasma oscillations, the frequencies $\omega_{\rm pl}^a$ of which are determined in the case $p \to 0$:

$$(\omega_{\rm pl}^a)^2 = \frac{n_a e_a^2}{\varepsilon_0 m_a}. \tag{4.116}$$

In a general situation, the spectral function of the screened potential (4.93) must be analysed carefully or, in other words, the spectral function of the inverse dielectric function $\varepsilon^{-1}(p, z)$. Of course, all properties of (4.93) are well defined if we use (4.68) and the formulae derived from it. However, for practical purposes (4.93) should be simplified by some approximations.

Let us discuss the nondegenerate situation only. In this case the real and imaginary parts of the polarization function are given by (4.71), (4.72). One main difficulty is the sum over the species by which the algebraic expressions become much more complicated than in the one-component case. In analogy to (4.108) we get now for the plasmon dispersion, i.e., for the possible frequencies of the plasma oscillations,

$$\omega^2(p) = \omega_{\text{pl}}^2 (1 + 3p^2/\bar{\varkappa}^2) \tag{4.117}$$

where

$$\bar{\varkappa}^2 = \omega_{\text{pl}}^4 / (\sum_c n_c e_c^2 k_B T / \varepsilon_0 m_c^2)$$

and

$$\omega_{\text{pl}}^2 = \sum_c \frac{n_c e_c^2}{\varepsilon_0 m_c} . \tag{4.118}$$

(4.117) coincides, up to p^2, with (4.108), in the case of the electron gas. Eq. (4.117) is, in the approximation given, a solution of the equation

$$\varepsilon'(p, \omega) = \operatorname{Re} \varepsilon(p, \omega) = 1 + \sum_c V_{cc}(p) \operatorname{Re} \Pi_c(p, \omega) = 0 . \tag{4.119}$$

Under the condition (4.119) the spectral function (4.93) has a sharp maximum, and taking into account $\operatorname{Im} \Pi_c \approx 0$, we may write instead of (4.93)

$$\begin{aligned} \operatorname{Im} \varepsilon^{-1}(p, \omega + i\varepsilon) &= \pi \, \delta(\varepsilon') \\ &= -\frac{\pi}{2} \omega(p) \{\delta(\omega - \omega(p)) - \delta(\omega + \omega(p))\} , \end{aligned} \tag{4.120}$$

where $\omega(p)$ is the solution of (4.119) and is given by (4.117). Here use was made of the fact that the real part of the dielectric function reads according to (4.86)

$$\varepsilon'(p, \omega) = 1 - \sum_a \frac{e_a^2 n_a}{\varepsilon_0 m_a \omega^2} \left(1 + 3 \frac{k_B T p^2}{m_a \omega^2}\right) \tag{4.121}$$

and that a solution of the equation (4.119) is found from (4.121) by iteration, i.e., if the ω^2 in the second term of the brackets is replaced by

$$\omega^2(p) \approx \sum_a \frac{n_a e_a^2}{\varepsilon_0 m_a} .$$

The plasmon dispersion is then given by eq. (4.117).

Using (4.120) and (4.117) we may determine the dielectric function, namely

$$\varepsilon^{-1}(p, z) = 1 - \int_{-\infty}^{+\infty} \frac{\operatorname{Im} \varepsilon^{-1}(p\omega)}{z - \omega} \frac{d\omega}{2\pi} \tag{4.122}$$

$$= 1 - \frac{\omega(p)}{2} \left(\frac{1}{\omega(p) - z} - \frac{1}{\omega(p) + z}\right) = \frac{z^2}{z^2 - \omega^2(p)} , \tag{4.123}$$

and $(z \to \omega + i0)$

$$\operatorname{Re} \varepsilon(p, \omega) = 1 - \frac{\omega^2(p)}{\omega^2}. \tag{4.124}$$

We have to underline now that eqs. (4.120), (4.123) and (4.124) are valid under the condition (4.119) only, if the quantity $\omega(p)$ is taken to be the solution of (4.119); i.e, the relations mentioned are valid near the plasmon pole of $\varepsilon^{-1}(pz)$ and thus are refered to as plasmon pole approximations.

Because the expressions (4.120), (4.123) and (4.124) are sufficiently simple and contain one essential physical effect it is desirable to retain the shape of (4.120), i.e. the structure of delta functions, but to generalize the expression (4.117) in such a way that it retains the property (approximately) to fulfil eq. (4.119), and gives with a spectral function of type (4.120) improved dielectric functions as compared with (4.123) and (4.124) which are valid not only near the plasmon pole. For this reason let us consider „*pole approximations*" of the dielectric function which agree with certain limiting situations of the exact result.

Let us thus collect some limiting formulae for the (real part of the) dielectric function. For zero frequency and small momenta we have after (4.77)

$$\lim_{p \to 0} \varepsilon(p, 0) = 1 + \frac{\varkappa_0^2}{p^2} \approx \frac{\varkappa_0^2}{p^2}, \tag{4.125}$$

where

$$\varkappa_0^2 = \sum_c \frac{n_c e_c^2}{\varepsilon_0 k_B T}$$

and, consequently

$$\lim_{p \to \infty} \varepsilon^{-1}(p, 0) = \frac{p^2}{\varkappa_D^2}. \tag{4.126}$$

For small momenta and finite ω we have after (4.86)

$$\varepsilon(0, \omega) = 1 - \frac{\omega_{pl}^2}{\omega^2} \tag{4.127}$$

with ω_{pl} after (4.118) and

$$\varepsilon^{-1}(0, \omega) = \frac{\omega^2}{\omega^2 - \omega_{pl}^2}, \tag{4.128}$$

$$\lim_{\omega \to \infty} \varepsilon^{-1}(0, \omega) = 1 + \frac{\omega_{pl}^2}{\omega^2}. \tag{4.129}$$

Approximation (4.129) is in agreement with the sum rule (4.51), namely the imaginary part of the DF reads now (cf. (4.120))

$$\operatorname{Im} \varepsilon^{-1}(0, \omega) = -\frac{\pi \omega_{pl}}{2} \left(\delta(\omega - \omega_{pl}) - \delta(\omega + \omega_{pl})\right).$$

Another limiting situation which should be met by improved approximations refers to the spectral function, or, to the imaginary part of the dielectric function itself.

Under the condition

$$p^3 \gg \frac{ne^2}{\varepsilon_0 \hbar} \sqrt{\frac{m_e}{k_B T}} \tag{4.130}$$

we get from (4.71), (4.72) that

$$\operatorname{Im} \varepsilon^{-1}(p, \omega) = -\frac{\varepsilon''(p, \omega)}{\varepsilon'^2 + \varepsilon''^2} \approx -\operatorname{Im} \varepsilon(p, \omega) \tag{4.131}$$

and, according to the relation

$$\lim_{a \to 0} \frac{1}{\sqrt{\pi a}} e^{-\frac{x^2}{a}} = \delta(x) \tag{4.132}$$

we may write the sharply peaked exponentials of (4.72) in the shape of delta functions, i.e., with (4.132) we have for a system of electrons (e) and ions (i)

$$\operatorname{Im} \varepsilon^{-1}(p, \omega) = -\pi \frac{ne^2}{\varepsilon_0 \hbar p^2} [\delta(\omega - \omega_{pe}) - \delta(\omega + \omega_{pe}) + \delta(\omega - \omega_{pi}) - \delta(\omega + \omega_{pi})] . \tag{4.133}$$

The quantities $\pm\omega_{pe}$ and $\pm\omega_{pi}$ determine the zeros of the exponents in (4.72) and are given as

$$\hbar \omega_{pa} = \frac{\hbar^2 p^2}{2m_a}, \qquad a = e, i, \tag{4.134}$$

and lead, in (4.71), to the replacement ${}_1F_1(1, \frac{3}{2}, x) = 1$.

In the sense discussed above we get a so-called "*single-pole approximation*", if we take instead of the plasmon dispersion (4.117) the function

$$\omega^2(p) = \omega_{pl}^2 \left(1 + \frac{p^2}{\varkappa_D^2}\right) + \left(\frac{\hbar p^2}{2m}\right)^2 \tag{4.135}$$

and for the spectral function

$$\operatorname{Im} \varepsilon^{-1}(p, \omega) = -\pi \frac{\omega_{pl}^2}{\omega(p)} (\delta(\omega - \omega(p)) - \delta(\omega + \omega(p))) . \tag{4.136}$$

The single-pole approximation (4.135), (4.136) is of advantage especially for a one-component system and fulfils (4.126)—(4.129) and (4.133) (CARINI et al., 1980; COUBLE and BOERCKER, 1983). The dielectric function has then the form

$$\varepsilon(p, \omega) = 1 - \frac{\omega_{pl}^2}{\omega^2 - \omega^2(p) + \omega_{pl}^2} . \tag{4.137}$$

The single-pole approximation may be applied also to two-component systems if the two species have nearly the same mass. (LUNDQUIST, 1967; ZIMMERMANN, 1976; HAUG and TRAN THOAI, 1978; ZIMMERMANN et al., 1978). The dispersion relation has then to be taken in the form

$$\omega^2(p) = \omega_{pl}^2 \left(1 + \frac{p^2}{\varkappa_D^2}\right) + \left(\frac{1}{2} \frac{\hbar p^2}{2\mu_{12}}\right)^2 \tag{4.138}$$

with

$$\omega_{pl}^2 = \frac{ne^2}{\varepsilon_0 \mu_{12}}, \qquad \mu_{12} = \frac{m_1 m_2}{m_1 + m_2} .$$

It should be mentioned that the expressions (4.135) and (4.138) are no solutions of (4.119).

As is to be seen, e.g., from (4.133) it is not possible to meet all demands discussed above by a single-pole approximation if the system consists of species with strongly different masses. For this reason RÖPKE (1981) proposed a double pole approximation by the following generalization of eq. (4.137)

$$\varepsilon(p, N\omega) = 1 - \frac{\omega_{pe}^2}{\omega^2 + \omega_{pe}^2 - \omega_e^2(p)} - \frac{\omega_{pi}^2}{\omega^2 + \omega_{pi}^2 - \omega_i^2(p)} . \tag{4.139}$$

Here new quantities were introduced

$$\omega_{p,c}^2 = \frac{n_c e_c^2}{\varepsilon_0 m_c}, \quad \omega_c^2(p) = \omega_{pc}^2\left(1 + \frac{p^2}{\varkappa_c^2}\right) + \left(\frac{\hbar p^2}{2m_c}\right)^2,$$

$$\varkappa_c^2 = \frac{n_c e_c^2}{\varepsilon_0 k_B T_c}, \quad \omega_{pe}^2 + \omega_{pi}^2 = \omega_{pl}^2, \quad \beta_c = 1/k_B T_c . \tag{4.140}$$

The (different) temperatures β_c are choosen in such a way that now the dispersion relation of the plasmons may be fulfilled (eq. (4.119)), namely

$$\beta_e + \beta_i = 2\beta = \frac{2}{k_B T}, \qquad \frac{1}{\beta_e m_e^2} + \frac{1}{\beta_i m_i^2} = \frac{3}{\beta}\left(\frac{1}{m_e^2} + \frac{1}{m_i^2}\right). \tag{4.141}$$

From (4.139) we have then ($z \equiv \omega + i\varepsilon$)

$$\varepsilon^{-1}(p, z) = \frac{(z^2 - \omega_e^2(p) + \omega_{pe}^2)(\omega^2 - \omega_i^2(p) + \omega_{pi}^2)}{z^4 - z^2(\omega_e^2(p) + \omega_i^2(p)) - \omega_{pe}^2\omega_{pi}^2 + \omega_e^2(p)\,\omega_i^2(p)} . \tag{4.142}$$

The roots of the denominator, i.e., the poles of $\varepsilon^{-1}(q, z)$ are located at ω_1 and ω_2, and have the shape

$$\omega_{1,2} = \frac{\omega_e^2(p) + \omega_i^2(p)}{2} \pm \sqrt{\omega_{pe}^2\omega_{pi}^2 + (\omega_e^2(p) - \omega_i^2(p))^2/4} . \tag{4.143}$$

The dispersion of the two branches has the properties:

(i) For large momenta we have

$$\lim_{p\to\infty} \omega_1^2 = \omega_e^2(p), \ \lim_{p\to\infty} \omega_2^2 = \omega_i^2(p)$$

and

(ii) for small momenta (with (4.140), (4.141))

$$\omega_1^2 = \omega_{pl}^2 + \frac{p^2}{\omega_{pl}^2}\left(\frac{\omega_{pe}^4}{\varkappa_e^2} + \frac{\omega_{pi}^4}{\varkappa_i^2}\right)$$

or

$$\omega_1^2 = \omega_{pl}^2\left(1 + 3\frac{p^2}{\bar{\varkappa}^2}\right) \tag{4.144}$$

(in agreement with (4.117))

and

$$\omega_2^2 = \frac{p^2}{\omega_{pl}^2}\omega_{pe}^2\omega_{pi}^2\left(\frac{1}{\varkappa_e^2} + \frac{1}{\varkappa_i^2}\right). \tag{4.145}$$

This latter expression is characterized by

$$\omega_2 \sim p$$

and may be refered to as the acoustic mode.

From (4.142) we get for the spectral function

$$\begin{aligned}\operatorname{Im} \varepsilon^{-1}(p, \omega) &= \pi[\delta(\omega + \omega_1) - \delta(\omega - \omega_1)] \frac{1}{2\omega_1} \\ &\cdot \frac{\omega_1^2 \omega_{pl}^2 + 2\omega_{pe}^2 \omega_{pi}^2 - \omega_{pe}^2 \omega_i^2(p) - \omega_{pi}^2 \omega_e^2(p)}{\omega_1^2 - \omega_2^2} \\ &+ \pi[\delta(\omega + \omega_2) - \delta(\omega - \omega_2)] \frac{1}{2\omega_2} \\ &\cdot \frac{\omega_2^2 \omega_{pl}^2 + 2\omega_{pe}^2 \omega_{pi}^2 - \omega_{pe}^2 \omega_i^2(p) - \omega_{pi}^2 \omega_e^2(p)}{\omega_2^2 - \omega_1^2} . \end{aligned} \tag{4.146}$$

Eq. (4.146) coincides for large p with (4.136). The resulting dielectric function according to (4.122) is in agreement with (4.125) and (4.126) for $\omega \to 0$ and $p \to 0$, for small momenta and finite or large frequencies it coincides with (4,127)—(4.129), respectively. RÖPKE (1981) discussed further the case of $m_e/m_i \ll 1$ and $m_1 \approx m_2$.

We want to refer to some recent papers in which the dielectric function is applied in various connections. There especially higher order contributions are taken into account. (For the inclusion of bound states see Section 4.5.) UTSUMI and ICHIMURA (1980) discussed the screening length, the structure factor, the correlation function and the correlation energy. The structure factor was dealt with by GREEN, LOWY and SZYMANSKI (1982), and MIYAGI (1981) determined the correlation energy of simple metals. The energy loss in metals which is closely connected with the dielectric function is the subject of a paper by STURM (1982).

Two-dimensional problems were already mentioned in the introduction and are not dealt with in this monograph. Here we want to refer, e.g., only to a paper by ISIHARA and IORIATTI (1980), in which the exchange energy and the specific heat are determined using the dielectric function.

In a large number of papers, people try to express the properties of systems consisting originally of fermions by the creation and annihilation operators of composite particles, i.e., by Boson operators. Here also the dielectric function plays an essential role; see, e.g., ARPONEN and PAJANNE (1975, 1979, 1982), and PAJANNE (1982).

4.3. Single-Particle Excitations

4.3.1. Quasi-Particle Concept

In the framework of many-particle physics the motion of a single particle is described by the single-particle Green's function. The latter must be determined from the Dyson equation as outlined in Chapter 3. In the Dyson equation the influence of the surrounding medium is fully taken into account by the self energy Σ. The self energy, however may be determined up to some approximation only, what is equivalent to the fact that the hierarchy of (Green's functions) equations must be truncated somehow.

For many practical purposes it is convenient, to condense the influence of the medium (at least approximately) into the single-particle properties (energy, mass) so that one has a collective of (approximately) noninteracting particles which are refered to as quasi-particles (PINES and NOZIÈRES, 1966; MATTUCK, 1967).

For this reason one has to determine some approximation of the self energy operator Σ which gives, as a result, the possibility of the determination of the quasi-particle energy, the quasi-particle damping (inverse life time of a single-particle state) and an effective mass. Generally, one should mention that the quasiparticle concept is useful only if the quasiparticle damping remains small as compared with the quasi-particle energy.

From the Dyson equation

$$G_a(p, z_\nu) = \left(\hbar z_\nu - \frac{\hbar^2 p^2}{2m_a} - \Sigma_a(p, z_\nu)\right)^{-1} \tag{4.147}$$

one should, in principle, determine all single-particle properties; however one must be careful, and the denominator of (4.147) does *not* give in a simple manner the single-particle properties as its zeros (poles of the Green's function).

Before we proceed with the discussion of the analytic properties of the Green's function $G(pz)$ we will show that the denominator of (4.147) does not have zeros. For this reason we write instead of Σ in the denominator its spectral representation, and we have thus instead of (4.147)

$$G_a(p, z) = \left(\hbar z - E_{\mathrm{HF}}(p) - \int \frac{\mathrm{d}\overline{\omega}}{2\pi} \frac{\Gamma_a(p, \overline{\omega})}{z - \overline{\omega}}\right)^{-1}. \tag{4.148}$$

Let us take the case of small Γ, and let $z = \omega + i\varepsilon$, $\varepsilon \to 0$. Then we have

$$G(p, \omega + i\varepsilon) = \left(\omega + i\varepsilon - E_{\mathrm{HF}}(p) - P\int \frac{\mathrm{d}\overline{\omega}}{2\pi} \frac{\Gamma(p, \overline{\omega})}{\omega - \overline{\omega}} + \frac{i}{2}\Gamma(p, \omega)\right)^{-1}. \tag{4.149}$$

While the quasiparticle energy may be approximately determined to be

$$\mathrm{Re}\, \tilde{E}_a(p) = E^a_{\mathrm{HF}}(p) + P\int \frac{\mathrm{d}\overline{\omega}}{2\pi} \frac{\Gamma_a(p, \overline{\omega})}{\omega - \overline{\omega}}, \tag{4.150}$$

the imaginary part of the denominator of (4.149) does not vanish.

As discussed in Chapter 3., the complex function (4.148) is analytic in the upper half plane (G_{I}) and does not have poles. This property follows from the Cauchy type representation of $G_a(p, z)$

$$G_a(p, z) = \int \frac{A_a(p, \omega)}{z - \omega} \frac{\mathrm{d}\omega}{2\pi}. \tag{4.151}$$

According to Chapter 3. the function $G_{\mathrm{I}}(pz)$ may be continued to the lower half plane, and reads then

$$G_{\mathrm{I}}(p, z) \to G_{\mathrm{II}}(p, z) + A(p, z) \quad (\mathrm{Im}\, z < 0) \tag{4.152}$$

where G_{II} is analytic and has no poles in the lower half plane. The continuation of G_{I} (4.152) to the lower half plane has poles which are that of $A(p, z)$. In order to determine the quasi-particle properties one has thus to determine the poles of the analytic continuation of the spectral function in the lower half plane. This means, one has to consider

$$A_a(p, z) = \Gamma_a(p, z) \left\{(z - E^a_{\mathrm{HF}}(p) - \mathrm{Re}\, \overline{\Sigma}_a(p,z))^2 + \frac{\Gamma^2_a(p, z)}{4}\right\}^{-1}$$

$$= i\left\{\left(z - E^a_{\mathrm{HF}}(p) - \mathrm{Re}\, \overline{\Sigma}_a(p, z) + \frac{i}{2}\Gamma_a(p, z)\right)^{-1}\right.$$

$$\left.-\left(z - E^a_{\mathrm{HF}}(p) - \mathrm{Re}\, \overline{\Sigma}_a(p, z) - \frac{i}{2}\Gamma_a(p, z)\right)^{-1}\right\}, \qquad \Gamma_a = 2\,\mathrm{Im}\, \Sigma_a\,. \quad (4.153)$$

It is seen that (4.153) has a pole in the lower half plane which is refered to as the single-particle energy (quasiparticle energy). While its real part is given approximately by (4.150), the damping is given by

$$\mathrm{Im}\, \widetilde{E}_a(p) = \left.\frac{2\,\mathrm{Im}\, \overline{\Sigma}_a(p, \omega)}{1 + \frac{\partial}{\partial\omega}\mathrm{Re}\, \overline{\Sigma}_a(p, \omega) \cdot \frac{1}{\hbar}}\right|_{\hbar\omega = \mathrm{Re}\, \widetilde{E}_a(p)} \quad (4.154)$$

While the determination of the poles of (4.153) leads to complex quasiparticle energies, which are considered to be the proper quantities, one may approximately also discuss the spectral function $A_a(p, \omega)$ which is sharply peaked if $\mathrm{Im}\, \overline{\Sigma}$ is sufficiently small. The position of the peak determines the quasiparticle energy, while the width is connected with the quasiparticle damping (inverse life time).

For practical cases and in order to have quasiparticles the damping must be (assumed to be) small. Alternatively, the complex quasiparticle energy may be written

$$\widetilde{E}_a(p) = \mathrm{Re}\, \widetilde{E}_a - i\,\mathrm{Im}\, \widetilde{E}_a(p) = \frac{\hbar^2 p^2}{2m_a^{\mathrm{eff}}} - i\,\mathrm{Im}\, \widetilde{E}_a(p)\,. \quad (4.155)$$

Here m_a^{eff} is the effective mass (Mattuk, 1967) which accounts for the influence of the plasma. For the evaluation of higher order contributions to the mass operator see Stolz and Zimmermann (1979).

4.3.2. Self-Energy in V^s-Approximation

Now let us discuss the simplest approximation which leads to a finite life time of single-particle excitations, i.e.,

$$\overline{\Sigma}_a(12) = i V^s_{aa}(12)\, G_a(12)\,. \quad (4.156)$$

Here we have to mention that $\overline{\Sigma}_a$ does not include the Hartree self energy which was used to construct the effective external field in order to construct the properly screened Green's functions, see Chapter 3. However, (4.156) includes the Hartree-Fock-self-energy yet, which reads

$$\Sigma_a^{\mathrm{HF}}(12) = i V_{aa}(12)\, G_a(12) \quad (4.157)$$

and has the momentum Matsubara frequency representation

$$\Sigma_a^{\mathrm{HF}}(p) = -\int \frac{dk}{(2\pi)^3} \frac{e_a^2}{\varepsilon_0 k^2} f_a(p - k)\,. \quad (4.158)$$

The essential feature of the Hartree-Fock self energy is that it is only momentum dependent and thus real valued. In this way this contribution does not lead to a finite life time of the single particle excitations, but only influences, e.g., the effective mass. Σ_a^{HF} has no classical limit as it arises from exchange effects. In the nondegenerate limit we may do the integral in (4.158); the result is

$$\Sigma_a^{\mathrm{HF}}(p) = -\frac{2e_a^2}{\varepsilon_a(s+1)} n_a \Lambda_a^2 \, {}_1F_1\left(1, \frac{3}{2}; -\frac{\hbar^2 p^2}{2m_a k_{\mathrm{B}} T}\right). \tag{4.159}$$

The difference between (4.156) and (4.157) is in the literature refered to as the correlation part of Σ_a,

$$\Sigma_a^{\mathrm{corr}}(12) = i\left(V_{aa}^{s}(12) - V_{aa}(12)\right) G_a(12) .$$

According to Chapter 3. we have, for the spectral density of the self energy

$$2 \operatorname{Im} \Sigma_a^{\mathrm{corr}}(k, \omega - i0) = \Gamma_a(k, \omega)$$

$$= \hbar 4\pi \int \frac{\mathrm{d}\boldsymbol{p}'}{(2\pi)^3} \int_{-\infty}^{+\infty} \frac{\mathrm{d}\omega'}{2\pi} \left(\delta\left(\hbar\omega - \hbar\omega' - \frac{(\boldsymbol{k}+\boldsymbol{p}')^2 \hbar^2}{2m_a}\right)\right)$$

$$\cdot V_{aa}(p') \left(1 - f(\boldsymbol{p}' + \boldsymbol{k}) + n_{\mathrm{B}}(\omega')\right) \operatorname{Im} \varepsilon^{-1}(p', \omega' + i0) . \tag{4.160}$$

The spectral weight function $\frac{1}{2}\Gamma_a(k, \omega)$ is connected with the single particle life time τ_a (in the lowest approximation):

$$\Gamma_a\left(k, \frac{\hbar^2 k^2}{2m_a}\right) = \hbar \tau_a^{-1}(k) . \tag{4.161}$$

The real part of the self energy reads in the V^s approximation

$$\operatorname{Re} \Sigma_a^{\mathrm{corr}}(k, \omega) = -P\hbar \int \frac{\mathrm{d}\boldsymbol{p}'}{(2\pi)^3} \int_{-\infty}^{+\infty} \frac{\mathrm{d}\omega'}{2\pi} \cdot 2V_{aa}(p')$$

$$\cdot \frac{\left(1 - f(\boldsymbol{p}' + \boldsymbol{k}) + n_{\mathrm{B}}(\omega')\right) \operatorname{Im} \varepsilon^{-1}(p', \omega' + i0)}{\hbar(\omega - \omega') - \dfrac{\hbar^2}{2m_a}(\boldsymbol{p}' + \boldsymbol{k})^2 .} \tag{4.162}$$

A further analytic evaluation of (4.160), (4.162) is possible only in limiting situations. Let us discuss the cases of weak and strong degeneracy and let $\operatorname{Im} \varepsilon^{-1}$ be the RPA version discussed in the previous Section.

Let us first consider the *nondegenerate limit*. We take into account that (for Coulomb systems)

$$V_{aa}(p) \operatorname{Im} \varepsilon^{-1}(p, \omega + i0) = |V_{aa}^{s}(p, \omega)|^2 \sum_b \operatorname{Im} \Pi_b(p, \omega + i0) . \tag{4.163}$$

If now, moreover, in a simplified version the modulus of the dynamically screened potential is replaced by a static (Debye) one, we get from (4.160) in the simplified nondegenerate limit

$$\Gamma_a\left(k, \frac{\hbar^2 k^2}{2m_a}\right) = \sum_c \frac{e_a^2 (k_{\mathrm{B}}T)^2 m_c^2 \Lambda_c^3}{\varepsilon_0 16\pi^3 \hbar^4} \, {}_1F_1\left(1, \frac{3}{2}; -\frac{\hbar^2 m_c k^2}{m_a^2 k_{\mathrm{B}} T}\right). \tag{4.164}$$

However, the damping Γ_a according to (4.164) has no classical limit. On the other hand it does not exhibit the sharp maximum which is due to the excitation of plasmons described by the correct expression for ε^{-1} involved in (4.160). Such discussions may be carried out only numerically (FENNEL and WILFER, 1975; see Fig. 4.2).

The mean value of $\operatorname{Re} \Sigma_a$ is related (approximately) to the chemical potential μ_a of species a. This may be shown as follows. In the nondegenerate case we introduce

$$\left\langle \Sigma_a\left(\boldsymbol{k}, \frac{\hbar^2 k^2}{2m_a}\right)\right\rangle = \int \Sigma_a\left(\boldsymbol{k}, \frac{\hbar^2 k^2}{2m_a}\right) \exp\left(-\beta\hbar^2 k^2/2m_a\right) \Lambda_a^3 \frac{\mathrm{d}\boldsymbol{k}}{(2\pi)^3}.$$

The factor Λ_a^3 is the remainder of the normalization integral, the spin factor $(2s_a + 1)$ cancels. As known from thermodynamics in the nondegenerate situation, see Ebeling et al. (1976, 1979), it is sufficient to determine the following integral:

$$\begin{aligned}\langle \Sigma_a\rangle &= \sum_b n_b \int \frac{\mathrm{d}\boldsymbol{k}\,\mathrm{d}\boldsymbol{k}'\,\mathrm{d}\boldsymbol{p}}{(2\pi)^9} \Lambda_a^3 \Lambda_b^3 V_{ab}(\boldsymbol{k}-\boldsymbol{p})\, V_{ab}^{\mathrm{s}}(\boldsymbol{k}-\boldsymbol{p})\, 2m_{ab}/\hbar^2 \\ &\cdot \exp\left\{-\beta\hbar^2\left(\boldsymbol{k}'\frac{m_{ab}}{m_b} - \frac{m_{ab}}{m_a}\boldsymbol{p}\right)^2\Big/ 2m_{ab}\right) - \left(\beta\hbar^2(\boldsymbol{p}+\boldsymbol{k}')^2/2(m_a+m_b)\right)\right\} \\ &\cdot \left[\left(\frac{m_{ab}}{m_b}\boldsymbol{k}' - \frac{m_{ab}}{m_a}\boldsymbol{p}\right)^2 - \left(k - \frac{m_{ab}}{m_b}(\boldsymbol{p}+\boldsymbol{k}')\right)^2\right]^{-1}.\end{aligned}$$

This equation follows from (4.162), if in the screening equation, $V^{\mathrm{s}} = V + V\Pi V^{\mathrm{s}}$, only in the integral term the full dynamics is retained in the polarization function Π, while the V^{s} (in the integral term) is replaced by a static (Debye) potential, see EBELING et al. (1976, 1979).

All further evaluation may be carried out exactly using some variable substitutions. As an intermediate result we have

$$\begin{aligned}\langle \Sigma_a\rangle &= -\sum_b n_b \Lambda_b^3 \Lambda_a^3 \frac{16 m_{ab}\pi^{3/2}}{(2\pi)^3\hbar^2}\left[\frac{(m_a+m_b)\,k_{\mathrm{B}}T}{2\hbar^2}\right]^{3/2} \\ &\cdot \int \frac{\mathrm{d}\boldsymbol{k}\,\mathrm{d}\boldsymbol{l}}{(2\pi)^3} \frac{1}{2kl_x} V_{ab}(l)\, V_{ax}^{\mathrm{s}}(l)\, \mathrm{e}^{-\frac{\beta\hbar^2}{2m_{ab}}\left(k-\frac{l}{2}\right)^2}, \qquad x = \frac{\boldsymbol{l}\cdot\boldsymbol{k}}{lk}.\end{aligned}$$

The final results reads

$$\langle \Sigma_a\rangle = -\sum_b \frac{n_b e_a^2 e_b^2}{8kT_{\mathrm{B}}\varepsilon^2\varkappa} G(\varkappa\lambda_{ab}),$$

$$G(y) = \frac{\sqrt{\pi}}{y}\left\{1 - \exp\left(\frac{y^2}{4}\right)\left[1 - \Phi\left(\frac{y}{2}\right)\right]\right\} = 1 - \frac{\sqrt{\pi}}{4}y + O(y^2).$$

The expansion begins with the limiting law

$$\langle \Sigma_a\rangle = -\frac{e_a^2\varkappa}{8\pi\varepsilon}.$$

Evidently, this is the chemical potential in Debye approximation. It may be shown easily that the equality between $\langle \Sigma_a\rangle$ and μ_a, the chemical potential of species a holds also in higher approximations. One has to take into account that

$$\mu_a = \left(\frac{\partial F}{\partial N_a}\right), \qquad F - F_{\text{id}} = \int_0^1 \frac{d\lambda}{\lambda} \langle V \rangle_\lambda .$$

For the mean interaction potential we get in Montroll-Ward approximation (EBELING et al., 1976, 1979)

$$\langle V \rangle = -\frac{\Omega k_B T}{2\varkappa} \sum_{ab} n_a n_b \left(\frac{e_a e_b}{4\pi\varepsilon k_B T}\right)^2 G(\varkappa\lambda_{ab}) .$$

Consequently we get

$$\mu_a - \mu_a^{\text{id}} = -\frac{1}{2\varkappa} \cdot \frac{e_a^2}{4\pi\varepsilon} \sum_b \frac{n_b e_b^2}{\varepsilon k_B T} G(\varkappa\lambda_{ab}) = \langle \Sigma_a \rangle . \tag{4.165}$$

The approximate equality between the averaged self energy and the interaction part of the chemical potential will be of some importance for the thermodynamical discussions in Chapter 6.

For several purposes one needs also the non-averaged self energy. The evaluation of the real part, given by eq. (4.162), may be carried out approximately if we use the expansion

$$n_B(\omega) = \frac{k_B T}{\hbar\omega} - \frac{1}{2} - \frac{\hbar\omega}{12 k_B T} + \dots$$

and further approximations of the types (4.78)—(4.79). Moreover the spectral representation of the screened potential must be applied (Chapter 3., eq. (3.248)).

For $\hbar\omega \to \dfrac{\hbar^2 k^2}{2m_a}$ we get from (4.162) (KRAEFT and STOLZMANN, 1982; KRAEFT, 1982)

$$\begin{aligned}
\operatorname{Re} \widetilde{E}_a(k) \equiv E_a(k) = & -\varkappa_D \frac{e_a^2}{4\pi\varepsilon_0}\left(\frac{1}{2} + \frac{1}{4}\sum_c \frac{m_c}{2m_a} - \frac{\pi}{32}\left(\sum_c \frac{1}{2}\left(\frac{m_c}{m_a}\right)^{1/2}\right)^2\right) \\
& + \frac{\hbar^2 k^2}{2m_a}\left(1 + \frac{\varkappa_D e_a^2}{4\pi\varepsilon_0 k_B T}\left[\frac{1}{12}\sum_c \frac{m_c^2}{2m_a^2} - \frac{1}{16}\left(\sum_c \frac{m_c}{2m_a}\right)^2\right.\right. \\
& - \frac{\pi}{48}\left(\sum_c \frac{1}{2}\left(\frac{m_c}{m_a}\right)^{1/2}\right)^2 + \frac{1}{6}\sum_c \frac{m_c}{2m_a} - \frac{\pi}{64}\left(\sum_c \frac{1}{2}\left(\frac{m_c}{m_a}\right)^{3/2}\right)\left(\sum_c \frac{1}{2}\left(\frac{m_c}{m_a}\right)^{1/2}\right. \\
& \left.\left.\left. - \frac{\pi^2}{64}\cdot\frac{5}{16}\left(\sum_c \frac{1}{2}\left(\frac{m_c}{m_a}\right)^{1/2}\right)^4\right]\right) + \dots ; \qquad \varkappa_D^2 = \sum_c \frac{e_c^2 n_c}{\varepsilon_0 k_B T} .
\end{aligned} \tag{4.166}$$

For equal masses, from (4.166) follows

$$E_a(k) = -\frac{\varkappa_D e_a^2}{4\pi\varepsilon_0}\left(\frac{3}{4} - \frac{\pi}{32}\right) + \frac{\hbar^2 k^2}{2m_a}(1 + \gamma_a) , \qquad \gamma_a = 0.0248 \frac{\varkappa_D e_a^2}{4\pi\varepsilon_0 k_B T} . \tag{4.167}$$

In contrast to the ideally free particles, the quasiparticles have, at zero momentum, a negative single-particle energy. Though the contribution just mentioned is a classical one, classical calculations (ICHIMARU, 1973) give, instead of ours

$$\frac{\varkappa_D e_a^2}{4\pi\varepsilon_0}\left(\frac{3}{4} - \frac{\pi}{32}\right) \to \frac{\varkappa_D e_a^2}{4\pi\varepsilon_0} \tag{4.168}$$

while from KUDRIN (1974) follows

$$\frac{\varkappa_D e_a^2}{4\pi\varepsilon_0} \cdot \frac{1}{2}. \tag{4.169}$$

The numerical evaluation of (4.162) is in agreement with the momentum independent part of (4.167) (ZIMMERMANN and RÖSLER, 1982).

From (4.167) we may define an effective mass according to

$$\frac{1}{m_a^{\text{eff}}} = \frac{\partial^2}{\partial(\hbar k)^2} E_a(k) = \frac{1+\gamma}{m_a}.$$

From ICHIMARU (1973) follows

$$\Delta m_a^{\text{eff}} = \frac{1}{3}\left(1 - \frac{\pi}{8}\right) \frac{\varkappa_D e_a^2 m_a}{4\pi\varepsilon_0 k_B T} = m_a^{\text{eff}} - m_a \,. \tag{4.170}$$

Now let us consider the case of *high degeneracy*. In the low temperature (high density) limit one may apply the Sommerfeld expansion of the Fermi function according to (4.87) and use (4.91), (4.92). From (4.160) then follows for momenta near the Fermi surface (KREMP, KRAEFT and FENNEL, (1972) for an electron gas

$$\Gamma(k) = \left(\frac{k}{k_F} - 1\right)^2 \frac{e^4 m_e k_F^3}{(4\pi\varepsilon_0)^2 k_{TF}^3 \hbar^2} + \frac{e^4 m_e^3}{\varepsilon_0^2 \hbar^6 k_{TF}^4} \frac{\pi}{6} (kT)^2 \left(\frac{\pi k_{TF}}{4k_F} - 2\left(\frac{k}{k_F} - 1\right)\right). \tag{4.171}$$

Here $k_F = (3\pi^2 n_e)^{1/3}$ is the Fermi wave number and k_{TF} is given according to (4.90). Often the abbreviations are applied $\left(\text{with } 2m_e = \frac{e^2}{2} = \hbar = 1\right)$

$$k_F^{-1} = \alpha r_s; \quad \alpha^3 = 4/9\pi; \quad r_s = d/a_0; \quad d^3 = 3/4\pi n; \quad a_0 = \hbar^2/me^2 \,.$$

Eq. (4.171) reads then (KREMP, KRAEFT and FENNEL, 1972)

$$\Gamma(k) = \frac{1}{4}\left(\frac{\pi}{\alpha r_s}\right)^{3/2} \left(\frac{k}{k_F} - 1\right)^2 + \frac{\pi^3}{24} (\alpha\pi)^{5/2} (k_B T)^2 \left(r_s^{5/2} - \frac{4r_s^2}{\sqrt{\alpha\pi}}\left(\frac{k}{k_F} - 1\right)\right). \tag{4.172}$$

This result is for $T = 0$ in agreement with that of QUINN and FERRELL (1958) and shows that at $k = k_F$ there exist undamped single-particle excitations. At $T \neq 0$, however, even at $k = k_F$ the quasi-particles are of finite life time. For intermediate degeneracy see the paper by FENNEL and WILFER (1975).

The real part of the quasi-particle energy follows analogously from eq. (4.162), and the result is (KREMP, KRAEFT and FENNEL, 1972; FENNEL, KRAEFT and KREMP, 1974)

$$\begin{aligned} E(p) = {} & \frac{\hbar^2 p^2}{2m_e} + \frac{e^2}{2\pi}\left\{(p_F - p)\left(\ln r_s + 2 + \ln\frac{\alpha}{\pi}\right) - 2p_F\right\} \\ & + \frac{\pi^2}{12} \frac{(k_B T)^2}{\varepsilon_F}\left[\left(\frac{p}{p_F} - 1\right)\left(\frac{9}{4} - \frac{1}{0.66 r_s}\right.\right. \\ & \left.\left. + \frac{1}{2}(\ln r_s - 1.8)\right) - \frac{1}{2} - \frac{1}{2}(\ln r_s - 1.8)\right]. \end{aligned} \tag{4.173}$$

The $T = 0$ contribution is in agreement with that of QUINN and FERRELL (1958).

4.4. Two-Particle Properties in a Plasma

4.4.1. Bethe-Salpeter Equation for a Two-Particle Cluster

The properties of a plasma in a special region of the density-temperature plane are determined also by the formation of bound states, see Chapter 2. In this Section we consider the formation of two-particle bound states (i.e., the formation of H-atoms from electrons and protons or the formation of excitons from electrons and holes, f.i.). Generalizations to higher complexes as molecules or biexcitons are straightforeward. The main result of this Section 4.4. will be that a self-energy of the two-particle complex can be introduced in analogy to the one-particle self-energy, and different approximations are discussed to evaluate the properties of a two-particles system embedded in a surrounding plasma.

In this Section 4.4.1. we repeat some formulas which also have been given in Chapter 3. An appropriate basis for the discussion of two-particle states in a interacting many-body system is the two-particle Green's function G_2 defined according to (cf. (3.73))

$$G_2(12, 1'2') = \frac{1}{(i\hbar)^2} \langle T\{a(1)\, a(2)\, a^+(2')\, a^+(1')\}\rangle \tag{4.174}$$

where 1 denotes species, spin and momentum, We are especially interested in the particle-particle channel where $t_1 = t_2 = t$, $t_1' = t_2' = t'$, so that according to the Kubo-Martin-Schwinger boundary condition the resulting $G_2(t - t')$ may be represented by a Fourier series

$$G_2(t - t') = \frac{1}{-i\hbar\beta} \sum_\lambda e^{-i\Omega_\lambda(t-t')}\, G_2(\Omega_\lambda) \tag{4.175}$$

with

$$\hbar\Omega_\lambda = \frac{\pi\lambda}{-i\beta} + \mu_a + \mu_b\,, \qquad \lambda = 0, \pm 2, \pm 4, \dots,$$

being the Matsubara frequencies for the two-particle system (ab).

The Fourier-coefficients $G_2(\Omega_\lambda)$ are related to the spectral function $A_2(\omega)$ according to

$$G_2(\Omega_\lambda) = \int \frac{d\omega}{2\pi} \frac{A_2(\omega)}{\Omega_\lambda - \omega} \tag{4.176}$$

so that the spectral function $A_2(\omega)$ is determined by the analytical continuation of $G_2(\Omega_\lambda)$ from the discrete set of Matsubara frequencies Ω_λ into the whole complex Ω-plane according to

$$A_2(\omega) = i\{G_2(\omega + i0) - G_2(\omega - i0)\}\,.$$

The physical meaning of the spectral function has been discussed in Section 3.2.5.

The two-particle Green function $G_2(t - t')$ is determined by a Bethe-Salpeter equation (BSE) which is represented by the following diagrams:

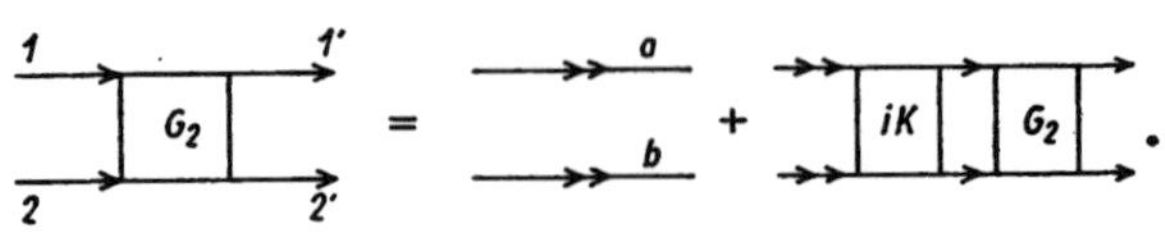

(4.177)

We consider the case $a \neq b$ where no exchange term arises. The one-particle propagator contained in eq. (4.177) obeys the Dyson equation

(4.178)

where Σ_a denotes the self energy. The kernel K in eq. (4.177) can be represented by the sum of all irreducible diagrams in the particle-particle channel and describes the effective interaction. Special approximations for the self energy Σ_a and the effective interaction K are considered in Sections 4.4.3. and 4.4.4.

In general, because of the dynamical character of K, the BSE (4.177) contains also on the right-hand side the three-time Green function (incoming lines at different times) $G_2(t_1, t_2, t_3, t_4)_{t_3=t_4}$ which yields after a Fourier transform the two-frequency Green function $G_2(z_\nu^a, \Omega_\lambda - z_\nu^a)$, $\hbar z_\nu^a = \dfrac{\pi\nu}{-i\beta} + \mu_a$, $\nu = \pm 1, \pm 3, \ldots$

To obtain a closed equation for $G_2(\Omega_\lambda)$ we must express $G_2(z_\nu^a, \Omega_\lambda - z_\nu^a)$ in terms of $G_2(\Omega_\lambda)$. For a statically (screened) potential V (see also eq. (4.210)) the effective interaction K depends only on Ω_λ, $K = K(\Omega_\lambda)$; in the general case its frequency dependence is more complicated. For a statical potential V, the BSE gives

$$G_2(121'2', z_\nu, \Omega_\lambda - z_\nu) = G_1(1, z_\nu)\, G_1(2, \Omega_\lambda - z_\nu) \,.$$
$$\left(\delta_{11'}\, \delta_{22'} + \int \mathrm{d}1''\, \mathrm{d}2''\, K(121''2'', \Omega_\lambda)\, G_2(1''2''1'2', \Omega_\lambda)\right) .$$

After a summation over ν we get the BSE for a statical potential in the usual form

$$G_2(\Omega_\lambda) = G_2^0(\Omega_\lambda)\left(1 + K(\Omega_\lambda)\, G_2(\Omega_\lambda)\right) .$$

The labels 1,1′ etc. remain unchanged.
Here we used (See section 3.2.8.)

$$G_2^0(\Omega_\lambda) = \sum_\nu \frac{1}{-i\beta}\, G_1(z_\nu)\, G_1(\Omega_\lambda - z_\nu) \,.$$

Using the definition

$$\frac{1}{-i\beta\hbar} \sum_\nu G_2(z_\nu, \Omega_\lambda - z_\nu) = G_2(\Omega_\lambda)$$

we get the following relation between the two-frequency function and the one-frequency function:

$$G_2(121'2', z_\nu^a, \Omega_\lambda - z_\nu^a) = \frac{G_a(1, z_\nu^a)\, G_b(2, \Omega_\lambda - z_\nu^a)}{\dfrac{1}{-i\beta\hbar} \sum\limits_\nu G_a(1, z_\nu^a)\, G_b(2, \Omega_\lambda - z_\nu^a)}\, G_2(121'2', \Omega_\lambda) \,. \qquad (4.179)$$

The same relation between $G_2(z_\nu, \Omega_\lambda - z_\nu)$ and $G_2(\Omega_\lambda)$ was also used for the case of a dynamical interaction (Shindo approximation, see Section 3.2.8.; Shindo, 1970; Kilimann and Kraeft, 1978). We argue that the resulting closed BSE is correct up to the first order in the retardation.

After some transformations, (4.177) may be written in the form

$$\{\varepsilon_a(1) + \varepsilon_b(2) + \Delta_{ab}^{\mathrm{eff}}(12, \Omega_\lambda) - \Omega_\lambda\}\, G_2(121'2', \Omega_\lambda)$$
$$+ \sum_{\bar{3}\bar{4}} V_{ab}^{\mathrm{eff}}(12\bar{3}\bar{4}, \Omega_\lambda)\, G_2(\bar{3}\bar{4}1'2', \Omega_\lambda) = i\delta_{11'}\, \delta_{22'}\, N_{ab}(12) \qquad (4.180)$$

with $\varepsilon_a(1) = \hbar^2 k_1^2/2m_a$,

$$N_{ab}(12) = \frac{1}{-\beta} \sum_\nu \{G_a(1, z_\nu^a) + G_b(2, \Omega_\lambda - z_\nu^a)\} \approx 1 - f_a(1) - f_b(2)$$

(phase space occupation), and

$$\Delta_{ab}^{\text{eff}}(12, \Omega_\lambda) = N_{ab}^{-1}(12) \frac{1}{-\beta} \sum_\nu \{\Sigma_a(1z_\nu^a) + \Sigma_b(2,\Omega_\lambda - z_\nu^a)\} \{G_a(1, z_\nu^a) + G_b(2, \Omega_\lambda - z_\nu^a)\},$$

$$V_{ab}^{\text{eff}}(12\bar{3}\bar{4}, \Omega_\lambda) = \frac{1}{i\beta} \sum_{\nu\bar{\nu}} \{G_a(1z_\nu^a) + G_b(2, \Omega_{\bar{\lambda}} - z_\nu^a)\} N_{ab}^{-1}(\bar{3}\bar{4})$$
$$\cdot K(12\bar{3}\bar{4}, z_\nu^a, \Omega_\lambda - z_\nu^a, z_{\bar{\nu}}^a) \{G_a(\bar{3}, z_{\bar{\nu}}^a) + G_b(\bar{4}, \Omega_\lambda - z_{\bar{\nu}}^a)\} \tag{4.181}$$

being the effective self-energy and the effective potential, respectively. Both these quantities are, in general, frequency dependent which expresses its dynamical character and are closely related which is shown from the relation valid, f.i., for the V^s-approximation (see 4.4.3.)

$$\Delta_{ab}^{\text{eff}}(12\Omega_\lambda) = \sum_{1'2'} \{ V_{ab}^{\text{eff}}(121'2'\Omega_\lambda)\, N_{ab}(1'2') - V(121'2')\} . \tag{4.182}$$

So, they have to be evaluated in the equivalent manner, and this equation should be valid after approximations in Δ_{ab}^{eff} and V_{ab}^{eff}.

Further we can verify from the expressions given in 4.4.3. and 4.4.4. that the following spectral representations for Δ_{ab}^{eff} and V_{ab}^{eff} are valid:

$$\begin{aligned} \Delta_{ab}^{\text{eff}}(\Omega_\lambda) &= \bar{\Delta}_{ab}^{\text{eff}} + \int \frac{d\omega}{\pi} \frac{\text{Im}\, \Delta_{ax}^{\text{eff}}(\omega - i0)}{\Omega_\lambda - \omega}, \\ V_{ab}^{\text{eff}}(\Omega_\lambda) &= \bar{V}_{ab}^{\text{eff}} + \int \frac{d\omega}{\pi} \frac{\text{Im}\, V_{ab}^{\text{eff}}(\omega - i0)}{\Omega_\lambda - \omega}. \end{aligned} \tag{4.183}$$

The real contributions $\bar{\Delta}_{ab}^{\text{eff}}$ and $\bar{V}_{ab}^{\text{eff}}$ contain those terms which are not dependent on Ω_λ.

As long as $N_{ab}(12) > 0$, the BSE (4.180) can be transformed into a symmetric Hermitean form by incorporating the denominator $N_{ab}^{-1}(\bar{3}\bar{4})$ into the Greens' functions according to

$$\bar{G}_2(121'2', \Omega_\lambda) = G_2(121'2', \Omega_\lambda)\, N_{ab}^{-1/2}(12)\, N_{ab}^{-1/2}(1'2') \tag{4.184}$$

so that we obtain the following form:

$$\{H_{ab}^{(0)} - \Omega_\lambda\}\, \bar{G}_2(\Omega_\lambda) + H_{ab}^{\text{pl}}(\Omega_\lambda)\, \bar{G}_2(\Omega_\lambda) = i \tag{4.185}$$

Here $H_{ab}^{(0)}$ denotes the Hamiltonian of the isolated two-particle system which has the following matrix elements in momentum representation:

$$\langle 12|\, H_{ab}^{(0)}\, |\bar{3}\bar{4}\rangle = \{\varepsilon_a(1) + \varepsilon_b(2)\}\, \delta_{1\bar{3}}\delta_{2\bar{4}} + V_{ab}(12\bar{3}\bar{4}) \tag{4.186}$$

with

$$V_{ab}(12\bar{3}\bar{4}) = \delta_{\sigma_1\bar{\sigma}_3}\, \delta_{\sigma_2\bar{\sigma}_4}\, \delta_{k_1+k_2, \bar{k}_3+\bar{k}_4}\, V(k_1 - \bar{k}_3) . \tag{4.187}$$

The influence of the surrounding matter on the two-particle complex is described by $H_{ab}^{\text{pl}}(\Omega_\lambda)$ which has the following matrix elements being symmetric in momentum representation, see (4.185):

$$\langle 12| H^{\text{pl}}_{ab}(\Omega_\lambda) |\bar{3}\bar{4}\rangle = \Delta^{\text{eff}}_{ab}(12, \Omega_\lambda)\, \delta_{1\bar{3}}\delta_{2\bar{4}}$$
$$+ N_{ab}^{-1/2}(12)\, V^{\text{eff}}_{ab}(12, \bar{3}\bar{4}, \Omega_\lambda)\, N_{ab}^{1/2}(\bar{3}\bar{4}) - V_{ab}(12\bar{3}\bar{4}). \tag{4.188}$$

Because of the spectral representations for Δ^{eff}_{ab} and V^{eff}_{ab} (4.183) we find also the spectral representation

$$H^{\text{pl}}_{ab}(\Omega_\lambda) = \bar{H}^{\text{pl}}_{ab} + \int \frac{d\omega}{\pi} \frac{\text{Im}\, H^{\text{pl}}_{ab}(\omega + i0)}{\Omega_\lambda - \omega} \tag{4.189}$$

with

$$\bar{H}^{\text{pl}}_{ab} = \bar{\Delta}^{\text{eff}}_{ab} + N_{ab}^{-1/2} \bar{V}^{\text{eff}}_{ab} N_{ab}^{1/2} - V_{ab}$$

and

$$\text{Im}\, H^{\text{pl}}_{ab}(\omega - i0) = \text{Im}\, \Delta^{\text{eff}}_{ab}(\omega - i0) + N_{ab}^{-1/2}\, \text{Im}\, V^{\text{eff}}_{ab}(\omega - i0)\, N_{ab}^{1/2}\,.$$

In eq. (4.189), the terms $\bar{H}^{\text{pl}}_{ab}$ and $\text{Im}\, H^{\text{pl}}_{ab}(\omega - i0)$ are Hermitean operators because of the relations

$$\bar{H}^{pl}_{ab}(12\bar{3}\bar{4}) = \bar{H}^{pl*}_{ab}(\bar{3}\bar{4}12), \quad \text{Im}\, H^{\text{pl}}_{ab}(12\bar{3}\bar{4}, \omega - i0) = \text{Im}\, H^{\text{pl}*}_{ab}(\bar{3}\bar{4}12, \omega - i0)\,.$$

Here $\text{Im}\, H^{\text{pl}}_{ab}$ depends also on the energy ω like on any real parameter. To determine the spectral function (4.189) we need $G_2(\omega \pm i0)$ which is determined by the BSE (4.185) after analytical continuation from Ω_λ to the complex Ω-plane:

$$\{H^{\text{eff}}_{ab} - \hbar(\omega \pm i0)\}\, \bar{G}_2(\omega \pm i0) \pm i\, \text{Im}\, H^{\text{pl}}_{ab}(\omega - i0)\, \bar{G}_2(\omega \pm i0) = i\,. \tag{4.190}$$

The effective Hamiltonian $H^{\text{eff}}_{ab}(\omega)$ is given by

$$H^{\text{eff}}_{ab}(\omega) = H^{(0)}_{ab} + \text{Re}\, H^{\text{pl}}_{ab}(\omega) = H^{(0)}_{ab} + \bar{H}^{pl}_{ab} + P \int \frac{d\bar{\omega}}{\pi} \frac{\text{Im}\, H^{\text{pl}}_{ab}(\bar{\omega} - i0)}{\omega - \bar{\omega}}\,. \tag{4.191}$$

In this way, we have related the BSE which depends on the Matsubara frequency Ω_λ to an equation which depends on the real parameter ω.

4.4.2. Solution of the Bethe-Salpeter Equation. Effective Wave Equation and Spectral Representations

Let us first discuss the case of an isolated two-particle system, i.e., we put $H^{\text{pl}}_{ab}(\Omega)$ equal to zero. In this case, we associate the homogeneous BSE

$$\{H^{(0)}_{ab} - \Omega\}\, \psi^{(0)}_{ab}(\Omega) = 0 \tag{4.192}$$

to the inhomogeneous BSE (4.185). This homogeneous BSE has the meaning of a wave equation and is identical with the Schrödinger equation for the two-particle system if we consider Ω as a real variable; normalized solutions $\psi^{(0)}_{nP} = \psi^{(0)}_{ab}(12, E^{(0)}_{nP})$ are obtained only for discrete values $E^{(0)}_{nP}$. With the help of the solutions of the homogeneous BSE, the Green's function $G_{2,0}(\Omega)$ for an isolated two-particle system may be represented by a bilinear expansion

$$G_{2,0}(121'2', \Omega) = \sum_{nP} \frac{\psi^{(0)}_{nP}(12)\, \psi^{(0)*}_{nP}(1'2')}{\hbar\Omega - E^{(0)}_{nP}} \tag{4.193}$$

and it is easily shown that at the values $\Omega = \Omega_\lambda$ this expression is the solution of the inhomogeneous BSE. The bilinear expansion has a similar structure as the free one-particle Green's function.

The bilinear expansion of the two-particle Green's function $G_2^{(0)}$ may be found in the following way: We use the complete orthonormalized set of eigenstates $\psi_{nP}^{(0)}$ with the corresponding eigenvalues $E_{nP}^{(0)}$ to give a representation of the BSE of the isolated two-particle problem:

$$\{E_{nP}^{(0)} - (\omega \pm i0)\}\, \overline{G}_2(n, P, n', P', \omega \pm i0) = i\delta_{nn'}\delta_{PP'}\,; \tag{4.194}$$

the solutions are the matrix elements of $\overline{G}_2$ which are diagonal in the representation according to $H_{ab}^{(0)}$.

$$\overline{G}_2(n, P, n', P', \omega \pm i0) = \frac{i\delta_{nn'}\delta_{PP'}}{E_{nP}^{(0)} - \hbar(\omega \pm i0)}\,. \tag{4.195}$$

From this expression we find the matrix elements of the spectral function, see Section 3.2.6.:

$$A(n, P, n', P', \omega) = \delta_{nn'}\delta_{PP'}\,\delta(\omega - E_{nP}^{(0)})\,. \tag{4.196}$$

After that, we can interpret $E_{nP}^{(0)}$ as the two-particle excitation energies of the system. Because we consider the case of an isolated two-particle system without any interactions with the surrounding, the spectral function consists of sharp δ-like peaks with zero bandwidth describing undamped quasiparticles.

A similar discussion is possible if we consider $H_{ab}^{\text{pl}}(\Omega)$ in Hartree-Fock approximation. The energy eigenvalues of the homogeneous wave equation which is associated to the inhomogeneous BSE are obtained in perturbation theory by

$$E_{nP} = E_{nP}^{(0)} + \langle nP|\, H_{ab}^{\text{pl}}(E_{nP})\, |nP\rangle \tag{4.197}$$

and we can give in this approximation the Green's function by the bilinear expansion

$$G_2(121'2', \Omega) = \sum_{nP} \frac{\psi_{nP}^{(0)}(12)\, \psi_{nP}^{(0)*}(1'2')}{\hbar\Omega - E_{nP}^{(0)} - \langle nP|\, H_{ab}^{\text{pl}}|nP\rangle}\,. \tag{4.198}$$

The spectral function consists in this approximation on δ-like peaks at the Hartree-Fock quasiparticle energies.

The concept of the bilinear expansion (4.198) breaks down in the case that $H_{ab}^{\text{pl}}(\Omega)$ contains an imaginary part. In this case the sign of the imaginary part in (4.197) has to be fixed correctly so that damped quasiparticles are described. However it seems to be adequate to avoid the bilinear expansion in this case and to work directly with the spectral function for $G_2(\Omega)$, see Kilimann, Kremp and Röpke (1983).

For this, we consider the homogeneous wave equation which is defined by the real part H_{ab}^{eff} (4.191) of the inhomogeneous BSE which depends on the energy ω as on a parameter:

$$H_{ab}^{\text{eff}}(\omega)\, |nP, \omega\rangle = E_{nP}(\omega)\, |nP, \omega\rangle\,. \tag{4.199}$$

Because the effective Hamiltonian $H_{ab}^{\text{eff}}(\omega)$ is Hermitean for any real value ω, the relations of orthonormality and completeness of the set of basis states hold

$$\sum |nP, \omega\rangle \langle \omega, nP| = 1\,, \qquad \langle \omega, P'n' \mid nP, \omega\rangle = \delta_{nn'}\delta_{PP'}$$

and we define the wave function according to

$$\langle p'P' \mid nP, \omega\rangle = \delta_{PP'}\, \psi_n(p', P; \omega)\,.$$

We point out that the eigenvalues $E_{nP}(\omega)$ of the Hamiltonian $H^{\text{eff}}_{ab}(\omega)$ do not represent immediately the spectrum of the two-particle excitations. The two-particle excitation spectrum is obtained below from the spectral function.

To solve the BSE (4.190) we consider the representation with respect to $H^{\text{eff}}_{ab}(\omega)$ which reads after taking into account the conservation of the total momentum P (only diagonal elements with respect to P)

$$\{E_{nP}(\omega) - \hbar(\omega \pm i0)\}\, \overline{G}_2(n, n', P, \omega + i0) \pm i \sum_m \langle\omega, Pn|\, \mathrm{Im}\, H^{\text{pl}}_{ab}(\omega - i0)\, |mP, \omega\rangle \cdot \overline{G}_2(m, n', P, \omega \pm i0) = i\delta_{nn} \,. \tag{4.200}$$

If we consider the non-diagonal matrix elements of $\mathrm{Im}\, H^{\text{pl}}_{ab}$ as small quantities, the matrix equation (4.200) has the approximative solution (no degenerate bound states)

$$\overline{G}_2(n, n', P, \omega \pm i0) = i \frac{\delta_{nn'}}{E_{nP}(\omega) - \hbar(\omega \pm i0) \pm i\,\Gamma_{nn}(P, \omega)} \mp \frac{\Gamma_{nn'}(P, \omega)\,(1 - \delta_{nn'})}{(E_{nP}(\omega) - \hbar(\omega \pm i0) \mp i\Gamma_{nn}(P, \omega))(E_{n'P}(\omega) - \hbar(\omega \pm i0) \pm i\Gamma_{n'n'}(P, \omega))}\,, \tag{4.201}$$

with

$$\Gamma_{nn'}(P, \omega) = \langle\omega, nP|\, \mathrm{Im}\, H^{\text{pl}}_{ab}(\omega - i0)\, |Pn', \omega\rangle \tag{4.202}$$

for bound states and

$$\Gamma_{pp'}(P, \omega) = \delta_{pp'}\; \mathrm{Im}\, \Delta^{\text{eff}}_{ab}\,(p, P, \omega - i0) \tag{4.203}$$

for scattering states. The original Green's function G_2 is then given by the operator

$$G_2(\omega \pm i0) = \sum_{nn'P} N^{1/2}_{ab}\, |nP\omega\rangle\, \overline{G}_2(n, n', P, \omega \pm i0)\, \langle\omega Pn'|\, N^{1/2}_{ab} \,. \tag{4.204}$$

In the non-degenerate case where $N_{ab} \approx 1$ and neglecting the non-diagonal elements $\Gamma_{nn'}$ $(n \neq n')$ we find for the spectral function A_2

$$A_2(n, P, \omega) = -2i \frac{\Gamma_{nn}(P, \omega)}{(E_{nP}(\omega) - \hbar\omega)^2 + \Gamma^2_{nn}(P\omega)} \,. \tag{4.205}$$

This expression represents the diagonal part of the spectral function. If the non-diagonal elements $\Gamma_{nn'}$ in (4.201) are taken into account which correspond to transitions between two-particle states, in (4.205) arise additional contributions which are connected with the incoherent part of the spectral function.

From the spectral function (4.205) we deduce the spectrum of the two-particle excitations by the solutions $\tilde{E}_{nP}$ of the dispersion relation

$$\hbar\omega = E_{nP}(\omega) \tag{4.206}$$

and the damping of these quasiparticle excitations by $\Gamma_{nP}(\omega = \tilde{E}_{nP})$.

Solving the wave equation (4.199) by perturbation technique, we find for the bound-state part of the spectrum

$$\tilde{E}_{nP} \approx E^{(0)}_{nP} + \langle Pn|\, \overline{H}^{\text{eff}}_{ab} + P \int \frac{\mathrm{d}\overline{\omega}}{\pi} \frac{\mathrm{Im}\, H^{\text{pl}}_{ab}\,(\overline{\omega} - i0)}{E^{(0)}_{nP} - \hbar\overline{\omega}}\, |nP\rangle \tag{4.207}$$

and for the scattering state part

$$\tilde{E}_{pP} \approx \frac{\hbar^2 p^2}{2m_{ab}} + \frac{P^2}{2M} + \mathrm{Re}\, \Delta^{\text{eff}}_{ab}\left(p, P, \frac{p^2}{2m_{ab}} + \frac{P^2}{2M}\right). \tag{4.208}$$

This different behaviour of bound states and of scattering state energy values leads to the decrease of the ionization energy and to the Mott transition as will be discussed below, see also Section 3.2.8.

The different behaviour of bound and scattering states can also be discussed from the homogeneous wave equation associated to the BSE (4.200):

$$\begin{aligned}(\varepsilon_a(1) + \varepsilon_b(2) - \hbar z)\,\psi_{ab}(p_1\,p_2, z) &- \sum_q V(q)\,\psi_{ab}(p_1 + q, p_2 - q, z)\\ = \sum_q V(q)\,\{&\underbrace{[N_{ab}(p_1, p_2) - 1]\,\psi_{ab}(p_1 + q, p_2 - q, z)}_{\text{(i)}} - \underbrace{[N_{ab}(p_1 + q, p_2 - q) - 1]\,\psi_{ab}(p_1, p_2, z)}_{\text{(ii)}}\\ + \sum_q \Delta V_{ab}^{\text{eff}}\,(p_1, p_2, q, z)\,\{&\underbrace{N_{ab}(p_1 p_2)\,\psi_{ab}(p_1 + q, p_2 - q, z)}_{\text{(iv)}} - \underbrace{N_{ab}(p_1 + q, p_2 - q)\,\psi_{ab}(p_1, p_2, z)\}}_{\text{(iii)}}\end{aligned} \tag{4.209}$$

where (4.188) has been used, and $\Delta V_{ab}^{\text{eff}} = V_{ab}^{\text{eff}} - V_{ab}$. This wave equation may be interpreted as the many-particle version of the two-particle Schrödinger equation The "isolated two-particle problem" is described by the left-hand side. Many-body effects are condensed into H_{ab}^{pl} which consists of four contributions:

(i) phase space occupation (statistical correlation),
(ii) exchange self-energy,
(iii) dynamical self-energy correction, and
(iv) dynamically screened effective potential.

It can easily be seen that a compensation acts between (i) and (ii) and between (iii) and (iv), since the curly brackets in (4.209) vanish both at $q = 0$. Thus, all of these terms must be taken into account for the determination of the two-particle spectrum. This means also in general that consistent approximations in the self energy Σ as well as in the effective interaction K have to be performed. The compensation acts especially for strongly bound states where the wave function is sharply localized in coordinate space so that its Fourier transform varies not strongly with q. This compensations do not occur for scattering states.

4.4.3. Two-Particle States in the Dynamically Screened Ladder Approximation

In the simplest approximation we take for K the dynamically screened potential,

$$iK \approx i V_{ab}^{\text{s}} = \tag{4.210}$$

where

$$V_{ab}^{\text{s}}(q, z) = \frac{V_{ab}(q)}{\varepsilon(q, z)} = V_{ab}(q)\left\{1 + \int \frac{d\overline{\omega}}{\pi}\,\frac{\operatorname{Im} \varepsilon^{-1}(q, \overline{\omega} - i0)}{z - \overline{\omega}}\right\} \tag{4.211}$$

so that

$$\operatorname{Im} V_{ab}^{\text{s}}(q, \omega - i0) = V_{ab}(q)\,\operatorname{Im} \varepsilon^{-1}(q, \omega - i\varepsilon)\,. \tag{4.212}$$

Then the self-energy Σ has to be taken in the V^{s}-approximation:

$$\Sigma = \quad = \Sigma^{\text{HF}} + \Sigma^{\text{MW}}\,, \tag{4.213}$$

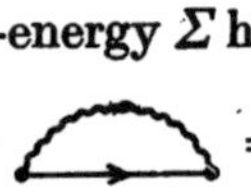

where the Hartree-Fock contribution is given by Σ^{HF} and Σ^{MW} denotes the Montroll-Ward contribution.

The evaluation of the self-energy yields, see 4.3.,

$$\Sigma_a(p, z) = \sum_q V(q) \left[f_a(p + q) + \int_{-\infty}^{\infty} \frac{d\omega}{\pi} \operatorname{Im} \varepsilon^{-1}(q, \omega) \frac{1 - f_a(p + q) + n_B(\omega)}{\hbar z - \hbar\omega - \varepsilon_a(p + q)} \right], \tag{4.214}$$

where the first term on the right-hand side is the exchange self-energy, whereas the second term is the correlation part.

Also the effective potential is calculated by carrying out the frequency summations, and after a lengthy calculation we obtain from (4.181) the analytical continuation

$$\Delta V_{ab}^{\text{eff}} (p_1, p_2, q, z) = V(q) \{[1 - f_a(p_1) - f_b(p_2)] [1 - f_a(p_1 + q) - f_b(p_2 - q)]\}^{-1}$$
$$\cdot \int_{-\infty}^{\infty} \frac{d\omega}{\pi} \operatorname{Im} \varepsilon^{-1}(q, \omega + i0) \left\{ \frac{n_B(\omega) [f_a(p_1 + q) - f_a(p_1) + f_a(p_1 + q) [1 - f_a(p_1)]}{\hbar\omega - \varepsilon_a(p_1 + q) + \varepsilon_a(p_1)} \right.$$
$$+ \frac{n_B(\omega) [1 - f_a(p_1) - f_b(p_2 - q)] + [1 - f_a(p_1)] [1 - f_b(p_2 - q)]}{\hbar z - \hbar\omega - \varepsilon_a(p_1) - \varepsilon_b(p_2 - q)}$$
$$\left. + (a \leftrightarrow b, p_1 \leftrightarrow p_2, q \leftrightarrow -q) \right\} . \tag{4.215}$$

This result may be discussed in the way that the first contribution describes the interactions with the density fluctuations during which the particle changes its momentum, and the second one describes the virtual excitation of the two-particle state into two free particle states. The numerators contain the Pauli occupation factors for the corresponding process.

Altogether, we have in the dynamically screened ladder approximation

$$H_{ab}^{\text{pl}}(12\bar{3}\bar{4}, \Omega_\lambda) = \delta_{\sigma_1\bar{\sigma}_3} \delta_{\sigma_2\bar{\sigma}_4} H_{ab}^{\text{pl}}(p_1, p_2, \bar{p}_3, \bar{p}_4, \Omega_\lambda) \tag{4.216}$$

with

$$\bar{H}_{ab}^{\text{eff}}(p_1, p_2, \bar{p}_3, \bar{p}_4) = [\Sigma_a^{\text{HF}}(p_1) + \Sigma_b^{\text{HF}} (p_2)] \delta_{p_1\bar{p}_3} \delta_{p_2\bar{p}_4}$$
$$- N_{ab}^{1/2}(p_1, p_2) V_{aa}(p_1, p_2, \bar{p}_3, \bar{p}_4) N_{ab}^{1/2}(\bar{p}_3, \bar{p}_4) - V_{aa}(p_1, p_2, \bar{p}_3, \bar{p}_4) \tag{4.217}$$

and

$$\operatorname{Im} H_{ab}^{\text{pl}}(p_1, p_2, \bar{p}_3, \bar{p}_4, \omega - i0) = \delta_{p_1\bar{p}_3} \delta_{p_2\bar{p}_4} N_{ab}^{-1}(p_1, p_2) \sum_q [\operatorname{Im} V_{aa}^{s}(q, \omega$$
$$- \varepsilon_a(p_1 - q) - \varepsilon_b(p_2)) (1 + n_B(\omega - \varepsilon_a(p_1 - q) - \varepsilon_b(p_2))) + (a \leftrightarrow b, p_1 \leftrightarrow p_2)]$$
$$- \delta_{p_1+p_2, \bar{p}_3+\bar{p}_4} N_{ab}^{-1/2}(p_1, p_2) N_{ab}^{-1/2}(\bar{p}_3, \bar{p}_4) [\operatorname{Im} V_{aa}^{s}(p_1 - \bar{p}_3, \omega - \varepsilon_a(p_1) - \varepsilon_b(\bar{p}_4)]$$
$$\cdot (1 + n_B(\omega - \varepsilon_a(p_1) - \varepsilon_b(\bar{p}_4))) + (a \leftrightarrow b, p_1 \leftrightarrow \bar{p}_3, p_2 \leftrightarrow \bar{p}_4)] . \tag{4.218}$$

The first contributions in (4.217) and (4.218) are obtained from the self energy in Δ_{ab}^{eff} and the last ones from V_{ab}^{eff}.

A further evaluation may be performed in the plasmon-pole approximation for $\operatorname{Im} \varepsilon^{-1}(q, \omega)$, see 4.2. Our aim is to be correct up to $n^{-1/2}$.

$$\operatorname{Im} \varepsilon^{-1}(q, \omega + i0) = -\pi \frac{\omega_{\text{pl}}^2}{2\omega q} \{\delta(\omega - \omega_q) - \delta(\omega + \omega_q)\} ,$$
$$\omega_{\text{pl}}^2 = \frac{ne^2}{\varepsilon_0 m_{ab}}, \quad \omega_q^2 = \omega_{\text{pl}}^2(1 + q^2/\varkappa^2) + \left(\frac{\hbar}{2} q^2/2m_{ab}\right)^2 . \tag{4.219}$$

In this approximation the ω-integral can be performed, replacing the integrand by a suitably chosen mean value.

In the non-degenerate case, the Fermi functions can be neglected with respect to unity, and we arrive at

$$\Delta V^{\text{eff}}_{ab}(p_1p_2qz) = V(q)\,\frac{\omega_{\text{pl}}^2}{2\omega_q}\left[\frac{1+n_{\text{B}}(\omega_1)}{z-\omega_1-\varepsilon_a(p_1)-\varepsilon_b(p_2-q)}\right.$$
$$\left.+\frac{n_{\text{B}}(\omega)}{z+\omega_q-\varepsilon_a(p_1)-\varepsilon_b(p_2-q)}+(a\leftrightarrow b,\,p_1\leftrightarrow p_2,\,q\leftrightarrow -q)\right]. \tag{4.220}$$

Physically, the expressions entering (4.220) may be interpreted in terms of virtual processes: The first term corresponds to the decay of a two-particle state (z) into a free pair with stimulated emission of a plasmon. The inverse process is the absorption of a thermally excited plasmon, again with a free final state (second term). A more elaborate version of the theory which shows that also the final state should be a true two-particle state (bound or scattering state) is given in the next Section, see RÖPKE et al. (1980). Numerical results are given in Section 4.4.5. As pointed out in 4.1.4., the approximation discussed here can be interpreted as the V^{S}-approximation for the two-particle self energy (4.28).

4.4.4. Two-Particle States in Surrounding Medium in First Born Approximation

The dynamically screened ladder approximation described in the previous Section can be improved in that way that all contributions in first Born approximation are taken into account which describe collisions between the considered two-particle system and free as well as bound states contained in the surrounding matter, so that all terms linear in the density of free particles and of bound states are taken into account.

The correct evaluation of Σ_a and K to obtain all terms linear in the free particle und bound state densities would require the solution of the three-body and the four-body problem, respectively. We restrict ourselves to a first Born approximation which means that besides the ladder T-matrix (in the backward going lines, which give the density of the corresponding particle)

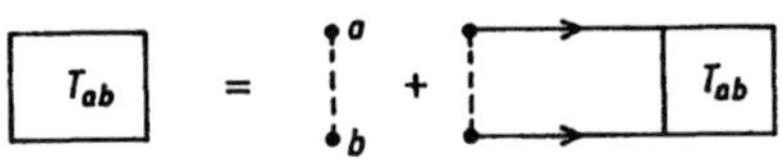

(4.221)

all diagrams with at most one explicit interaction line are taken into acoount:

$\Sigma_a =$

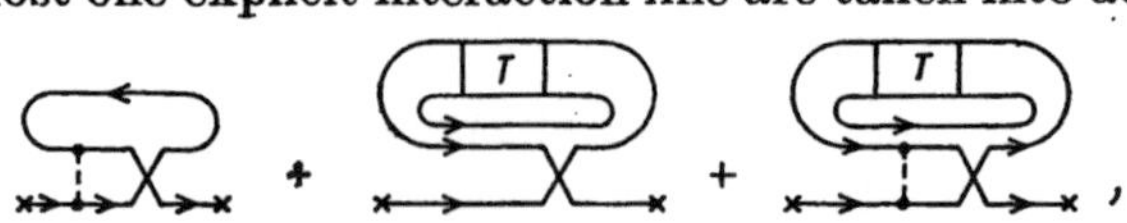

(4.222)

$K =$

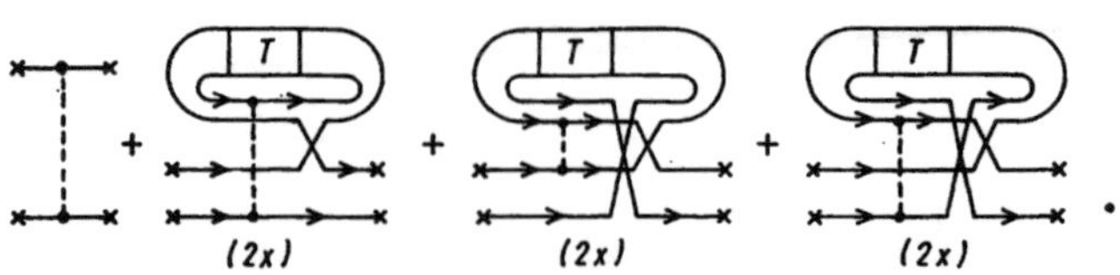

Such diagrams exceed the ordinary Born approximation; in the frame of the chemical picturet he backward going lines include composites. Furthermore we will replace the bare Coulomb potential by the dynamically screened Coulomb potential V^{s}. So we arrive at the following approximations for Σ and K up to first Born approximation with respect to V^s (in topological equivalence to (4.222)):

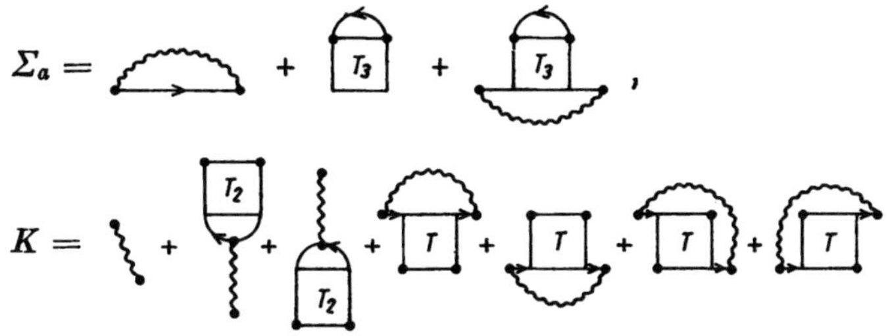

(4.223)

where $T_2(T_3)$ starts with two (three) interaction lines. By the introduction of the dynamically screened potential the influence of collective excitations as, e.g., plasma oscillations, on the two-particle system is taken into account. This means a partial summation of higher-order contributions which must be done to avoid the divergency arising in the second Born approximation due to the long-range behaviour of the Coulomb potential.

The T-matrices occurring in (4.222) and (4.223) may be expressed by unperturbed antisymmetrized two-particle states $\psi_{nP}^{(0)}(p_1, p_1 + p_2)$ and two-particle energies $E_{nP}^{(0)}$:

$$T_{ab}(p, P - p, p + q, P - p - q, \Omega_\lambda)$$
$$= \sum_n \frac{\psi_{nP}^{(0)}(p, P)\, \psi_{nP}^{(0)*}(p + q, P)}{\hbar\Omega_\lambda - E_{nP}^{(0)}} \left(E_{nP}^{(0)} - \varepsilon_a(p) - \varepsilon_b(P - p)\right) \left(\hbar\Omega_\lambda - \varepsilon_a(p + q)\right.$$
$$\left. - \varepsilon_b(P - p - q)\right). \tag{4.224}$$

The evaluation of the contributions presented in (4.224) is performed straightforward by standard methods. For the self energy we obtain

$$\Sigma_a(p, z_\nu^a) = - \sum_q V_{aa}(q) \left\{ f_a(p + q) + \int_{-\infty}^{\infty} \frac{d\hbar\omega}{\pi} \operatorname{Im} \varepsilon^{-1}(q, \omega + i0) \frac{1 + n_B(\hbar\omega)}{\hbar z_\nu^a - \hbar\omega - \varepsilon_a(p + q)} \right\}$$
$$+ \sum_{nPb\sigma_b} |\psi_{nP}^{(0)}(p, P)|^2 \frac{(\varepsilon_a(p) + \varepsilon_b(P - p) - E_{nP}^{(0)})^2}{\hbar z_\nu^a - E_{nP}^{(0)} + \varepsilon_b(P - p)} \{g_{ab}(E_{nP}^{(0)}) + f_b(P - p)\}$$
$$+ \sum_{nPqb\sigma_b} |\psi_{nP}^{(0)}(p + q, P)|^2\, V_{aa}(q)\, g_{ab}(E_{nP}^{(0)}), \tag{4.225}$$

whereas we get for the effective interaction (total momentum equal to zero)

$$V_{ab}^{\text{eff}}(p, -p, p + q, -p - q, \Omega_\lambda) = V_{ab}(q) \left\{ 1 + \int_{-\infty}^{\infty} \frac{d\hbar\omega}{\pi} \operatorname{Im} \varepsilon^{-1}(q, \omega + i0) \right.$$
$$\left. \cdot \left[\frac{1 + n_B(\hbar\omega)}{\hbar\Omega_\lambda - \hbar\omega - \varepsilon_a(p + q) - \varepsilon_b(-p)} + \frac{1 + n_B(\hbar\omega)}{\hbar\Omega_\lambda - \hbar\omega - \varepsilon_a(p) - \varepsilon_b(-p - q)} \right] \right\}$$
$$- \sum_{nPc\sigma_c} V_{cb}(q)\, g_{ac}(E_{nP}^{(0)})\, \psi_{nP,ac}^{(0)}(pP)\, \psi_{nP,ac}^{(0)*}(p + q, P)$$
$$- \sum_{nPc\sigma_c} V_{ac}(q)\, g_{cb}(E_{nP}^{(0)})\, \psi_{nP,cb}^{(0)}(P + p, P)\, \psi_{nP,cb}^{(0)*}(P + p + q, P)$$

$$- \sum_{nP} V_{ab}(P) \left\{ g_{ab}(E_{nP}^{(0)}) + \int_{-\infty}^{\infty} \frac{d\hbar\omega}{\pi} \operatorname{Im} \varepsilon^{-1}(P, \omega + i0) \frac{1 + n_B(\hbar\omega) + g_{ab}(E_{nP}^{(0)})}{\hbar\Omega_\nu - t\omega - E_{nP}^{(0)}} \right\}$$

$$\cdot \{\psi_{nP,ab}^{(0)}(P + p, P) - \psi_{nP,ab}^{(0)}(p, P)\} \{\psi_{nP,ab}^{(0)}(P + p + q, P) - \psi_{nP,ab}^{(0)*}(p + q, P)\}$$

$$+ \sum_{q'} V_{ab}(q') \int_{-\infty}^{\infty} \frac{d\hbar\omega}{\pi} \operatorname{Im} \varepsilon^{-1}(q', \omega + i0) (1 + n_B(\hbar\omega))$$

$$\cdot \left\{ \frac{\delta_{q,0} - \delta_{q,-q'}}{\hbar\Omega_\lambda - \hbar\omega - \varepsilon_a(p - q') - \varepsilon_b(-p)} + \frac{\delta_{q,0} - \delta_{q,-q'}}{\hbar\Omega_\lambda - \hbar\omega - \varepsilon_a(p) - \varepsilon_b(-p - q')} \right\}. \quad (4.226)$$

From these expressions for Σ_a and V_{ab}^{eff} we can derive the plasma Hamiltonian for $p_1 = -p_2 = p$, $\bar{p}_3 = -\bar{p}_4 = p + q$; see KILIMANN, KREMP and RÖPKE (1983) and (4.188):

$$\bar{H}_{ab}^{\text{pl}}(p, -p, p + q, -p - q) = \delta_{q,0}[\Sigma_a^{\text{HF}}(p) + \Sigma_b^{\text{HF}}(-p) + 4 \sum_{n\bar{P}qcd} |\psi_{n\bar{P},cd}^{(0)}(p + \bar{q})|^2$$

$$\cdot V_{aa}(\bar{q})\, g_{cd}(E_{n\bar{P}}^{(0)})] + N_{ab}^{1/2}(p, -p)\, V_{ab}(q)\, N_{ab}^{1/2}(p + q, -p - q) - V_{ab}(q)$$

$$+ \sum_{n\bar{P}} V_{ab}(\bar{P})\, g_{ab}(E_{n\bar{P}}^{(0)}) [\psi_{n0,ab}^{(0)}(p + \bar{P}) - \psi_{n0,ab}^{(0)}(p)] [\psi_{n0,ab}^{(0)*}(\bar{P} + p + q) - \psi_{n0,ab}^{(0)*}(p + q)]$$

$$-2 \sum_{nPc} V_{cb}(q)\, g_{ac}(E_{nP}^{(0)})\, \psi_{n0,ac}^{(0)}(p)\, \psi_{n0,ac}^{(0)*}(p + q)$$

$$-2 \sum_{nPc} V_{ac}(q)\, g_{cb}(E_{nP}^{(0)})\, \psi_{n0,cb}^{(0)}(p + P)\, \psi_{n0,cb}^{(0)*}(P + p + q) \,. \quad (4.227)$$

For the imaginary part of H_{ab}^{pl} we have

$$\operatorname{Im} H_{ab}^{\text{pl}}(p, -p, p + q, -p - q, \omega - i0) = \delta_{q,0} \operatorname{Im} \Delta_{ab}^{\text{eff}}(p, \omega - i0)$$

$$+ N_{ab}^{1/2}(p, -p) \operatorname{Im} V_{ab}^{\text{eff}}(p, -p, p + q, -p - q; \omega + i0)\, N_{ab}^{1/2}(p + q; p - q) \quad (4.228)$$

with

$$\operatorname{Im} \Delta_{ab}^{\text{eff}}(p, \omega - i0) = [\operatorname{Im} \Sigma_a(p; \omega - \varepsilon_a(p) - i0) + \operatorname{Im} \Sigma_b(-p, \omega - \varepsilon_b(p) - i0)]$$

$$\cdot N_{ab}^{-1}(p, -p) \,,$$

$$\operatorname{Im} \Sigma_a(p, \omega - i0) = \sum_{\bar{q}} \operatorname{Im} V_{aa}^{s}(\bar{q}, \omega + \varepsilon_a(p + \bar{q}) - i0) [1 + n_B(\omega + \varepsilon_a(p + \bar{q}))]$$

$$+ 2\pi \sum_{nPc} (\psi_{n0,ac}^{(0)}(p, 0))^2 (\varepsilon_a(p) + \varepsilon_c(P - p) - E_{nP}^{(0)})^2 [g_{ac}(E_{nP}^{(0)}) + f_c(P - p)] \quad (4.229)$$

and

$$\operatorname{Im} V_{ab}^{\text{eff}}(p, -p, p + q, -p - q; \omega - i0) = \operatorname{Im} V_{ab}^{s}(q, \omega - \varepsilon_a(p + q) - \varepsilon_b(-p))$$

$$\cdot [1 + n_B(\omega - \varepsilon_a(p + q) - \varepsilon_b(-p))] + (a \leftrightarrow b, p \leftrightarrow p + q)$$

$$- \sum_{q'} \operatorname{Im} V_{ab}^{s}(q', \hbar\omega - \varepsilon_a(p - q') - \varepsilon_b(p)) \{1 + n_B(\hbar\omega - \varepsilon_a(p - q') - \varepsilon_b(-p)\}$$

$$\cdot \{\delta_{q,0} - \delta_{q,-q'}\} - \sum_{q'} \{a \leftrightarrow b, p \leftrightarrow p + q'\} \,. \quad (4.230)$$

The effect of the surrounding plasma is described by an effective plasma Hamiltonian which contains the Fermi and Bose distribution functions for free and bound states, respectively. Instead of the chemical potentials μ_a, μ_b, we can introduce the densities of free pairs n_f and of bound states n_{ab}, which are determined by a mass action law (see Chapter 6.) We will consider the low-density relations

$$f_a(p) = \tfrac{1}{2}\, n_f\, \Lambda_a^3 \exp\left(-\beta\, \hbar p^2/2m_a\right), \qquad g_{ab}(E_{nP}) = \tfrac{1}{4}\, n_{ab}\, \Lambda_{ab}^3 \exp\left(-\beta\hbar^2 P^2/2M\right) \delta_{n,0} \tag{4.231}$$

with

$$M = m_a + m_b\,, \quad \Lambda_a = \hbar\sqrt{2\pi\beta/m_a}\,, \qquad \Lambda_{ab} = \hbar\sqrt{2\pi\beta/M}\,.$$

All bound states are assumed to be in the ground state $n = 0$.

We derive an expression for the ground state energy shift

$$\Delta E_0 = \Delta E_{0,\,\text{band filling}} + \Delta E^{(\text{HF})}_{0,\,\text{self energy}} + \Delta E^{(\text{HF})}_{0,\,\text{eff. int.}} + \Delta E_{0,\,\text{plasmon.}} \tag{4.232}$$

In the first contribution to ΔE_0 which corresponds to band filling effects, we expand the one-particle Green's function for small densities (i.e. small Σ):

$$\Delta E_{0,\,\text{band filling}} = \sum_{pq} V(q)\, \psi_0(p)\, \psi_0(p+q)\, \{f_a(p) + f_b(-p)\}$$
$$+ 2 \sum_{Ppq} V(q)\, \psi_0(p)\, \psi_0(p+q) \left\{\left|\psi_0\left(p - \frac{m_a}{M} P\right)\right|^2 + \left|\psi_0\left(p + \frac{m_b}{M} P\right)\right|^2\right\} g_{ab}(E_{nP}). \tag{4.233}$$

This contribution may be interpreted in the following way: Many-body effects such as phase space occupation (statistical correlations) are caused not only by free particles but also by bound states which are distributed over the momentum space according to their wave functions.

For the Hartree-Fock self-energy contribution we obtain

$$\Delta E^{(\text{HF})}_{0,\,\text{self energy}} = -\sum_{pq} V(q)\, |\psi_0(p)|^2\, \{f_a(p-q) + f_b(-p-q)\}$$
$$+ \sum_{pq} V(q)\, |\psi_0(p)^2| \left\{\psi_0\left(p - \frac{m_a}{M} P\right) \psi_0\left(p + q - \frac{m_a}{M} P\right)\right.$$
$$\left. + \psi_0\left(p + \frac{m_b}{M} P\right) \psi_0\left(p + q + \frac{m_b}{M} P\right)\right\}$$
$$\cdot 2 g_{ab}(E_{nP}) - \sum_{Ppq} V(q)\, |\psi_0(p)|^2 \left\{\left|\psi_0\left(p + q - \frac{m_a}{M} P\right)\right|^2\right.$$
$$\left. + \left|\psi_0\left(p + q + \frac{m_b}{M} P\right)\right|^2\right\} 2 g_{ab}(E_{nP})\,, \tag{4.234}$$

and the Hartree-Fock contributions of the effective interaction are immediately derived from (4.227):

$$\Delta E^{(\text{HF})}_{0,\,\text{eff. int.}} = -\sum_{Ppq} V(q)\, \psi_0(p)\, \psi_0(p+q)\, 2 g_{ab}(E_{nP})$$
$$\cdot \left\{\psi_0\left(p - \frac{m_a}{M} P\right) \psi_0\left(p + q - \frac{m_a}{M} P\right) + \psi_0\left(p + \frac{m_b}{M} P\right) \psi_0\left(p + q + \frac{m_b}{M} P\right)\right\}$$
$$- \sum_{Ppq} V(-P)\, \psi_0(p)\, \psi_0(p+q)\, g_{ab}(E_{nP})$$
$$\cdot \left\{\psi_0\left(p - \frac{m_a}{M} P\right) - \psi_0\left(p + \frac{m_b}{M} P\right)\right\} \left\{\psi_0\left(p + q - \frac{m_a}{M} P\right) - \psi_0\left(p + q + \frac{m_b}{M} P\right)\right\}. \tag{4.235}$$

The plasmon contributions to the self energy and the effective interaction are combined into

$$\Delta E_{0,\,\mathrm{plasmon}} = -\sum_{nPpq} V(-P)\,\psi_0(p)\,\psi_0(p+q) \int_{-\infty}^{\infty} \frac{\mathrm{d}\hbar\omega}{\pi}\,\mathrm{Im}\,\varepsilon^{-1}(-P, \hbar\omega + i0)$$

$$\cdot \frac{1 + n_\mathrm{B}(\hbar\omega) + g_{ab}(E_{nP})}{E_0 - \hbar\omega - E_{nP}}$$

$$\cdot \left\{\psi_n\left(p - \frac{m_a}{M}P\right) - \psi_n\left(p + \frac{m_b}{M}P\right)\right\}\left\{\psi_n\left(p + q - \frac{m_a}{M}P\right) - \psi_n\left(p + q + \frac{m_b}{M}P\right)\right\}. \tag{4.236}$$

Using the single plasmon pole approximation (4.236), see Section 4.2., we arrive at

$$\Delta E_{0,\,\mathrm{plasmon}} = -\sum_{npqq'} V(q')\,\psi_0(p)\,\psi_0(p+q)\,\frac{\omega_\mathrm{pl}^2}{2\omega_{q'}}$$

$$\cdot \left\{\frac{n_\mathrm{B}(\hbar\omega_{q'}) - g_{ab}(E_{nq'})}{E_0 + \hbar\omega_{q'} - E_{nq'}} + \frac{1 + n_\mathrm{B}(\hbar\omega_{q'}) + g_{ab}(E_{nq'})}{E_0 - \hbar\omega_{q'} - E_{nq'}}\right\}$$

$$\cdot \left\{\psi_n\left(p + \frac{m_a}{M}q'\right) - \psi_n\left(p - \frac{m_b}{M}q'\right)\right\}\left\{\psi_n\left(p + q + \frac{m_a}{M}q'\right) - \psi_n\left(p + q - \frac{m_b}{M}q'\right)\right\}. \tag{4.237}$$

This expression may be considered as describing the processes of spontaneous and induced emission and absorption of thermal plasmons, where the statistical weights of the final states for the direct and inverse processes

$$\left(1 + n_\mathrm{B}(\hbar\omega_{q'})\right)\left(1 + g_{ab}(E_{nq'})\right) - n_\mathrm{B}(\hbar\omega_{q'})\,g_{ab}(E_{nq'}) \tag{4.238}$$

and

$$n_\mathrm{B}(\hbar\omega_{q'})\left(1 + g_{ab}(E_{nq'})\right) - \left(1 + n_\mathrm{B}(\hbar\omega_{q'})\right) g_{ab}(E_{nq'}) \tag{4.239}$$

occur in the numerator.

The result (4.237) differs from the corresponding expression in dynamically screened ladder approximation in replacing transitions from the ground state (E_0) to free particle states by transitions into two-particle states.

An alternative approach is given by Stolz, Zimmermann and Röpke (1983); there, an effective Hamiltonian is constructed from a variational principle which yields the correct Schrödinger equation for the two-particle problem in the Hartree-Fock approximation. Both approaches give identical results.

The generalization of the Hartree-Fock energy shifts of bound states in a surrounding clustered medium to systems with arbitrary cluster size has been considered for hot nuclear matter by Röpke et al. (1983).

4.4.5. Numerical Results and Discussion of the Two-Particle States

The shift and the damping of two-particle states may be evaluated by perturbation theory or by using a variational procedure to minimize the real part of the Hamiltonian

$$H(\omega, \varphi) = \langle\varphi|\, H_{ab}^{\mathrm{eff}}(\omega)\,|\varphi\rangle \tag{4.240}$$

with a trial function φ. Afterwards the energy of the lowest bound state E_0 is found by solving the self-consistent equation

$$\omega = H(\omega, \varphi_{\min}) \,. \tag{4.241}$$

The 1s-type function

$$\varphi(p) = \frac{8\sqrt{\pi a^3}}{(1 + (pa)^2)^2}\,, \qquad \varphi(r) = \frac{1}{\sqrt{\pi a^3}}\,\mathrm{e}^{-r/a} \tag{4.242}$$

is taken as trial function with the variational parameter a. For the unperturbed system, (4.242) is the exact eigenfunction with $a = a_\mathrm{B}$. In the screened ladder approximation (4.227) with (4.242) we find for zero center-of-mass momentum ($p_1 + p_2 = 0$)

$$\begin{aligned} H(\omega, a) = {} & \frac{\hbar^2}{2a^2 m_{ab}} - \frac{e^2}{4\pi\varepsilon_0 a} - \sum_{pp'} V_{aa}(p - p')\,|\varphi(p) - \varphi(p')|^2\,\frac{\hbar\omega_\mathrm{pl}^2}{2\omega_{p-p'}} \\ & \cdot \left\{ \frac{n_\mathrm{B}(\hbar\omega_{p-p'}) + 1}{\hbar^2 p'^2/2m_a + \hbar^2 p^2/2m_b + \hbar\omega_{p-p'} + \Delta_a + \Delta_b - \hbar\omega} \right. \\ & \left. + \frac{n_\mathrm{B}(\hbar\omega_{p-p'})}{\hbar^2 p'^2/2m_a + \hbar^2 p^2/2m_b - \hbar\omega_{p-p'} + \Delta_a + \Delta_b - \hbar\omega} \right\}. \end{aligned} \tag{4.243}$$

The sum of quasi-particle shifts, $\Delta_a = \Sigma_a(0, \Delta_a)$, is obtained from the self energy (4.214) in plasmon pole approximation:

$$\Delta_a + \Delta_b = -\sum_q V_{aa}(q)\,\frac{\omega_\mathrm{pl}^2}{2\omega_q} \left\{ \frac{n_\mathrm{B}(\hbar\omega_q) + 1}{\hbar^2 q^2/2m_a + \hbar\omega_q} + \frac{n_\mathrm{B}(\hbar\omega_q)}{\hbar^2 q^2/2m_a - \hbar\omega_q} + (a \leftrightarrow b) \right\}. \tag{4.244}$$

The plasma correction in (4.243) contains a typical wave function difference squared. The quadratic terms $\varphi^2(p)$ and $\varphi^2(p')$ are due to the effective self-energy shift (iii) in (4.209), whereas the mixed term, $-2\varphi(p)\,\varphi(p')$, stems from the effective potential (iv). For a bound state, both contributions compensate each other to a large extent. In the limit of small densities, one finds from (4.243)

$$E_0 = E_0^{(0)} + 0(na_\mathrm{B}^3) \,. \tag{4.245}$$

The position of the continuum edge (shift of continuum states) can be obtained by inserting the asymptotic scattering wave function $\varphi^{(0)}(p) \to \delta_{p,0}$ into (4.243) (which is equivalent to $a \to \infty$). Then, obviously, only the self-energy terms contribute, and one ends up with

$$H(\omega = \Delta_a + \Delta_b, a \to \infty) = \Delta_a + \Delta_b \tag{4.246}$$

when comparing with (4.244). Thus, the continuum shift equals the quasiparticle shift as expected. The limiting behaviour for small densities

$$\Delta_a + \Delta_b \propto -(na_b^3)^{1/2} \tag{4.247}$$

shows that the continuum shifts more rapidly than the bound state energy (4.245). The omitted factor in (4.247) does not reproduce exactly the well known Debye shift $-e^2\varkappa$ in the high-temperature limit which is due to the single plasmon pole approximation.

A similar discussion can also be applied to the Hartree-Fock self-energy and the phase space occupation term (Pauli blocking), (i) and (ii), in (4.209).

Numerical calculations were performed by ZIMMERMANN et al. (1978) for $m_a = m_b$. The equation

$$\hbar\Omega = \langle \varphi_{\min} | H^{(0)} + H^{\mathrm{pl}}(\Omega) | \varphi_{\min} \rangle \tag{4.248}$$

was solved for complex Ω. Results for the temperature $k_B T = E_0$ are shown in Fig. 4.6. As expected, the bound state energy shifts downwards only slightly, whereas the

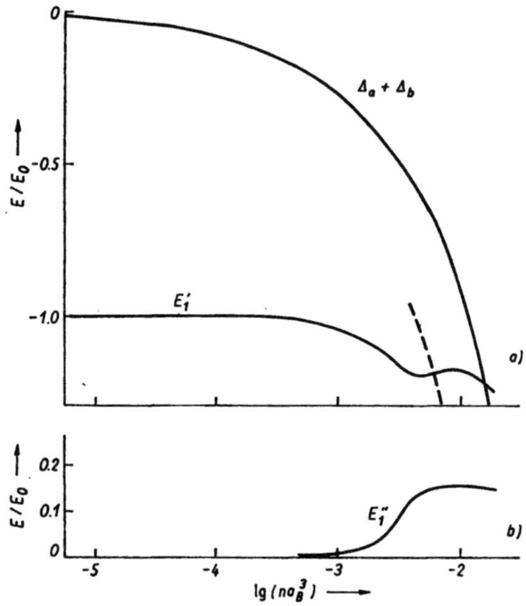

Fig. 4.6. Two-particle energies versus plasma density n at temperature $k_B T = E_0$
a) Continuum edge $\Delta^a + \Delta^b$ and real part E_1' of the lowest bound state energy. Dashed curve: Threshold for plasmon absorption, b) imaginary part E_1''

continuum moves down rapidly. The imaginary part E_0'' (damping) remains very small until a transition from the bound state into the continuum becomes possible by absorbing a thermally exicted plasmon. The dashed curve in Fig. 4.6a) marks the corresponding threshold energy $\Delta_a + \Delta_b - \omega_{\mathrm{pl}}$. The curves could be traced into the continuum which corresponds to a relative minimum above the absolute one for $a \to \infty$. Being without physical relevance, this finding provides clear evidence that the lowest bound state vanishes at a finite density (Mott density) and does not approach asymptotically the continuum edge as claimed by HAUG and TRAN THOAI (1978). Curves for $k_B T = 0.1 E_0$ are displayed in Fig. 4.7. E_0' deviates from the low-density value E_0 less than 1% until the bound state merges into the continuum which is in accordance with experimental findings in strongly excited semiconductors. Due to the low temperature, plasmon damping is very small.

The results for the Mott density at different temperatures given by ZIMMERMANN et al. (1978) are compared with the simple Mott criterion

$$\varkappa^{-1} = r_D = 0.84 a_B \tag{4.249}$$

in Fig. 4.8. In view of the sophisticated treatment the differences are rather small if dynamical screening of the interaction as well as in the self energy is taken into account.

A more detailed discussion of the shift of the ground state energy E_0 in dependence on temperature, density and mass ratio $\gamma = m_a/m_b$ may be given in the first Born approximation after Section 4.4.4. The different contributions to ΔE_0 given there may be rearranged in the following way:

$$\Delta E_0 = (\delta_\text{f}^{(1)} + \delta_\text{f}^{(2)} + \delta_\text{f}^{(3)})\, n_\text{f} + (\delta_{ab}^{(1)} + \delta_{ab}^{(2)})\, n_{ab}\ ; \tag{4.250}$$

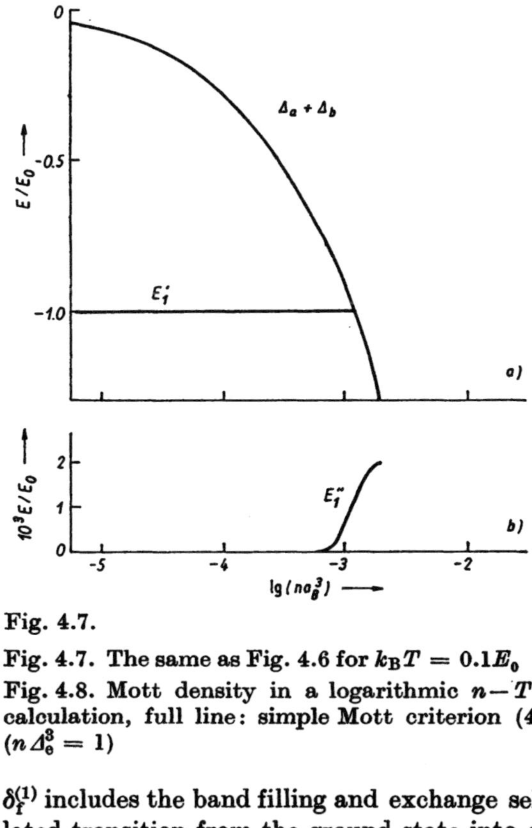

Fig. 4.7.

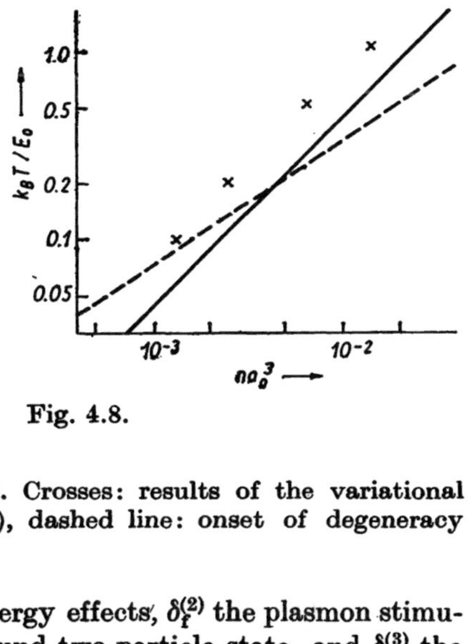

Fig. 4.8.

Fig. 4.7. The same as Fig. 4.6 for $k_\text{B}T = 0.1E_0$

Fig. 4.8. Mott density in a logarithmic $n-T$-plot. Crosses: results of the variational calculation, full line: simple Mott criterion (4.249), dashed line: onset of degeneracy ($n\Lambda_e^3 = 1$)

$\delta_\text{f}^{(1)}$ includes the band filling and exchange self-energy effects, $\delta_\text{f}^{(2)}$ the plasmon stimulated transition from the ground state into a bound two-particle state, and $\delta_\text{f}^{(3)}$ the transitions into scattering two-particle states where the scattering states may be approximated by the product of free particle wave functions; $\delta_{ab}^{(1)}$ contains exchange-type contributions arising in (4.233)—(4.235), $\delta_{ab}^{(2)}$ describes a Coulomb-type contribution which comes from the last term on the right-hand side of (4.235).

Explicite expressions for the different contributions in (4.250) are given by Röpke et al. (1980). The behaviour of $\delta_\text{f}^{(1)}$ in the region $T = 0.1\,E_0, \dots, E_0$ is shown in Fig. 4.9 for different mass ratios. For $T \to \infty$ this contribution goes to zero whereas for $T = 0$ the value 24π is approached.

The contributions $\delta_\text{f}^{(2)}$ and $\delta_\text{f}^{(3)}$ are evaluated within the single plasmon pole approximation for $\varepsilon(q, \omega)$ (4.120), for the mass ratio $\gamma = 1$. The sum over all bound states in $\delta_\text{f}^{(2)}$ is reduced to the contribution of only the first excited state ($2p$-type). This approach works well because for $\gamma = 1$ transitions to bound s-states give no contributions, and, on the other hand, transitions to higher excited states give only small contributions (the $1s-3p$ contribution is of about 10% of the $1s - 2p$ one). A graph of

$\delta_f^{(2)}$ is presented in Fig. 4.10. $\delta_f^{(2)}$ tends to zero if $T \to \infty$ and goes to the value -36.66 at $T = 0$. The behaviour of $\delta_f^{(3)}$ is similar to that of $\delta_f^{(2)}$ as seen in Fig. 4.10. For $T = 0$ the value $-41,5$ is taken.

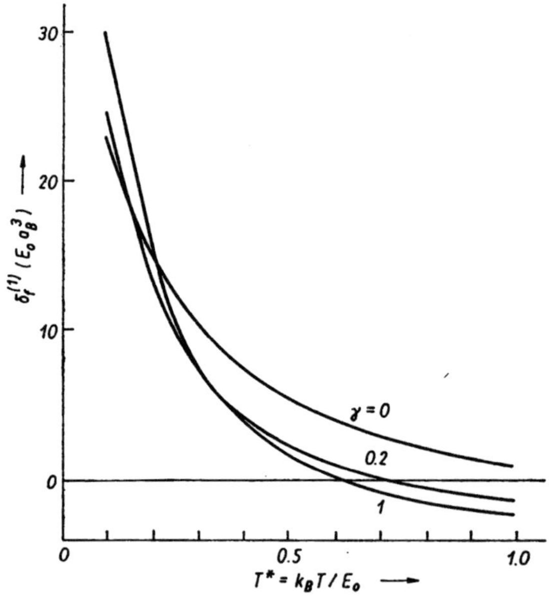

Fig. 4.9. Band filling and exchange self-energy effects ($\delta_f^{(1)}$, eq. (4.250)) versus temperature T^* at different mass ratios $\gamma = 0, 0.2, 1$

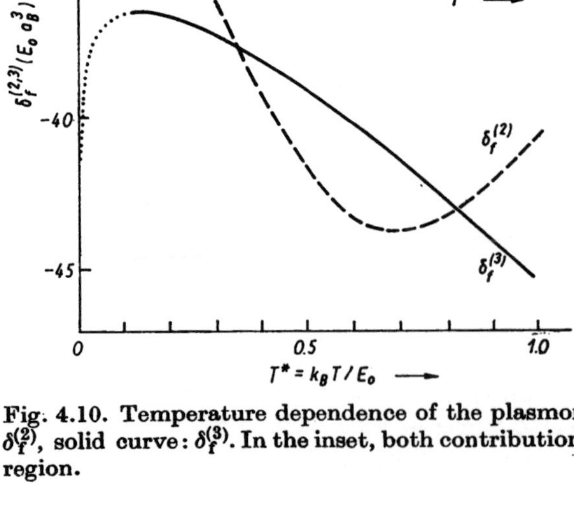

Fig. 4.10. Temperature dependence of the plasmon contributions ($\gamma = 1$). Dashed curve: $\delta_f^{(2)}$, solid curve: $\delta_f^{(3)}$. In the inset, both contributions are presented for a larger temperature region.

The contributions $\delta_{ab}^{(1)}$ and $\delta_{ab}^{(2)}$ are shown in Fig. 4.11. The contribution $\delta_{ab}^{(1)}$ takes the value $26\pi/3$ at $T = 0$ and can be compared with the expressions found by HANAMURA (1974), $\delta_{ab}^{(2)}$ is zero if $\gamma = 0$ or 1 and remains small for other values of the mass ratio.

To get more reliable results for the ground state energy shift we have to improve the Born approximation allowing, f.i., formation of molecules

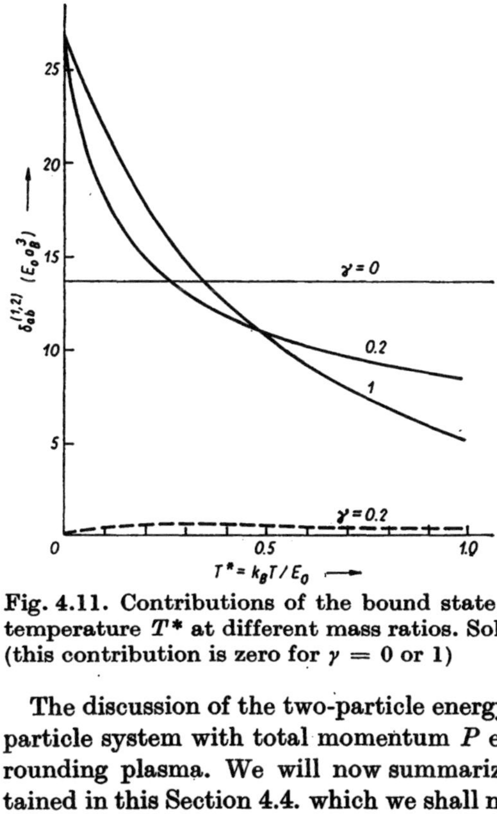

Fig. 4.11. Contributions of the bound state bound state interaction as a function of the temperature T^* at different mass ratios. Solid curves: $\delta_{ab}^{(1)}$, dashed curve: $\delta_{ab}^{(2)}$ for $\gamma = 0.2$ (this contribution is zero for $\gamma = 0$ or 1)

The discussion of the two-particle energy spectrum given up to now refers to a two-particle system with total momentum P equal to zero in the rest frame of the surrounding plasma. We will now summarize the most important physical results obtained in this Section 4.4. which we shall need in the following Chapters:

(i) The discrete energy states of a pair of oppositely charged particles $e_b = -e_a$ imbedded into a rare plasma are quite near to the unperturbed levels

$$E_{sl}(n, T) = E_{sl}^{(0)} + O(n) . \tag{4.251}$$

We note that this result refers only to symmetrical plasmas $Z = 1$. In the case of unsymmetrical plasmas $Z \neq 1$ a shift of order $(Z - 1)^{3/2} \varkappa e^2/2$ may appear as shown by KUDRIN (1974) (see also RÖPKE et al. (1978)). The total density of elementary particles n can be split into the density of free particles n_f and bound states n_{ab}. The bound state energy shift due to free particle density is described quite well in the dynamically screened ladder approximation, see Section 4.4.3., whereas the energy shift due to bound states in the surrounding plasma is given in Born approximation in Section 4.4.4.

(ii) The edge of the continuum in a rare plasma is shifted down by the so-called Debye shift

$$\Delta(P) = \Delta_a(P/2) + \Delta_b(P/2) . \tag{4.252}$$

We explicitly point out the dependence on the total momentum P of the two-particle system. For a two-particle system with $P \neq 0$ the shifts due to the interaction with the surrounding plasma are decreasing, and the most representative shifts of the continuum edge area obtained for $\hbar^2 P^2 \approx M k_B T$. Instead of the continuum shift at $P = 0$, mean properties are determined by the thermal averaged shift which is given approximatively by the chemical potential of the plasma

$$\Delta = (2\pi M k_B T)^{-3/2} \int \mathrm{d}\boldsymbol{P}\, \Delta(P) \exp\left(-P^2/2Mk_BT\right)$$
$$\approx \mu(n, T) - \mu_{\mathrm{id}}(n, T) = \mu_{\mathrm{xc}}(n, T) = \mu_a^{\mathrm{xc}} + \mu_b^{\mathrm{xc}} \,. \tag{4.253}$$

The interaction part of the chemical potential is sometimes called exchange-correlation part.

(iii) The discrete energy levels E_{sl} merge into the continuum at certain critical density n_{sl} which is defined by the condition

$$E_{sl}(P, n, T) - \Delta(P, n, T) = 0 \,. \tag{4.254}$$

It is interesting to note that disappearance of lines depends on the total momentum of the atom. The strongest effect of the surrounding plasma on the bound state levels is observed for atoms at rest and on the other hand the influence of the plasma on atoms moving with very high velocity relative to the plasma is extremely weak. Therefore, in principle all possible bound states of an atom exist in the plasma since at least a few atoms have high relative velocities, however fast atoms are exponentially rare due to the Maxwell distribution. In practice the optical observability of bound states and the corresponding lines is given by those atoms which are present with the highest probability, i.e., by those having thermal momenta

$$\hbar P_{\mathrm{th}} \sim \sqrt{M k_B T} \,.$$

In this way the dissappearance of lines is practically given by the condition

$$E_{sl}(P_{\mathrm{th}}, n, T) - \Delta(P_{\mathrm{th}}, n, T) = 0 \,.$$

It seems to be more convenient to average eq. (4.254) with a Maxwellian distribution which gives

$$E_{sl}(n, T) - \Delta(n, T) = 0 \tag{4.255}$$

where the mean shifts may be approximated by the interaction part of the chemical potential after eq. (4.253).

The Mott density $n_M(T) = n_{10}(T)$ corresponds to the disappearance of the ground state level

$$E_{10}(n_M, T) - \Delta(n_M, T) = 0 \,. \tag{4.256}$$

An estimate is given by

$$E_{10} \approx -(e^2/8\pi\varepsilon a_B) \,, \qquad \Delta \approx -(e^2/4\pi\varepsilon r_D)$$

which corresponds to the familiar Mott criterion

$$r_D \approx a_B \,.$$

In Fig. 4.12 a set of curves — calculated in Debye approximation — are drawn, which show the densities where the 1s-level disappears. In order to give a physical application let us consider the surface of the sun (photosphere) with a temperature

$T_s = 5785$ K and the density 10^{-7} g/cm³ (ALFVEN and ARRHENIUS, 1976) corresponding to an electron density of about $10^{16} - 10^{17}$ cm^{-3}. Using the Saha equation, eq. (6.75), this would lead to a free electron density of only about 10^{13} cm^{-3}. From Fig. 4.12 one would expect the lines to merge between $s = 100$ and $s = 1000$. However, one has to take into account that states with high quantum numbers have only low occupation rates and short life times. In fact, the maximum hydrogen Balmer line observed in the solar photosphere is $s = 17$ (MOORE, MINNAERT and HOUTGAST, 1966; ROUSE, 1983). For a more correct estimate we would need better calculations of $\Delta(n_e, T)$ or at least of the interaction part of the chemical potential $\mu_{xc}(n, T_e)$ (see Chapter 6. and EBELING et al., 1985).

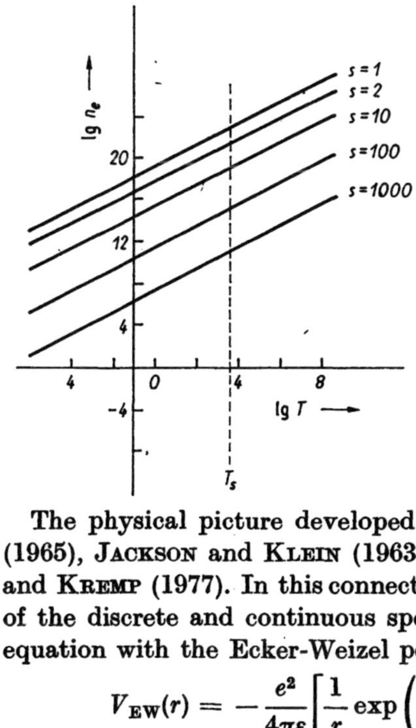

Fig. 4.12. Characteristic lines in the density-temperature plane of hydrogen where the levels with the main quantum number s disappear due to the merging in the continuum

The physical picture developed above goes back in part to ECKER and WEIZEL (1965), JACKSON and KLEIN (1963, 1969), KUDRIN (1974) and to EBELING, KRAEFT and KREMP (1977). In this connection we note that a qualitative correct description of the discrete and continuous spectra is given by the solution of the Schrödinger equation with the Ecker-Weizel potential

$$V_{EW}(r) = -\frac{e^2}{4\pi\varepsilon}\left[\frac{1}{r}\exp\left(-\frac{r}{r_D}\right) + \frac{1}{r_D}\right]. \tag{4.257}$$

Further we note that a related potential has been treated by ROUSE (1967, 1968, 1969, 1971).

The correct foundation of the picture given in this Section on the basis of the BSE is due to KILIMANN et al. (1977), ZIMMERMANN et al. (1978) and RÖPKE et al. (1978, 1980, 1981).

4.5. Dielectric Function Including Bound States

4.5.1. Extended RPA Dielectric Function for a Partially Ionized Plasma

As already discussed in Sections 3.3., 4.2., the dielectric function $\varepsilon(q, \omega)$ is defined according to

$$\varepsilon(q, \omega) = \lim_{\delta \to +0} \{1 - \sum_{ab} V_{ab}(q)\, \Pi_{ab}(q, \omega + i\delta)\}\,. \tag{4.258}$$

The polarization function $\Pi_{ab}(q, z)$ is to be taken as the analytical continuation of $\Pi_{ab}(q, \omega_\mu)$ defined at the Bose-type Matsubara frequencies ω_μ into the complex z-plane. The polarization function $\Pi_{ab}(q, \omega_\mu)$ can be evaluated within the many-body approach by using a perturbation theoretical expansion as described in Section 3. so that $\Pi_{ab}(q, \omega_\mu)$ may be represented by the sum of all irreducible diagrams which do not disintegrate into two separate parts by cutting an internal interaction potential line:

$\Pi_{ab}(q, \omega_\mu) = q, \omega_\mu, a$  $q, \omega_\mu, b =$ 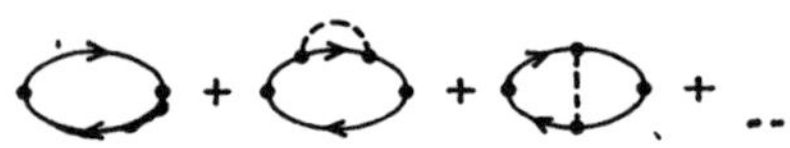

(4.259)

This diagram technique may be used to construct different approximations for the dielectric function in a systematical way. The use of the Green's function approach allows to derive approximation schemes for the dielectric function which can be compared with equivalent approximations for thermodynamic, transport and optical properties of Coulomb systems.

The use of the Green's function technique gives the dielectric function in the grand canonical ensemble, i.e. as function of temperature $T = 1/k_B\beta$ and the chemical potentials μ_a of the different species a. To transform this function into a function of another set of thermodynamic parameters, as, for instance, pressure or the densities n_a of the constituents of the plasma, we have to use the thermodynamic equation of states which will be presented in detail in the next Chapter. In this Section 4.5., we apply only simple versions of the equation of state to obtain the dielectric function in dependence of temperature and particle densities and to discuss some limiting cases.

The first diagram of the right-hand side of eq. (4.259) gives the well known random phase approximation

$$\Pi_{ab}^{\mathrm{RPA}}(q, z) = 2\delta_{ab} \frac{1}{\Omega} \sum_{p,a} \frac{f_a(E_p^a) - f_a(E_{p-q}^a)}{E_p^a - \hbar z - E_{p-q}^a}. \tag{4.260}$$

The resulting RPA dielectric function $\varepsilon^{\mathrm{RPA}}(q, \omega)$ was discussed in Section 4.2. For non-degenerate systems, especially, we have in the limit $\omega = 0$, $q \to 0$

$$\lim_{q \to 0} [\varepsilon^{\mathrm{RPA}}(q, 0) - 1] = \frac{\varkappa^2}{q^2}, \qquad \varkappa^2 = \frac{1}{\varepsilon_0} \beta \sum_a n_a e_a^2. \tag{4.261}$$

This result means that all charged particles contribute to the screening of a static potential. On the other hand, if bound states of charged particles occur, these two-particle bound states should contribute to the Debye screening length only due to their total charge. Especially, in the case of a symmetric plasma, the charged particles which are constituents of bound states (atoms or excitons, e.g.) should drop out from (4.259) in a simple approximation. Within the chemical picture where a bound state corresponds to a new composite particle such a result would be evident. In accordance with this picture, screening effects often are treated in a modified RPA (4.260) for the dielectric function where the total number of charged particles is replaced by the number of "free" carriers. The number of free carriers is determined from a mass action law.

However, within a rigorous treatment of the many-body problem the effect of the occurrence of bound states should be accounted for in a consequent way. Only within this approach is it possible to treat dense Coulomb systems and, especially, the Mott effect where a simple chemical picture becomes inapplicable.

Improvements of the RPA dielectric function (4.260) have been widely discussed in the case of strong degeneracy as, for instance, electrons in metals, c.f. BISHOP and LÜHRMANN (1978). In that case, the occurrence of bound states is not of importance. The influence of bound states on the dielectric function was considered in plasma physics by KLIMONTOVICH (1975, 1980) and also in the case of impurities in semiconductors (TAKESHIMA, 1978; HUGON and GHAZALI, 1976) describing the system by a model Hamiltonian. A Green's function approach to the dielectric function which takes into account the formation of bound states introducing the corresponding T-matrix expressions in the polarization function has been given by RÖPKE and DER (1979), see also RÖPKE, SEIFERT and KILIMANN (1981).

The main idea how to find approximations for the polarization function $\Pi_{ab}(q, z)$ where free particles and bound states are taken into account on the same footing, was given in Section 4.1.4. The extended RPA dielectric function is given explicitly by the following contributions to the polarization function:

(4.262)

where the ladder T operator equation (see Section 3.2.8. and (4.21))

$$T_{ab}(z) = \ldots + \ldots = \sum_{nP} (E_{nP}^{(0)} - K) |nP\rangle \frac{1}{\hbar z - E_{nP}^{(0)}} \langle nP| (\hbar z - K) \tag{4.263}$$

was introduced; K is the kinetic energy operator, $E_{nP}^{(0)}$ and $|nP\rangle$ are the eigenvalues and eigenstates, respectively, of the isolated atom with total momentum P and internal quantum number n. The last diagram on the right-hand side of eq. (4.262) cancels the corresponding diagram contained in the last but one term.

In the non-degenerate case where the Fermi function may be neglected with respect to unity, the diagrams (4.262) are evaluated (RÖPKE and DER, 1979) and give

$$\varepsilon(q, \omega) = 1 + 4\pi\alpha_1(q, \omega) + [4\pi\alpha_2(q, \omega) - (4\pi\alpha_1(q, \omega))^2] \tag{4.264}$$

with the polarizabilities

$$\alpha_1(q, \omega) = -\frac{2}{\varepsilon_0 \Omega q^2} \sum_a e_a^2 \sum_p \frac{f_a(E_p^a) - f_a(E_{p-q}^a)}{E_p^a - \hbar(\omega + i\delta) - E_{p-q}^a},$$

$$\alpha_2(q, \omega) = -\frac{4}{\varepsilon_0 \Omega q^2} \sum_{nn'P} |M_{nn'}(q)|^2 \frac{g_{ab}(E_{nP}^{ab}) - g_{ab}(E_{n'P-q}^{ab})}{E_{nP}^{ab} - \hbar(\omega + i\delta) - E_{n'P-q}^{ab}}$$
$$+ \frac{4}{\varepsilon_0 \Omega q^2} \sum_{p_1 p_2, ab} e_a^2 \frac{g_{ab}(E_{p_1}^a + E_{p_2}^b) - g_{ab}(E_{p_1-q}^a + E_{p_2}^b)}{E_{p_1}^a - \hbar(\omega + i\delta) - E_{p_1-q}^a},$$

$$M_{nn'}(q) = \int d^3p\, \psi_n^*(p) \left\{ e_a \psi_{n'}\left(p - \frac{m_a}{M} q\right) + e_b \psi_{n'}\left(p + \frac{m_b}{M} q\right) \right\}. \tag{4.265}$$

In order to compare this result with the result given by KLIMONTOVICH (1976, 1980), we decompose the spectrum of two-particle states into the bound (b) and the scattering (sc) part and obtain the following structure for (4.264):

$$\varepsilon(q, \omega) = 1 + 4\pi(\alpha_f + \alpha_{b,b} + \alpha_{b,sc} + \alpha_{sc,b} + \alpha_{sc,sc} - \alpha_{f,f} - 4\pi\alpha_f^2), \tag{4.266}$$

which is in contrast to KLIMONTOVICH (1975, 1980), where a decomposition of $\varepsilon(q, \omega)$

$$\varepsilon(q, \omega) = 1 + 4\pi(\alpha_{f,f} + \alpha_{f,b} + \alpha_{b,f} + \alpha_{b,b}) \tag{4.267}$$

has been proposed approximating two-particle scattering states by free particle states. Our expression (4.266) gives also the correct dependence on the free particle number, because the divergent terms in $\alpha_{sc,sc}$ and $\alpha_{f,f}$ cancel in (4.266).

The extended RPA-expression (4.264) for the dielectric function can be interpreted in a rather simple way: The single particle energies and distribution functions are replaced by the corresponding two particle quantities, and the coupling to the Coulomb potential is given by the charge e_a in the case of free particles, and by the matrix element $M_{nn'}(q)$ in the case of a two-particle state. The matrix element $M_{nn'}(q)$ can be represented by a multipole expansion (power series expansion with respect to q), and in the long wave limit we have

$$\lim_{q \to 0} M_{nn'}(q) = i\boldsymbol{q} \cdot \boldsymbol{d}_{nn'} \tag{4.268}$$

where

$$\boldsymbol{d}_{nn'} = \int \mathrm{d}\boldsymbol{r}\, \psi_n^*(r)\, e\boldsymbol{r}\psi_{n'}(r)$$

is the dipole matrix element for the transition $n \to n'$.

The extended RPA dielectric function (4.264), (4.265) can be compared with the result which is obtained for the polarizability $\hat{\alpha}_n(q, \omega)$ of a bound state n of a single atom (LANDAU and LIFSHITS, 1967; KUBO, 1973)

$$\hat{\alpha}_n(q, \omega) = -\frac{1}{q^2} \sum_{n'} \left(\frac{\langle n| \varrho_q |n'\rangle \langle n'| \varrho_{-q} |n\rangle}{E_n - \omega - E_{n'}} + \frac{\langle n| \varrho_{-q} |n'\rangle \langle n'| \varrho_q |n\rangle}{E_n + \omega - E_{n'}} \right), \tag{7.269}$$

$$\varrho_q = \int \mathrm{d}\boldsymbol{r} \sum_a e_a n_a(\boldsymbol{r}) \exp(i\boldsymbol{q} \cdot \boldsymbol{r})\,.$$

For a two-particle state $|nP\rangle$ we obtain

$$\langle nP| \varrho_q |n'P'\rangle = \sum_{p_1p_2p_1'p_2'} \Omega^{-2} \int \mathrm{d}\boldsymbol{r}_1\, \mathrm{d}\boldsymbol{r}_2 \langle nP \mid p_1p_2 \rangle\, \mathrm{e}^{i(\boldsymbol{p}_1\boldsymbol{r}_1 + \boldsymbol{p}_2\boldsymbol{r}_2)}$$

$$\cdot (e_a\, \mathrm{e}^{i\boldsymbol{q}\boldsymbol{r}_1} + e_b\, \mathrm{e}^{i\boldsymbol{q}\boldsymbol{r}_2})\, \mathrm{e}^{-i(\boldsymbol{p}_1'\boldsymbol{r}_1 + \boldsymbol{p}_2'\boldsymbol{r}_2)} \langle p_1'p_2' \mid n'P'\rangle$$

$$= \sum_{p_1p_2} \langle nP| p_1p_2\rangle \{e_a\langle \boldsymbol{p}_1 + \boldsymbol{q}, \boldsymbol{p}_2 \mid n'P'\rangle + e_b \langle \boldsymbol{p}_1, \boldsymbol{p}_2 + \boldsymbol{q} \mid n'P'\rangle\}\,. \tag{4.270}$$

We conclude that the extended RPA-dielectric function (4.264, 4.65) contains the polarizability of bound two-particle states weighted by their distribution function.

The extended RPA dielectric function may also be compared with results found from linear response theory. For instance, for electrons bound to a periodic lattice, which are described by a band index l and a quasi-momentum $\boldsymbol{k}$, the following expression for the dielectric function is obtained by EHRENREICH and COHEN (1959):

$$\varepsilon(\boldsymbol{q}, \omega) = 1 - \frac{e^2}{\varepsilon_0 \Omega q^2} \sum_{kll'} |\langle \boldsymbol{k}l \mid \boldsymbol{k} + \boldsymbol{q}l'\rangle|^2 \frac{f(\boldsymbol{k} + \boldsymbol{q}, l') - f(\boldsymbol{k}, l)}{E_{l'}(\boldsymbol{k} + \boldsymbol{q}) - E_l(\boldsymbol{k}) - \hbar(\omega + i\delta)}\,,$$

$$\langle \boldsymbol{k}l| \boldsymbol{k} + \boldsymbol{q}l'\rangle = \int \psi_{kl}^*(\boldsymbol{r})\, \mathrm{e}^{-i\boldsymbol{q}\boldsymbol{r}}\, \psi_{\boldsymbol{k}+\boldsymbol{q},l'}(\boldsymbol{r})\, \mathrm{d}\boldsymbol{r}\,. \tag{4.271}$$

This expression can be obtained from the general formula (4.264) in the adiabatic approximation for a periodic lattice. Thus the extended RPA-dielectric function reflects the occurrence of bound states in a correct way in the low density limit. Some special cases will be discussed in the next Section. Furthermore, this dielectric function may serve as a starting point to derive the thermodynamic properties of Coulomb systems which show the correct low-density limit (ZIMMERMANN, 1983), and to derive transport and optical properties, see Chapter 8. Improvements may be obtained in a straightforward manner by taking further diagrams for $\Pi_{ab}(q, \omega_\mu)$, see below Sections 4.5.3., 4.5.4.

4.5.2. Limiting Behaviour of the Extended RPA Dielectric Function

The limiting behaviour of the RPA dielectric function for $\omega \to 0$, $\omega \to \infty$ and $q \to 0$ has been discussed in Section 4.2. The same discussion can be applied to the term $\alpha_2(q, \omega)$ in the extended RPA dielectric function (4.265). However, if we introduce the densities of the constituents n_a instead of the corresponding chemical potentials μ_a, we have also to evaluate the equation of state $\mu_a = \mu_a(\beta, n_a, n_b)$. This equation of state can be found from the relation

$$n_a = \frac{2}{\Omega} \frac{1}{-i\beta} \sum_p \sum_\nu G_a^<(p, z_\nu^a) , \tag{4.272}$$

see Section 6.2. and STOLZ and ZIMMERMANN (1979). Improvements in the determination of the one-particle Green's function must be performed on the same footing as the improvement of the polarization function (4.262). The influence of bound states on the one-particle Green's function is taken into account in the following approximation:

$$G_a(p, z_\nu^a) = \tag{4.273}$$

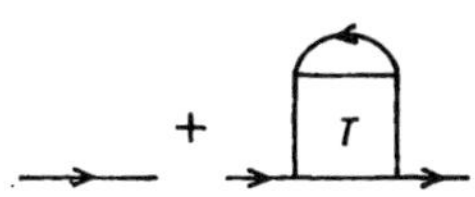

with the result

$$n_a = \frac{2}{\Omega} \sum_p f_a(E_p^a) + \frac{4}{\Omega} [\sum_{nP} g_{ab}(E_{nP}^{ab}) - \sum_{pq} g_{ab}(E_{p+q}^a + E_p^b)] \tag{4.274}$$

($a \neq b$), see STOLZ and ZIMMERMANN (1979).

The spectrum of two-particle states $|nP\rangle$ can be decomposed into the bound (b) and scattering (sc) part. Whereas the bound state wave functions are introduced according to

$$\langle p_1 p_2 | nP \rangle^{(b)} = \delta(p_1 + p_2 - P) \sqrt{\frac{(2\pi)^3}{\Omega}}\, \psi_n(p) , \tag{4.275}$$

$$p = (m_b p_1 - m_a p_2)/(m_a + m_b)$$

the wave function of scattering states (incoming momenta p_a, p_b) can be given by a power series expansion with respect to the potential V:

$$\langle p_1 p_2 | \, p_a = p_1 + q, p_b = p_2 - q \rangle = \delta(q, 0) \tag{4.276}$$

$$+ \frac{1}{\Omega} \frac{\langle p_1 p_2 | V_{ab}^\delta | p_1 + q, p_2 - q \rangle}{E_{p_1+q}^a + E_{p_2-q}^b - E_{p_1}^a - E_{p_2}^b + i0} + \dots$$

The Coulomb potential $V_{ab}(q) = e_a e_b / \varepsilon_0 q^2$ was replaced by the potential $V_{ab}^\delta(q) = e_a e_b / \varepsilon_0 (q^2 + \delta^2)$ with δ arbitrarily small to avoid the divergencies of the Coulomb potential. Taking only the first term of the right-hand side of (4.276) the scattering part and the free two-particle part cancel in (4.266), and in the low density limit considered here we obtain

$$n_a = n_a^* + \sum_m{}^{(b)} n_{ab,m}^* ,$$
$$e^{\beta\mu_a} = (2\pi\beta\hbar^2/m_a)^{3/2}\, n_a^*/2 ,$$
$$e^{\beta(\mu_a+\mu_b-E_m^{ab})} = (2\pi\beta\hbar^2/(m_a+m_b))^{3/2}\, n_{ab,m}^*/4 . \tag{4.277}$$

These expressions are equivalent to a simple mass action law, the densities of free particles and of bound states are denoted by asterisks.

Correspondingly, the limiting behaviour of the RPA-contribution $\alpha_1(q, \omega)$ is determined by the free particle densities:

$$\lim_{\omega\to\infty} \omega^2\alpha_1(q, \omega) = -(\omega_{\rm pl}^{\rm RPA})^2 = -\sum_a n_a^* e_a^2/\varepsilon_0 m_a ,$$
$$\lim_{\omega\to 0} \omega^2\alpha_1(0, \omega) = -(\omega_{\rm pl}^{\rm PRA})^2 , \tag{4.278}$$
$$\lim_{q\to 0} q^2\alpha_1(q, 0) = \varkappa^2 = \sum_a \frac{1}{\varepsilon_0} n_a^* e_a^2\beta ,$$

Discussing the bound state contribution $\alpha_2(q, \omega)$ to the extended RPA dielectric function, we treat the scattering states also by a Born series expansion (4.276) and find that the contributions of occupied two-particle scattering states, due to intermediate transitions into two-particle scattering states, are at least of second order with respect to the potential. (The zeroth and first orders with respect to the potential are canceled by the last terms on the right-hand side of (4.264) and (4.265)). In the long wave limit we obtain, see RÖPKE et al. (1979), ($m_{ab}^{-1} = m_a^{-1} + m_b^{-1}$)

$$\lim_{q\to 0} \alpha_2(q, \omega) = \sum_{a \neq b} \sum_n{}^{(b)} \alpha_n(\omega) \int \frac{4\mathrm{d}\boldsymbol{P}}{(2\pi)^3}\, g_{ab}(E_{nP}^{ab})$$
$$- \sum_{a\neq b} \sum_n{}^{(b)} e^2 \int \mathrm{d}p \left(\frac{\partial}{\partial p_z}\psi_n(p)\right)^2 \left\{\frac{1}{\hbar^2p^2/2m_{ab} - \hbar(\omega + i\delta) - E_n}\right.$$
$$\left. + \frac{1}{\hbar^2p^2/2m_{ab} + \hbar(\omega + i\delta) - E_n}\right\} \int \frac{4\mathrm{d}\boldsymbol{P}}{(2\pi)^3}\, g_{ab}(\hbar^2p^2/2m_{ab} + \hbar^2p^2/2(m_a + m_b))$$
$$- \sum_{a\neq b} 2\hbar^4 \left[\frac{e_a}{m_a} - \frac{e_b}{m_b}\right]^2 \int \frac{\mathrm{d}\boldsymbol{k}}{(2\pi)^3}\, k_z^2 (V_{ab}(k))^2 \int \frac{\mathrm{d}\boldsymbol{p}_1}{(2\pi)^3} \int \frac{\mathrm{d}\boldsymbol{p}_2}{(2\pi)^3}$$
$$\cdot \frac{1}{(E_{p_1+k}^a + E_{p_2-k}^b - E_{p_1}^a - E_{p_2}^b)^4}\, \frac{g_{ab}(E_{p_1+k}^a + E_{p_2-k}^b) - g_{ab}(E_{p_1}^a + E_{p_2}^b)}{E_{p_1+k}^a + E_{p_2-k}^b - \hbar(\omega + i\delta) - E_{p_1}^a - E_{p_2}^b}$$
$$- \sum_{a,b} 16\beta e_a e_b \omega^{-2} \int \frac{\mathrm{d}\boldsymbol{k}}{(2\pi)^3}\, |V_{ab}|^2\, S_{ab}(k) \tag{4.279}$$

with

$$S_{ab}(k) = \int \frac{dp_1}{(2\pi)^3} \int \frac{dp_2}{(2\pi)^3} \frac{g_{ab}(E^a_{p_1} + E^b_{p_2})}{E^a_{p_1} + E^b_{p_2} - E^a_{p_1+k} - E^b_{p_2-k}} \frac{1}{p_{1z}/m_a - p_{2z}/m_b}$$
$$\cdot \left\{\frac{\beta}{2} [p_{1z}/m_a)^3 - (p_{2z}/m_b)^3] - p_{1z}/m_a^2 + p_{2z}/m_b^2 - \frac{1}{2m_{ab}} \frac{(p_{1z}/m_a)^2 - (p_{2z}/m_b)^2}{p_{1z}/m_a - p_{2z}/m_b}\right.$$
$$\left. + \frac{k_z}{m_{ab}} \frac{(p_{1z}/m_a{}^2) + (p_{2z}/m_b)^2}{E^a_{p_1} + E^b_{p_2} - E^a_{p_1+k} - E^b_{p_2-k}}\right\},$$

and the polarizabilities

$$\alpha_n(\omega) = e^2 \sum_{n'} |\langle n| z |n'\rangle|^2 \left\{\frac{1}{E_n - \hbar(\omega + i\delta) - E_{n'}} + \frac{1}{E_n + \hbar(\omega + i\delta) - E_{n'}}\right\}. \tag{4.280}$$

From this expression we derive the limiting behaviour at high frequencies

$$\lim_{\omega\to\infty} \omega^2\alpha_2(0, \omega) = -\frac{8}{\hbar} \sum_{a\neq b} \frac{e^2}{4\pi\varepsilon_0} \sum_n^{(b)} \frac{1}{\Lambda^3_{ab}} e^{-\beta(E_n-\mu_a-\mu_b)}$$
$$\cdot \sum_{n'} (\langle n| z |n'\rangle)^2 (E_{n'} - E_n) + O(V^2) \tag{4.281}$$
$$= -\frac{2}{\hbar^2} \sum_{a\neq b} \sum_n^{(b)} \frac{e^2}{4\pi\varepsilon_0} n^*_{ab,n} \sum_{n'} |\langle n| z |n'\rangle|^2 (E_{n'} - E_n), \tag{4.281}$$

if terms of the order V^2 are neglected, which are proportional to n_a^{*2}. Using the commutation relations of quantum mechanincs, we find that

$$\lim_{\omega\to\infty} \omega^2(\varepsilon(0, \omega) - 1) = -\sum_a \frac{n^*_a e^2_a}{\varepsilon_0 m_a} + \sum_{a\neq b} \sum_n^{(b)} \frac{e^2 n^*_{ab,n}}{\varepsilon_0 m_{ab}} = \frac{e^2 n_p}{\varepsilon_0 m_{ab}} \tag{4.282}$$

is determined by the total number of charged particles (f-sum rule).

On the other hand, in the low-frequency case, transitions between bound states or between bound and scattering states, respectively, give only a constant contribution to the real part of $\varepsilon(0, \omega)$ and contributions of the order ω^{-2} arise from the RPA-contribution and transitions between scatering states:

$$\lim_{\omega\to 0} \omega^2\varepsilon(0, \omega) = -(\omega_{\text{pl}}^{\text{RPA}})^2 - \sum_{ab} 64\pi\beta e_a e_b \int \frac{dk}{(2\pi)^3} |V_{ab}(k)| S_{ab}(\text{L}). \tag{4.283}$$

In first order in n^*_a the singularity of $\varepsilon(0, \omega)$ at $\omega \to 0$ is determined only by the number of free particles n^*_a.

Similarly, in the first order with respect to the densities n^*_a, $n^*_{ab,m}$ we have in the limit $\omega = 0$, $p \to 0$

$$\varepsilon(q, 0) \approx 1 + \frac{e^2}{k_B T \varepsilon_0 q^2} \sum_a n^*_a - \frac{\hbar^2 e^2}{12\,\varepsilon_0} \beta^2 \sum_a \frac{1}{m_a} n^*_a$$
$$- \frac{2}{\varepsilon_0} e^2 \sum_m^{(b)} \int dp \left|\frac{\partial}{\partial z} \psi_m(p)\right|^2 n^*_{ab,m} \frac{1 - e^{\beta(E^{ab}_m - \hbar^2 p^2/2m_{ab})}}{E^{ab}_m - \hbar^2 p^2/2m_{ab}}$$
$$- \frac{1}{\varepsilon_0} e^2 \sum_m^{(b)} \sum_{m'}^{(b)} (\langle m| z |m'\rangle)^2 (n^*_{ab,m} - n^*_{ab,m'}) \frac{1}{E^{ab}_m - E^{ab}_{m'}}. \tag{4.284}$$

In addition to the free charged particles, which contribute in the order q^{-2}, the two-particle bound states contribute to the dielectric function in the order q^0 due to their polarizability. The coefficient of the singular part at $q \to 0$ is reduced if the number of free particles is reduced because of the formation of bound states.

Up to the order q^0, no imaginary part is obtained from the RPA dielectric function. Therefore, in the long wave limit, the imaginary part of the extended RPA dielectric function is determined only by the contributions beyond RPA, which give the following expression according to (4.279):

$$\begin{aligned}\lim_{q\to 0} \operatorname{Im} \varepsilon(q, \omega) &= \sum_{a \neq b} \pi \frac{\mathrm{e}^2}{\varepsilon_0} \sum_n^{(b)} \sum_{n'} |\langle n| z |n'\rangle|^2 \frac{4}{\Lambda_{ab}^3} \mathrm{e}^{-\beta(E_n - \mu_a - \mu_b)} \\ &\cdot \{\delta(E_n + \hbar\omega - E_{n'}) - \delta(E_{n'} + \hbar\omega - E_n)\} \\ &- \sum_{a \neq b} \pi \frac{e^2}{\varepsilon_0} \sum_n^{(b)} \int \mathrm{d}\boldsymbol{p} \left(\frac{\partial}{\partial p_z} \psi_n(p)\right)^2 \frac{4}{\Lambda_{ab}^3} \mathrm{e}^{-\beta(\hbar^2 p^2/2m_{ab} - \mu_a - \mu_b)} \\ &\cdot \{\delta(E_n + \hbar\omega - \hbar^2 p^2/2m_{ab}) - \delta(\hbar^2 p^2/2m_{ab} + \hbar\omega - E_n)\} \\ &+ \sum_{a \neq b} \frac{2\pi e^2}{\varepsilon_0 m_{ab}^2 \omega^4} \int \frac{\mathrm{d}\boldsymbol{k}}{(2\pi)^3} \frac{k^2}{3} |V_{ab}(k)|^2 \int \frac{\mathrm{d}\boldsymbol{p}_1}{(2\pi)^3} \int \frac{\mathrm{d}\boldsymbol{p}_2}{(2\pi)^3} \mathrm{e}^{-\beta(E_{p_1}^a + E_{p_2}^b - \mu_a - \mu_b)} \\ &\cdot \delta(E_{p_1}^a + E_{p_2}^b - E_{p_1+k}^a - E_{p_2-k}^b + \hbar\omega)\,(1 - \mathrm{e}^{-\beta\hbar\omega})\,. \end{aligned} \tag{4.285}$$

We may restrict ourselves to positive frequencies $\omega > 0$ only. Expression (4.285) consists of three part: The first contribution on the right-hand side of (4.285) describes transitions from bound states to other two-particle states by emission or absorption of the energy $\hbar\omega$; especially it contains the line spectrum in the approximation of Doppler-broadened transitions between sharp energy levels as discussed more in detail in Chapter 8. The second contribution describes transitions from scattering to bound states by emission of the energy $\hbar\omega$ (recombination). The last contribution describes transitions between scattering states. Especially, in the limit $\omega \to 0$, the dc conductivity may be obtained from this term, see RÖPKE et al. (1979), but this limit needs special considerations given in Chapter 7. regarding the formulation of linear response theory.

4.5.3. Self-Energy and Vertex Corrections to the Extended RPA Dielectric Function

As far as the extended RPA dielectric function was constructed, the polarization function has been approximated by bubble diagrams of the free particle propagator as well as the isolated two-particle system. This approximation is applicable in the low density limit as well as in the high density limit; but, in the intermediate region the interaction with the surrounding matter becomes of importance. Improvements of the extended RPA dielectric function can be found along different lines.

In this Section we discuss improvements which are essential in the case of a high degree of ionization so that the interaction with free particles is most important. In the next Section 4.5.4., however, we consider the case where the interaction with bound states dominates.

Corrections to the polarization function may be constructed treating the single-particle propagator and the two-particle propagator in an equivalent way as pointed

out in Section 4.1.4. So we have

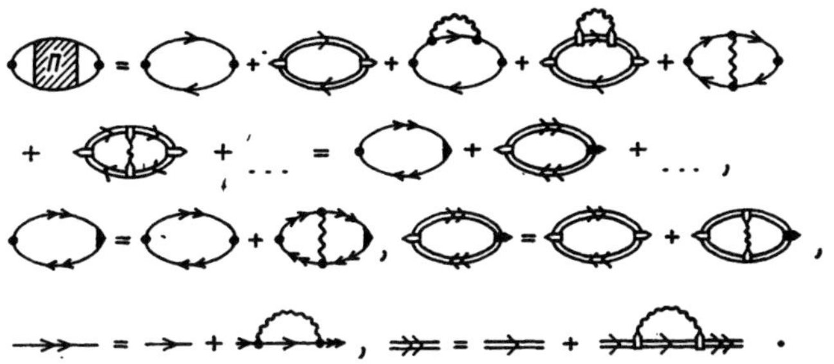

(4.286)

These approximations mean that the propagators of free complexe are replaced by the propagators of quasiparticles which follow from the solution or the Dyson equation or the Bethe-Salpeter equation, respectively, with given approximations for the self-energy and the effective interaction. The vertex corrections can be considered as the exchange term to the potential in the equation for the screened potential (3.232). Note that by summing up classes of diagrams in this way, we have to avoid double counting of diagrams by subtracting the corresponding diagrams which are of lower order with respect to the screened interaction. As long as we consider only the bound state part of the two-particle propagator, we can omit such terms of lower order with respect to the interaction.

The improvements of the single particle contribution $\Pi_1(q, \omega_\mu)$ to the polarization function are well known from literature. For instance, the dielectric function of EHRENREICH and COHEN (1959) eq. (4.271) is obtained in the quasiparticle picture, and the self-energy effects are taken into account introducing the Bloch states. Vertex corrections to the single particle contribution can be derived, for instance, with the aid of time-dependent Hartree-Fock calculations, see HEDIN and LUNDQUIST (1971). In this way, the expressions for the dielectric function due to HUBBARD (1957, 1958, 1967) or due to SINGWI et al. (1968) may be derived.

Let us consider in more detail the improvements of the bound state contribution in $\Pi_2(q, \omega_\mu)$ to the polarization function. Instead of the isolated atom, we introduce the two-particle system imbedded in a surrounding plasma which comes out from the effective wave equation given in Section 4.4. With this approximation, which accounts also for the Mott effect, we construct the polarization function $\Pi_2(q, \omega_\mu)$ in quasiparticle representation (self-energy corrections) if we take

$$G_2(n, n', P, z) = -\int_{-\infty}^{\infty} \frac{d\omega}{2\pi} \frac{A(n, P, \omega)}{\omega - z} \delta_{nn'},$$

$$A(n, P, \omega) = -2i \frac{\Gamma_{nP}(\omega)}{[E_{nP} - \hbar\omega + \Delta_{nP}(\omega)]^2 + \Gamma_{nP}^2(\omega)}, \tag{4.287}$$

see also Section 4.1.4. The spectral function can be approximated by a Lorentzian in the low density limit; the energy shift $\Delta_{nP}(E_{nP})$ and the width $\Gamma_{nP}(E_{nP})$ of the energy level E_{nP} are obtained from the perturbative solution of the wave equation.

With the improved two-particle Green's function (4.205) we evaluate the bound part of the polarization function

$$\Pi_2^{(b)}(q,\omega) = \frac{1}{-i\beta} \sum_{nn'P\lambda} |M_{nn'}(q)|^2 \int \frac{d\omega'}{2\pi} \int \frac{d\omega''}{2\pi} \frac{A(n,P,\omega')\,A(n',P-q,\omega'')}{(\omega'-\Omega_\lambda)\,(\omega''-\Omega_\lambda-\omega_\mu)}$$
$$= \sum_{nn'P} |M_{nn'}(q)|^2 \int \frac{d\hbar\omega'}{2\pi} \int \frac{d\omega''}{2\pi} \frac{A(n,P,\omega')\,A(n',P-q,\omega'')}{\omega''-\omega'-\omega_\mu}$$
$$\cdot \{g_{ab}(\hbar\omega') - g_{ab}(\hbar\omega'')\}\,. \tag{4.288}$$

From this follows

$$\operatorname{Im}\varepsilon(q,\omega) = \operatorname{Im}\varepsilon^{\mathrm{RPA}}(q,\omega) + V(q) \sum_{nn'P} |M_{nn'}(q)|^2$$
$$\cdot \int_{-\infty}^{\infty} \frac{d\omega'}{2\pi} \{g_{ab}(\hbar\omega') - g_{ab}(\hbar(\omega-\omega'))\}\, A(n,P,\omega')\, A(n',P-q,\omega-\omega')\,, \tag{4.289}$$

i.e., the imaginary part of the dielectric function is given by the convolution integral of the spectral functions. Further evaluation is possible if a Lorentzian form is assumed and $g_{ab}(\hbar\omega')$ is replaced by $g_{ab}(E^{ab}_{nP})$,

$$\Pi_2^{(b)}(q,\omega+i\delta) = \sum_{nn'P} |M_{nn'}(q)|^2 \{g_{ab}(E^{nP}_{ab}) - g_{ab}(E^{ab}_{n'P-q})\}$$
$$\cdot (E^{ab}_{nP} + \Delta_{nP} + i\Gamma_{nP} - E^{ab}_{n'P-q} - \Delta_{n'P-q} + i\Gamma_{n'P-q} - \hbar\omega)^{-1}\,. \tag{4.290}$$

In this approximation, the dielectric function is determined by the effective two-particle properties in a dense plasma. The absorption spectrum is obtained from the imaginary part of $\varepsilon(q,\omega)$ and describes pressure broadening and spectral line shift as described by RÖPKE, SEIFERT and KILIMANN (1981), see also Chapter 8.

The vertex corrections follow from the integral equation for the vertex function $\Gamma(1,2,\omega_\mu,q)$ with $1 = (n_1, P_1, \omega_1)$:

$$\sum_{12} M_{12}(q) \frac{A(1)\,A(2)}{\omega_1-\omega_2-\omega_\mu} \{g(\hbar\omega_1) - g(\hbar\omega_2)\}\, \Gamma(1,2,\omega_\mu,q) = \sum_{12} |M_{12}(q)|^2$$
$$\cdot \frac{A(1)\,A(2)}{\omega_1-\omega_2-\omega_\mu} \{g(\hbar\omega_1) - g(\hbar\omega_2)\} + \sum_{1234} M_{12}(q) \int \frac{d\bar{q}}{(2\pi)^3} \int \frac{d\bar{\omega}}{2\pi} V(\bar{q}) \operatorname{Im}\varepsilon^{-1}(\bar{q},\bar{\omega})$$
$$\cdot (1+n_{\mathrm{B}}(\hbar\bar{\omega}))M_{23}(\bar{q})\, M_{41}(\bar{q}) \frac{A(1)A(2)A(3)A(4)\,\{g(\hbar\omega_4) - g(\hbar\omega_3)\}}{(\omega_1-\omega_2-\omega_\mu)\,(\omega_4-\omega_3-\omega_\mu)} \Gamma(4,3,\omega_\mu,q)\,. \tag{4.291}$$

This integral equation can be solved analytically only in certain approximations which will be presented in Chapter 8. The resulting corrections for the spectral line profiles, for instance, are of the same order with respect to the density as the self-energy corrections are.

4.5.4. Local Field Effects and Enhancement of the Dielectric Function

In dense Coulomb system a more or less sharp transition from a metallic state to an isolating, dielectric state (Mott transition) is possible because of the abrupt change of the ionization degree. In the metallic region, the screening by free carriers is the most important process which is expressed by the screened potential. Approaching the Mott density from the metallic state, screening of the Coulomb potential becomes more ineffective with lowering of the density, and correlations and bound states are established.

On the other hand, approaching the Mott density from the dielectric state, the static dielectric function is enhanced with increasing density so that a singularity occurs if the metallic state is achieved. The enhancement of the dielectric function in heavily doped semiconductors has been described, f.i., by CASTNER et al. (1975) and has been further discussed by HUGON and GHAZALI (1976). A unified treatment of enhancement and screening in the dielectric function was attempted by TAKESHIMA (1978). However, up to now, there has been no systematic investigation of the enhancement of the dielectric function near the Mott transition which should be based on a many-particle approach. In this Section we outline some steps in evaluating the dielectric function for a dense dielectric system.

For this we discuss three topics: The local field effects (Clausius-Mossotti relation), the self-consistent determination of the bound-state polarizabilities, and the influence of screening by free particles on the self-consistent treatment of local field effects.

The local field effects are obtained from the bound state contribution to the polarization function if the self-energy and vertex contributions are taken into account as given in the expression (4.22) and (4.28). At low temperatures, only the lowest bound state is occupied,

$$g_{ab}(E^{ab}_{mP}) \approx \tfrac{1}{4}\, n\Lambda^3_{ab}\, \mathrm{e}^{-\beta\hbar^2P^2/2M}\, \delta_{m,0}\,, \tag{4.292}$$

and furthermore the long-wave limit is considered where only dipole matrix elements contribute to the vertex function $M_{nn'}(q)$,

$$M_{nn'}(q) \approx i\boldsymbol{q}\cdot\boldsymbol{d}_{nn'}\,. \tag{4.293}$$

In contrast to the previous Section where the terms proportional to $\mathrm{Im}\,\varepsilon^{-1}(q,\omega)$ were important in the self-energy and the vertex contribution, see eq. (4.291), now the Hartree-Fock terms are taken because the interaction between bound neutral states does not show the divergent behaviour of the Coulomb interaction. Hence we find for the bound state self energy

$\Sigma^{\mathrm{HF}}_2(n, n', P, \Omega_\lambda) =$  $= -\sum_{q_1} (\boldsymbol{q}_1 ; \boldsymbol{d}_{n0})\,(-\,\boldsymbol{q}_1\cdot\boldsymbol{d}_{0n'})\,\frac{1}{q_1^2}\,\frac{n\Lambda^3_{ab}}{\varepsilon_0^4}\,\mathrm{e}^{-\beta\frac{\hbar^2(P-q_1)^2}{2M}}\,.$ (4.294)

The integral equation for the vertex function has the form

$$\begin{aligned}\Gamma(n, n', P, \Omega_\lambda, q, \omega_\mu) &= \sum_{\bar n\bar n'} G^{\mathrm{HF}}_2(n, \bar n; P, \Omega_\lambda)\, G^{\mathrm{HF}}_2(n', \bar n', P-q, \Omega_\lambda-\omega_\mu)\,(\boldsymbol{q}\cdot\boldsymbol{d}_{\bar n\bar n'})\\ &+ \sum_{q_1\mu_1 n_1' \bar n' \bar n'} G^{\mathrm{HF}}_2(n, \bar n, P, \Omega_\lambda)\, G^{\mathrm{HF}}_2(n', \bar n', P-q, \Omega_\lambda-\omega_\mu)\,(\boldsymbol{q}_1\cdot d_{\bar n n_1})\\ &\cdot(-\,\boldsymbol{q}_1\cdot d_{\bar n\, n_1'})\,\frac{4\pi}{q_1^2}\,\Gamma(n_1, n_1', P-q_1, \Omega_\lambda-\omega_{\mu_1}, q, \omega_\mu)\,.\end{aligned} \tag{4.295}$$

The summations over the internal frequencies can be performed, see RÖPKE and DER (1979), and for $\omega = 0$ the ground state polarizability

$$\hat\alpha = 2\sum_n (d^z_{0n})^2\,(E_n - E_0)^{-1} \tag{4.296}$$

appears. The summation over the internal momentum can be performed in the saddle-point approximation, and using rotational invariance $(q_1^z)^2$ is replaced by $q_1^2/3$, and in the limit $q^2 \to 0$ we arrive at

$$\Pi_2^{(b)}(0, 0) = q^2 \frac{n_{ab}\hat{\alpha}}{1 - 4\pi n_{ab}\hat{\alpha}/3}, \tag{4.297}$$

which is the well-known Clausius-Mossotti relation. For vanishing denominator, i.e. at the bound state density $n_{ab} = 3/4\,\pi\hat{\alpha}$, the static dielectric function $\varepsilon(0, 0)$ diverges. However, this divergency is not related to the Mott transition for which other mechanisms should also be taken into account.

Further improvements are obtained if the potential $V(q)$ in the Hartree-Fock self energy as well as in the integral equation for the vertex function is replaced by the screened potential. Furthermore, replacing the Coulomb interaction by the screened interaction, also the two-particle properties (energy levels and wave functions) are modified, which leads to a modified ground state polarizability.

Within a more phenomenological approach, we start from the static dielectric function given for low bound state densities by

$$\varepsilon_{ab}(q, 0) = 1 + n_{ab} \frac{\hat{\alpha}}{\varepsilon_0}(q, 0). \tag{4.298}$$

The ground state polarizability $\hat{\alpha}(q, 0)$ can be derived from the general expression (Hugon and Ghazali, 1976)

$$\hat{\alpha}(q, 0) = \frac{e^2}{q^2} \sum_n |\langle n| \, e^{i\boldsymbol{q}\cdot\boldsymbol{r}} \, |0\rangle|^2 \frac{g_{ab}(E_0^{ab})}{E_n^{ab} - E_0^{ab}}, \tag{4.299}$$

where the energy values E_i and the wave functions $|i\rangle$ are determined approximatively by the solution of the Schrödinger equation with the statically screened potential

$$V_{ab}^s(q, 0) = V_{ab}(q)/\varepsilon_{ab}(q, 0) \tag{4.300}$$

so that eqs. (4.298) and (4.299) should be solved self-consistently. Because of the q-dependence of $\varepsilon_{ab}(q, 0)$, this problem must be solved numerically. If we replace $\varepsilon_{ab}(q, 0)$ by $\varepsilon_{ab}(0, 0)$, we have a hydrogen-like system with the enhanced polarizability

$$\hat{\alpha}(0, 0) = 18\pi a_B^3 \varepsilon_0^2 (\varepsilon_{ab}(0, 0)/\varepsilon_0)^4 \tag{4.301}$$

and $\varepsilon_{ab}(0, 0)$ is given by the solution of (4.298):

$$\varepsilon_{ab}(0, 0) = \varepsilon_0 + 18\pi n_{ab} a_B^3 \varepsilon_0 (\varepsilon_{ab}(0, 0)/\varepsilon_0)^4. \tag{4.302}$$

For $n_{ab}^{1/3} a_B > 0.123$ no solution of (4.302) is found. This behaviour indicates a non-stability of the dielectric phase of the plasma for higher densities n_{ab} than a critical one. However, screening effects are overestimated in these considerations, and the q-dependence as well as the dynamical character of the screening would change the actual value of the singularity in $\varepsilon_{ab}(0, 0)$.

A unified description which takes also into account the formation of free particles was presented by Takeshima (1978), see also Schmidt, Haronska and Röpke (1981). In this approach the dielectric function $\varepsilon(q, 0)$ is expanded in a power series near $q = 0$:

$$\varepsilon(q, 0) = \frac{A}{q^2} + B + O(q^2). \tag{4.303}$$

The coefficients A, B are connected with the free particle contribution and the bound state contribution to the susceptibility of the plasma:

$$A = \varkappa^2 \varepsilon_{ab}(0, 0) ,$$

$$B = 1 + \frac{n_{ab} \cdot \hat{\alpha}(0, 0)}{1 - \frac{4\pi}{3} n_{ab} \hat{\alpha}(0, 0)} = \varepsilon_{ab}(0, 0) . \tag{4.304}$$

Binding energy, polarizability and free particle density must be determined in a self-consistent way, and the bound state contribution to the dielectric function as well as the shift of the bound state energy levels shows a singular behaviour at a critical density which will be temperature dependent. More details are given by SCHMIDT, HARONSKA and RÖPKE (1981), where also the screening of the local field by free carriers is considered with the following result for the bound state contribution to the polarizability

$$\Pi_2^{(b)}(0, 0) = q^2 \frac{n_{ab} \hat{\alpha}}{1 - \frac{4\pi}{3} n_{ab} \hat{\alpha}[1 - F(\varkappa)]} ,$$

$$F(\varkappa) = \hbar^2 \varkappa^2 \beta / M + \sqrt{\pi/2}\, \hbar^3 \varkappa^3 \sqrt{\beta/M} \exp(\beta \hbar^2 \varkappa^2 / M) [1 - \Phi \sqrt{(\beta \hbar^2 \varkappa^2 / 2M)}] , \tag{4.305}$$

Φ denotes the error integral.

We pointed out some possible improvements of the extended RPA dielectric function. These improvements can be based on a systematic many-particle approach. We were mainly concerned with diagrams which can be interpreted as self-energy or vertex corrections. Another way to construct further approximations consists of the incorporation of higher complexes into the extended RPA dielectric function as molecules, biexcitons and more general clusters, see Section 4.1.

5. Equilibrium Properties in Classical and Quasiclassical Approximation

5.1. The One-Component Plasma Model

The justification for including a chapter on classical and quasiclassical approximations into a book devoted to quantum statistics comes from the well known observation that many physical properties of Coulombic systems are determined by purely classical physics. To the class of problems which are mainly determined by classical effects belong, e.g., the equilibrium structure (Baus and Hansen, 1980; Ichimaru, 1982), long-range correlations and screening (Gruber et al., 1980, 1981; Blum et al., 1981, 1982), charge and field fluctuations (Jancovici, 1981, 1982; Lebowitz, 1983) as well as the response behaviour (Golden and Kalman, 1976, 1982). It is not the place here to discuss the classical theory in detail, the interested reader may get some orientation from the references given above, however it seems to be meaningful to explain some basic ideas and to point out the links between the classical and the quantum statistical theory. Therefore we shall concentrate on the question how quantum effects may be included into the classical theories and in what places quantum corrections to the classical behaviour appear.

We start our considerations with the one-component plasma, a system of identical point charges, interacting exclusively through the Coulomb potential, and which are immersed in a rigid, uniform background of opposite charge to ensure overall charge neutrality. This system may be described entirely classically, there are no classical divergencies which would enforce a quantummechanical treatment. Quantum effects play a role only at sufficiently high densities, where the mean distance between the charges may get smaller than the thermal De Broglie wavelength. At lower densities the conditions for a classical or semiclassical description are always fulfilled (see also 2.1.).

The one-component plasma (OCP) is the simplest possible model for a Coulombic system: N positive point charges (ze) are immersed in a uniform neutralizing negative background. In other words we assume that the electrons are not discrete, but their total charge ($-Nze$) is assumed to be continuously smeared out over the total volume V. A comprehensive survey of the theory of the OCP has recently been given by Baus and Hansen (1980). In the OCP the nuclei are considered as classical particles but the electron gas is assumed to be so highly degenerate that it can be assimilated to a rigid continuum in which the classical nuclei move. In spite of its approximative feature the OCP has proved to be very useful for the description of many properties of real plasmas.

For the classical OCP the equilibrium statistical mechanics reduces to the calculation of the configuration integral

$$Q_N = \int d\boldsymbol{r}_1 \dots d\boldsymbol{r}_N \exp\left[-\beta U_N(r_1, \dots, r_N)\right] \tag{5.1}$$

which is connected with the free energy by

$$F = F_{\text{id}}^{\text{B}} - kT \ln Q_N\,, \qquad \beta = 1/k_{\text{B}}T\,, \tag{5.2}$$

or the mean interaction energy

$$U_{\text{int}} = \langle U_N \rangle = \int d\boldsymbol{r}_1 \dots d\boldsymbol{r}_N\, U_N \exp(-\beta U_N) / \int d\boldsymbol{r}_1 \dots d\boldsymbol{r}_N \exp(-\beta U_N)\,. \tag{5.3}$$

Here we have used the notation

$$\begin{aligned} F_{\text{id}}^{\text{B}} &= \sum_a N_a kT\left[\ln(n_a \Lambda_a^3) - 1\right], \\ \Lambda_a &= h[2\pi m_a kT]^{-1/2}\,, \qquad n_a = N_a/V\,, \\ U_N &= \sum_{1\leq i<j\leq N} V_{ij}(\boldsymbol{r}_i \boldsymbol{r}_j)\,, \\ V_{ij} &= e_i e_j/4\pi\varepsilon_0 r\,, \qquad r = |\boldsymbol{r}_i - \boldsymbol{r}_j|\,. \end{aligned} \tag{5.4}$$

The mathematical problem of the calculation of the thermodynamic function reduces to the calculation of 3 N-dimensional integrals ($N \sim 10^{23}$). Analytical results are known only for the case of very small densities by using the methods of cluster expansions (ABÉ, 1959; FRIEDMAN, 1962; EBELING et al. 1966; CZERWON, 1972). One finds, e.g.,

$$\begin{aligned} P_{\text{int}}/nkT &= -\tfrac{1}{6}\mu - \tfrac{1}{12}\mu^2\left(\ln\mu + 2C + \ln 3 - \tfrac{4}{3}\right) \\ &\quad - \tfrac{1}{8}\mu^3(\ln\mu + 0.4064)\,, \\ F_{\text{int}}/NkT &= -\tfrac{1}{3}\mu - \tfrac{1}{12}\mu^2\left(\ln\mu + 2C + \ln 3 - \tfrac{11}{6}\right) \\ &\quad - \tfrac{1}{12}\mu^3(\ln\mu + 0.0731) \end{aligned} \tag{5.5}$$

where

$$\mu = (e^2/4\pi\varepsilon_0 r_{\text{D}} k_{\text{B}}T), \qquad C = 0.577216, \qquad r_{\text{D}} = [ne^2/\varepsilon_0 k_{\text{B}}T]^{-1/2}\,. \tag{5.6}$$

Unfortunately these results are of little use since the OCP works only at high densities due to the condition that the electrons should be degenerate. Therefore what one really needs are calculations of the integrals (5.1)—(5.3) at high densities. For this region the methods of Monte Carlo calculation (MC) of molecular dynamics (MD) and of nonlinear integral equations (hyper-netted-chain, Percus-Yevick — HNC, PY etc.) are available now. The MC method introduced by METROPOLIS, ROSENBLUTH and TELLER (1953) and worked out for Coulomb systems by BRUSH, SAHLIN and TELLER (1966) is a sampling method based on a certain Markov process. The phase space average is reduced to a time average

$$\bar{A} = \frac{1}{M}\sum_{\nu=1}^{M} A(\boldsymbol{R}_\nu), \qquad \boldsymbol{R}_\nu = (\boldsymbol{r}_{1\nu}, \dots, \boldsymbol{r}_{N\nu})\,. \tag{5.7}$$

One constructs a random walk of point through the space via a Markov process such that the probability $P(\boldsymbol{R})$ tends toward the canonical distribution $P_{\text{eq}}(\boldsymbol{R})$ as $M \to \infty$ (BINDER, 1979; ZAMALIN, NORMAN and FILINOV, 1977).

This Markov process is defined by specifying a transition probability $W(\boldsymbol{R}_\nu \to \boldsymbol{R}_\nu')$ from one configuration $\boldsymbol{R}_\nu$ to another $\boldsymbol{R}_\nu'$. In order that the Markov process has the desired property that $P(\boldsymbol{R}_\nu)$ converges toward $P_{eq}(\boldsymbol{R}_\nu)$, it is sufficient to impose the detailed balance condition

$$P_{eq}(\boldsymbol{R}_\nu)\, W(\boldsymbol{R}_\nu \to \boldsymbol{R}_\nu') = P_{eq}(\boldsymbol{R}_\nu')\, W(\boldsymbol{R}_\nu' \to \boldsymbol{R}_\nu) \tag{5.8}$$

which means that the ratio of transition probabilities depends on the change in the potential energy only:

$$W(\boldsymbol{R}_\nu \to \boldsymbol{R}_\nu') : W(\boldsymbol{R}_\nu' \to \boldsymbol{R}_\nu) = \exp\{-[U_N(\boldsymbol{R}_\nu) - U_N(\boldsymbol{R}_\nu')]/k_B T\}\,. \tag{5.9}$$

The choice of W commonly used is

$$W(\boldsymbol{R}_\nu \to \boldsymbol{R}_\nu') = \begin{cases} \tau_s^{-1} \exp(-\delta U_N/k_B T) & \text{if} \quad \delta U_N > 0\,, \\ \tau_s^{-1} & \text{otherwise} \end{cases} \tag{5.10}$$

where τ_s is some time scaling factor. Next we consider the question of the way in which the MC scheme is practically realized. One starts by specifying an initial condition $\boldsymbol{r}_k(0)$ for the set of coordinates of the ions. Then one selects one particle i for which the coordinates will be changed randomly from $\boldsymbol{r}_i$ to $\boldsymbol{r}_i'$. Next one computes the change in the potential energy δU_N produced by the trial move, and computes the transition probability W. Then one selects a random number z in the interval $0<z<1$. If $\tau_s W < z$ the move is rejected and the old state is counted once more as a "new" state in the averaging. If $\tau_s W > z$; the move is accepted, i.e., the state with the new coordinates $\boldsymbol{r}_i'$ is taken as the new configuration in the averaging. This procedure is repeated very often, and thus a Markov chain of M events is generated which serves as a basis for the calculation of the average.

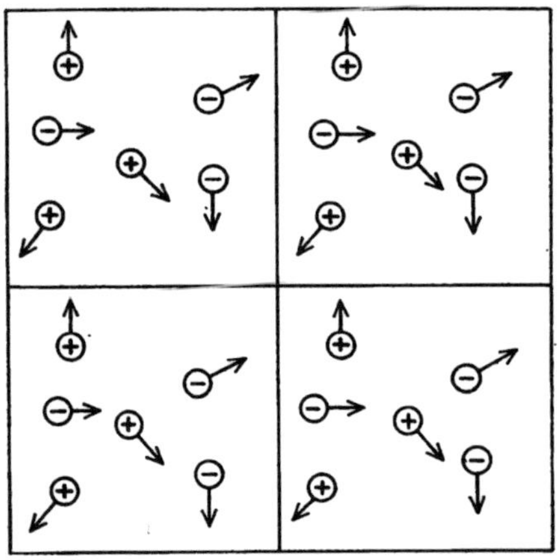

Fig. 5.1. Elementary cells and periodic boundary conditions in Monte Carlo calculations

In the MC method, one studies finite systems of N particles $N \sim 10^3 - 10^4$ while one is really interested in the bulk properties of very large systems ($N \to \infty$). In order to be able to extrapolate to the thermodynamic limit ($N \to \infty$, $V \to \infty$, N/V-finite) special boundary conditions have been introduced. The usual boundary condition of a small system to simulate infinite systems is the periodic boundary condition. We assume that the N particles are enclosed in a hypercubic box with the linear dimensions L_1, L_2, L_3. Now we compose an infinite system by the combination of many hypercubic boxes assuming that in each of these elementary boxes happens exactly the same (Fig. 5.1). Some special problems appear in the case of Coulomb interactions since, due to their long range nature, an ion in a given cell interacts with the ions in many surrounding cells. In other words, account must be taken not only of the electrostatic interactions of a given ion with the $N - 1$ ions in the same box, but also with all the periodic replicas of these particles. The resulting infinite sums are poorly convergent; the difficulty can be overcome by using the Ewald summation technique similar to those used in solid state physics to compute Coulombic lattice sums (BINDER, 1972). MC calculations for dense OCP, i.e.

$$\Gamma \gg 1\,, \quad \Gamma = \frac{z^2e^2}{4\pi\varepsilon kTd}\,, \qquad d = \left(\frac{3}{4\pi n_i}\right)^{1/3} \tag{5.11}$$

are extremely computer time consuming. The pioneering MC work on the dense OCP is due to BRUSH, SAHLIN and TELLER (1966) and to HANSEN (1973). Hansens computation covered the range $1 \lessapprox \Gamma \lessapprox 300$ for systems of 16, 54, 128 and 250 ions and generated about $5 \cdot 10^5$ configurations per run. His interaction energies were fitted by DE WITT (1976, 1978), within the statistical errors by the very simply function

$$U_{\text{int}}/NkT = -\,0.896434\,\Gamma + 0.86185\,\Gamma^{1/4} - 0.5551\,. \tag{5.12}$$

At high densities the interaction energy approaches within a few percent the static *bcc* lattice energy

$$U_{\text{int}}/Nk_{\text{B}}T = -0.895292\,\Gamma\,. \tag{5.13}$$

The MC calculations performed by POLLOCK and HANSEN (1973) for the crystalline phase may be fitted by the expression (DE WITT, 1976)

$$\begin{aligned} &U_{\text{int}}/Nk_{\text{B}}T = a\Gamma + b\Gamma^{1/2} + c\ln\Gamma + d\,,\\ &a = -0.90047, \qquad b = 0.2688263\,,\\ &c = 0.0719925, \quad d = 0.0537919\,. \end{aligned} \tag{5.14}$$

The Helmholtz free energies, obtained from integrating the equations of state are for the fluid phase represented by

$$F_{\text{int}}/Nk_{\text{B}}T = -0.896434\Gamma + 3.4474\Gamma^{1/4} - 0.5551\ln\Gamma - 2.809 \tag{5.15}$$

and for the crystalline phase by

$$F_{\text{int}}/Nk_{\text{B}}T = -0.90047\Gamma + 0.5376526\Gamma^{1/2} + \dots \tag{5.16}$$

The free energies for both phases intersect at

$$\Gamma_{\text{melt}} = 155 \pm 10$$

being the locus of the phase transition liquid-solid. The OCP model has been generalized to treat mixtures of point ions by HANSEN, TORRIE and VIEILLEFOSSE (1977). These authors have performed MC computations for H^+—He^{++} mixtures which are of course of special importance for the treatment of fusion plasmas. To model electrons in a plasma as a rigid uniform background on which the ions move is reasonable only in the limit of extremely high densities. At lower densities the polarization of the electron background by the ions should be taken into account. Neglecting the dynamical screening effects, the Coulombic potential energy must then be modified by the introduction of the static dielectric function $\varepsilon(k)$ of the electron background. In the Fourier representation follows (see also Section 4.2.)

$$U_N(r_1, \dots, r_N) = \frac{1}{2V} \sum_k{}' \frac{z^2e^2}{\varepsilon_0 k^2} \left[\varrho_k \varrho_{-k} \frac{1}{\varepsilon(k)} - N \right],$$

$$\varrho(k) = \frac{1}{\sqrt{N}} \sum_{j=1}^{N} \exp(-i\boldsymbol{k} \cdot \boldsymbol{r}_j) . \tag{5.17}$$

Hubbard and Slattery (1971) as well as De Witt and Hubbard (1976) have carried out MC computations for H^+ plasmas and for H—He plasmas in a responding electron background using for $\varepsilon(k)$ the Lindhard dielectric function and considering two densities corresponding to the Brueckner parameters $r_s = 0.1$ and $r_s = 1.0$. The main result of their calculations is that the deviation of the free energy from its $r_s \to 0$ limit is essentially linear in r_s and relatively small up to $r_s = 1$. For H-plasmas the correction term is given approximately by

$$\delta F = Nk_BT\, r_s[0.0579\Gamma + 0.971\Gamma^{1/4} - 0.343] . \tag{5.18}$$

These results are in quite good agreement with a simple perturbation theory around the original OCP model with a rigid background which starts from (Galam and Hansen, 1976)

$$\delta F = Nk_BT \frac{1}{3\pi} \int_0^\infty \mathrm{d}k\, k^2\, S(k)\, w(k) ,$$

$$w(k) = 3\Gamma k^{-2} [\varepsilon^{-1}(k) - 1],\ S(k)\text{-structure factor.} \tag{5.19}$$

Introducing here the binary correlation function from the rigid background OCP, δF may be easily calculated.

5.2. Many-Component Systems. Slater Sums

5.2.1. Partition Functions and Effective Potentials

In the canonical ensemble which will be used in this Chapter we have

$$F = -k_BT \ln Z$$

where in coordinate representation

$$Z = \sum_{\sigma_i} \int \mathrm{d}\boldsymbol{r}_1 \dots \mathrm{d}\boldsymbol{r}_N \langle \sigma_1 r_1 \dots \sigma_N r_N | \exp(-\beta H_N) | \sigma_N r_N \dots \sigma_1 r_1 \rangle . \tag{5.20}$$

In the case of ideal particles $H_N = \Sigma p_i^2/2m$ and Boltzmann statistics we have

$$Z = (V^N/N!\, \Lambda^{3N}) , \quad \Lambda = (\hbar/\sqrt{2\pi mkT}) . \tag{5.21}$$

Therefore, in the general case we may write

$$F = F_{\mathrm{id}}^{\mathrm{B}} - k_BT \ln \{N!\, \Lambda^{3N}\, V^{-N}$$
$$\cdot \sum_{\sigma_i} \int \mathrm{d}\boldsymbol{r}_1 \dots \mathrm{d}\boldsymbol{r}_N \langle \sigma_1 r_1 \dots \sigma_N r_N | \exp(-\beta H_N) | \sigma_N r_N \dots \sigma_1 r_1 \rangle\} . \tag{5.22}$$

Let us define Slater sums by

$$S^{(N)}(r_1, \dots, r_N) = N!\, \Lambda^{3N} \sum_{\sigma_i} \langle \dots |\exp(-\beta H_N)| \dots \rangle \,.$$

This is the counterpart of the classical Boltzmann factor. Remember that in classical statistics

$$F = F_{\text{id}}^{\text{B}} - k_{\text{B}}T \ln \int d\boldsymbol{r}_1 \dots d\boldsymbol{r}_N \exp\big(-\beta U_N(r_1, \dots, r_N)\big)\, V^{-N} \tag{5.23}$$

where

$$U_N(r_1, \dots, r_N) = \sum_{i<j} V_{ij}(r_i, r_j)\,.$$

In the quantum mechanical case $S^{(N)}$ may be expressed by the energy eigen function

$$S^{(N)}(r_1, \dots, r_N) = N!\, \Lambda^{3N} \sum_{n,\sigma_i} \exp(-\beta E_n) \cdot |\psi_n(r_1, \sigma_1, \dots, r_N \sigma_N)|^2\,. \tag{5.24}$$

In the classical limit we have

$$\lim_{\hbar \to 0} S^{(N)}(r_1, \dots, r_N) = \exp\big(-\beta \sum_{i<j} V_{ij}\big)\,. \tag{5.25}$$

Therefore it seems to be useful to introduce near to the classical limit the representation if $\hbar \neq 0$

$$S^{(N)}(r_1, \dots, r_N) = \exp\big[-\beta \sum_{i<j} u_{ij}(r_i r_j) - \beta \sum_{i<j<k} u_{ijk}(r_i, r_j, r_k) + \dots\big]\,. \tag{5.26}$$

We have introduced here higher order interactions, beside two-particle interactions, since quantum effects create correlations between particles. Formally the representation (5.26) is correct. By specialization to $N = 2$, $N = 3$, etc. we find

$$u_{ij} = -k_{\text{B}}T \ln S^{(2)}(r_i, r_j)$$

$$u_{ijk} = -k_{\text{B}}T \ln \{S^{(3)}(r_i, r_j, r_k)/S^{(2)}(r_i, r_j)\, S^{(2)}(r_i, r_k)\, S^{(2)}(r_j, r_k)\}\,. \tag{5.27}$$

We note that the N-particle canonical distribution which is defined as the diagonal element density matrix

$$F_N(r_1, \dots r_N) = \langle r_1 \dots r_N |\exp(-\beta H_N)| r_N \dots r_1 \rangle \tag{5.28}$$

is in analogy to the classical case given by

$$F_N(r_1, \dots, r_N) = \frac{\exp(-\beta U_N)}{\int dr_1 \dots dr_N \exp(-\beta U_N)}\,. \tag{5.29}$$

Having introduced the effective potentials u_{ij}, u_{ijk} etc. we are able to calculate all the equilibrium properties via the classical route (Zelener, Norman and Filinov, 1981). For the internal energy one finds, e. g.,

$$U/Nk_{\text{B}}T = (U_{\text{id}} + \langle V_N \rangle)/Nk_{\text{B}}T\,, \tag{5.30}$$

$$\langle V_N \rangle = \int d\boldsymbol{r}_1 \dots d\boldsymbol{r}_N\, V_N F_N(r_1, \dots, r_N)\,, \tag{5.31}$$

$$V_N = \sum_{1 \le i < j \le N} \frac{e_i e_j}{4\pi\varepsilon\, |\boldsymbol{r}_i - \boldsymbol{r}_j|}\,. \tag{5.32}$$

We note that the potential energy V_N is different from the effective potential U_N.

It seems to be promising to start MC calculations on the basis of these expressions. In other words the idea is, to construct again a Markov process such that the probability $P(R)$ tends toward the Slater sum (5.26) as $M \to \infty$. Then the phase average

(5.31) is replaced by a time average (or sampling average)

$$\bar{V}_N = \frac{1}{M} \sum_{\nu=1}^{M} V_N(\boldsymbol{R}_\nu) \,. \tag{5.33}$$

Such an approach has been developed by ZELENER, NORMAN and FILINOV (1981). However the MC method for quantum plasmas is still at the very beginning. Still there are many difficulties to be solved, connected mainly with the existence of many-particle effective interactions. The progress which has been reached recently in calculating the ground state energies of quantum systems by MC methods is very encouraging (CEPERLEY and KALOS, 1979, CEPERLEY and ALDER, 1980, 1981). Looking at the very promising results for low-temperature systems we believe that the method proposed in the investigations of Zelener, Norman and Filinov for high-temperature plasmas has good perspectives. However further progress will depend of course on better results in the calculation of the higher order effective potentials. Since the essential part of the many particle correlations comes from the ideal gas contributions we propose to write the canonical distribution in the following way

$$F_N(r_1, \dots, r_N) = F_N^0(r_1, \dots, r_N)\, C \exp\Big[-\beta \sum_{i<j} v_{ij}(r_i, r_j) - \beta \sum_{i<j<k} v_{ijk}(r_i, r_j, r_k) - \dots\Big] \,. \tag{5.34}$$

The density operator of ideal systems F_N^0 is exactly known and has been given also in the previous Chapter.

We note that the representation given above is very similar to the ground state wave function used in MC calculations (CEPERLEY and KALOS, 1979; CEPERLEY and ALDER, 1980, 1981)

$$\psi(r_1, \dots, r_N) = \psi^0(r_1, \dots, r_N) \exp\Big[-\tfrac{1}{2} \sum_{i<j} v_{ij}(r_i, r_j)\Big] \tag{5.35}$$

where ψ^0 is the ideal gas wave function. Such a form has been proposed by Bijl, Mott, Dingle and Jastrow and is nowadays being used by many authors working on the numerical study of quantum many-body systems.

How does a MC simulation for quantum distribution functions (5.34) work? The Metropolis algorithm is again a biased random walk in configuration space. Each particle is moved one after another to a new position uniformly distributed inside a cube of side s. That move is either accepted or rejected depending on the magnitude of the distribution function F_N. Suppose $\boldsymbol{R}$ is the old position and $\boldsymbol{R}'$ the new. Then if

$$F_N(\boldsymbol{R}') \geqq F_N(\boldsymbol{R}) \tag{5.36}$$

the new point is accepted. Otherwise the new point is accepted with probability q where

$$q = F_N(\boldsymbol{R}')/F_N(\boldsymbol{R}) \,. \tag{5.37}$$

The sample average of the potential energy (or another quantity)

$$\bar{V}_N = \frac{1}{M} \sum_{\nu=1}^{M} V_N(\boldsymbol{R}_\nu) \tag{5.38}$$

will approach the ensemble average $\langle V_N \rangle$ as M increases. Periodic boundary condition may be used to eliminate surface effects. The long-range problem posed by the Coulomb tail of the two particle potential u_{ij} may be solved again by the use of the Ewald image potential. Since many particle interactions are very difficult to calculate, one may also introduce variational parameters $\alpha_1, \alpha_2, \dots$ into v_{ij}, i. e.

$$F_N(r_1, \dots, r_N; \alpha_1, \alpha_2, \dots) = F^0(r_1, \dots, r_N)\, C \exp\left[-\beta \sum v_{ij}(r_i, r_j; \alpha_1, \alpha_2, \dots)\right]. \tag{5.39}$$

Afterwards the free parameters $\alpha_1, \alpha_2, \dots$ are found by minimizing the free energy:

$$F(N, V, T; \alpha_1, \alpha_2, \dots) = \text{Min.} \tag{5.40}$$

There are many problems where the multi-component description may be reduced to a pseudo-one-component description by using the asymmetry of masses (Brovman and Kagan, 1967, 1969; Brovman et al., 1972; Onishtshenko and Krasny, 1973; Galam and Hansen, 1976; Baus and Hansen, 1980; Ebeling et al., 1984).

5.2.2. Calculation of Slater Sums and Effective Potentials

Let us consider first the case of two particles with the charges e_a, e_b, the masses m_a, m and the spins s_a, s_b interacting via Coulomb's law

$$V_{ab} = \frac{e_a e_b}{4\pi\varepsilon\, |\boldsymbol{r}_2 - \boldsymbol{r}_1|} \tag{5.41}$$

and having the Hamiltonian

$$H_{ab} = -\frac{\hbar^2}{2m_a}\Delta_1 - \frac{\hbar^2}{2m_b}\Delta_2 + V_{ab}. \tag{5.42}$$

Let us then consider the two-particle Slater sum which is still relatively simple to calculate. By using the momentum eigenfunctions we get the representation

$$\begin{aligned} S_{ab}(r_1, r_2) &= \int \mathrm{d}\boldsymbol{k}_1\, \mathrm{d}\boldsymbol{k}_2 \\ &\cdot \sum_{\sigma_1\sigma_2} \psi^*_{k_1k_2\sigma_1\sigma_2}(r_1, r_2) \exp(-\beta H_{ab})\, \psi_{k_1k_2\sigma_1\sigma_2}(r_1, r_2) \end{aligned} \tag{5.43}$$

where the wave functions with the appropriate symmetry are

$$\begin{aligned} \psi_{k_1k_2\sigma_1\sigma_2}(r_1, r_2) &= \text{const}\,\{\exp(i\boldsymbol{k}_1\boldsymbol{r}_1 + i\boldsymbol{k}_2\boldsymbol{r}_2)\, \chi_1(\sigma_1)\, \chi_2(\sigma_2) \\ &+ (-1)^{2s_a} \exp(i\boldsymbol{k}_1\boldsymbol{r}_2 + i\boldsymbol{k}_2 \cdot \boldsymbol{r}_1)\, \chi_1(\sigma_2)\, \chi_2(\sigma_1)\}. \end{aligned} \tag{5.44}$$

By introducing relative and mass center coordinates and

$$H_{ab} = -\frac{\hbar^2}{2m_{ab}}\Delta + V_{ab}, \qquad m_{ab} = \frac{m_a m_b}{m_a + m_b}, \tag{5.45}$$

we get after performing the spin summations and one $\boldsymbol{k}$-integration

$$\begin{aligned} S_{ab}(r) &= \lambda_{ab}^3 \pi^{-3/2} \int \mathrm{d}\boldsymbol{k}\, [\exp(-i\boldsymbol{k} \cdot \boldsymbol{r}) + \delta_{ab}(-1)^{2s_a}(2s_a + 1)^{-1} \\ &\cdot \exp(+i\boldsymbol{k} \cdot \boldsymbol{r})]\, W_{ab}(k, r, \beta), \\ W_{ab}(k, r, \beta) &= \exp(-\beta H_{ab}) \exp(i\boldsymbol{k} \cdot \boldsymbol{r}), \\ \lambda_{ab} &= \hbar(2\pi m_{ab} kT)^{-1/2}. \end{aligned} \tag{5.46}$$

By differentiation with respect to β one gets the Bloch equation

$$\frac{\partial W_{ab}}{\partial \beta} + H_{ab} W_{ab} = 0 \tag{5.47}$$

or more explicitely

$$\frac{\partial W_{ab}}{\partial \beta} = -V_{ab} W_{ab} + \frac{\hbar^2}{2m_{ab}} \Delta W_{ab} . \tag{5.48}$$

Substituting here

$$W_{ab} = \exp\left[i\boldsymbol{k}\cdot\boldsymbol{r} - \frac{\beta\hbar^2}{2m_{ab}} k^2\right] v_{ab}(k, r, \beta) \tag{5.49}$$

we get finally

$$\frac{\partial v_{ab}}{\partial \beta} = -V_{ab} v_{ab} + \frac{\hbar^2}{2m_{ab}} [\Delta v_{ab} + 2i\boldsymbol{k}\cdot\nabla v_{ab}] . \tag{5.50}$$

The first method for the solution of this equation uses an $\hbar^2$-expansion. By substituting

$$v_{ab} = \exp\{-\beta V_{ab}^{(0)} - \beta\hbar^2 V_{ab}^{(1)} - \beta\hbar^4 V_{ab}^{(2)} - \beta\hbar^6 V_{ab}^{(3)} - \dots\} \tag{5.51}$$

we get finally (Hoffman and Kelbg, 1966, 1967; Rohde et al., 1968)

$$\begin{aligned} S_{ab}(r) = \exp\Bigg\{&-\beta \frac{e_a e_b}{4\pi\varepsilon_0 r} + \frac{\beta^3\hbar^2}{24 m_{ab} r^4}\left(\frac{e_a e_b}{4\pi\varepsilon_0}\right)^2 \\ &+ \frac{\beta^4\hbar^4}{60 m_{ab}^2 r^6}\left(\frac{e_a e_b}{4\pi\varepsilon_0}\right)^2 - \frac{\beta^5\hbar^4}{120 m_{ab}^2 r^7}\left(\frac{e_a e_b}{4\pi\varepsilon_0}\right)^3 + \frac{3\beta^5\hbar^6}{224 m_{ab}^3 r^8}\left(\frac{e_a e_b}{4\pi\varepsilon_0}\right)^6 \\ &- \frac{89\beta^6\hbar^6}{7560 m_{ab}^3 r^9}\left(\frac{e_a e_b}{4\pi\varepsilon_0}\right)^3 + \frac{13\beta^7\hbar^6}{5040 m_{ab}^3 r^{10}}\left(\frac{e_a e_b}{4\pi\varepsilon_0}\right)^4 + O(\hbar^8)\Bigg\} \\ &+ \text{exchange terms.} \end{aligned} \tag{5.52}$$

The investigation of this series shows convergence only in the case of large r-values; here the exchange terms may be neglegted since they decrease exponentially with increasing r. In the classical limit $\hbar \to 0$, the Boltzmann factor is obtained. We note that the exponent in eq. (5.52) represents for large r the effective two-particle potential. A second solution method is based on a perturbation expansion with respect to e^2 which reads

$$v_{ab}(k, r, \beta) = 1 + v_{ab}^{(1)} + v_{ab}^{(2)} + \dots , \qquad v_{ab}^{(n)} = O(e^{2n}) , \tag{5.53}$$

or for the Fourier transforms

$$v_{ab}(k, t, \beta) = (2\pi)^3 \delta(\boldsymbol{t}) + \tilde{v}_{ab}^{(1)} + \tilde{v}_{ab}^{(2)} + \dots \tag{5.54}$$

By iteration we get easily the n-th order term

$$\begin{aligned} v_{ab}^{(n)}(k, t, \beta) = -\frac{1}{(2\pi)^3}\int_0^\beta \mathrm{d}\beta' \exp\Bigg[&-\frac{\hbar^2}{2m_{ab}}(t^2 - 2\boldsymbol{k}\cdot\boldsymbol{t})(\beta - \beta') \\ &\cdot \int \mathrm{d}\boldsymbol{t}'\, \tilde{V}_{ab}(\boldsymbol{t} - \boldsymbol{t}')\, \tilde{v}_{ab}^{(n-1)}(k, t', \beta')\Bigg] . \end{aligned} \tag{5.55}$$

In this way we get in the linear order in e^2 for the Fourier transform of the Slater sum (Kelbg, 1964; Ebeling et al., 1967)

$$\begin{aligned}\tilde{S}_{ab}(t) &= (2\pi)^3\,\delta(t) - (\beta\, e_a e_b/\varepsilon t^2)\exp\left(-\tfrac{1}{4}\lambda_{ab}^2 t^2\right)\\ &\cdot {}_1F_1\left(\tfrac{1}{2},\tfrac{3}{2};\tfrac{1}{4}\lambda_{ab}^2 t^2\right) + \delta_{ab}(-1)^{2s_a}(2s_a+1)^{-1}\\ &\cdot \left\{\lambda_{ab}^3\pi^{3/2}\exp\left(-\tfrac{1}{4}\lambda_{ab}^2 t^2\right) - \tfrac{1}{2}(\beta e_a e_b/\varepsilon)\,\lambda_{ab}^2\exp\left(-\tfrac{1}{4}\lambda_{ab}^2 t^2\right)\right.\\ &\left.\cdot {}_2F_2\left(\tfrac{1}{2},\tfrac{1}{2},\tfrac{3}{2},\tfrac{3}{2};\tfrac{1}{4}\lambda_{ab}^2 t^2\right)\right\} + O(e^4)\,. \end{aligned} \tag{5.56}$$

From this result we get for the Fourier transform of the effective potential in the first oder

$$\begin{aligned} w_{ab}(t) &= (e_a e_b/\varepsilon)\exp\left(-\tfrac{1}{4}\lambda_{ab}^2 t^2\right)\left\{t^{-2}\,{}_1F_1\left(\tfrac{1}{2},\tfrac{3}{2};\tfrac{1}{4}\lambda_{ab}^2 t^2\right)\right.\\ &\left.+ \tfrac{1}{2}\,\delta_{ab}(-1)^{2s_a}(2s_a+1)^{-1}\lambda_{ab}^2\,{}_2F_2\left(\tfrac{1}{2},\tfrac{1}{2},\tfrac{3}{2},\tfrac{3}{2};\tfrac{1}{4}\lambda_{ab}^2 t^2\right)\right\}\,. \end{aligned} \tag{5.57}$$

For particles of different species $a \neq b$ this is easily inverted to the coordinate space and yields the so-called Kelbg potential (Kelbg, 1964, 1972; Klimontovich and Kraeft, 1974)

$$w_{ab}(r) = \frac{e_a e_b}{4\pi\varepsilon_0 r}\left[1 - \exp\left(-\frac{r^2}{\lambda_{ab}^2}\right) + \sqrt{\pi}\left(\frac{r}{\lambda_{ab}}\right)\left(1 - \Phi\left(\frac{r}{\lambda_{ab}}\right)\right)\right]. \tag{5.58}$$

We note that w_{ab} is finite for $r \to 0$, i.e., the Coulomb divergence disappears due to Heisenberg's uncertainty effect (Fig. 5.2).

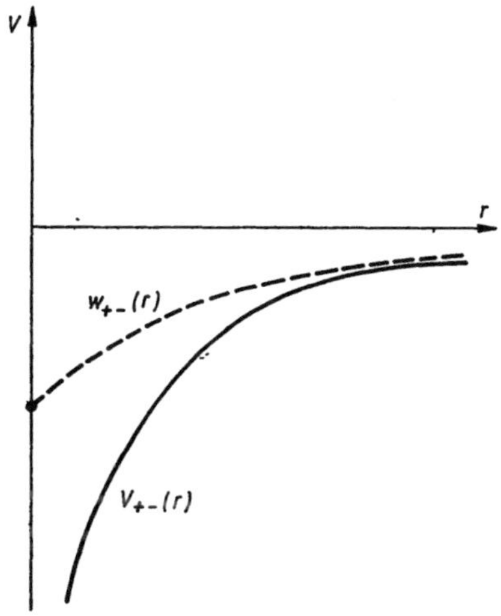

Fig. 5.2. Interaction energy between different charges and the corresponding effective potential after Kelbg (1964)

Now we shall give a representation of S_{ab} by means of the energy eigenfunctions (Morita, 1959; Ebeling, 1968, 1969). By starting from eq. (5.24) we write

$$S_{ab}(r) = (1+\delta_{ab})!\,8\pi^{3/2}\lambda_{ab}^3\sum_\alpha \exp(-\beta E_\alpha)\,|\psi_\alpha(r,\vartheta,\varphi)|^2 \tag{5.59}$$

where ψ_α are the eigen functions and E_α the eigen energies for the relative motion (including the discrete as well as the continuous part of the spectrum).

By using a summation theorem for the spherical functions

$$\sum_{m=-l}^{+l} |Y_{lm}(\vartheta,\varphi)||^2 = \frac{1}{4\pi}(l+1) \tag{5.60}$$

we obtain finally

$$S_{ab}(r) = 2\lambda_{ab}^3 \sqrt{\pi} \{\theta(-e_a e_b) \sum_{s=1}^{\infty} \sum_{l=0}^{\infty} (2l+1) \exp(-\beta E_s) |R_{sl}(r)|^2$$

$$+ \sum_{l=0}^{\infty} (2l+1) [1 + \delta_{ab}(-1)^{2s_a+l} (2s_a+1)^{-1}]$$

$$\cdot \int_0^{\infty} dk \frac{dn_{lk}}{dk} \exp(-\lambda_{ab}^2 k^2) |R_{kl}(r)|^2 k , \quad (5.61)$$

where $R(r)$ are the radial parts of the wave function and dn_{lk}/dk — density of states. Since $R(r)$ is analytically given for Coulomb systems, $S_{ab}(r)$ may be calculated from this formula. (BARKER, 1968, 1971; STORER, 1968; EBELING et al., 1968; ROHDE et al., 1968; ZELENER, NORMAN and FILINOV, 1972, 1973, 1974, 1975, 1981).

The calculation of the wavefunctions and performance of all the summtions and integrations is a very time consuming procedure. Therefore we shall use here an interpolation method which is based on Taylor expansions for small r. By splitting the two-particle Slater sum into an Heisenberg term and a symmetry term,

$$S_{ab}(r) = S_{ab}^{\mathrm{H}} + S_{ab}^{\mathrm{S}} , \quad (5.62)$$

we get by expansion of eq. (5.61) with respect to r

$$S_{ab}^{\mathrm{H}} = [4\sqrt{\pi}\, \xi_{ab} J_1(\xi_{ab}) + \theta(-e_a e_b) \sqrt{\pi}\, \xi_{ab}^3 Z_3(\xi_{ab})] \cdot \exp[-\xi_{ab} r/\lambda_{ab} + O(r^2/\lambda_{ab}^2)] , \quad (5.63)$$

$$S_{ab}^{\mathrm{S}} = \delta_{ab}(-1)^{2s_a} (2s_a+1)^{-1} 4\sqrt{\pi} \xi_{ab} J_1(\xi_{ab}) \cdot \exp[-\xi_{ab} r/\lambda_{ab} + O(r^2/\lambda_{ab}^2)] \quad (5.64)$$

where (EBELING et al., 1968)

$$J_n(\xi) = \int_0^{\infty} dx\, x^n [1 - \exp(-\pi\xi/x)]^{-1} ,$$

$$Z_n(\xi) = \sum_{s=1}^{\infty} s^{-n} \exp[\xi^2/4s^2] . \quad (5.65)$$

We need here only J_1 and Z_3 which are given in Tab. 5.1. In the higher Taylor coefficient of n-th order appear the corresponding functions J_n and Z_{n+2}. The following

Tab. 5.1. $I_1^{\mathrm{T}}(B)$ and $Z_3^{\mathrm{T}}(B)$ as a function of B.
$I_1^{\mathrm{T}}(\pi\xi) = 4\sqrt{\pi}\, \xi I_1(\xi)$; $Z_3^{\mathrm{T}}(\pi\xi) = Z_3(\xi)$.

B	$I_1^{\mathrm{T}}(B)$	$Z_3^{\mathrm{T}}(B)$
20	2.257 +1	2.514 +4
15	1.693 +1	2.993 +2
10	1.129 +1	1.929 +1
5	5.712	2.111
1	1.701	1.229
.5	1.320	1.209
0	1.000	1.202
−.5	7.556 −1	
−1.0	5.730 −1	
−5.0	7.035 −2	
−10	6.449 −3	
−15	7.161 −4	
−20	9.186 −5	

Pade approximations have the correct r-dependence for small as well as for large distances:

$$
\begin{aligned}
S^{\mathrm{H}}_{ab}(r) &= \exp\left[\frac{A_{ab} - \xi_{ab}\varrho + 12A_{ab}\xi_{ab}^{-1}\varrho^3}{1 - \varrho^2 + 12A_{ab}\xi_{ab}^{-2}\varrho^4}\right], \\
S^{\mathrm{S}}_{ab}(r) &= \delta_{ab}(-1)^{2s_a}(2s_a + 1)^{-1}\exp\left[A_{ab} - \xi_{ab}\varrho - \varrho^2\right], \\
A_{ab} &= A(\xi_{ab}) = \ln\left\{4\pi\xi_{ab}J_1(\xi_{ab}) + \theta(-\xi_{ab})\sqrt{\pi}\,\xi_{ab}^3 Z_3(\xi_{ab})\right\}, \\
\varrho &= r/\lambda_{ab}\,.
\end{aligned}
\tag{5.66}
$$

5.3. The Pair Distribution Function

5.3.1. Basic Equations and Hierarchy

In Section 3.1.4. we have introduced the reduced s-particle density operators (eq. (3.36))

$$
F_s(1, \dots, s) = V^s \operatorname*{Tr}_{(s+1)\dots N} \varrho_N(1,\dots, N)\,. \tag{5.67}
$$

Starting from the canonical ensemble with

$$
\varrho_N = Z^{-1}\exp(-\beta H_N)
$$

we now introduce the diagonal part of the coordinate representation of the reduced density operator by

$$
\begin{aligned}
F_s(r_1, \dots, r_s) &= Q^{-1}\int d\boldsymbol{r}_{s+1}\dots d\boldsymbol{r}_N\, S^{(N)}(r_1, \dots, r_N)\, V^s \\
&= V^sQ^{-1}\int d\boldsymbol{r}_{s+1}\dots d\boldsymbol{r}_N \exp\left[-\beta\sum u_{ij} - \beta\sum u_{ijk} - \dots\right].
\end{aligned}
\tag{5.68}
$$

By differentiation with respect to the first coordinate and using (5.68) follows (Bogolyubov, 1969, 1979, 1971)

$$
\begin{aligned}
\frac{\partial}{\partial \boldsymbol{r}_1}F_s &= -\beta F_s\frac{\partial}{\partial \boldsymbol{r}_1}U_s \\
&\quad - \beta n\int d\boldsymbol{r}_{s+1}\,F_{s+1}\frac{\partial}{\partial \boldsymbol{r}_1}U_{s+1}\,.
\end{aligned}
\tag{5.69}
$$

Taking into account explicitly the dependence on the species to which the s particles belong we may write

$$
\begin{aligned}
\frac{\partial}{\partial \boldsymbol{r}_1}F_{a_1\dots a_s} &= -\beta F_{a_1\dots a_s}\frac{\partial}{\partial \boldsymbol{r}_1}U_{a_1\dots a_s} \\
&\quad - \beta\sum_{a_{s+1}} n_{a_{s+1}}\int d\boldsymbol{r}_{s+1}\,F_{a_1\dots a_{s+1}}\frac{\partial}{\partial \boldsymbol{r}_1}U_{a_1\dots a_{s+1}}\,.
\end{aligned}
\tag{5.70}
$$

We proceed now to a formal solution of this hierarchy by density expansions (Bogolyubov, 1946, 1969; Petrucci, 1971)

$$
F_{a_1\dots a_s} = \sum_{m=1}^{\infty} F^{(m)}_{a_1\dots a_s}\,, \qquad F^{(m)}_{a_1\dots a_s} = O(n^m)\,.
$$

Substituting the expansion into the hierarchy we obtain

$$\frac{\partial}{\partial \boldsymbol{r}_1} F^{(m)}_{a_1\ldots a_s} = -\beta F^{(m)}_{a_1\ldots a_s} \frac{\partial}{\partial \boldsymbol{r}_1} U_{a_1\ldots a_s} - \beta \sum_{a_{s+1}} n_{a_{s+1}} \int \mathrm{d}\boldsymbol{r}_{s+1} F^{(m-1)}_{a_1\ldots a_{s+1}} \frac{\partial}{\partial \boldsymbol{r}_1} U_{a_1\ldots a_{s+1}} \tag{5.71}$$

where

$$F^{(0)}_{a_1\ldots a_s} \to 1 \quad \text{if} \quad |\boldsymbol{r}_k - \boldsymbol{r}_l| \to \infty\,,$$

$$F^{(m)}_{a_1\ldots a_s} \to 0 \quad \text{if} \quad |\boldsymbol{r}_k - \boldsymbol{r}_l| \to \infty\,, \quad m \geqq 1\,.$$

In the zeroth order we obtain

$$F^{(0)}_{a_1\ldots a_s} = \exp\left(-\beta\, U_{a_1\ldots a_s}\right). \tag{5.72}$$

The solution to the first order equation is given by

$$F^{(1)}_{a_1\ldots a_s} = \sum n_{a_{s+1}} \int \mathrm{d}\boldsymbol{r}_{s+1} \{\exp\left[-\beta U_{a_1\ldots a_{s+1}}\right] - \exp\left[-\beta U_{a_1\ldots a_s}\right] (1 + \sum_{i=1}^{s} (\exp\left[-\beta U_{a_i a_{s+1}}\right] - 1))\}\,. \tag{5.73}$$

In an analogous way one can find the other coefficients of our density expansion. Bogolyubovs density expansion diverges in the limit of infinite volumes $V \to \infty$ what is due to the long range nature of the Coulomb forces. In order to get convergent terms only one has to perform some rearrangement of the terms in the series. Formally this corresponds to ring summations. We shall use here a quite simple elimination procedure (Schmitz, 1966, 1968; Ebeling, 1966; Falkenhagen and Ebeling, 1971). The long range potential

$$V_{ab} = \frac{e_a e_b}{4\pi\varepsilon r} \tag{5.74}$$

is eliminated by means of the following integral equation which defines the Debye correlation function g_{ab}:

$$g_{ab} = -\beta V_{ab} - \sum_c n_c \int \mathrm{d}\boldsymbol{r}_3\, V_{ac} g_{bc}\,. \tag{5.75}$$

It is well known that the solution of this equation is given by

$$g_{ab} = -\frac{e_a e_b}{4\pi\varepsilon r} \exp\left(-\varkappa r\right), \qquad \varkappa = r_{\mathrm{D}}^{-1} = [\sum_a n_a e_a^2/\varepsilon k_{\mathrm{B}} T]^{1/2}\,. \tag{5.76}$$

We may express the potential V_{ab} by means of g_{ab} using an iteration of the integral equation (5.75) (Schmitz, 1966, 1968; Petrucci, 1971):

$$\beta V_{ab} = -g_{ab} + \sum n_c \int \mathrm{d}\boldsymbol{r}_3\, g_{ac} g_{bc} - \sum n_c n_d \int \mathrm{d}\boldsymbol{r}_3\, \mathrm{d}\boldsymbol{r}_4\, g_{ac} g_{bc} g_{cd} + \ldots$$

Substituting this into the formal density expansion and collecting all terms with equal prefactors $n_a n_b n_c \ldots$ etc. we get convergent density expansions. For example one finds for the pair distribution function in the first order (EBELING, KELBG and ROHDE, 1968)

$$F_{ab}(r_1, r_2) = \exp\,(g_{ab} - \beta V'_{ab})\,\{1 + \sum_c n_c \int \mathrm{d}r_3$$
$$\cdot [\Phi_{ac}g_{bc} + \Phi_{bc}g_{ac} + \Phi_{ac}\Phi_{bc} + \exp\,(g_{ac} + g_{bc} - \beta V'_{ac} - \beta V'_{bc})$$
$$\cdot (\exp\,(-\beta u_{abc}) - 1)] + O(n^2)\}\,, \tag{5.77}$$
$$\Phi_{ab} = \exp\,(g_{ab} - \beta V'_{ab}) - 1 - g_{ab}\,.$$

In other notation by using Slater sums we get for the pair distribution function

$$F_{ab}(r_1, r_2) = S^{(2)}_{ab}(r_1, r_2) \cdot \exp\,(9\mathrm{ab} + \beta V_{ab}) \cdot \{1 + \sum_c n_c \int \mathrm{d}r_3$$
$$\cdot [\Phi_{ac}g_{bc} + g_{ac}\Phi_{bc} + \Phi_{ac}\Phi_{bc} + (S^{(3)}_{abc}/S^{(2)}_{ab} - S^{)2)}_{ac}S^{(2)}_{bc})\,.$$
$$\cdot \exp\,(g_{ac} + g_{bc} + \beta V_{ac} + \beta V_{bc})] + O(n^2)\} \tag{5.78}$$

(See also KRAEFT and KREMP (1968).

5.3.2. Discussion of the Pair Distribution

In zeroth order the eq. (5.78) for the pair distribution function simplifies to (KREMP and KRAEFT, 1968)

$$F_{ab}(r) = S^{(2)}_{ab}(r) \exp\left[\frac{e_a e_b}{4\pi\varepsilon k_B T r}\,(1 - \mathrm{e}^{-\varkappa r})\right] + O(n)\,. \tag{5.79}$$

This formula however is restricted to low density nearly ideal plasmas where the correlations are weak. Of more interest are nonideal plasmas where the properties are strongly influenced by the correlations between the charged particles (EBELING, 1983). Of special importance are of course the pair correlations. In dense plasmas we may assume again, in a first approximation, that local thermal equilibrium holds. Let us consider a two-component plasma consisting of electrons with the density n_e and mass m_e and ions with the density n_i and mass m_i where $n_e = n_i = n$, $m_e \ll m_i$. The region of electron and ion correlations respectively is given by

$$\Gamma_e = \frac{e^2}{4\pi\varepsilon d\theta_e} \gtrsim 1\,, \qquad \Gamma_i = \frac{e^2}{4\pi\varepsilon d\theta_i} \gtrsim 1\,, \qquad d = (3/4\pi n)^{1/3}$$

where θ_k are the mean kinetic energies per degree of freedom for the corresponding Fermi gases. For a fixed temperature $T \lesssim 10^5$ K, electron and ion correlations are created simultaneously with increasing density. The plasmas become strongly correlated; at a density of about 10^{24} cm^{-3} the electron correlations disappear due to the kinetic energy of degenerate electron gases, the ion correlations however are still present. At temperatures $T \gtrsim 10^6$ K the electrons behave like ideal gases in the whole density region, the ions however are strongly correlated in the density region between about 10^{27} cm^{-3} and 10^{34} cm^{-3}. Let us consider first the region of nondegenerate particles with strong correlations. Here the pair correlation function F_{ab} may be expressed by the Slater sum S_{ab} and by the classical pair function which is assumed to be known.

$$F_{ab}(r) = F^{\mathrm{cl}}_{ab}(r) \exp\,(V_{ab}/k_B T)\, S_{ab}(r) \tag{5.80}$$

where V_{ab} is the interaction potential. The Slater sum may in principle be calculated from the known wave functions, however this is a very time consuming procedure. In Section 5.2.2. we have calculated the following representation for large distances using $\hbar^2$-expansions:

$$\ln S_{ab}(r) = -\frac{\beta e_a e_b}{4\pi\varepsilon r} - \frac{\hbar^2\beta^4 e_a^2 e_b^2}{(4\pi)^2\, 24 m_{ab} r^4} + c_6 r^6 + c_7 r^7 + c_8 r^8 + c_9 r^9 + \dots$$

Furthermore, we have given Taylor representations for small distances using known properties of the wave functions. Thus given

$$\ln S_{ab}(r) - = A(\xi_{ab}) + \xi_{ab}\frac{r}{\lambda_{ab}} + a_2 r^2 + a_3 r^3 + a_4 r^4 + a_5 r^5 + a_6 r^6 + \dots,$$

$$\xi_{ab}' = -(e_a e_b / 4\pi\varepsilon k_B T \lambda_{ab}),$$

all the coefficients given above are known.

The function $A(x)$ is given by

$$A(x) = \ln\left[\theta(x)\sqrt{\pi}\sum_{s=1}^{\infty} s^{-3}\exp(x^2/4s^2)\right.$$

$$\left. + 4\sqrt{\pi}\, x\int_0^\infty dy y\,\{1 - \exp(-x/y)\}^{-1}\right].$$

Based on this knowledge Padé approximations may be constructed which work in the whole region of distances. The simplest possibility is

$$S_{ab}(r) = \exp\left[\frac{A - \xi R + 12A\xi^{-1}R^3}{1 - R^2 + 12A\xi^{-2}R^3}\right] \pm \frac{1}{2}\delta_{ab}\exp\left[A - \xi R - R^2\right], \quad (5.81)$$

$$\xi = \xi_{ab}, \qquad R = r/\lambda_{ab}.$$

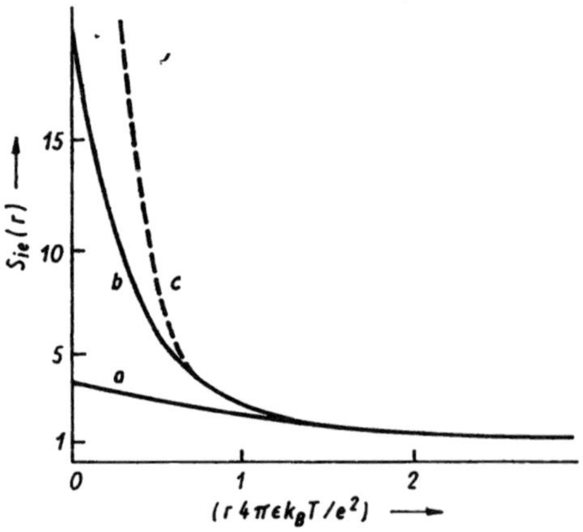

Fig. 5.3.

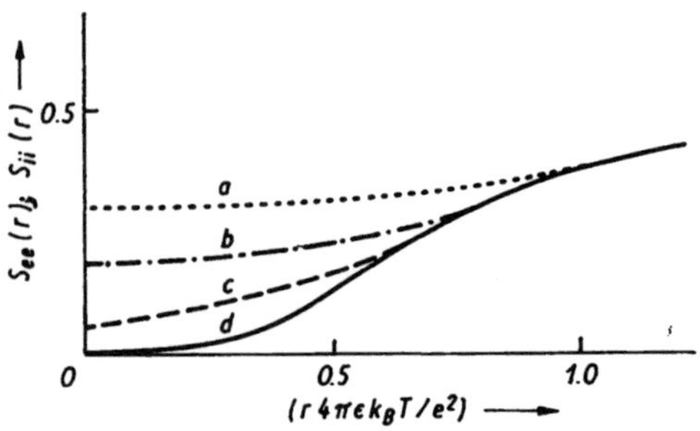

Fig. 5.4.

Fig. 5.3. The Slater sum for electron-ion pairs
a — $T = 1.5 \cdot 10^5$ K, b — $T = 2.5 \cdot 10^5$ K, c — $T = 0$ K (classical Boltzmann factor)

Fig. 5.4. Slater sums for electron-electron and for ion-ion-pairs
a — $T_e = 3.1 \cdot 10^6$ K, $T_i = 5.7 \cdot 10^9$ K,
b — $T_e = 7.8 \cdot 10^5$ K, $T_i = 1.4 \cdot 10^9$ K,
c — $T_e = 1.2 \cdot 10^5$ K, $T_i = 2.3 \cdot 10^8$ K,
d — $T_e = 0$ K, $T_i = 0$ K (classical Boltzmann factor)

In Figs. 5.3 and 5.4 several examples for the binary Slater sum are shown which demonstrate the influence of temperature on the correlations at small distances. Eq. (5.80) is correct only in the case of densities where classical screening still works.

A more extended range of validity has the formula (EBELING, 1983; EBELING and GRIGO, 1984)

$$F_{ab}(r) = F_{ab}^0 + F_{ab}^1 + \alpha^2 \exp(\beta\mu_{xc}) \{S_{ab}(r) - S_{ab}^0(r) - S_{ab}^1(r)\} . \tag{5.82}$$

Where the terms with the upper index 0 or 1 respectively denote the zeroth or first order of perturbation theory with respect to e^2 including degeneracy effect. Furthermore the degree of ionization is calculated from a mass action law (EBELING and SÄNDIG, 1973; EBELING and GRIGO, 1980, 1982, 1984) and μ_{xc} is the interaction part of the chemical potential of the plasma which we express by Padé approximations (EBELING et al., 1981; EBELING and RICHERT, 1982) the simplest of which is (see Chapter 6.)

$$\mu_{xc} = \frac{a_0(T)(\alpha n)^{1/2} + a_1(T)\, a_2(T)\, (\alpha n)^{3/2}}{1 + (a_1(T)/a_0(T))\, (\alpha n)^{1/2} + a_1(T)\, a_0(T)\, (\alpha n)^{7/6}} .$$

The temperature functions are choosen in such a way that the Debye law, the quantum correction at low density, the Hartree-Fock contribution and the lattice energy at high density are represented in the right way. We note that there are also other approaches to the pair distribution function based on e^2 expansions (SCHMITZ and KREMP, 1968; ISIHARA and MONTROLL, 1971; ISIHARA and WADATI, 1969, 1972). The knowledge of the pair distribution function for two ions (protons) at short distance is of importance as it governs the rate of nuclear reaction which is proportional to the probability that two nuclei approach one another (GRABOSKE et al., 1973; JANCOVICI, 1977; HAUBOLD and JOHN, 1982).

In dense plasmas two effects increase this probability: (1) screening effects, (2) quantum effects which both lead to an enhancement of the rate of nuclear reactions. The second effect is shown in Fig. 5.4. which demonstrates that Coulomb repulsion looses its significance at temperatures above 10^8 K. Futhermore the knowledge of the pair correlation functions given above may be used for the calculation of correlation contributions to the transport coefficients of nonideal plasmas (EBELING, 1981, 1982; EBELING et al., 1983, 1984).

5.4. Thermodynamic Functions

5.4.1. Cluster Expansions of the Free Energy

As shown in Section 3.1.6. the free energy may be expressed by the mean value of the interaction part of the Hamiltonian by using a charging process:

$$F = F_{id} + \int_0^1 \frac{d\lambda}{\lambda} \langle V(\lambda) \rangle \tag{5.83}$$

where

$$V(\lambda) = \lambda \sum_{i<j} \frac{e_i e_j}{4\pi\varepsilon |\boldsymbol{r}_i - \boldsymbol{r}_j|} . \tag{5.84}$$

The mean value may be expressed by two-particle density operators (compare Section 3.1.6.):

$$\langle V\rangle = \frac{1}{2} \operatorname*{Tr}_{12} \left(V(12)\, F(12) \right).$$

By using the pair distribution functions we get in this way finally

$$F = F_{\text{id}} + \frac{1}{2} \sum_{ab} n_a n_b \int_0^1 \frac{d\lambda}{\lambda} \, d\boldsymbol{r}_1 \, d\boldsymbol{r}_2 / V_{ab}(\lambda) \, F_{ab}(r_1, r_2; \lambda) \,. \tag{5.85}$$

Substituting here for the pair distribution function the cluster expansion eq. (5.78) as well as eq. (5.75) and reordering with respect to the density we get the cluster expansion (EBELING, 1967, 1968, 1969)

$$F = F_{\text{id}}^{\text{B}} - k_{\text{B}} T V \left\{ \tfrac{1}{12} \varkappa^3 + S^{(2)} + S^{(3)} + \dots \right\}, \tag{5.86}$$

$$S^{(2)} = \tfrac{1}{2} \sum_{ab} n_a n_b \int d\boldsymbol{r}_2 \left[\Phi_{ab} - \tfrac{1}{2} g_{ab}^2 \right], \tag{5.87}$$

$$S^{(3)} = \tfrac{1}{6} \sum_{abc} n_a n_b n_c \int d\boldsymbol{r}_2 \, d\boldsymbol{r}_3 \, [\Phi_{ab} \Phi_{ac} \Phi_{bc}$$

$$+ (S_{abc} - S_{ab} S_{ac} S_{bc}) \exp (g_{ab} + g_{ac} + g_{bc} + \beta V_{ab} + \beta V_{ac} + \beta V_{bc})] \,, \tag{5.88}$$

$$\Phi_{ab} = S_{ab}(r) \exp [g_{ab} + \beta V_{ab}] - 1 - g_{ab} \,. \tag{5.89}$$

The orders of magnitudes of the cluster integrals are (FRIEDMAN, 1962)

$$\begin{aligned}
S^{(2)} &= \begin{cases} O(n^2) & \text{if} \quad \mu_3 = 0 \,, \\ O(n^2 \ln n) & \text{if} \quad \mu_3 \neq 0 \,, \end{cases} \\
S^{(3)} &= \begin{cases} O(n^3 \ln n) & \text{if} \quad \mu_3 = 0 \,, \\ O(n^{5/2}) & \text{if} \quad \mu_3 \neq 0 \,, \end{cases} \\
S^{(4)} &= \begin{cases} n^3 & \text{if} \quad \mu_3 = 0 \,, \\ n^{5/2} & \text{if} \quad \mu_3 \neq 0 \,, \end{cases} \\
S^{(5)} &= \begin{cases} n^{7/2} & \text{if} \quad \mu_3 = 0 \,, \\ n^3 & \text{if} \quad \mu_3 \neq 0 \,. \end{cases}
\end{aligned} \tag{5.90}$$

The third order moment used here is given by

$$\mu_3 = \sum_a n_a e_a^3 \,.$$

We note that $\mu_3 = 0$ for charge-symmetrical systems.

5.4.2. Density Expansions of Free Energy

The explicit calculation of the cluster integrals is a rather difficult task. A numerical evaluation of $S^{(2)}$ using interpolation formulae for the Slater sums has been given by ROHDE et al. (1968). An analytical calculation is possible at small densities using the following ideas. First one isolates the divergencies by writing the identity

$$S^{(2)} = 2\pi \sum_{ab} n(n_b \lim_{R\to\infty} \lim_{u\to 0} \left\{ \int_u^R \mathrm{d}\boldsymbol{r}\, r^2 \left[S_{ab}(r) \right.\right.$$

$$\cdot \exp\left(\frac{\beta e_a e_b}{4\pi\varepsilon r}(1-\mathrm{e}^{-\varkappa r})\right) - 1 + \frac{\beta e_a e_b}{4\pi\varepsilon r} - \frac{1}{2}\left(\frac{\beta e_a e_b}{4\pi\varepsilon r}\mathrm{e}^{-\varkappa r}\right)^2$$

$$+ \frac{1}{6}\left(\frac{\beta e_a e_b}{4\pi\varepsilon r}\mathrm{e}^{-\varkappa r}\right)^3\Bigg] - \frac{1}{6}\left(\frac{\beta e_a e_b}{4\pi\varepsilon}\right)^3 [\mathrm{Ei}(-3\varkappa R) - \mathrm{Ei}(-3\varkappa u)]$$

$$\left.\left. + \frac{1}{24}\left(\frac{\beta e_a e_b}{4\pi\varepsilon}\right)^4 \left[\frac{1}{R}\exp(-4\varkappa R) - \frac{1}{u}\exp(-4\varkappa u) + 4\varkappa \mathrm{Ei}(-4\varkappa R) - 4\varkappa\, \mathrm{Ei}(-4\varkappa u)\right]\right\}.$$

By expansion with respect to $\varkappa$ we obtain

$$S^{(2)} = 2\pi \sum_{ab} n_a n_b \left\{ \frac{1}{6}\left(\frac{\beta e_a e_a}{4\pi\varepsilon}\right)^3 \left[\left(1 + \frac{\beta e_a e_b \varkappa}{4\pi\varepsilon}\right)\ln \varkappa\lambda_{ab} - \frac{\beta e_a e_b \varkappa}{4\pi\varepsilon}\left(1 - \ln\frac{4}{3}\right)\right]\right.$$

$$\left. + \left(1 + \frac{\beta e_a e_b \varkappa}{4\pi\varepsilon}\right)\lambda_{ab}^3 K_0(\xi_{ab}; s_a) + O(\varkappa^2)\right\} \tag{5.91}$$

where the so-called quantum virial function introduced by EBELING (1967, 1968, 1969) is given by

$$\lambda_{ab}^3 K_0(\xi_{ab}; s_a) = \lim_{R\to\infty} \left\{ \int_0^R \mathrm{d}\boldsymbol{r}\, r^2 \left[S_{ab}(r) - 1 - \frac{\beta e_a e_b}{4\pi\varepsilon r} - \frac{1}{2}\left(\frac{\beta e_a e_b}{4\pi\varepsilon r}\right)^2\right]\right.$$

$$\left. + \frac{1}{6}\left(\frac{\beta e_a e_b}{4\pi\varepsilon}\right)^3 \left[C + \ln\frac{3R}{\lambda_{ab}}\right]\right\}. \tag{5.92}$$

By splitting the two-particle Slater sum into a Heisenberg term and an exchange term after (5.46) or (5.61), (5.62) we may write

$$K_0(\xi_{ab}; s_a) = Q(\xi_{ab}) + \frac{(-1)^{2s_a}}{2s_a + 1} E(\xi_{ab}) \tag{5.93}$$

where

$$\xi_{ab} = -\frac{\beta e_a e_b}{4\pi\varepsilon\lambda_{ab}}, \qquad \lambda_{ab} = \frac{\hbar}{\sqrt{2m_{ab}k_B T}}. \tag{5.94}$$

The functions $Q(x)$ and $E(x)$ are analytical in the interaction parameter and may be represented by Taylor series

$$Q(x) = \sum_{n=0}^{\infty} q_n x^n, \qquad E(x) = \sum_{n=0}^{\infty} e_n x^n. \tag{5.95}$$

The first coefficients up to $n = 3$ are most easily calculated by using the Fourier representations of the Slater sums after eqs. (5.46)–(5.50) and (5.53)–(5.57). In this way follows

$$q_0 = 0\,, \qquad q_1 = -\tfrac{1}{6}, \qquad q_2 = -\tfrac{1}{8}\sqrt{\pi}\,,$$
$$e_0 = \tfrac{1}{4}\sqrt{\pi}\,, \quad e_1 = \tfrac{1}{2}\,, \qquad e_2 = \tfrac{1}{4}\sqrt{\pi}\ln 2\,. \tag{5.96}$$

These coefficients have been obtained in several independent ways (VEDENOV and LARKIN, 1959; DE WITT, 1962, 1966; TRUBNIKOV and ELESIN, 1964; KELBG, 1964; YUKHNOVSKII and BLAZHIEVSKII, 1966; EBELING et al., 1967; KREMP and SCHMITZ, 1967; KREMP and KRAEFT, 1968; YUKHNOVSKII and HETZHEIM, 1968; KLIMONTOVICH and EBELING, 1972; EBELING et al., 1970, 1976, 1979). The third order leads to quite difficult integrals which have been solved analytically (HOFFMANN and EBELING, 1968) and numerically (KRAEFT and KREMP, 1968); the result is

$$q_3 = -\tfrac{1}{6}\left[\tfrac{1}{2}C + \ln 3 - \tfrac{1}{2}\right], \quad C = 0.577216\,\ldots\,,$$
$$e_3 = \tfrac{1}{72}\pi^2\,. \tag{5.97}$$

The higher coefficients are most easily obtained by using the methods of scattering theory (EBELING, 1969; EBELING et al., 1970, 1976, 1979); the result is ($n \geqq 4$)

$$q_n = \sqrt{\pi}\,\zeta(n-2)\,2^{-n}/\Gamma(\tfrac{1}{2}n+1)\,,$$
$$e_n = \sqrt{\pi}(1-2)^{2-n}\,\zeta(n-1)\,2^{-n}/\Gamma(\tfrac{1}{2}n+1)\,. \tag{5.98}$$

The result obtained independently by KOPYSHEV (1968) differs from this in some minor points as the sign of e_n and the appearance of a term $\ln(\varkappa l)$ instead of $\ln(\varkappa\lambda_{ab})$ in eq. (5.91). Taking into account that there now exist several independent derivations using quite different methods we may suppose that the virial functions $Q(x)$ and

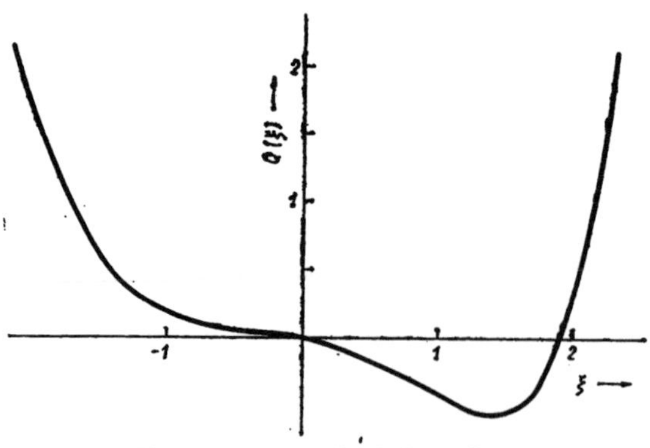

Fig. 5.5. The quantum virial function, Heisenberg contribution

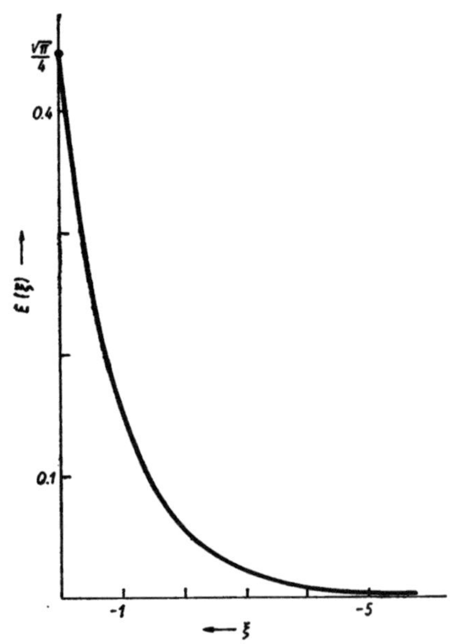

Fig. 5.6. The quantum virial function, exchange contribution

$E(x)$ which determine the second virial coefficient are exactly known. Figs. 5.5 and 5.6 show a graphical representation of these functions. We note that the exchange part $E(x)$ disappears exponentially for $(-x) \to \infty$ and $Q(x)$ behaves for $|x| \to \infty$ like (EBELING, 1969)

$$Q(x) = 2\sqrt{\pi}\,\theta(x)\left[\sigma(x) - \frac{1}{8}x^2\right]$$

$$-\frac{1}{6}x^3\left[\ln|x| + 2C + \ln 3 - \frac{11}{6}\right] - \frac{1}{12}x - \frac{1}{60x} + O(x^{-3}),$$

$$\sigma(x) = \sum_{s=1}^{\infty} s^2\left[\exp\left(\frac{x^2}{4s^2}\right) - 1 - \left(\frac{x^2}{4s^2}\right)\right],$$

$$\theta(x) = \begin{cases} 1 & \text{if} \quad x > 0, \\ 0 & \text{if} \quad x < 0. \end{cases} \tag{5.99}$$

Little is known about the behaviour of the higher virial coefficients. The calculations of CZERWON (1972) and KRIENKE et al. (1975) yield (see also EBELING, 1977)

$$S^{(3)} = -2\pi^2 \sum_{abc} n_a n_b n_c a_1 \varkappa^{-1} (\beta/4\pi\varepsilon)^5 (e_a^4 e_b^3 e_c^3)$$

$$-\tfrac{1}{12}\pi^4 \sum n_a n_b n_c (\beta/4\pi\varepsilon)^6 (e_a e_b e_c)^4 \ln(\varkappa\lambda_{ab})$$

$$-\tfrac{4}{3}\pi^2 (\ln 2) \sum n_a n_b n_c (\beta/4\pi\varepsilon)^6 (e_a^5 e_b^3 e_c^4) \ln(\varkappa\lambda_{ab}) + O(n^3),$$

$$S^{(4)} = a_2 \varkappa^{-3} \sum n_a n_b n_c n_d (\beta/4\pi\varepsilon)^6 (e_a e_b e_c e_d)^3 + O(n^3),$$

$$a_1 = 0.543\ldots, \qquad a_2 = 10.13\ldots \tag{5.100}$$

By collecting these results one obtains the low density expansion of the free energy given in Section 2.3. without proof.

6. Quantum Statistical Calculations of Equilibrium Properties

6.1. Equation of State in the Screened Ladder Approximation

6.1.1. The Second Virial Coefficient

We want to determine thermodynamic functions. For this purpose we start from eq. (3.235) which expresses the pressure as a function of the fugacities.

In a diagrammatic version the pressure is given by the general expression

$$p - p_0 = -\frac{1}{2\Omega}\sum_{ab}\int_0^1 \frac{d\lambda}{\lambda}$$

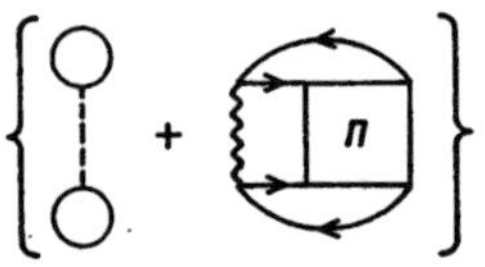

(6.1)

which is an abbreviation for

$$p - p_0 = \frac{1}{2}\sum_{ab}{}'\int_0^1 \frac{d\lambda}{\lambda}\{\lambda V_{ab}(12)\, G^{H}_{ab}(121^{++}2) + V^{s}_{ab}(12\lambda)\, \Pi_{ab}(121^{++}2^{+}\lambda)\}\, d2\,.$$

The first contribution in the curly brackets, the Hartree contribution, vanishes for electro-neutral systems. The second contribution includes the polarization function discussed in Section 3.3. and the dynamically screened potential. All quantities depend on the coupling parameter λ. Further thermodynamic information may be deduced from eq. (6.1) by known thermodynamic relations.

For practical purposes it is necessary to choose some approximation for the polarization function Π and to carry out the integral over the coupling parameter in some reasonable way. Let us choose, for the polarization function Π, the ladder approximation which reads (see (3.236))

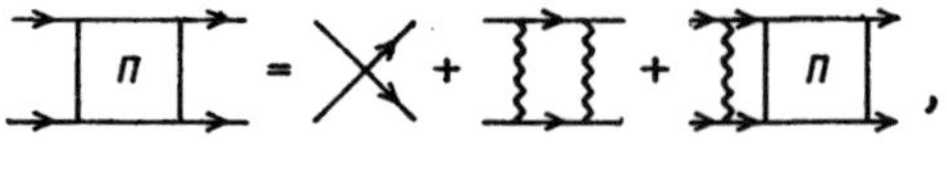

(6.2)

$$\Pi_{ab}(1234) = -\delta_{ab}G_a(14)\, G_a(23)$$
$$- \int G_a(11')\, G_a(1'\bar{1})\, G_a(\bar{1}3)\, V^{s}_{ab}(1'2')\, V^{s}_{ab}(\bar{1}\bar{2})\, G_b(22')\, G_b(2'\bar{2})\, G_b(\bar{2}4)\, d1'd2'd\bar{1}d\bar{2}$$
$$+ i\int G_a(11')\, G_b(22')\, V^{s}_{ab}(1'2')\, \Pi_{ab}(1'2'34)\, d1'2d'\,. \qquad (6.3)$$

The derivation of the screened ladder equation (6.2) is given by Kraeft, Kremp and Kilimann (1973); see also Ebeling, Kraeft and Kremp (1976, 1979) and Stolz and Zimmermann (1979). Let us scetch briefly the derivation.

We start from eq. (3.228) and replace the derivative of $\bar{\Sigma}$ according to the rule

$$\frac{\delta \bar{\Sigma}_a(12)}{\delta U_b^{\text{eff}}(1'2')} = \sum_c \int d3\, d4 \frac{\delta \bar{\Sigma}_a(12)}{\delta G_c(34)} \frac{\delta G_c(34)}{\delta U_b^{\text{eff}}(1'2')}. \tag{6.4}$$

With the approximation

$$\frac{\delta G_c(34)}{\delta U_b^{\text{eff}}(1'2')} = \delta_{bc} G_b(31')\, G_b(2'4) \tag{6.5}$$

we get instead of (3.228)

$$\Pi_{ab}(12'1'2) = -\delta_{ab} G_a(12)\, G_a(2'1')$$
$$-\int G_a(1\bar{1}) \frac{\delta \bar{\Sigma}_a(\bar{1}\bar{2})}{\delta G_b(34)} G_a(\bar{2}1')\, G_b(32)\, G_b(2'4)\, d3 d4 d\bar{1} d\bar{2}\,. \tag{6.6}$$

Now the appropriate approximation for $\delta\bar{\Sigma}/\delta G$ must be looked for. We start from (3.229) and have to take from both sides the functional derivative with respect to G; before the relation (6.4) and the approximation (6.5) are applied to (3.229). As a result the integral term involves three Green's function. The functional derivative of V_s gives (taking into account (6.5)) two contributions.

The functional derivative of eq. (3.229) is then taken in the following approximation: The first right-hand side term of (3.229) gives one term coming from G_a and one coming from V_{aa}^s (the second one coming from V_{aa}^s is neglected). The second right-hand side term of (3.229) gives a number of contributions from which only that coming from one of the three Green's functions is retained. The resulting equation for the determination of $\delta\bar{\Sigma}/\delta G$ is then (Kraeft, Kremp and Kilimann, 1973)

$$\frac{\delta \bar{\Sigma}_a(11')}{\delta G_b(22')} = \delta_{ab} V_{aa}^s(12')\, \delta(12)\, \delta(1'2') - G_a(11')\, G_b(2'2)\, V_{ab}^s(12')\, V_{ab}^s(21')$$
$$+\int d3' d4'\, V_{ab}^s(12')\, G_a(13')\, G_b(2'4') \frac{\delta \bar{\Sigma}_a(3'1')}{\delta G_b(24')}. \tag{6.7}$$

While (3.229) is a functional integral equation we have now in our approximation (6.7) only an integral equation, which includes all ladder diagrams (exchange ones as well), except the first order direct contribution. This fact is due to the renormalization in connection with the screening procedure as a result of which the Hartree contribution was used to construct the effective external potential (see Section 3.3.).

For some purposes (see also Section 5.4.) it is useful to introduce the T-matrix (which includes all ladder type diagrams) by

$$T_{ab}(1234) = \frac{\delta(\bar{\Sigma}_a 13)}{\delta G_b(42)} + V_{ab}(12)\, \delta(13)\, \delta(24) \tag{6.8}$$

If the ladder type quantity $\delta\bar{\Sigma}/\delta G$ according to eq. (6.7) is inserted into (6.6), we have a ladder type approximation for Π_{ab}, which includes all direct and exchange ladder type diagrams, however without the free pair of Green's functions (which is not included in the definition of Π) and without the first order (with respect to V^s) diagram, which is not included in $\delta\bar{\Sigma}/\delta G$ according to (6.7).

In this way eq. (6.3) is derived. That means, as a result of the screening renormalization, the diagram  does not occur in (6.2), and, in the ladder approximation, Π differs from the corresponding two-particle Green's function by the following relation:

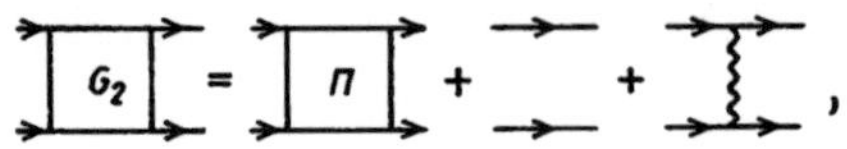

(6.9)

while the ladder approximation of the two-particle Green's function is defined by

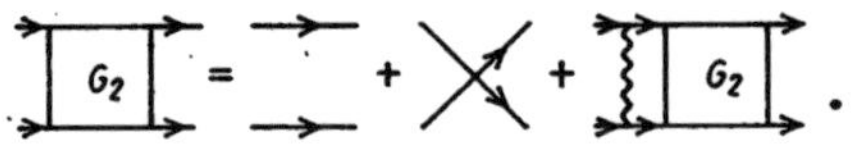

(6.10)

In the ladder approximation, eq. (6.2) may be decomposed as follows:

$$p - p_0 = -\frac{1}{2\Omega}\sum_{ab}\int_0^1 \frac{d\lambda}{\lambda}$$

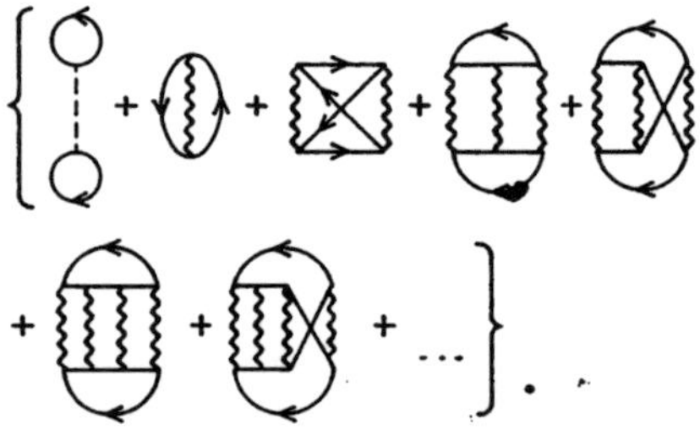

(6.11)

Here the solid lines represent single particle Green's functions, while the wavy lines are (dynamically) screened potentials.

Such expansions coincide in the low degeneracy limit with virial expansions; the right-hand side of (6.11) is, apart from the special Coulomb problem of screening, the second virial coefficient approximation. See also Section 3.2.9.

This equation is called the screened ladder approximation of the pressure. It includes, beside the Hartree contribution, one diagram with one wavy line only. This diagram consists of the Hartree-Fock (HF) and the Montroll-Ward (MW) contributions which read  and  according to the equation for the screened potential in random phase approximation (Sections 3.3. and 4.2.). The diagram with two wavy lines is the second order exchange diagram. For any higher order, $n \geqq 3$, we have one direct and one exchange diagram.

Let us denote the contributions to the pressure including n wavy lines by p_n, so that

$$p = p_0 + p_{\mathrm{H}} + p_1 + p_2 + p_3 + p_4 + \dots \tag{6.12}$$

As already mentioned, we have $p_{\mathrm{H}} = 0$ for electroneutral systems, and

$$p_1 = p_{\mathrm{HF}} + p_{\mathrm{MW}}\,. \tag{6.13}$$

The explicit evaluation of the different contributions p_n is possible only for p_0, p_{H} and p_{HF}. For p_{MW} only a numerical evaluation is possible (see 6.1.3.).

6.1.2. Evaluation of the Higher Order Contributions

For the higher order terms, $n \geqq 4$, an essential approximative simplification is possible by replacing the screened potential by the Coulomb potential, because divergencies occur only up to $n = 3$. To this end let us consider an expression for thermodynamic quantities in the ladder approximation which is valid for short range potentials.

For short range potentials it is not necessary to do some renormalization which is connected with the screening procedure. As a starting point we have in this case (cf. eq. (6.1)) (see KREMP et al., 1984a)

$$p - p_0 = \frac{1}{2} \sum_{ab} \int_0^1 \frac{d\lambda}{\lambda} \lambda V_{ab}(12)\, G_{ab}(121^{++}2^{+}\lambda)\, d2\,. \tag{6.14}$$

From (6.14) we can derive the second virial coefficient, if G_{ab} is taken in the ladder approximation; see Section 3.2.9. and eq. (3.220).

As is well known G_{ab} may be expanded into a Fourier series. In operator notation we have

$$p - p_0 = \frac{1}{2\Omega} \frac{1}{i\beta} \sum_{ab} \sum_{\nu} \int_0^1 \frac{d\lambda}{\lambda} \mathrm{Tr}^{-}(V_{ab} G_{ab}(\omega_\nu^{ab}))_\lambda\,. \tag{6.15}$$

Here Ω is the volume, ω_ν^{ab} is the Boson Matsubara frequency,

$$\omega_\nu^{ab} = \frac{\pi\nu}{-i\beta} + \mu_a + \mu_b\,, \qquad \nu \text{ even}\,.$$

Tr^{-}refers to the anti-symmetric Hilbert space. The frequency summation may be carried out in the usual manner (cf. Section 3.2.8. and KADANOFF and BAYM, 1962); the result is (the subscript λ will be droped henceforth)

$$p - p_0 = -\frac{1}{2\Omega} \int_0^1 \frac{d\lambda}{\lambda} \sum_{ab} \int_C \mathrm{Tr}^{-}\, g_B^{ab}(z)\, (V G_{ab}(z)) \frac{dz}{2\pi}\,, \tag{6.16}$$

where

$$g_B^{ab}(z) = \{\exp[\beta(z - \mu_a - \mu_b)] - 1\}^{-1}$$

is the Bose function. The contour C encircles the singularities of $G_{ab}(z)$ being located at the real axis.

Instead of (6.16) we may write

$$p - p_0 = \frac{1}{2\Omega} \int_0^1 \frac{d\lambda}{\lambda} \sum_{ab} \int_{-\infty}^{+\infty} \frac{d\omega}{\pi} \mathrm{Tr}^{-} g_B^{ab}(\omega)\, \mathrm{Im}(V_{ab} G_{ab}(\omega + i\varepsilon))\,. \tag{6.17}$$

After a partial integration we get

$$p - p_0 = -\frac{1}{2\beta\Omega} \int_0^1 \frac{d\lambda}{\lambda} \sum_{ab} \int_{-\infty}^{+\infty} \frac{d\omega}{\pi} \mathrm{Tr}^{-} \ln |1 - e^{-\beta(\omega - \mu_a - \mu_b)}|$$

$$\cdot \mathrm{Im}\left(V_{ab} \frac{dG_{ab}(\omega + i\varepsilon)}{d\omega}\right). \tag{6.18}$$

Now we introduce the ladder approximation for G_{ab} which is defined by (6.10) and reads in matrix notation (KREMP et al., 1984a)

$$G_{ab}(z) = G_{ab}^0(z) + G_{ab}^0(z)\, V_{ab} G_{ab}(z)\,, \tag{6.19}$$

where the free two-particles Green's function is given by

$$G^0_{ab}(p_1, p_2, p_1', p_2', z) = \delta(p_1, p_1')\, \delta(p_2, p_2') \cdot \frac{N_{ab}(p_1, p_2)}{z - \varepsilon_a(p_1) - \varepsilon_b(p_2)} \tag{6.20}$$

with

$$N_{ab}(p_1, p_2) = 1 - f_a(p_1) - f_b(p_2) \tag{6.21}$$

being the Pauli blocking corresponding to the phase space occupation (see also Section 4.4.).

From (6.19), (6.20) we get (with H^0 being the Hamiltonian of free particles)

$$(z - H^0 - \lambda V N)\, G(z;\lambda) = N \tag{6.22}$$

and

$$NV \frac{\mathrm{d}G(z;\lambda)}{\mathrm{d}z} = - \frac{\mathrm{d}G(z;\lambda)}{\mathrm{d}\lambda}. \tag{6.23}$$

With (6.23) the λ-integration in (6.18) may be carried out; the result is

$$p - p_0 = \sum_{ab} \frac{k_B T}{2\Omega} \int\limits_{-\infty}^{+\infty} \frac{\mathrm{d}\omega}{\pi} \operatorname{Tr}^- \ln |\, 1 - \mathrm{e}^{-\beta(\omega - \mu_a - \mu_b)}\, | \cdot \operatorname{Im} N_{ab}^{-1} \{G_{ab}(\omega + i\varepsilon) - G^0_{ab}(\omega + i\varepsilon)\}\,. \tag{6.24}$$

For questions of practical evaluations it is useful to introduce the T-matrix,

$$T_{ab}(z) = V_{ab} + V_{ab}\, G^0_{ab}(z)\, T_{ab}(z)\,. \tag{6.25}$$

Due to the property of $T_{ab}(z)$ to remain finite for the scattering spectrum, we may deduce from (6.20) for frequencies belonging to the scattering spectrum of $T_{ab}(z)$ (see also Chapter 3. and KREMP et al., 1984a)

$$\lim_{\varepsilon \to 0} \operatorname{Im} G^0_{ab}(\omega + i\varepsilon) = - \pi\delta(\omega - H^0_{ab})\, N_{ab}\,, \tag{6.26}$$

while we must retain in the case of frequencies belonging to the bound state spectrum of $T_{ab}(z)$ (here $T_{ab}(z)$ has simple poles)

$$\operatorname{Im} G^0_{ab}(\omega + i\varepsilon) = \varepsilon \frac{\mathrm{d}G^0_{ab}(\omega)}{\mathrm{d}\omega}. \tag{6.27}$$

With (6.24) — (6.27) we get for the equation of state (cf. KREMP et al., 1984)

$$\begin{aligned} p - p_0 = - \frac{k_B T}{2\Omega} \sum_{ab} \Bigg\{ & \sum_{nP} \ln |1 - \mathrm{e}^{-\beta(E^{ab}_{nP} - \mu_a - \mu_b)}|\, \sigma_{ab,n} \\ & - \sum_{p_1 p_2} N_{ab}(p_1, p_2) \Bigg[\frac{1}{k_B T}\, g_B(\varepsilon_a(p_1) + \varepsilon_b(p_2)) \\ & \cdot \operatorname{Re} \langle p_1 p_2 |\, T_{ab}(\varepsilon_a(p_1) + \varepsilon_b(p_2))\, | p_2 p_1 \rangle^- \\ & - \ln |\, 1 - \exp[-\beta(\varepsilon_a(p_1) + \varepsilon_b(p_2) - \mu_a - \mu_b)] | \\ & \cdot \sum_{p_1' p_2'} 2\pi\delta(\varepsilon_a(p_1) + \varepsilon_b(p_2) - \varepsilon_a(p_1') - \varepsilon_b(p_2')) \\ & \cdot \operatorname{Im} \{ \langle p_1 p_2 |\, T'^*_{ab}(\varepsilon_a(p_1) + \varepsilon_b(p_2) + i\varepsilon) |\, p_2' p_1' \rangle \\ & \cdot N_{ab}(p_1', p_2')\, \langle p_1' p_2' |\, T_{ab}(\varepsilon_a(p_1) + \varepsilon_b(p_2) + i\varepsilon)\, | p_2 p_1 \rangle^- \} \Bigg] \Bigg\}. \end{aligned} \tag{6.28}$$

E_{nP} is the eigen value of the Schrödinger equation of relative motion (set of quantum numbers n for bound states) and includes the kinetic energy of the center of mass, $\hbar^2P^2/2M$. Further $\sigma_{ab,n}$ takes into account the exchange effects and reads

$$\sigma_{ab,n} = (2s_a + 1)(2s_b + 1)\left\{1 - \frac{\delta_{ab}}{(2s_b + 1)} e^{i\varphi_n}\right\},$$

where s_a, s_b — spin, and the phase factor is in the case of a spherically symmetric potential simply $-(1)^l$. The symbol $\langle\ \rangle^-$ means antisymmetric matrix elements with the following meaning:

$$\langle p_2p_1| A_{ab} |p_1p_2\rangle^- = (2s_a + 1)(2s_b + 1) \cdot \left\{\langle p_2p_1| A_{ab} |p_1p_2\rangle - \frac{\delta_{ab}}{2s_b + 1} \langle p_2p_1| A_{ab} |p_2p_1\rangle\right\}.$$

Disregarding bound states, in the case $T = 0$ results of such type were first derived by Galitskii and Migdal (1958). Eq. (6.28) is a generalization of equations of state given by Baumgartl (1967) and Kraeft, Ebeling and Kremp (1969) and Kremp, Kraeft and Ebeling (1971). The result is similar to that given by Boercker and Dufty (1979) and may also be compared to that of Dashen, Ma and Bernstein (1969) and Kilimann (1978). Eq. (6.28) is an equation of state in the ladder approximation which accounts for bound states and for degeneracy effects (density effects). The degeneracy occurs at different places; namely we have (i) Bose-type shapes of the two particle sums over states both in the bound state and in the scattering state contributions to the pressure; (ii) Pauli blocking expressed by the functions $N_{ab}(p_1, p_2)$ which describe a phase space occupation of the interacting fermions; (iii) a density dependence of all two-particles quantities such as E^{ab}_{nP}, T_{ab} etc., as a result of the dependence on the Pauli blocking function $N_{ab}(p_1, p_2)$ of the solution of the corresponding "many-particle modification" of the two-particle Schrödinger equation (Bethe-Salpeter equation, see Section 5.4.).

It is also possible to rewrite eq. (6.28) in terms of the on-shell S-matrix which is closely connected to the on-shell-T-matrix, see Dashen, Ma and Bernstein (1969) and Kremp et al. (1984a).

Let us consider the equation of state (6.28) under the further approximation that the degeneracy of the Fermi particles is neglected, while the boson character of the bound states is retained as well as that of the two-particle states. The interparticle potential is assumed to be spherically symmetric. Then we may write (in angular momentum quantities)

$$p - p_0 = -\frac{k_B T}{2} \sum_{ab} \int \frac{dP}{(2\pi)^3} \sum_{l=0}^{\infty} (2s_a + 1)(2s_b + 1)(2l+1)\left(1 - \delta_{ab}\frac{(-1)^l}{2s_b + 1}\right) \cdot \left\{\sum_n \ln\left|1 - \exp\left[-\beta\left(E^{ab}_n + \frac{\hbar^2P^2}{2M} - \mu_a - \mu_b\right)\right]\right| + \frac{1}{\pi}\int_0^{\infty} dE \ln\left|1 - \exp\left[-\beta\left(E + \frac{\hbar^2P^2}{2M} - \mu_a - \mu_b\right)\right]\right| \frac{d\delta^{ab}_l(E)}{dE}\right\}. \tag{6.29}$$

Equation (6.29) is a slight modification of earlier results (KRAEFT, EBELING and KREMP, 1969; KREMP, KRAEFT, and EBELING, 1971); see also STOLZ and ZIMMERMANN (1979).

Instead of (6.29) one may write a complex contour integral which reads (EBELING, KRAEFT and KREMP 1976, 1979)

$$p - p_0 = -\frac{k_B T}{2} \sum_{ab} \int \frac{\mathrm{d}\boldsymbol{P}}{(2\pi)^3} \sum_{l=0}^{\infty} (2s_a + 1)(2s_b + 1)\left(1 - \delta_{ab}\frac{(-1)^l}{2s_b + 1}\right)$$

$$\cdot (2l+1) \cdot \int\limits_C \frac{\mathrm{d}z}{2\pi i} \ln\left|1 - \exp\left[-\beta\left(z + \frac{\hbar^2 P^2}{2M} - \mu_a - \mu_b\right)\right]\right| \cdot \frac{\mathrm{d}\ln D_l^{ab}(z)}{\mathrm{d}z}. \quad (6.30)$$

$D_l^{ab}(z)$ is the Jost function. The contour C encircles the singularities of the Hamiltonian, i.e., bound and scattering states as well. Eq. (6.30) is of special interest if the general properties of thermodynamic functions have to be considered.

It is, especially, possible to show that the second virial coefficient is analytic with respect to the coupling parameter, and there is no singularity of the sum of bound states; the latter is, if at all, compensated by that of the continuum contribution (KREMP and KRAEFT, 1972; KREMP, KRAEFT and EBELING, 1971).

Let us return now to the Coulomb problem. The equation of state (6.12) may be written in the form

$$p = p_0 + p_H + p_1 + p_2 + p_3 + p_L''$$

where p_L'' is given by the sum of ladders with more than three rungs, i.e. by eqs. (6.28)—(6.30) without the contributions of order V, V^2 and V^3. We remember again that the terms contained in p_L'' need not necessarily be screened due to their convergent character. Therefore the procedure developed above may be applied to the calculation of p_L''. For illustration let us consider the case of nondegenerate plasmas. Then we get from eq. (6.29)

$$p_L'' = k_B T \, 4\pi^{3/2} \sum_{ab} z_a z_b \lambda_{ab}^3 \left[\theta(-e_a e_b) \sum_{s=1}^{\infty} \sum_{l=0}^{s-1} (2l + 1)\left(e^{-\beta E_{sl}} - 1 + \beta E_{sl}\right)\right.$$

$$\left. + \sum_{l=0}^{\infty} (2l + 1)\left[1 + \delta_{ab}\frac{(-1)^{2s_a+l}}{2s_a + 1}\right] \int\limits_0^{\infty} \mathrm{d}E \, e^{-\beta E} \frac{1}{\pi} \frac{\mathrm{d}}{\mathrm{d}E} \delta_l''(E)\right]; \quad (6.31)$$

$\theta(x)$ — Heaviside function. Here the two-particle quantities E_{sl} and $\delta_l(E)$ are to be determined from the effective wave equation. Further $\delta_l''(E)$ denotes the phase shifts without the contributions of order V, V^2 and V^3. We see that the problem of divergency of the partition function of the two-particle bound states has found a quite natural solution. The partition function remains finite even at the case of zero density due to the subtraction procedure in eq. (6.31). Moreover at finite densities the effective Schrödinger equation has only a finite number of bound states. Therefore no convergence problems appear in our theory.

6.1.3. Evaluation of the Hartree-Fock and the Montroll-Ward Contributions

Let us now return to the eq. (6.12) and give explicit results for the Hartree-Fock and Montroll-Ward contributions to pressure.

It is worthwhile to remark that for some applications of physical interest it is not necessary to take into account the higher order contributions with respect to the coupling parameter and to consider only the correction (6.13) to the ideal pressure. This is, especially, possible in those regions of the density-temperature plane in which bound states do not play an essential role. We have such situations in weakly degenerate systems at sufficiently high temperatures, and in highly degenerate plasmas as well.

We will now give expressions and numerical date for p_{HF} and p_{MW} which are valid at any degree of degeneracy. In addition we will give the ideal pressure p_0 which is simply given by Fermi (Bose) integrals and reads

$$p_0 = \sum_a \frac{(2s_a + 1)}{\Lambda_a^3} k_B T \, I_{3/2}(\mu_a/k_B T) \,, \tag{6.32}$$

where

$$I_\nu(\alpha) = \sum_{r=1}^{\infty} (\pm 1)^{r+1} r^{-\nu-1} \exp(r\alpha), \quad \nu > 1 \,, \tag{6.33}$$

with the property

$$I_{\nu-1}(\alpha) = \frac{\mathrm{d}}{\mathrm{d}\alpha} I_\nu(\alpha) \,.$$

The upper sign in (6.33) refers to Bose, the lower sign to Fermi systems. From eq. (6.1) we get the following expression for the Hartree-Fock contribution of the pressure as a function of the fugacity:

$$p_{HF} = \sum_a \int \frac{\mathrm{d}\boldsymbol{p}\, \mathrm{d}\boldsymbol{q}}{(2\pi)^6} V_{aa}(\boldsymbol{p} - \boldsymbol{q}) f_a(p) f_a(q) \,. \tag{6.34}$$

Here f_a are Fermi functions, and V_{aa} is the Fourier transform of the Coulomb potential

$$V_{aa}(p) = \frac{e_a^2}{\varepsilon_r \varepsilon_0 q^2}$$

(ε_0 — vacuum dielectric constant, ε_r relative dielectric constant).

According to De Witt (1961), eq. (6.34) may be rewritten

$$p_{HF} = \sum_a \frac{(2s_a + 1)\, e_a^2}{4\pi\varepsilon_r\varepsilon_0\Lambda_a^4} \int_{-\infty}^{\alpha_a} \mathrm{d}\alpha' \, I_{-1/2}^2(\alpha') = \sum_a \frac{(2s_a + 1)\, e_a^2}{4\pi\varepsilon_r\varepsilon_0} \mathcal{J}_{HF}(\alpha_p) \,. \tag{6.35}$$

The quantity $I_{-1/2}$ is the Fermi integral defined by (6.33),

$$I_\nu(\alpha) = \frac{1}{\Gamma(\nu + 1)} \int_0^\infty \mathrm{d}x \, \frac{x^\nu}{\mathrm{e}^{x-\alpha} + 1} \,; \tag{6.36}$$

$\Gamma(\frac{1}{2}) = \sqrt{\pi}$, $\alpha_a = \mu_a/k_B T$ — degeneracy parameter, μ_a — chemical potential, s_a — spin projection, $\Lambda_a = h/(2\pi m_a k_B T)^{1/2}$.

An analytic evaluation of (6.35) is possible in the low degeneracy limit, cf. Kraeft and Stolzmann (1976) and Filinov (1975); the result is

$$p_{HF}(\alpha \ll -1) = \sum_a \frac{(2s_a + 1)\, e_a^2}{4\pi\varepsilon_r\varepsilon_0\Lambda_a^4} \left[\frac{\mathrm{e}^{2\alpha_a}}{2} - \frac{2}{3\sqrt{2}} \mathrm{e}^{3\alpha_a} \right]. \tag{6.37}$$

In the case of high degeneracy, we may derive from (6.35) the asymptotic expansion (Berg, 1968; Kraeft and Stolzmann, 1979)

$$p_{\mathrm{HF}}(\alpha \gg 1) = \sum_a \frac{(2s_a + 1)\, e_a^2}{4\pi\varepsilon_r\varepsilon_0 \Lambda_a^4} \left[\frac{2}{\pi}\alpha_a^2 - \frac{\pi}{3}\ln\alpha_a + C\right]. \tag{6.38}$$

This result follows also if the Fermi integrals occuring in (6.35) are replaced by their Sommerfeld expansions (Kraeft and Stolzmann, 1976). However, the constant C may be determined exactly only if the "Hartree-Fock inetgral" $\mathcal{J}_{\mathrm{HF}}(\alpha_a)$ defined by (6.35) is evaluated numerically (Kraeft and Stolzmann, 1979). Different analytic procedures to determine C give entirely different results, cf. Kraeft and Stolzmann (1976).

The numerical evaluation of $\mathcal{J}_{\mathrm{HF}}(\alpha_a)$ yielded (Kraeft and Stolzmann, 1979)

$$C = 0.5040 + O(\alpha_a^{-2}) . \tag{6.39}$$

The numerical evaluation of (6.35) gives the Hartree-Fock integral tabulated in Kraeft and Stolzmann (1979).

In order to give a connection to a real physical situation, the fugacity was replaced in (6.35) by the density according to the ideal relation

$$n_a = \frac{2s_a + 1}{\Lambda_a^3} I_{1/2}(\alpha_a) . \tag{6.40}$$

In Fig. 6.1 the quantity p_{HF} was drawn for an electron gas for a fixed density. It is seen that the low degeneracy result (6.37) fits quite well to the numerical evaluation,

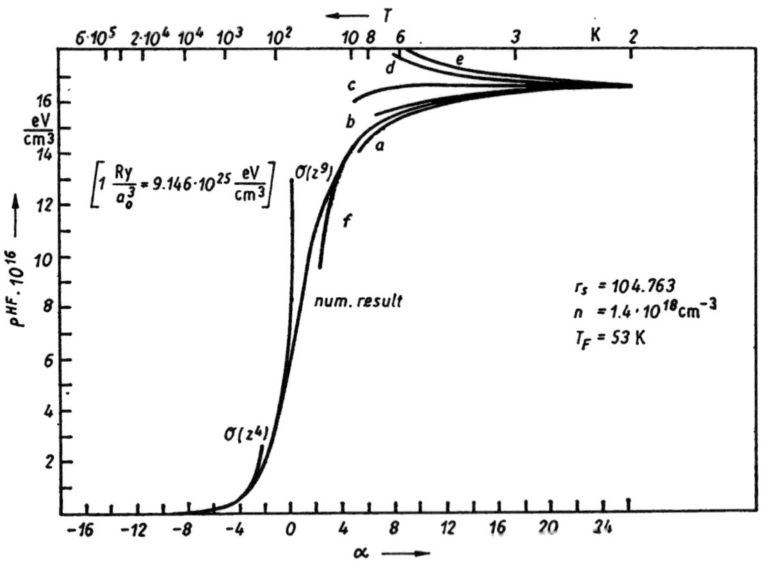

Fig. 6.1. Numerically determined values of the Hartree-Fock pressure p_{HF} as a function of α. Temperature according to (6.40) for a fixed density. $O(z^4)$ and $O(z^9)$ — fugacity expansions according to Stolzmann (1976)
a—f: Approximations in the highly degenerate region. a — Wassermann, Buckholtz, De Witt (1970), b — Horovitz and Thieberger (1974), c — Isihara and Kojima (1974), d — Fennel, Kraeft and Kremp (1974), e — Glasser (1975), f — equation (6.38) with (6.39).

while among the different high degeneracy expressions (De Witt, 1961; Horovitz and Thieberger, 1974; Isihara and Kojima, 1974; Glasser, 1975; Fennel, Kraeft and Kremp, 1974) only (6.38) and (6.39) coincide satisfactorily with the exact result.

Let us now proceed to the evaluation of the direct contributions of the order e^4. The corresponding contribution to the pressure is usually refered to as the Montroll-Ward approximation. The latter is defined by the following expression:

$$p_{\mathrm{MW}} = -\hbar \int \frac{\mathrm{d}\boldsymbol{p}}{(2\pi)^3} \int_0^\infty \frac{\mathrm{d}\omega}{2\pi} \coth \frac{\beta\hbar\omega}{2}$$

$$\cdot \left[\arctan \frac{\sum_c V_{cc}(p) \operatorname{Im} \Pi_c(p\omega)}{1 - \sum_b V_{bb}(p)\, \Pi_b(p\omega)} - \sum_d V_{dd}(p) \operatorname{Im} \Pi_d(p\omega) \right]. \tag{6.41}$$

Here we have to use

$$\operatorname{Im} \Pi_c(p, \omega) = \frac{(2s_c + 1)\, k_B T m_c^2}{4\pi\hbar^4 p} \ln \left| \frac{\mathrm{e}^{-\alpha_c} + \mathrm{e}^{-\frac{\beta}{2m_c}\left(\frac{m_c\omega}{p} + \frac{\hbar p}{2}\right)^2}}{\mathrm{e}^{-\alpha_c} + \mathrm{e}^{-\frac{\beta}{2m_c}\left(\frac{m_c\omega}{p} - \frac{\hbar p}{2}\right)^2}} \right| \tag{6.42}$$

and

$$\operatorname{Re} \Pi_c(p, \omega) = -(2s_c + 1) \int \frac{\mathrm{d}\boldsymbol{p}'}{(2\pi)^3} \frac{f_c\left(\boldsymbol{p}' + \frac{\boldsymbol{p}}{2}\right) - f_c\left(\boldsymbol{p}' - \frac{\boldsymbol{p}}{2}\right)}{\hbar\omega + \frac{\hbar^2}{2m}\left(\left(\boldsymbol{p}' - \frac{\boldsymbol{p}}{2}\right)^2 - \left(\boldsymbol{p}' + \frac{\boldsymbol{p}}{2}\right)^2\right)}. \tag{6.43}$$

For numerical purposes it is convenient to carry out a partial integration. We get with dimensionless variables and for

$m_c = m_e$, $|e_c| = e$, $\alpha_c = \alpha$, from (6.41)—(6.43)

$$p_{\mathrm{MW}} = -\frac{|\alpha k_B T|^{5/2}\, (2m_e)^{3/2}}{4\pi^3\, \hbar^3} \int_0^\infty \mathrm{d}x\, x^2$$

$$\cdot \int_0^\infty \mathrm{d}y\, \frac{\mathrm{e}^{|\alpha| y} + 1}{\mathrm{e}^{|\alpha| y} - 1} \left[\arctan \frac{P(x, y)}{1 - Q(x, y)} - P(x, y) \right], \tag{6.44}$$

where the functions P, Q are defined by (for later purposes we retain the subscripts)

$$P(x, y) = \sum_c \frac{e_c^2 (2\, m_c)^{1/2}}{4\pi\varepsilon_r\varepsilon_0 \cdot 2\hbar (k_B T)^{1/2} |\alpha_c|^{3/2}\, x^3} \cdot \ln \frac{\mathrm{e}^{-\alpha_c} + \mathrm{e}^{-\frac{|\alpha_c|}{4}\left(\frac{y^2}{x^2} + 2y + x^2\right)}}{\mathrm{e}^{-\alpha_c} + \mathrm{e}^{-\frac{|\alpha_c|}{4}\left(\frac{y^2}{x^2} - 2y + x^2\right)}}, \tag{6.45}$$

$$Q(x, y) = \sum_b \frac{e_b^2(2m_b)^{1/2}}{4\pi\varepsilon_r\varepsilon_0\hbar 2\pi(k_B T)^{1/2}\,|\alpha_b|^{3/2}\,x^3} \cdot \int_0^\infty dt \frac{L(x, y, t, \alpha_b)}{\left[e^{\frac{1}{2}(t-\alpha_b)} + e^{-\frac{1}{2}(t-\alpha_b)}\right]^2}, \tag{6.46}$$

$$L(x, y, t, \alpha) = (a_-^2 - b^2)\ln\left|\frac{a_- - b}{a_- + b}\right| - (a_+^2 - b^2)\ln\left|\frac{a_+ - b}{a_+ + b}\right| - 2bx\,,$$

$$a_\pm = y/2x \pm x/2; \qquad b = (t/|\alpha|)^{1/2}\,. \tag{6.47}$$

In the nondegenerate limit we get from (6.41) (see EBELING, KRAEFT and KREMP 1976, 1979)

$$p_{MW} = \frac{K^2 k_B T}{12\pi} - \sum_{ab} \frac{\pi^3}{4} \frac{e_a^2 e_b^2 \lambda_{ab}}{(4\pi\varepsilon_r\varepsilon_0)^2\,k_B T} \frac{z_a z_b (2s_a + 1)\,(2s_b + 1)}{\Lambda_a^3\,\Lambda_b^3}$$

$$\left(\varkappa^2 = \sum_a \frac{e_a^2 z_a (2s_a + 1)}{\varepsilon_r\varepsilon_0 k_B T \Lambda_a^3}, \qquad z_a = e^{\beta\mu_a}, \qquad \lambda_{ab} = \frac{\hbar}{\sqrt{2m_{ab}\,k_B T}}\right). \tag{6.48}$$

In the degenerate situation, we have for the one component (electron) plasma

$$p_{MW} = -n_e(0.0622 \ln r_s - 0.142 - 0.0054 r_s \ln r_s - 0.015 r_s) \cdot \mathrm{Ry}/a_B^3 \tag{6.49}$$

(see GELL-MANN and BRUECKNER, 1957; CARR and MARADUDIN, 1964).

$$\mathrm{Ry} = \frac{m_e e^4}{2\hbar^2(4\pi\varepsilon)^2}, \qquad a_B = \frac{\hbar^2 4\pi\varepsilon}{m_e e^2},$$

$$r_s = \left(\frac{3}{4\pi n_e}\right)^{1/3} \cdot \frac{1}{a_B}, \qquad \varepsilon = \varepsilon_r\varepsilon_0\,.$$

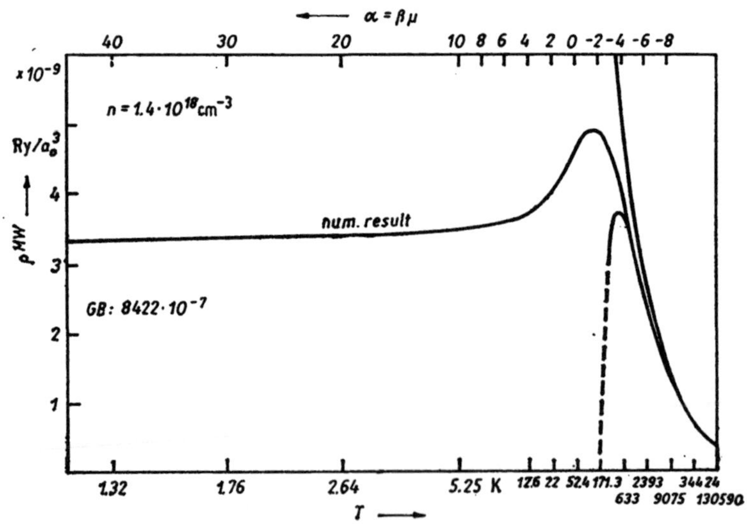

Fig. 6.2. Numerically determined values of the Montroll-Ward pressure p_{MW} as a function of α. Temperature according to (6.40) for a fixed density. GB — Gell-Mann-Brueckner value of p_{MW}. Dashed line — high temperature approximation (6.48), full line at high temperatures — first contribution of (6.48) only

Tab. 6.1. Montroll-Ward pressure after eq. (6.44) at fixed temperature in Ry a_B^{-3}. Density according to eq. (6.40).

α	$T = 5 \cdot 10^2$ K	
	lg n	p^{MW}
−6.0	17.13	1.4985 −10
−5.0	17.56	6.3096 −10
−4.0	18.00	2.5671 − 9
−3.0	18.42	9.8986 − 9
−2.0	18.84	3.5330 − 8
−1.0	19.25	1.1209 − 7
1.0	19.93	6.7453 − 7
2.0	20.18	1.2509 − 6
3.0	20.39	2.0266 − 6
4.0	20.55	3.9893 − 6
5.0	20.68	4.1226 − 6
6.0	20.79	5.4261 − 6
8.0	20.97	8.5185 − 6
10.0	21.11	1.2202 − 5
15.0	21.38	2.3862 − 5
20.0	21.56	3.8508 − 5
25.0	21.71	5.6406 − 5
30.0	21.83	7.7044 − 5
35.0	21.93	1.0016 − 4
40.0	22.01	1.2608 − 4

α	$T = 3.4291 \cdot 10^3$ K		α		
	lg n	p^{MW}		lg n	p^{MW}
−8.0	17.51	2.3512 −10	5.5	21.99	1.4688 −4
−7.5	17.73	4.9316 −10	6.0	22.05	1.6621 −4
−7.0	17.95	1.0333 − 9	6.5	22.09	1.8653 −4
−6.5	18.16	2.1642 − 9	7.0	22.14	2.0772 −4
−6.0	18.38	4.5115 − 9	8.0	22.23	2.5366 −4
−5.5	18.60	9.3744 − 9	10.0	22.37	2.5603 −4
−5.0	18.81	1.9380 − 8	12.0	22.49	4.7500 −4
−4.5	19.03	3.9838 − 8	14.0	22.59	6.0553 −4
−4.0	19.25	8.1105 − 8	15.0	22.63	6.7464 −4
−3.5	19.46	1.6320 − 7	16.0	22.67	7.5094 −4
−3.0	19.67	3.2390 − 7	18.0	22.75	9.0703 −4
−2.5	19.88	6.3040 − 7	20.0	22.82	1.0786 −3
−2.0	20.10	1.1977 − 6	22.0	22.88	1.2591 −3
−1.5	20.30	2.2083 − 6	24.0	22.93	1.4485 −3
−1.0	20.50	3.9129 − 6	25.0	22.96	1.5511 −3
−0.5	20.69	6.6285 − 6	26.0	22.99	1.6642 −3
0.0	20.87	1.0646 − 5	28.0	23.03	1.8823 −3
0.5	21.04	1.6160 − 5	30.0	23.08	2.1119 −3
1.0	21.18	2.3426 − 5	32.0	23.12	2.3444 −3
1.5	21.32	3.2058 − 5	34.0	23.16	2.5778 −3
2.0	21.44	4.2202 − 5	35.0	23.18	2.6985 −3
2.5	21.54	5.3709 − 5	36.0	23.20	2.8192 −3
3.0	21.64	6.6455 − 5	38.0	23.23	3.1206 −3
3.5	21.72	8.0324 − 5	39.0	23.25	3.2602 −3
4.0	21.80	9.5367 − 5	40.0	23.27	3.4038 −3
4.5	21.87	1.1145 − 4	41.0	23.28	3.5393 −3
5.0	21.93	1.2863 − 4	42.0	23.30	3.6861 −3

Tab. 6.1, cont

α	$T = 6.0 \cdot 10^3$ K		$T = 8.0 \cdot 10^3$ K	
	lg n	P^{MW}	lg n	P^{MW}
−8.0	17.87	6.2176 −10	18.06	1.0279 −9
−7.5	18.09	1.3180 − 9	18.28	2.1808 −9
−7.0	18.31	2.7633 − 9	18.50	4.5834 −9
−6.5	18.53	5.7977 − 9	18.72	9.6184 −9
−6.0	18.74	1.2107 − 8	18.93	2.0106 −8
−5.5	18.96	2.5210 − 8	19.15	4.1906 −8
−5.0	19.17	5.2239 − 8	19.36	8.6934 −8
−4.5	19.39	1.0766 − 7	19.58	1.7933 −7
−4.0	19.61	2.1995 − 7	19.80	3.6724 −7
−3.5	19.82	4.4452 − 7	20.01	7.4352 −7
−3.0	20.03	8.8557 − 7	20.22	1.4846 −6
−2.5	20.24	1.7339 − 6	20.43	2.9130 −6
−2.0	20.45	3.3103 − 6	20.64	5.5752 −6
−1.5	20.66	6.1260 − 6	20.85	1.0344 −5
−1.0	20.86	1.0914 − 5	21.05	1.8460 −5
−0.5	21.05	1.8515 − 5	21.24	3.1387 −5
0.0	21.23	2.9822 − 5	21.42	5.0560 −5
0.5	21.40	4.5339 − 5	21.59	7.6839 −5
1.0	21.54	6.5329 − 5	21.73	1.1044 −4
1.5	21.68	8.9285 − 5	21.87	1.5111 −4
2.0	21.79	1.1697 − 4	21.98	1.9762 −4
2.5	21.90	1.4823 − 4	22.09	2.4972 −4
3.0	22.00	1.8260 − 4	22.19	3.0664 −4
3.5	22.08	2.1967 − 4	22.27	3.6878 −4
4.0	22.16	2.5938 − 4	22.35	4.3499 −4
4.5	22.23	3.0250 − 4	22.42	5.0357 −4
5.0	22.29	3.4762 − 4	22.48	5.8015 −4
5.5	22.35	3.9451 − 4	22.54	6.5946 −4
6.0	22.40	4.4583 − 4	22.59	7.4258 −4
6.5	22.45	4.9789 − 4	22.64	8.2770 −4
7.0	22.51	5.5597 − 4	22.70	9.2056 −4
8.0	22.58	6.7363 − 4	22.77	1.1152 −3
10.0	22.72	9.3992 − 4	22.91	1.5521 −3
12.0	22.84	1.2453 − 3	23.03	2.0418 −3
15.0	22.99	1.7620 − 3	23.18	2.8967 −3
20.0	23.17	2.7991 − 3	23.36	4.5699 −3
25.0	23.32	4.0330 − 3	23.51	6.5716 −3
30.0	23.44	5.4361 − 3	23.63	8.8376 −3
35.0	23.54	6.9852 − 3	23.73	1.1369 −2
40.0	23.62	8.7273 − 3	23.81	1.4099 −2

α	$T = 10^4$ K	
	lg n	P^{MW}
−8.0	18.20	1.5161 −9
−7.5	18.42	3.2285 −9
−7.0	18.64	6.7984 −9
−6.5	18.86	1.4241 −8
−6.0	19.07	2.9798 −8
−5.5	19.29	6.2141 −8
−5.0	19.50	1.2900 −7
−4.5	19.72	2.6621 −7
−4.0	19.94	5.4613 −7
−3.5	20.75	1.1075 −6

Tab. 6.1. cont

α	$T = 10^4$ K	
	ln n	p^{MW}
−3.0	20.36	2.2154 −6
−2.5	20.68	4.3543 −6
−2.0	20.80	8.3526 −6
−1.5	21.00	1.5523 −5
−1.0	21.20	2.7756 −5
−0.5	21.38	4.7227 −5
0.0	21.56	7.6155 −5
0.5	21.73	1.1570 −4
1.0	21.88	1.6623 −4
1.5	21.00	2.2683 −4
2.0	22.12	2.9676 −4
2.5	22.23	3.7431 −4
3.0	22.34	4.5863 −4
3.5	22.41	5.0566 −4
4.0	22.50	6.4814 −4
4.5	22.56	7.5056 −4
5.0	22.63	8.6161 −4
5.5	22.68	9.7739 −4
6.0	22.74	1.0995 −3
6.5	22.78	1.2269 −3
7.0	22.84	1.3609 −3
8.0	22.92	1.6453 −3
10.0	23.06	2.2811 −3
12.0	23.18	3.0106 −3
15.0	23.33	4.2500 −3
20.0	23.51	6.6756 −3
25.0	23.65	9.5771 −3
30.0	23.78	1.2860 −2
35.0	23.87	1.6576 −2
40.0	23.95	2.0553 −2

α	$T = 2 \cdot 10^4$ K		$T = 5 \cdot 10^4$ K	
	lg n	p^{MW}	lg n	p^{MW}
−6.0	19.53	9.8719 −8	20.13	4.8654 −7
−5.0	19.95	4.3912 −7	20.56	2.1966 −6
−4.0	20.40	1.8658 −6	21.00	9.4712 −6
−3.0	20.81	7.6584 −6	21. 42	3.9257 −5
−2.0	21.24	2.9203 −5	21.84	1.5145 −4
−1.0	21.65	9.7934 −5	22.25	5.1541 −4
1.0	22.33	5.8968 −4	22.93	3.1284 −3
2.0	22.58	1.0457 −3	23.18	5.4990 −3
3.0	22.79	1.5992 −3	23.39	8.3154 −3
4.0	22.95	2.2356 −3	23.55	1.1466 −2
5.0	23.08	2.9393 −3	23.68	1.4892 −2
6.0	23.20	3.7221 −3	23.79	1.8625 −2
8.0	23.37	5.4996 −3	23.97	2.7047 −2
10.0	23.51	7.5488 −3	24.11	3.6644 −2
15.0	23.78	1.3856 −2	24.38	6.5511 −2
20.0	23.96	2.1608 −2	24.56	1.0112 −1
25.0	24.10	3.0921 −2	24.71	1.4315 −1
30.0	24.23	4.1090 −2	24.83	1.9040 −1
35.0	24.32	5.3129 −2	24.93	2.4270 −1
40.0	24.40	6.5266 −2	25.01	2.9926 −1

The evaluation of (6.44) was carried out numerically in a time consuming procedure (KRAEFT and JAKUBOWSKI, 1978; KRAEFT and STOLZMANN, 1979; STOLZMANN and KRAEFT, 1980). The result is shown for fixed density in Fig. 6.2. The connection to the density was fixed according to (6.40). It is seen that the numerical evaluation is sufficiently well represented, in the region of low degeneracy, by (6.48), while the high degeneracy result (6.49) fits to the numerical curve only if $r_s \ll 1$. The Tabls. 6.1 and 6.2 give the numerical evaluation for various temperatures at fixed density, and for various densities at fixed temperatures. Again (6.40) was applied; of course the connection between p_{MW} and α does not suffer from approximation (6.40).

Tab. 6.2. Montroll-Ward pressure after eq. (6.44) at fixed density (units of p_{MW} in Ry a_B^{-3}). Temperature according to eq. (6.40)

$n = 1.4 \cdot 10^{18}$ cm^{-3}

α	T, K	p_{MW}	α	T, K	p_{MW}
−8.0	9.0728 +3	1.2800 −9	6.0	8.6233	3.5407 −9
−7.5	6.5012 +3	1.5167 −9	6.5	7.9877	3.5210 −9
−7.0	4.6587 +3	1.7720 −9	7.0	7.4375	3.5065 −9
−6.5	3.3387 +3	2.0646 −9	8.0	6.5343	3.4754 −9
−6.0	2.3927 +3	2.3899 −9	9.0	5.8254	3.4632 −9
−5.5	1.7151 +3	2.7496 −9	10.0	5.2530	3.4453 −9
−5.0	1.2297 +3	3.1377 −9	11.0	4.7823	3.4261 −9
−4.5	8.8201 +2	3.5355 −9	12.0	4.3885	3.4209 −9
−4.0	6.3307 +2	3.9231 −9	14.0	3.7673	3.4128 −9
−3.5	4.5486 +2	4.2811 −9	15.0	3.5173	3.4086 −9
−3.0	3.2739 +2	4.5712 −9	16.0	3.2996	3.4000 −9
−2.5	2.3629 +2	4.7787 −9	18.0	2.9350	3.3959 −9
−2.0	1.7129 +2	4.8866 −9	20.0	2.6428	3.3884 −9
−1.5	1.2501 +2	4.8962 −9	22.0	2.4034	3.3818 −9
−1.0	9.2131 +1	4.8402 −9	24.0	2.2037	3.3765 −9
−0.5	6.8838 +1	4.7184 −9	25.0	2.1153	3.3733 −9
0.0	5.2357 +1	4.5661 −9	26.0	2.0346	3.3721 −9
0.5	4.0677 +1	4.4027 −9	28.0	1.8896	3.3698 −9
1.0	3.2347 +1	4.2629 −9	30.0	1.7638	3.3645 −9
1.5	2.6336 +1	4.1002 −9	32.0	1.6534	3.3609 −9
2.0	2.1924 +1	3.9888 −9	34.0	1.5563	3.3583 −9
2.5	1.8622 +1	3.8837 −9	35.0	1.5119	3.3576 −9
3.0	1.6099 +1	3.8099 −9	36.0	1.4700	3.3562 −9
3.5	1.4129 +1	3.7332 −9	38.0	1.3927	3.3542 −9
4.0	1.2561 +1	3.6745 −9	39.0	1.3571	3.3538 −9
4.5	1.1289 +1	3.6263 −9	40.0	1.3232	3.3511 −9
5.0	1.0241 +1	3.5933 −9	41.0	1.2909	3.3485 −9
5.5	9.3652 +0	3.5639 −9	42.0	1.2602	3.3468 −9

6.2. Density and Chemical Potential in the Screened Ladder Approximation

6.2.1. Bound State and Quasiparticle Contributions

In the preceding Section we have considered the thermodynamic properties on the basis of the charging procedure, that means we have determined primarily the equation of state for the pressure. This method is not very convenient if we take into account in the first step of our approximation effective one-particle and two-particle energies and if we replace the Coulomb potential by a screened one. All this quantities are functions of λ and the integration over λ is difficult. But it is very important to use these effective quantities because we obtain in this way non-perturbative expressions for the thermodynamic functions. In order to obtain thermodynamic properties it is therefore more convenient to start with the well known relation (6.50) (STOLZ and ZIMMERMANN, 1979). From the equation

$$n_a(\mu, T) = \int f_a(\omega)\, A_a(p, \omega) \frac{\mathrm{d}\omega}{2\pi} \frac{\mathrm{d}\boldsymbol{p}}{(2\pi)^3} \tag{6.50}$$

we obtain all thermodynamic information as for example the chemical potential $\mu(n, T)$ by inversion of (6.50). The equation of state for the pressure is given by

$$p = \Sigma \int\limits_{-\infty}^{\mu_a} n_a(T, \mu_a)\, \mathrm{d}\mu_a\,.$$

In the expression (6.50) $A_a(p, \omega)$ denotes the one-particle weight function given by

$$A_a(p, \omega) = \frac{\Gamma_a(p, \omega)}{\left[\omega - \dfrac{p^2}{2m_a} - \mathrm{Re}\, \Sigma_a(p, \omega)\right]^2 + \left[\dfrac{\Gamma_a(p, \omega)}{2}\right]^2} \tag{6.51}$$

and

$$f_a(\omega) = [\mathrm{e}^{-\beta(\omega - \mu_a)} \pm 1]^{-1}\,.$$

Because the spectral function is very complicated we use the approximation (3.145) from Chapter 3.2.5. (KREMP, KRAEFT and LAMBERT, 1984)

$$A_a(p, \omega) = 2\pi\delta(\omega - \varepsilon_a(p)) \left\{1 + \frac{\partial}{\partial \hbar\omega} \mathrm{Re}\, \Sigma_a(p, \omega)|_{\omega=\varepsilon_a}\right\}$$

$$- 2\, \mathrm{Im}\, \Sigma_a(p, \omega) \frac{\mathrm{d}}{\mathrm{d}\hbar\omega} \frac{P}{\omega - \varepsilon_a(p)}\,. \tag{6.52}$$

As we have shown in Chapter 3.2.4. the $\varepsilon_a(p)$ are the quasiparticle energies which follow from the dispersion relation

$$\varepsilon_a(p) = E_a(p) + \mathrm{Re}\, \Sigma_a\, (p, \omega)\,|_{\omega=\varepsilon_a(P)} \tag{6.53}$$

$(E_a(p)$ = Hartree-Fock self energy)

and Γ_a is given by

$$\Gamma_a(p, \omega) = \Sigma_a^{>}(p, \omega) \mp \Sigma_a^{<}(p, \omega)\,. \tag{6.54}$$

Finally let us write down the dispersion relation between $\mathrm{Re}\, \Sigma_a$ and Γ_a

$$\mathrm{Re}\, \Sigma_a(p, \omega) = \Sigma_a^{\mathrm{HF}}\, (p, \omega) + P \int^{\mathrm{U}} \frac{\Gamma(p, \bar{\omega})}{\omega - \bar{\omega}} \frac{\mathrm{d}\bar{\omega}}{2\pi}\,. \tag{6.55}$$

Using these relations we find from (6.50) (KREMP, KRAEFT and LAMBERT, 1984)

$$n_a(\mu_a, T) = \int \frac{\mathrm{d}\boldsymbol{p}}{(2\pi)^3} f(\varepsilon(p)) + \int \frac{\mathrm{d}\omega\, \mathrm{d}\boldsymbol{p}}{(2\pi)^4} 2\pi\delta(\hbar\omega - \varepsilon_a(p))\, f_a(\omega) \tag{6.56}$$

$$\cdot \frac{\partial}{\partial\omega} \mathrm{Re}\, \Sigma_a(p, \omega)\,|_{\omega=\varepsilon_a} - \int\limits_{-\infty}^{\infty} \frac{\mathrm{d}\omega}{2\pi} \int \frac{\mathrm{d}\boldsymbol{p}}{(2\pi)^3} \frac{\mathrm{d}}{\mathrm{d}\omega} \frac{P}{\hbar\omega - \varepsilon_a(p)} \Gamma_a(p, \omega)\, f_a(\omega)$$

or with (6.55) (Stolz and Zimmermann, 1979; Röpke, Münchow, and Schulz, 1982)

$$n_a = \int \frac{dp}{(2\pi)^3} f(\varepsilon_a(p)) + \int_{-\infty}^{\infty} \frac{d\omega}{2\pi} \int \frac{dp}{(2\pi)^3} \frac{\partial}{\partial\omega} P \frac{\Gamma_a(p,\omega)}{\hbar\omega - \varepsilon_a(p)} \{f_a(\omega) - f_a(\varepsilon(p))\} . \tag{6.57}$$

The physical meaning of this expression is the following: The first term is the contribution of the ideal Fermi gas of quasiparticles. The correlations between quasiparticles are described by the further terms. This term can be interpreted as a generalization of the second virial coefficient to quasiparticles.

For the further consideration we have to write down approximations for the self energy $\varepsilon_a(p)$. In the case of the long range Coulomb interaction the self energy is given by $\Sigma(p, \omega) = \Sigma_a^{\mathrm{H}}(p) + \overline{\Sigma}_a(p, \omega)$. The screened self energy $\varepsilon_a(p)$ is given by equation (3.229):

$$\overline{\Sigma}_a(1\bar{1}) = i\hbar\, V_{aa}^{s}(11') G_a(11') + i \sum_b \int d\bar{1}\, d\bar{2}\, V_{ab}^{s}(1\bar{2})\, G_a(1\bar{1})\, \delta\, \Sigma_a(\bar{1}1')/\delta U_b^{\mathrm{eff}}(2\bar{2}) ; \tag{6.58}$$

using (6.4) and the approximation (6.5) we find

$$\overline{\Sigma}_a(11') = i\hbar\, V_{aa}^{s}(11')\, G_a(11') + i \sum_b \int d\bar{1}\, d\bar{2}\, d5\, d6$$

$$\cdot\, V_{ab}^{s}(1\bar{2})\, G_a(1\bar{1}) \frac{\delta\overline{\Sigma}_a(\bar{1}1')}{\delta G_b(56)} G_b(\bar{2}6)\, G_b(5\bar{2}) . \tag{6.59}$$

Now we use the ladder approximation (6.7) for $\delta\Sigma/\delta G$ and introduce the T-Matrix (6.8). Then we find the screened ladder approximation for the self energy

$\Sigma_a(11') =$ 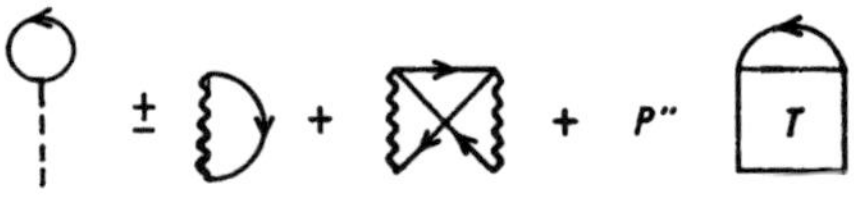 (6.60)

$$= \Sigma^{\mathrm{H}}(11') + \Sigma^{\mathrm{RPA}}(11') + \Sigma_{\mathrm{exch}}^{2}(11') + \Sigma_a^{\mathrm{ladd}}(11') ,$$

with

$\Sigma_a^{\mathrm{H}} =$ , $\Sigma_a^{\mathrm{RPA}} = \pm$ , (6.61)

$\Sigma_{\mathrm{exch}}(11') =$ , $\Sigma_a^{\mathrm{ladd}}(11') = P''$ .

The corresponding expression for $\Gamma_a(p, \omega)$ is given by

$$\Gamma_a(p, \omega) = \Gamma_a^{\mathrm{RPA}}(p, \omega) + \Gamma_{2a}^{\mathrm{ex}}(p, \omega) + \Gamma_a^{\mathrm{ladd}}(p, \omega) . \tag{6.62}$$

The contribution Γ_a^{RPA} is considered in detail in Chapters 3.3.3. and 5.3.2. and is given by the expressions (3.269) and (5.163). Using (5.163) and (3.258) it is easy to show that the contribution Γ_a^{RPA} to the correlation part of n_a vanishes. In the same manner $\Gamma_{2a}^{\mathrm{ex}}(p, \omega)$ gives no contribution. The contribution $\Gamma_a^{\mathrm{ladd}}(p)$ was discussed in

3.2.9., there we found

$$\Gamma_a(p, \omega) = \sum_b \int \frac{dp_2}{(2\pi)^3} 2 \operatorname{Im} \langle p_1 p_2 | T_{ab}(\hbar\omega + \varepsilon_b(p_2)) | p_2 p_1 \rangle \cdot \{n_B(\hbar\omega + \varepsilon_b(p_2)) + f_a(\varepsilon_b(p_2))\} . \tag{6.63}$$

For the further consideration it is more convenient to use the on-shell T-Matrix. After putting $T(\omega)$ on-shell that means taking $\hbar\omega = \varepsilon(p_1) + \varepsilon(p_2)$ only the pure scattering spectrum is included, and the bound states are lost. To study the effects of bound states we separate first the bound state contributions to n_a. Bound states are contained in the third term of (6.56). In ladder approximation we obtain for this term

$$\sum_b \int_{-\infty}^{\infty} \frac{d\omega}{2\pi} \int \frac{dp_1\, dp_2}{(2\pi)^6} \left(\frac{d}{d\omega} \frac{1}{\hbar\omega - \varepsilon_a(p_1) - \varepsilon_b(p_2)} \right) 2 \operatorname{Im} \langle p_1 p_2 | T_{ab}(\omega) | p_2 p_1) \cdot g_{ab}^{B}(\omega) \left(1 - f_a(\varepsilon_a(p_2))\right) . \tag{6.64}$$

Here we have used the substitution $\hbar\omega \to \hbar\omega + \varepsilon(p_2)$ and the relation

$$\{g_B^{ab}(\omega) + f_b(\varepsilon_b(p_1))\} f_a(\hbar\omega - \varepsilon_b(p_1)) = g_B^{ab}(\omega) \left(1 - f_b(\varepsilon_b(p_1))\right) . \tag{6.65}$$

In order to obtain the bound state part we consider the ω-integration in the interval $(-\infty, 0)$. Using the bilinear expansion (3.215) for the T-Matrix it is easy to show that

$$\sum_b \int_{-\infty}^{0} \frac{dp_1\, dp_2}{(2\pi)^6} \frac{d}{d\omega} \left(\frac{1}{\hbar\omega - \varepsilon_b(p_2) - \varepsilon_a(p_1)} \right) \langle p_1 p_2 | T_{ab}(\omega + i\delta) | p_2 p_1 \rangle = \sum_b \sum_{nP} \pi\delta(\hbar\omega - E_{nP}^{ab}) . \tag{6.66}$$

Then we obtain for the density (Kremp, Kraeft and Lambert, 1984)

$$\begin{aligned} n_a = & \int \frac{dp_1}{(2\pi)^3} f_a(\varepsilon_a(p_1)) + \frac{P''}{2V} \sum_b \sum_{nP} \left(e^{\beta(E_{nP} - \mu_a - \mu_b)} - 1\right)^{-1} \\ & + P'' \int \frac{dp}{(2\pi)^3} 2\pi f_a(\varepsilon_a(p)) \frac{\partial}{\partial \hbar\omega} \operatorname{Re} \Sigma_a^c(p\omega)|_{\omega=\varepsilon_a} - \sum_b \int_0^{\infty} \frac{d\omega}{2\pi} \int \frac{dp_1\, dp_2}{(2\pi)^6} P'' \\ & \cdot \left\{ \left(\frac{d}{d\omega} \frac{P}{\hbar\omega - \varepsilon_a(p_1) - \varepsilon_b(p_2)} \right) 2 \operatorname{Im} \langle p_1 p_2 | T_{ab}(\omega) | p_2 p_1 \rangle g_B^{ab}(\omega) \{1 - f_a(\varepsilon_b(p_1))\} \right\} . \end{aligned} \tag{6.67}$$

Considering the bound state contribution in this equation the following interpretation is possible: Evidently there is no essential difference between the bound state and the quasipartical contribution. For this reason it is convenient in statistical mechanics to use the following interpretation (Ebeling, 1974): bound states = new (composite) particles.

Therefore instead of our system consisting of two species of elementary Fermi particles, we consider now a system of many components which are the free particles a and b and the composite particles (ab). For the description of such a system it is necessary to introduce fugacities of all particles. For the new ones, it is obvious to define in eq. (6.67) that

$$\mu^n_{ab} = \mu_a + \mu_b - \beta E^{ab}_n\,, \qquad z^n_{ab} = z_a z_b\, \mathrm{e}^{-\beta E^{ab}_n}\,. \tag{6.68}$$

With this substitution the bound state part contributes to the density as a system of ideal Bose particles. In this way we describe the properties of the system in the chemical picture corresponding to an ionisation equilibrium:

$$a + b \rightleftharpoons (ab)\,. \tag{6.69}$$

6.2.2. The Mass Action Law

The condition for the chemical equilibrium is given by eq. (6.68). Applying this chemical picture we can divide the total density into a contribution of free particles a with the density n_a^{free} and a contribution where a is bound n_a^{bound}:

$$n_a = n_a^{\text{free}} + n_a^{\text{bound}}$$

with

$$n_a^{\text{free}} = \int \frac{\mathrm{d}p}{(2\pi)^3}\, f_a(\varepsilon_a(p)) + \frac{P''}{\hbar} \int \frac{\mathrm{d}p}{(2\pi)^3}\, 2\pi\, f_a(\varepsilon_a(p))$$
$$\cdot \frac{d}{d\hbar\omega} \operatorname{Re} \Sigma_a^{\text{corr}}(p\,\omega)|_{\omega=\varepsilon_a(p)} - P'' \int\limits_0^\infty \frac{\mathrm{d}\omega}{2\pi} \int \frac{\mathrm{d}p_1\, \mathrm{d}p_2}{(2\pi)^6}\, g_{\mathrm{B}}^{ab}(\omega)$$
$$\cdot \{1 - f_a(\varepsilon_a(p_1))\}\, 2 \operatorname{Im} \langle p_1 p_2 |\, T_{ab}(\omega)\, | p_1 p_2\rangle \frac{\mathrm{d}}{\mathrm{d}\omega} \frac{P}{\hbar\omega - \varepsilon_a(p_1) - \varepsilon_b(p_2)} \tag{6.70}$$

and

$$n_a^{\text{bound}} = \frac{P''}{2V} \sum_b \sum_n \int \frac{\mathrm{d}P}{(2\pi)^3} \left(\mathrm{e}^{\beta\left(\hbar^2 \frac{P^2}{2M} - \mu^n_{ab}\right)} - 1\right)^{-1}. \tag{6.71}$$

By using the eqs. (6.68) and (6.70), (6.71) we are able to describe the chemical equilibrium and to determine the composition of the system with respect to bound and free states. In this sense this equation just mentioned plays the role of a mass action law (MAL). Let us explain this in the more simple case of nondegenerate plasmas. For simplicity we neglect the correlation part in n_a^{free}. Further let us introduce the degree of ionisation by

$$\alpha = \frac{n_a^{\text{free}}}{n_a}\,, \qquad \frac{n_a^{\text{bound}}}{n_a^{\text{free}}} = \frac{1-\alpha}{\alpha} \tag{6.72}$$

and by using (6.70) and (6.71) (KREMP, KRAEFT and LAMBERT, 1984):

$$\frac{1-\alpha}{\alpha^2} = n_a$$
$$\frac{\displaystyle\int \frac{\mathrm{d}P}{(2\pi)^3}\, P'' \sum_{nl} \exp\left[-\beta \frac{\hbar^2 P^2}{2M} - \beta E^{ab}_{nl}(P)\right]}{\displaystyle\int \frac{\mathrm{d}p_a}{(2\pi)^3} \exp\left[-\beta \frac{\hbar^2 p_b^2}{2m_a} - \beta \operatorname{Re} \Sigma_a(p_a, \omega)\right] \int \frac{\mathrm{d}p_b}{(2\pi)^3} \exp\left[-\beta \frac{\hbar^2 p_b^2}{2m_b} - \beta \operatorname{Re} \Sigma_b(p_b, \omega)\right]}. \tag{6.73}$$

This equation is a generalization of the Saha equation to non-ideal plasmas; we note that the energy levels E_{nl}^{ab} are the solutions of the effective wave equation discussed in Chapter 5.4. The usual Saha equation may be obtained if we replace Re Σ by Re $\overline{\Sigma}^{RPA}$ with

$$\mathrm{Re}\,\overline{\Sigma}_a^{RPA}(p,\omega) = \int \frac{\mathrm{d}p}{(2\pi)^3}\,\mathrm{e}^{-\beta\frac{\hbar^2p^2}{2m_a}}\,\mathrm{Re}\,\Sigma_a^{RPA}(p,\varepsilon_a(p))\Lambda_a^3\,. \tag{6.74}$$

Using the relations (3.269) and (6.55) we find in the case of nondegenerate plasmas

$$\overline{\Sigma} = -\frac{\varkappa e^2}{2}, \qquad \varkappa = \frac{1}{r_D}.$$

Further we neglect excited states and replace the sum of states by exp $(-\beta E_{10})$ Then the Saha equation takes the well known form

$$\frac{1-\alpha}{\alpha^2} = n_e\,\Lambda_{ab}^3\,\mathrm{e}^{\beta(|E_{10}|-\varkappa e^2)}.$$

The usual Saha equation therefore is a very simplified form of the general MAL (6.74). The expression $\varkappa e^2$ is the so-called Debye shift and $|E_{10}|$ the ground state energy of the H-Atom. The difference $|E_{10}| - \varkappa e^2$ can be interpreted as an effective ionisation energy and $\varkappa e^2$ as the lowering of the ionisation energy.

In the primitive approximation given above we have considered only E_{10} as a bound state. In more refined calculations also the lower excited states have to be considered as bound ones. However — this is a peculiarity of Coulombic systems — we have to introduce an upper border ε^* of bound states, i.e., only the discrete states with

$$E_n^{ab} < \varepsilon^* < 0$$

are considered per definitionem as bound. However the discrete states above this border

$$\varepsilon^* < E_n^{ab} < 0$$

which have quite extended wave functions and low stability are considered as quasi-free and are to be treated together with the free states. Following out earlier work (Ebeling et al., 1976, 1979) we take the thermal energy as the appropriate border between bound and quasi-free discrete states:

$$\varepsilon^* = \zeta k_B T\,, \qquad \zeta \approx 1\,.$$

In order to specify this physical definition of a bound two-particle state we propose two possible choices of the operator P'' which selects the bound states from the sum over states; physically both these choices are equivalent:

(i) Planck-Brillouin-Larkin formula

$$P'' \sum_n \int \frac{\mathrm{d}\boldsymbol{P}}{(2\pi)^3} \exp\left[-\frac{\beta\hbar^2P^2}{2M} - \beta E_n^{ab}(P)\right] = \sum_n \int \frac{\mathrm{d}\boldsymbol{P}}{(2\pi)^3}\,\mathrm{e}^{-\frac{\beta\hbar^2P^2}{2M}}\left[\mathrm{e}^{-\beta E_n^{ab}} - 1 + \beta E_n^{ab}\right];$$

(ii) Riewe-Rompe convention

$$P'' \sum_n \int \frac{\mathrm{d}\boldsymbol{P}}{(2\pi)^3} \exp\left[-\frac{\beta\hbar^2P^2}{2M} - \beta E_n^{ab}(P)\right] = \sum_{E_n^{ab}<\varepsilon^*} \int \frac{\mathrm{d}\boldsymbol{P}}{(2\pi)^3}\,\mathrm{e}^{-\frac{\beta\hbar^2P^2}{2M} - \beta E_n^{ab}}.$$

An improved Saha equation may be obtained by using the approximation (KRAEFT et al., 1983; KREMP et al., 1984)

$$\int \frac{\mathrm{d}p}{(2\pi)^3} \exp\left[-\beta\left(\frac{\hbar^2 p^2}{2m_a} + \operatorname{Re} \Sigma_a\right)\right] \approx \int \frac{\mathrm{d}p}{(2\pi)^3} \exp\left[-\frac{\beta\hbar^2 p^2}{2m_a}\right] \cdot [1 - \beta \operatorname{Re} \Sigma_a] \approx \exp[-\beta\mu_a] .$$

In this way we obtain from eq. (6.73)

$$n_{ab} = (2s_a + 1)(2s_b + 1) \Lambda_{ab}^{-3} \sigma(T, n_a, n_b) \exp[\beta\mu(n_a, n_b, T)] ,$$
$$\mu(n_a, n_b, T) = \mu_a + \mu_b , \qquad (6.75)$$
$$\sigma(T, n_b, n_a) = \Lambda_{ab}^3 \int \frac{\mathrm{d}P}{(2\pi)^3} P'' \sum_{sl} (2l + 1) \exp\left(-\beta E_{sl}^{ab} - \beta\hbar^2 \frac{P^2}{2M}\right).$$

In the following $\mu(n_a, n_b, T)$ will be called the chemical plasma potential and $\sigma(T, n_a, n_b)$ the Brillouin-Planck-Larkin partition function; moreover we have assumed in the preceding text that the plasma consists of two species a and b only, which are oppositely charged. The Saha equation may be written in the equivalent form

$$\mu_{ab} = \mu_a + \mu_b \qquad (6.75\text{a})$$

with

$$\mu_{ab} = k_B T \ln [n_{ab} \Lambda_{ab}^3 / (2s_a + 1)(2s_b + 1) \sigma(T, n_a, n_b)]$$

being the chemical potential of the neutrals (the pairs ab). Eq. (6.75a) expresses the condition of chemical equilibrium (6.69). As a new and interesting result may be considered that in advanced theories the energy levels of the Bethe-Salpeter equation have to be introduced into the partition function and into the chemical potential of the bound pairs instead of the Schrödinger energy levels. In the Sections 6.3. — 6.5. this result will be shown of some importance for the behaviour of partially ionized plasmas.

6.3. One-Component Plasmas

6.3.1. Analytical Formulae for the Limiting Situations

In previous Sections the high- and low-density behaviour of the thermodynamic function was derived analytically (see 2.1., 2.3., 5.4., 6.1., 6.2.). On the other hand a few numerical points for the Montroll-Ward (MW) contribution to the pressure computed by KRAEFT and STOLZMANN (1979) and others (STOLZMANN and KRAEFT, 1979, 1980; ZIMMERMANN and RÖSLER, 1978) are available (see Tabs. 6.1 and 6.2).

The aim of the present Section is the construction of Padé approximations which coincide with the known analytical results in the highly and weakly degenerate regions, respectively, and give good approximations for the intermediate region.

The n-expansions at $n\Lambda^3 \ll 1$ and expansions according to inverse powers of n at $n\Lambda^3 \gg 1$ may be regarded as the first terms of a Taylor-like series. As is well known; the range of validity of such polynomial expressions may be improved by representing the functions $f(n)$ by Padé approximations, which have the same Taylor coefficients as the limiting cases.

In the case of the Montroll-Ward contribution we cannot hope to get a simple Padé approximation, since the region where limiting formulae are not valid, covers approximately 5 orders of magnitude.

The construction of analytical formulae for the thermodynamic functions of Coulombic systems is of great importance for the consideration of correlation effects in metals and in semiconductors. There are already many papers (see e.g. HUBBARD, 1958; PINES, 1958; CARR et al., 1961; CALLAWAY, 1964, 1969; ZIMMERMANN and RÖSLER, 1978; EBELING et al., 1981) which are devoted to the correlation energy of the electron fluid in the region of metallic densities $2 \lessapprox r_s \lessapprox 6$ where

$$r_s = (4\pi n a_B^3/3)^{-1/3}\,,$$

a_B — Bohr radius.

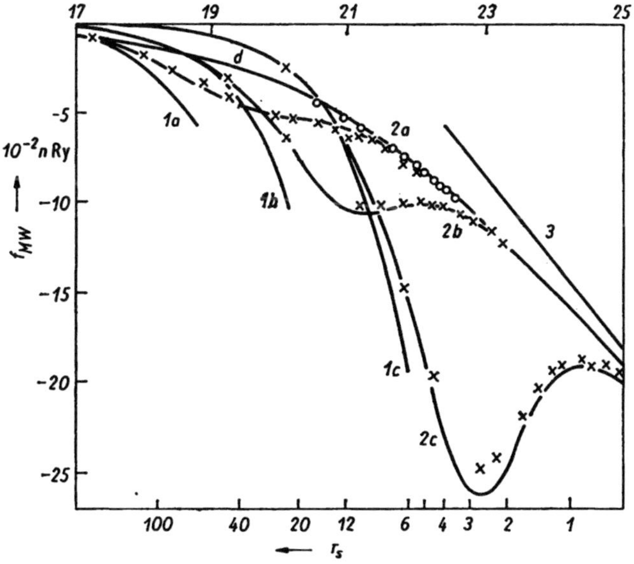

Fig. 6.3. Montroll-Ward contribution to the free energy density of an electron gas (with homogeneous positive background at different temperatures (a — 500 K, b — 3430 K, c — $5 \cdot 10^4$ K, d — 0 K) versus the Bruckner-parameter r_s. The curves 2 correspond to the Padé formula (6.96) and the curves 1 and 3 are the classical (leading term of eq. (6.86)) and the high degenerate (eq. (6.88)) limits, respectively. The crosses are the numerical points calculated by KRAEFT and STOLZMANN (1979) and the circles are RPA results by ZIMMERMANN and RÖSLER (1978). Upper abscissa — lg n.

In order to give an idea of the behaviour of the correlation part of the ground state energy of the electron gas, in Fig. 6.3. we gave the results of numerical calculations of the Montroll-Ward graph from KRAEFT and STOLZMANN (1979) and of RPA calculations by ZIMMERMANN and RÖSLER (1978). For comparison our Padé results are given too. We should mention here that the Montroll-Ward contribution coincides with the RPA calculations at $T = 0$ K. As we see from Fig. 6.3. the RPA values for the correlation part to the ground state energy correspond to the $T = 500$ K calculations for the Montroll-Ward contribution up to $r_s \sim 5.5$ in the case of an electron gas.

Let us now consider some possible applications. The magnitude of the ground state energy is of importance for many properties of the solid state. For example CALLAWAY (1964, 1969) calculated the cohesion energy of alkali metals which is equal to the difference of the ground state energies of valence electrons in the crystal and that for free atoms. In this work the temperature dependence of the exchange and correlation part is taken into acoount. The temperature behaviour of these two contributions is contrary indeed, but their sum is weakly temperature dependent. This does not influence the critical temperature much. However the critical density and the chemical potential are strongly influenced.

In the following let us assume that the diagrams up to the order e^4, i.e. the graphs with no more than 2 interaction lines give the dominant contribution to the thermodynamic functions of one-component plasmas.

As shown in the previous sections, the calculation of the pressure of a many-body system up to the order e^4 can be presented as

$$P = p_0 + p_{\mathrm{H}} + p_{\mathrm{HF}} + p^{(2)} = p_0 + p_{\mathrm{H}} + p_1 + p_2 ,$$

$$p_0 = \frac{2}{\beta \Lambda^3} I_{3/2}(\beta_e, \mu), \qquad p_{\mathrm{H}} = 0 ,$$

$$p_{\mathrm{HF}} = \frac{2e^2}{(4\pi\varepsilon)\, \Lambda^4} \int\limits_{-\infty}^{\beta\mu_e} \mathrm{d}y\, I_{-1/2}^2(y) ,$$

$$p^{(2)} = p_{\mathrm{MW}}(\mu_e, T) + p_{\mathrm{ex}}^{(2)}(\mu_e, T_e) . \tag{6.76}$$

We used $P_{\mathrm{H}} = 0$ in an electro-neutral system ($\sum n_a e_a = 0$). The arguments μ mean the chemical potentials. In the grand canonical ensemble the μ appear as independent variables. Furtheron $I_\nu(x)$ are the known Fermi integrals (see 6.1.).

In the limits of high and low densities, respectively, analytical formulae are valid. If we introduce generalized fugacities

$$z = 2\Lambda^{-3} \exp\,(\beta\mu_e)$$

which coincide with the densities in the classical limit we get in the low degenerate case

$$\beta p_{\mathrm{MW}} = (K^3/12\pi) - \tfrac{1}{4}\, \pi^{3/2} z^2 \lambda_{ee} (\beta e^2/4\pi\varepsilon)^2 + \dots ,$$

$$\beta p_{\mathrm{ex}}^{(2)} = -\tfrac{1}{4}\, \pi^{3/2} z^2 \lambda_{ee} (\beta e^2/4\pi\varepsilon)^2 \ln 2 - \dots$$

with

$$K^2 = \beta z e^2/\varepsilon , \qquad \lambda_{ee} = (\beta\hbar^2/m_e)^{1/2}, \qquad \varepsilon = \varepsilon_r \varepsilon_0 ,$$

ε_r — static dielectric constant of the medium.

The highly degenerate case is given after Gell-Mann and Brueckner (1957), Du Bois (1959), Du Bois and Kivelson (1969), Carr and Maradudin (1964) by

$$p_{\mathrm{MW}} = -nE_{\mathrm{R}}$$

(E_{R} — correlation energy per particle in the ring approximation),

$$E_{\mathrm{R}}/\mathrm{Ry} = 0.0622 \ln r_s - 0.142 ,$$
$$r_s = d/a_{\mathrm{B}}, \qquad \mathrm{Ry} = (e^2/8\pi\varepsilon a_{\mathrm{B}}) . \tag{6.77}$$

In order to get the complete e^4-contribution we take the e^4-exchange term into account and get finally

$$p^{(2)} = -nE^{(2)} ,$$
$$E^{(2)}/\mathrm{Ry} = 0.0622 \ln r_s - 0.096. \tag{6.78}$$

Let us consider now the thermodynamic functions in the canonical ensemble. As independent variables we have to use now the densities n_e or dimensionless densities

$$\bar{n} = n_e \Lambda_e^3 . \tag{6.79}$$

The density dependence of the pressure in the canonical representation follows by an inversion procedure (EBELING et al., 1976, 1979; STOLZMANN and KRAEFT, 1979, 1980) using the relation between the density and the chemical potential given by

$$n_e = \partial p / \partial \mu_e .$$

If we define $\alpha(n_e, T)$ as a function of the density by

$$n_e = 2\Lambda_e^{-3} I_{1/2}(\alpha) \tag{6.80}$$

we get, after an inversion, the chemical potential and the pressure (see 6.1. and EBELING et al., 1976, 1979)

$$\mu(\bar{n}, T) = \mu_e(n_e, T) = \beta^{-1}\alpha - \frac{e^2}{4\pi\varepsilon\Lambda} I_{-1/2}(\alpha) - \frac{\Lambda^3}{2I_{-1/2}(\alpha)} \frac{\partial}{\partial\alpha} p^{(2)} ,$$

$$p(\alpha(\bar{n}), T) = \frac{2}{\beta\Lambda^3} I_{3/2}(\alpha) + \frac{e^2}{2\pi\varepsilon\Lambda^4} [\mathcal{J}_{HF}(\alpha) - I_{1/2}(\alpha)\, I_{-1/2}(\alpha)]$$

$$+ p^{(2)} - [I_{1/2}(\alpha)/I_{-1/2}(\alpha)] \frac{\partial}{\partial\alpha} p^{(2)} , \tag{6.81}$$

$$\mathcal{J}_{HF}(\alpha) = \int_{-\infty}^{\alpha} dy\, I_{-1/2}^2(y) .$$

The free energy density is given by

$$f = F/V = N_e \mu_e - p ,$$
$$f = \frac{2}{\beta\Lambda^3} [\alpha I_{1/2} - I_{3/2}] - \frac{e^2}{4\pi\varepsilon\Lambda^4} e^{2\alpha} H(\alpha) - p^{(2)}(\beta^{-1}\alpha, T) . \tag{6.82}$$

In the low degenerate limit we get

$$\beta f^{(2)} = -\frac{\varkappa^3}{12\pi} \left[1 - \frac{3}{16}\sqrt{2}\, \varkappa\lambda(1 + \ln 2) + O(\varkappa^2\lambda^2)\right],$$

$$\varkappa^2 = \beta n_e e^2/\varepsilon , \quad \lambda^2 = \beta\hbar^2/m_e \tag{6.83}$$

and

$$\beta\mu^{(2)} = \beta e^2 \varkappa \,[1 - \tfrac{1}{4}\sqrt{\pi}\, \varkappa\lambda(1 + \ln 2) + O(\varkappa^2\lambda^2)] . \tag{6.84}$$

In the highly degenerate limit we find from eqs. (6.77) and (6.78)

$$f^{(2)}/n \text{ Ry} = E_4/n \text{ Ry} = 0.0622 \ln r_s - 0.096 ,$$
$$\mu^{(2)}/\text{Ry} = [E^{(2)} + n\, \partial E^{(2)}/\partial n]/\text{Ry} = 0.0622 \ln r_s - 0.1167 . \tag{6.85}$$

6.3.2. Padé Interpolations between the Degenerate and the Nondegeneraet Cases

For many applications (e.g., for the discussion of phase transition) approximate analytical formulae are desirable. In the following analytical interpolation formulae will be derived on the basis of the Padé approximations. A quantitative test to the approximations for the electron gas is the comparison with numerical values of P_{MW} given by KRAEFT and STOLZMANN (1979).

If we assume the border of the validity of the limiting laws to be at $\bar{n} \approx 1$, it is useful to write these laws in the form of n-expansions. The interpolation formulae are choosen to be based on two terms in each case both for the nongenerate and for the highly degenerate limit in order to keep the formulae simple enough. In the case of the electron gas, in the low degenerate limit $\bar{n} \ll 1$, the analytical formulae are as follows

$$\varphi_{\mathrm{MW}}/\mathrm{Ry} = f_{\mathrm{MW}}/n\,\mathrm{Ry} = -f_0\,\bar{n}^{1/2} - f_1\,\bar{n}\,,$$

$$f_0 = \frac{3}{2}\left(\frac{k_{\mathrm{B}}T}{\pi\,\mathrm{Ry}}\right)^{1/4}, \qquad f_1 = -\frac{1}{4\sqrt{2}}\,. \tag{6.86}$$

For the chemical potential we have

$$-\mu_{\mathrm{MW}}/\mathrm{Ry} = \mu_0\bar{n}^{1/2} + \mu_1\bar{n},$$

$$\mu_0 = \left(\frac{k_{\mathrm{B}}T}{\pi\,\mathrm{Ry}}\right)^{1/4}, \qquad \mu_1 = -\frac{1}{2\sqrt{2}}\,. \tag{6.87}$$

In the case $\bar{n} \gg 1$, with eqs. (6.77) and (6.85), we get

$$\varphi_{\mathrm{MW}}/\mathrm{Ry} = -f_2\ln\bar{n} - f_3\,, \tag{6.88}$$

$$f_2 = \frac{2}{3\pi^2}(1 - \ln 2), \qquad f_3 = -\frac{3}{2}f_2\ln\left[(36\pi)^{1/3}\,\beta\,\mathrm{Ry}\right] + 0.142\,, \tag{6.89}$$

$$-\mu_{\mathrm{MW}}/\mathrm{Ry} = \mu_2\ln\bar{n} + \mu_3\,, \tag{6.90}$$

$$\mu_2 = f_2\,, \qquad \mu_3 = f_2 + f_3\,. \tag{6.91}$$

The idea of Padé approximations is, to represent a function which is known only in certain regions of the arguments by rational functions. Padé approximations are being used now in many branches of theoretical physics and chemistry and have proved to be a very useful tool for the representation of functions which are known only in part. In our case the landscape formed by the thermodynamic functions, e.g., by f_{MW} over the density-temperature plane is to be guessed by using the available knowledge, here eqs. (6.86) and (6.88). The limiting behaviour suggests a Padé formula of the type

$$R_1(n, T) + R_2(n, T)\ln R_3(n, T)$$

where the R_i are certain rational functions. The following expression was shown to be adequate (EBELING et al., 1981):

$$-\varphi_{\mathrm{MW}}/\mathrm{Ry} = \frac{f_0\bar{n}^{1/2} + a_3\bar{n}^p\ln(1 + \bar{n}^q a_4)}{1 + a_1\bar{n}^{1/2} + a_2\bar{n}^p}\,, \tag{6.92}$$

$$a_3 = (a_2 f_2/q)\,, \qquad a_1 = -f_1/f_0\,, \qquad a_4 = \exp(f_3 q/f_2)\,, \tag{6.93}$$

p, q, a_2 to be fitted to numerical data. We used, e. g., $q = 1/4$, $p = 4/3$, $a_2 = 1/4$.

This structure yields with $p > 1/2$ in the nongenerate limit the interpolation formulae (EBELING and SÄNDIG, 1973; EBELING et al., 1976, 1979)

$$-\varphi_{\mathrm{MW}}/\mathrm{Ry} = \frac{f_0\bar{n}^{1/2}}{1 - (f_1/f_0)\,\bar{n}^{1/2}} \tag{6.94}$$

and in the $T = 0$ K limit (EBELING et al., 1981)

$$-\varphi_{\mathrm{MW}}/\mathrm{Ry} = \frac{f_2}{q} \ln\left[1 + \exp\left(0.142\, q/f_2\right) r_s^{-3q}\right]. \tag{6.95}$$

For finite temperatures eq. (6.92) has a maximum deviation of 8% from the numerical values given by KRAEFT and STOLZMANN (1979). Fig. 6.3. shows a comparison of the interpolation formulae (6.96) with both the numerical points (crosses) and the limiting laws. In the case of the high temperature limit the agreement is rather good. The agreement with the numerical data may be still improved if the exponent $q = 7/30$ is used instead of $q = 1/4$ (RICHERT, 1982; ZIMMERMANN et al., 1984). Then at $T = 0$ the deviations are smaller than 1% and at high temperatures smaller than 4%. Since the computer calculations are very difficult for high α-values, there are at present only values up to $\alpha = 40$ available. The chemical potential we may get by the relation

$$-\mu/\mathrm{Ry} = -(\varphi/\mathrm{Ry}) + \bar{n}\frac{\partial}{\partial \bar{n}}(-\varphi/\mathrm{Ry}). \tag{6.96}$$

For applications again Padé approximations are useful. We propose

$$-\mu_{\mathrm{MW}}/\mathrm{Ry} = \frac{\bar{\mu}_0 \bar{n}^{1/2} + \frac{3}{4q}\mu_2 \bar{n}^{3/2} \ln\left(1 + \exp\left(\mu_3/4\mu_2\right) \bar{n}^q\right)}{1 - (\mu_1/\mu_0)\, \bar{n}^{1/2} + \frac{3}{4}\, \bar{n}^{3/2}} \tag{6.97}$$

with $q = 1/4$ or $q = 7/30$ respectively. The μ_{MW}-curves are drawn in Fig. 6.4.

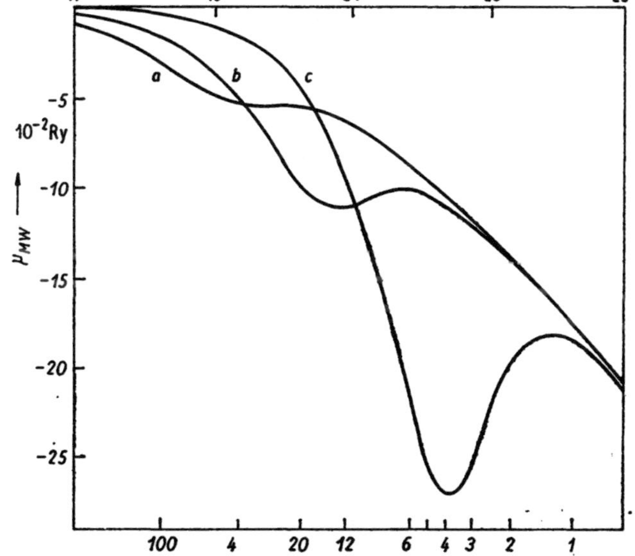

Fig. 6.4. The Montroll-Ward contribution to the chemical potential of an electron gas after eq. (6.97) vs. the electron density at three temperatures (a — 500 K, b — 3430 K, c — $5 \cdot 10^4$ K)

In the same way, as demonstrated above, we may develope the Padé approximations for the complete e^4-contribution. The e^4-exchange term at $\bar{n} \gg 1$ is a constant for f/n Ry as well as for μ/Ry (Du Bois, 1959; Onsager et al., 1966). This can be seen from the eqs. (6.85). So we are able to write the quantities $\varphi^{(2)}$ and $\mu^{(2)}$ in the same manner as the eqs. (6.88) and (6.90).

On the other hand the e^4-exchange term in the case of high temperatures can be calculated from eq. (6.83). It is of order n and influences the coefficients f_1 and μ_1. In the eqs. (6.83) and (6.84) the second terms are due to the MW term and the third terms are due to the e^4-exchange term. Finally we can write the limiting laws in the same way as eqs. (6.86) and (6.87). We should mention here that the e^4-exchange term coincides at $T = 0$ K with the well known Hubbard correction to the RPA results.

After these considerations and taking into account the resembling behaviour of the temperature and the density dependence of the direct and exchange e^4-contributions, there follows the possibility of application of the Padé approximations given above, if we use the limiting laws for the complete e^4-term. In this way one gets for the complete contribution of the order e^4 (Montroll-Ward and exchange term) the formulae

$$-\varphi^{(2)}/\mathrm{Ry} = \frac{f_0\bar{n}^{1/2} + f_2\bar{n}^{4/3}\ln\left(1 + \exp\left(\tilde{f}_3 q/f_2\right)\bar{n}^q\right)}{1 - (\tilde{f}_1/f_0)\,\bar{n} + \frac{1}{4}\,\bar{n}^{4/3}}, \tag{6.98}$$

$$-\mu^{(2)}/\mathrm{Ry} = \frac{\mu_0\,\bar{n}^{1/2} + 3\mu_2\,\bar{n}^{3/2}\ln\left(1 + \exp\left(\tilde{\mu}_3 q/\mu_2\right)\bar{n}^q\right.}{1 - (\tilde{\mu}_1\bar{n}^{1/2}/\mu_0) + \frac{3}{4}\,\bar{n}^{3/2}} \tag{6.99}$$

where

$$\tilde{f}_3 = f_3 - 0.04836\,, \qquad \tilde{f}_1 = f_1(1 + \ln 2)\,, \tag{6.100}$$

$$\tilde{\mu}_3 = \mu_3 - 0.04836\,, \qquad \tilde{\mu}_1 = \mu_1(1 + \ln 2)\,. \tag{6.101}$$

The constant $(\tilde{f}_3 - f_3)$ which is due to the e^4-exchange term has been calculated in a paper by Onsager, Mittag and Stephen (1966). One may ask of course whether it makes sense to take into account the e^4-exchange term neglecting all the other higher order contribution. We shall come back to this question in the next Section. The construction of Padé interpolations between the degenerate and the nondegenerate regions is of course not unique; there exist many possibilities which satisfy the limiting conditions. Between the choices which are under discussion we mention one proposed by Richert (1982) which makes use of the compensation effect between the temperature dependences of the Hartree-Fock term and the Montroll-Ward term (Zimmermann et al., 1984). As empirically known, the sum of the Hartree-Fock and the Montroll-Ward terms is nearly temperature-independent in a broad region beginning at $T = 0$. This suggests a treatment of both terms together. Richert (1982) proposes for the free energy

$$F_{\mathrm{xc}} = F_{\mathrm{HF}} + F_{\mathrm{MW}}$$

the Padé formula

$$\varphi_{\mathrm{xc}} = F_{\mathrm{xc}}(\bar{n}, T)/N_e = \frac{b\bar{n}\,\varepsilon_{\mathrm{xc}}(\bar{n}) - \mathrm{Ry}\left[f_0\bar{n}^{1/2} + f_5\bar{n}\right]}{1 + b\left[\bar{n}^2 + \ln\left(1 - (f_1/bf_0)\bar{n}^{1/2}\right)\right]},$$

$$f_5 = (e^2/16\pi\varepsilon\Lambda_e\,\mathrm{Ry}) \tag{6.102}$$

and f_0, f_1 have the meaning as above (eqs. (6.88), (6.89)).

The corresponding formula for the chemical potential of the electrons is

$$\mu_{xc} = \frac{b\bar{n}^2\mu_{xc}(n,\ T = 0) - \text{Ry}\,[\mu_0\bar{n}^{1/2} + \mu_5\bar{n}]}{1 + b[\bar{n}^2 + \ln\left(1 - (\mu_1/\mu_0 b)\ \bar{n}^{1/2}\right)]},$$

$$\mu_5 = \left(\frac{e^2}{8\pi\varepsilon\,\Lambda\,\text{Ry}}\right). \tag{6.103}$$

Here b is a fitting parameter which is estimated to be $b = 8$. Introducing here

$$\varepsilon_{xc}(n)/\text{Ry} = F_{xc}(n,\ T = 0)/N_e\,\text{Ry} = -0.9163\ r_s^{-1} - 0.08883 \ln\left(1 + f_3' r_s^{-7/10}\right),$$

$$\mu_{xc}(n,\ T = 0)/\text{Ry} = -0.12217 r_s^{-1} - 0.08883 \ln\left(1 + m_3' r_s^{-7/10}\right),$$

$$\text{without } e^4\text{-exchange: } f_3' = 4.9262,\ m_3' = 6.2208,$$

$$\text{with } e^4\text{-exchange: } f_3' = 2.8589,\ m_3' = 3.6094, \tag{6.104}$$

we get a very useful formula for the exchange-correlation contribution to the thermodynamic functions (RICHERT, 1982). We note that the formulae (6.104) for the ground state functions does not contain fitted parameters (except the exponent 7/10) and is in remarkable good agreement with numerical results by ZIMMERMANN and RÖSLER (1978); the deviations for the Montroll-Ward contribution are less than 1% in the region $0 < r_s \leq 12$. Therefore the ground state formula (6.104) seems to be advantageous in comparison with related approximations (HEDIN and LUNDQUIST, 1971;

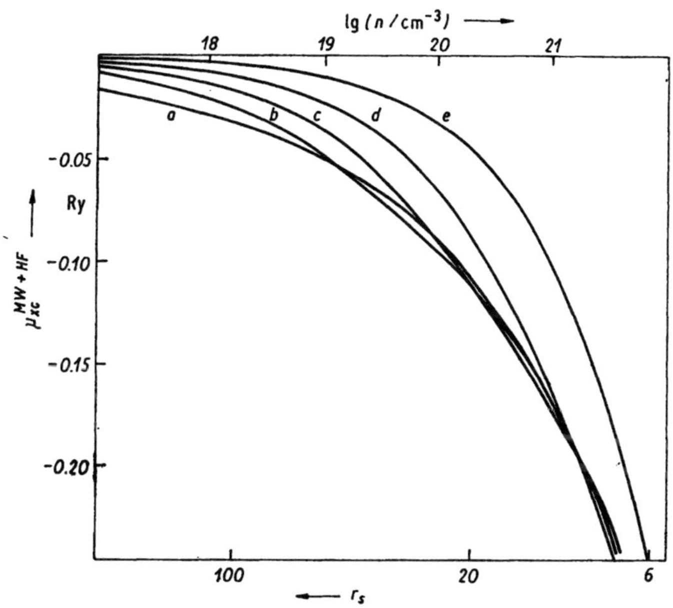

Fig. 6.5. The exchange-correlation contribution to the chemical potential according to eq. (6.103) (a — 0 K, b — 1000 K, c — 300 K, d — 10000 K, e — 50000 K)

VOSKO et al., 1980; WÜNSCHE, 1979, 1983). The density depedence of the exchange-correlation contribution to the chemical potential (Hartree-Fock and Montroll-Ward) at different temperatures is shown in Fig. 6.5. We note the weak temperature dependence below 3000 K.

Due to this effect the electronic contribution to the specific heat will be quite small at lower temperatures, We obtain the specific heat per volume by means of

$$c_v = -T\left(\frac{\partial^2}{\partial T^2} f\right)_v = -T\frac{\partial^2}{\partial T^2}[f_0 + f_{\mathrm{HF}} + f_{\mathrm{MW}}] = c_v^0 + c_v^{\mathrm{HF}} + c_v^{\mathrm{MW}} .$$

For the ideal term follows

$$c_v^0 = \frac{3k^{5/2}T^{3/2}(2m)^{3/2}}{16\pi^{3/2}\hbar^3}[5I_{3/2}(\alpha) - 3I_{1/2}^2(\alpha)/I_{-1/2}(\alpha)] . \tag{6.105}$$

In the low and in the highly degenerate limits we have

$$c_v^0 = \tfrac{3}{2}\, kn \quad \text{if } (-\alpha) \gg 1 ,$$

$$c_v^0 = \frac{2mk^2T}{6\hbar^2}(3\pi^2 n)^{1/3} \quad \text{if} \quad \alpha \gg 1 .$$

The exchange term in the first order is given by the Hartree-Fock specific heat (Kraeft and Stolzmann, 1979, 1984):

$$c_v^{\mathrm{HF}} = \left(\frac{e^2}{4\pi\varepsilon}\right)\frac{k^2T(2m)^2}{8\hbar^4\pi^2}\left[2\int\limits_{-\infty}^{\alpha} \mathrm{d}\alpha'\, I_{-1/2}^2(\alpha')\right.$$

$$\left. - \frac{9}{4} I_{1/2}(\alpha)\, I_{-1/2}(\alpha) + \frac{9}{4} I_{1/2}^2(\alpha)\, I_{-3/2}(\alpha)/I_{-1/2}(\alpha)\right] \tag{6.106}$$

with the limiting laws

$$c_v^{\mathrm{HF}} = \left(\frac{e^2}{4\pi\varepsilon}\right)\frac{\pi n^2\hbar^2}{kT^2 m} \quad \text{if} \quad (-\alpha) \gg 1 ,$$

$$c_v^{\mathrm{HF}} = \frac{k^2T(2m)^2}{4\hbar^4\pi^2}\left(\frac{e^2}{4\pi\varepsilon}\right)\left(c - \frac{\pi}{6} - \frac{\pi}{3}\ln\alpha\right) \quad \text{if} \quad \alpha \gg 1 ,$$

$$c = 0.5040 .$$

For the ring sum we obtained the Montroll-Ward specific heat (Kraeft and Stolzmann, 1984)

$$c_v^{\mathrm{MW}} = \frac{3}{4T}\left[3(I_{1/2}^2(\alpha)/I_{-1/2}^2(\alpha))\frac{\partial^2}{\partial\alpha^2} f^{\mathrm{MW}}(\alpha, T) + \frac{3}{4}T^2\frac{\partial^2}{\partial T^2} f^{\mathrm{MW}}(\alpha, T)\right.$$

$$+ (5(I_{1/2}(\alpha)/I_{-1/2}(\alpha)) - 3(I_{1/2}^2(\alpha)\, I_{-3/2}(\alpha)/I_{-1/2}^3(\alpha)))$$

$$\left. \cdot \frac{\partial}{\partial\alpha} f^{\mathrm{MW}}(\alpha, T) - 4T(I_{1/2}(\alpha)/I_{-1/2}(\alpha))\frac{\partial^2}{\partial T^2} f^{\mathrm{MW}}(\alpha, T)\right] \tag{6.107}$$

with

$$c_v^{\mathrm{MW}} = \frac{k_{\mathrm{B}}\pi^3}{16\pi} \quad \text{if} \quad (-\alpha) \gg 1 ,$$

$$c_v^{\mathrm{MW}} = \frac{k_{\mathrm{B}}^2 T}{6}(3\pi^2)^{1/3} r_s [0.0922 - 0.01667 \ln r_s] \quad \text{if} \quad \alpha \gg 1 ,$$

$$\varkappa^2 = (ne^2/\varepsilon k_{\mathrm{B}}T) , \qquad r_s = \frac{me^2}{\hbar^2 4\pi\varepsilon}\left(\frac{3}{4\pi n}\right)^{1/3} .$$

On the basis of the limiting formulae again, Padé approximation could be constructed. For details of calculations of the specific heat we refer to the papers of KRAEFT and STOLZMANN (1979, 1984). Further we refer to Padé approximations including the correct temperature dependence near $T = 0$ developed by KRAEFT et al. (1984).

6.3.3. Padé Approximations Including Higher Order Interaction Terms and Wigner Crystallization

So far we have discussed the case of weak and moderate interactions only (Fig. 6.6). Let us study now the effects of strong interactions. First we consider the case $T = 0$. Since WIGNER (1932, 1934, 1938, 1954) one knows that electrons form for low densities ($r_s \gtrsim 100$) a *bcc*-crystal with the correlation energy

$$\varepsilon_c(r_s)/\mathrm{Ry} = -\frac{a}{r_s} + O(r_s^{-3/2}), \qquad a = 0.8755 .$$

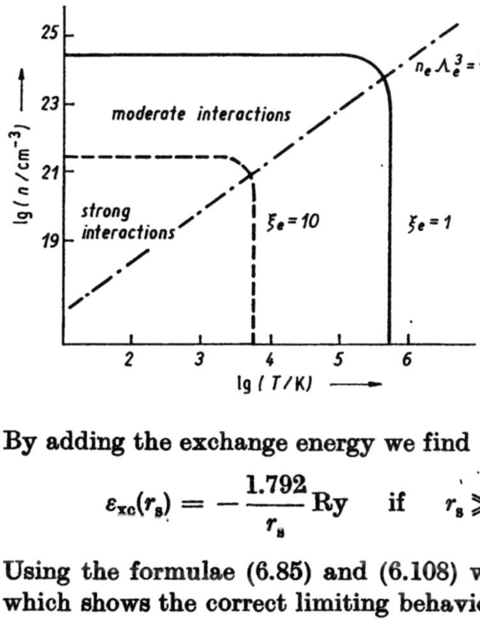

Fig. 6.6. The strong interaction region for the electron gas according to $\xi_e > 10$

By adding the exchange energy we find

$$\varepsilon_{xc}(r_s) = -\frac{1.792}{r_s}\,\mathrm{Ry} \quad \text{if} \quad r_s \gg 1, \quad T = 0 . \tag{6.108}$$

Using the formulae (6.85) and (6.108) we construct a Padé formula for ε_{xc} and μ_{xc} which shows the correct limiting behaviour (RICHERT and EBELING, 1984):

$$\eta_{xc}(r_s)/\mathrm{Ry} = -\eta_1 \ln\left[\frac{1 + \eta_2 r_s^{1/2} + \eta_3 r_s^{-1/2}}{1 + \eta_2 r_s^{1/2}}\right] - \frac{\eta_4}{r_s}, \tag{6.109}$$

$$\eta_{xc} = \varepsilon_{xc}, \mu_{xc}; \qquad \eta_i = \varepsilon_i, \mu_i; \qquad i = 1, 2, 3, 4 .$$

$$\varepsilon_1 = 0.1244, \qquad \mu_1 = 0.1244,$$

$$\varepsilon_2 = 0.3008, \qquad \mu_2 = 0.2664,$$

$$\varepsilon_3 = 2.117, \qquad \mu_3 = 2.501,$$

$$\varepsilon_4 = 0.9163, \qquad \mu_4 = 1.222.$$

We have compared the ε_{xc} values after (6.109) with numerical simulations of CEPERLEY (1981) and CEPERLEY and ALDER (1980, 1981). The deviations are neglegible in the region $r_s \geqq 20$. For smaller r_s-values the comparison is given in Tab. 6.3 and Fig. 6.7. The maximum deviation is less than about 10^{-3} Ry at $r_s \lessapprox 10$. Fig. 6.7 shows that the new formula proposed here gives much better agreement with the MC data than the interpolation formulae proposed in earlier papers by WIGNER (1934, 1938)

$$\varepsilon_c/\mathrm{Ry} = -\frac{0.88}{r_s + 7.8},$$

Tab. 6.3. Comparison of the Padé formulae (6.109) and (6.104) with the ground state energies after Ceperley and Alder

$1000\varepsilon_c$	r_s						
	1	2	5	10	20	50	100
CA	119.6	90.2	56.3	37.2	23.0	11.4	6.4
eq. (6.109)	120.2	89.3	55.8	36.7	22.9	11.4	6.4
eq. (6.104)	120.0	90.2	58.3	40.1	26.7	15.1	9.6

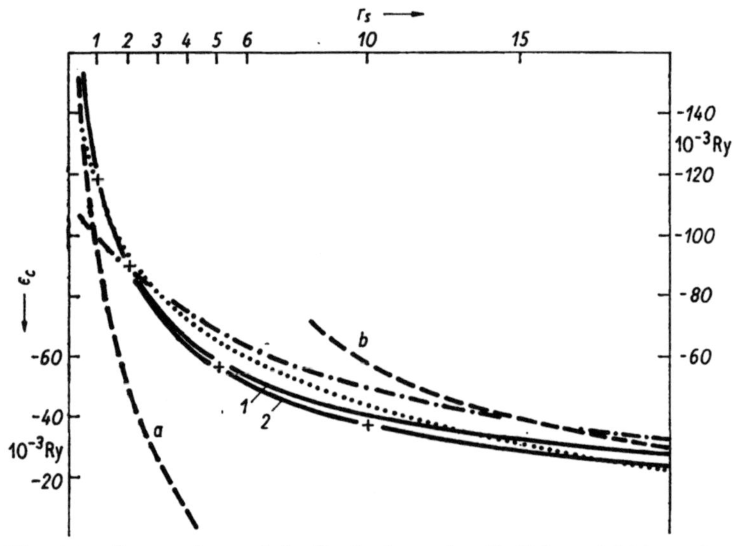

Fig. 6.7. Comparison of the Pade formulae (6.110 — 6.112) and of the limiting laws
The crosses are calculated by CEPERLY a d ALDER (1980);
a — GELL-MANN and BRUECKNER (1957), b — electrostatic limit, 1 — EBELING and RICHERT (1981), 2 — RICHERT and EBELING (1984), dash-dotted line — WIGNER (1934), dotted line — PINES and NOZIERES (1966)

by NOZIÈRES and PINES (1958, 1959)

$$\varepsilon_c/\mathrm{Ry} = -0.115 + 0.0313 \ln r_s,$$

and by EBELING, RICHERT, KRAEFT and STOLZMANN (1981) (see eq. (6.104)). In this way we have shown that, a consequent use of the available knowledge, condensed in Padé approximations, yields quite good results for $T = 0$ in the whole density range. In the following parts the formulae are extended to finite temperature. In the region of low densities and finite temperatures we introduce

$$\bar{n} = n_e \Lambda_e^3 = 6\sqrt{\pi}\, r_s^{-3} \tau^{-3/2} ,$$
$$\tau = k_B T/\mathrm{Ry}, \quad \Lambda^2 = 4\pi\tau^{-1} a_B^2 .$$

Using a representation by means of the plasma parameter

$$\mu = \frac{e^2 \varkappa}{4\pi\varepsilon} = \frac{2\sqrt{6}}{(r_s \tau)^{3/2}}$$

the general formulae for the correlation parts may be written as

$$\varphi_c(\mu, \tau)/\mathrm{Ry} = -\frac{1}{3}\mu\tau + \frac{\mu^2 \tau^{3/2}}{16}\sqrt{\frac{\pi}{2}} - \frac{1}{12}\mu^2 \tau\,[K_e(\xi) + \ln \mu] + O(\mu^2 \ln \mu) ,$$

$$\mu_c(\mu, \tau)/\mathrm{Ry} = -\frac{1}{2}\mu\tau + \frac{1}{8}\mu^2 \tau^{3/2}\sqrt{\frac{\pi}{2}} - \frac{1}{6}\mu^2 [K_e(\xi) + \ln \mu] + O(\mu^2 \ln \mu) \tag{6.110}$$

where the first term is due to Debye screening, the second stems from ring sum interaction (RPA) and all ladder terms are included in $K_e(\xi)$, the known temperature dependent virial function (see EBELING et al., 1976, 1979, and Section 5.4).:

$$K_e(\xi) = \frac{6}{\xi^3}\left\{Q_3(-\xi) - \frac{1}{2} E_2(-\xi)\right\} - \ln \xi ,$$

$$Q_3(\xi) = -\frac{\xi^3}{6}\left(\frac{1}{2} C_E + \ln 3 - \frac{1}{2}\right) + \sum_{n=4}^{\infty} \frac{\sqrt{\pi}\,\zeta(n-2)}{\Gamma\left(\frac{n}{2}+1\right)}\left(\frac{\xi}{2}\right)^n , \tag{6.111}$$

$$E_2(\xi) = \frac{\sqrt{\pi}}{4}(\ln 2)\,\xi^2 + \frac{\pi^2}{72}\xi^3 + \sum_{n=4}^{\infty} \frac{\sqrt{\pi}\,(1-2)^{2-n}\,\zeta(n-1)}{\Gamma\left(\frac{n}{2}+1\right)}\left(\frac{\xi}{2}\right)^n ,$$

$$\xi^2 = 2/\tau , \qquad C_E = 0.577216\ldots ,$$

$\zeta(x)$ — Riemann's function.

For explicit calculations we may use

$$K_e(\xi) \approx (0.879 - \xi^{-1} \cdot 2.251 - \ln \xi) \cdot \left[1 + 0.32\xi^{\left(\frac{2+6.4\xi}{1+2.6\xi}\right)}\right]^{-1} + 0.4197 + \tfrac{3}{4}\sqrt{\pi}\,\xi^{-1} .$$

The maximum deviation of this approximation from the correct function is less than 0.3%.

Now Padé approximations will be constructed. The following structure for φ_{xc} and μ_{xc} is assumed:

$$-\eta_{xc}(\mu, \tau) = \frac{\mathrm{Ry}\,[\eta_5 \mu\tau + \eta_6 (\mu\tau)^2] - \eta_9 \bar{n}^2 \eta_{xc}(r_s, 0)}{1 + \eta_7 \mu\tau^{1/2} + \eta_8 \mu\{\ln(\mu + h) - \ln \mu\} + \eta_9 \bar{n}^2} , \tag{6.112}$$

$$\varphi_5 = \frac{1}{3}, \quad \varphi_6 = \frac{1}{16}, \quad \varphi_7 = \frac{3}{16}\sqrt{\frac{\pi}{2}}, \quad \varphi_8 = \frac{1}{4},$$

$$\mu_5 = \frac{1}{2}, \quad \mu_6 = \frac{1}{8}, \quad \mu_7 = \frac{1}{4}\sqrt{\frac{\pi}{2}}, \quad \mu_8 = \frac{1}{3},$$

$$h = \exp(-K_e(\xi)) , \quad \varphi_9 = \mu_9 \approx 8 .$$

If one expands (6.112) for low μ, i.e. $\mu \ll 1$, eqs. (6.110) are obtained which is one of the conditions of consistency. For $\xi_e < 1$ the ladder contributions h become small and only the RPA part η_7 is to be taken into account, i.e., in RPA we put simply $h = 0$ and for $T = 0$ we have as in Section 6.3.2.

$$\begin{aligned} \eta_{xc}^{RPA}(r_s, 0) &= -0.08883 \ln\left[1 + \eta_{10} r^{-7/10}\right] - \eta_4 r_s^{-1}\,, \\ \varphi_{10} &= 4.9262\,, \quad \mu_{10} = 6.2208\,. \end{aligned} \tag{6.113}$$

The expansion for low temperatures gives in the first order $\eta_{xc}(r_s, 0)$ which is given by (6.109). The switch parameter $\bar{n}^2$ is chosen as to produce the lowest temperature correction proportional to τ^2 as it is expected from Hartree-Fock estimates (Kremp et al., 1972), this is easily to be seen if (6.112) is represented in r_s and τ only. By exclusion of μ by means of r_s and τ we obtain

$$\begin{aligned} \eta_{xc}(r_s, \tau) &= \frac{\eta_9 \eta_{xc}(r_s, 0) - \left(\dfrac{2}{3\pi}\,\eta_6 r_s^3 \tau^2 + \dfrac{\sqrt{6}}{18\pi}\,\eta_5 r_s^{9/2} \tau^{5/2}\right) \mathrm{Ry}}{\eta_9 + \dfrac{1}{36\pi}\, r_s^6 \tau^3 + \dfrac{\sqrt{6}}{18\pi}\,\eta_7 r_s^{9/2} \tau^2 L}\,, \\ L &= 1 + \frac{4}{3} \sqrt{\frac{2}{\pi\tau}} \ln\left[1 + \frac{h}{2\sqrt{6}}\,(r_s\tau)^{3/2}\right]. \end{aligned} \tag{6.114}$$

We want to note that eqs. (6.112) and (6.114) have no fitted parameters. All the values of the eight parameters $\eta_1, \ldots, \eta_8$ follow from the limiting laws. The parameter η_9 could still be fitted by using calculations of the temperature dependent corrections to the ground state energy beyond the Hartree-Fock term (Kraeft and Stolzmann, 1984; Kraeft et al., 1985). As in the case of $\tau = 0$ we would prefer to use (6.112) for the chemical potential as well as for the free energy instead of using differentiations.

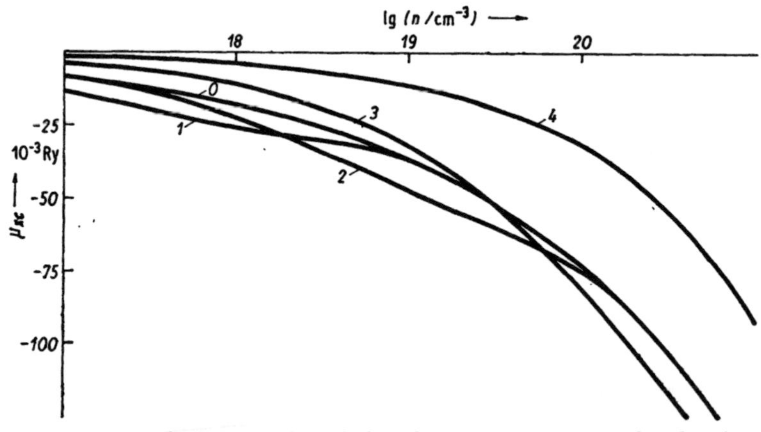

Fig. 6.8. Chemical potential of the electron-gas versus the density for five temperatures 0 — 0 K, 1 — 100 K, 2 — 500 K, 3 — 3430 K, 4 — 50000 K

As far as no reliable numerical calculations are available for $\tau \neq 0$ ($\xi \gtrsim 1$) and densities $\bar{n} \approx 1$, the quality of eq. (6.112) cannot be checked at this moment. However we assume (6.112) to be at least a first approximation for the whole n—T-plane, since in RPA the consistency with (6.102), (6.103) is quite good.

Fig. 6.8 shows the chemical potential after (6.112) for five temperatures. It is well visible that the curves change quite smoothly from the Debye curve to the $\tau = 0$ values.

In Fig. 6.9 the free energy φ_{xc} for 500 K is shown. Of special interest is the strong temperature dependence of exchange and correlation parts which is nearly cancelled due to the opposite sign. Therefore the sum φ_{xc} is near to the $\tau = 0$ curve up to quite low densities. This fact provides a justification for the broad use of ground state calculations in solid state theory. We underline however that it makes no sense to take into account temperature corrections in the Hartree-Fock term and to neglect then in the correlation contribution.

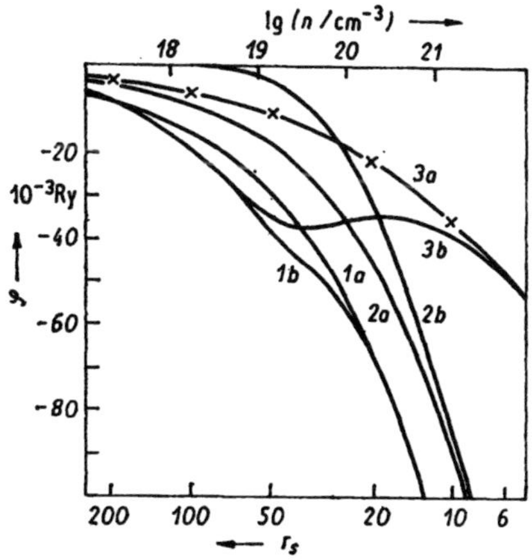

Fig. 6.9. Free energy of the electron-gas at $T = 500$ K (b) as a function of the density. The curves denoted by a are plotted for the ground state $T = 0$. 1 denotes the exchange-correlation sum, 2 the exchange and 3 the correlation contribution. The crosses are calculated by CEPERLEY and ALDER (1980). The strong temperature dependences of the curves 2 and 3 are canceled in the sum 1 up to relatively low densities

Let us further remark that the formulae proposed here could be improved in the low density range by differing between the solid and the fluid branch which has been neglected here for simplicity. Finally let us express the hope that reliable Monte Carlo calculations for finite temperatures will be available in the near future.

6.4. Electron-Hole Plasmas

6.4.1. Analytical Results for the Plasma Model

In a semiconductor an electron-hole (e—h) plasma may be generated by optical excitation. Estimations show that in a semiconductor with indirect gap the plasma is in thermodynamic equilibrium due to the lifetime of the excited states which is of the order of some microseconds. The density of the e—h-plasma may be varied in wide limits by varying the intensity of the optical excitation. In particular there exist density-temperature (n—T) states, for which the interactions in the e—h-plasma are essential. Thus the e—h-plasmas are interesting examples for non-ideal plasmas.

This non-ideality causes some properties of optically excited semiconductors, which are typical many-body effects. These effects are, e.g., bound states between electrons and holes, the excitons, and the corresponding ionization equilibrium $e + h \rightleftharpoons ex$ and the phase transition to an e—h-liquid. This phase transition was predicted by KELDYSH (1970) and experimentally verified by several authors (RICE et al., 1977, 1980). For this reason it is useful to investigate the thermodynamic properties of an e—h-plasma in a wide $n-T$-range by theoretical methods.

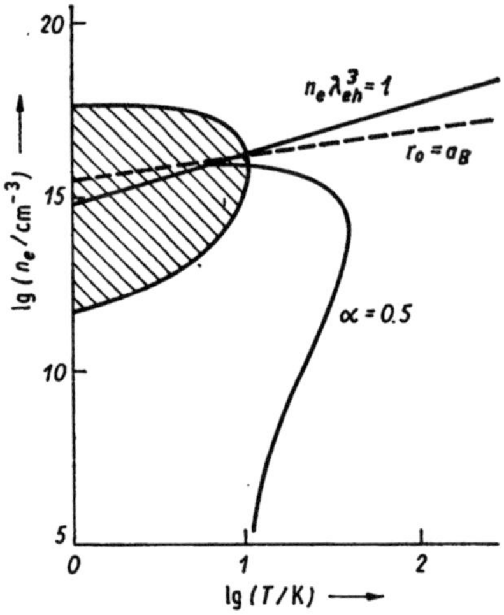

Fig. 6.10. Characteristic curves in the density-temperature plane for e—h-plasmas. In the hatched region coexistence between fully ionized droplets and a weakly ionized exciton gas is observed

Fig. 6.10 gives a survey of the behaviour of the e—h-plasma in the different regions of the n_e-T-plane. Let us introduce the following quantities: $\lambda_{eh} = \hbar \cdot (2m_{eh}kT)^{-1/2}$ — the thermal wavelength, $r_0 = [e^2(n_e^* + n_h^*)/\varepsilon k_B T]^{-1/2}$ — the (Debye) screening length and $a_B = 4\pi\varepsilon\hbar^2/m_{eh}e^2$ — the Bohr radius. Here n_e^*, $n_h^* = n$ represent the densities of free electrons and holes, respectively, m_{eh} is the reduced effective mass.

Now we define the curves $n_e^* \cdot \lambda_{eh}^3 = 1$, $r_0 = a_B$. These curves which are straight lines in the ($\log n^*$-$\log T$)-plane, divide the plane into regions, where the plasmas behave qualitatively different, as indicated in Fig. 6.10. In a charged many-body system the two-particle bound states are influenced by the other particles via screening. The effective two-particle Schrödinger equation (Bethe-Salpeter equation) has bound states only under the condition $r_0 \gtrsim a_B$. There are no bound states if $r_0 \lesssim a_B$ (cf. Sections 5.4. and 6.2.). Thus the existence of excitons is only possible in the lower part of the n_e-T-plane, where $r_0 \gtrsim a_B$ and here an ionization equilibrium must be discussed. The degree of ionization is given by $\alpha = n_e^*/n_e$ which must be calculated from the solution of a Saha equation. The curve $\alpha = 0.5$ divides the region below the straight line $r_0 = a_B$ into a region with a strongly ionized plasma (i.e. e, h dominate) and a region with a weakly ionized plasma (i.e. excitons dominate) (cf. Fig. 6.10).

The behaviour of the e—h-plasma in the thermodynamic equilibrium may be deduced from the thermodynamic potential, which must be calculated quantum statistically, in order to discuss phase transitions, ionization equilibria, etc. But it is not possible to calculate F analytically in the entire n_e-T-plane. In the degenerate e—h-plasma, where the relation $n_e\lambda_{eh}^3 \gg 1$ is valid (cf. Fig. 6.10) we may apply the theory of the degenerate electron gas in the random-phase approximation (RPA).

Calculations of this kind were carried out by several workers (ZIMMERMANN, 1971, 1976; ZIMMERMANN et al., 1978, 1984; COMBECSOT and NOZIERES, 1972; RICE et al., 1977, 1980; HAUG and TRAN THOAI, 1978). In the case $n_e\lambda_{eh}^3 \ll 1$ we have a non-degenerate e—h-plasma, in which the formation of excitons is essential. In this case we must use approximations, which contain all the screened ladder diagrams. But here we have other simplifications on account of $n_e\lambda_{eh}^3 \ll 1$.

The quantum statistics of e—h-plasmas was developed in earlier work; especially the full ladder sum was discussed and evaluated (KRAEFT et al., 1975; EBELING et al., 1976, 1981).

In this Section we want to develop the quantum statistics of e—h-plasmas, and we want to deal with (i) the calculation of the thermodynamic functions, (ii) the ionization equilibrium e + h → ex, and (iii) the stability of the thermodynamic functions and the phase diagram of e—h-plasmas.

As in the case of OCP let assume in the following that the dominant contributions to the thermodynamic potentials come from the diagrams up to the order e^4, i.e.

$$\begin{aligned} p &= p_0 + p_1 + p_2 = p_{\mathrm{id}} + p_{\mathrm{HF}} + p^{(2)}\,, \\ f &= f_{\mathrm{id}} + f_{\mathrm{HF}} + f^{(2)}\,, \\ \mu &= \mu_e + \mu_h = \mu_{\mathrm{id}} + \mu_{\mathrm{HF}} + \mu^{(2)} \end{aligned} \tag{6.115}$$

where in the canonical ensemble

$$\begin{aligned} \mu_{\mathrm{id}} &= \beta^{-1}(\alpha_e + \alpha_h)\,, \\ n_e &= 2\Lambda_e^{-3}\, I_{1/2}(\alpha_e)\,, \qquad n_h = 2\Lambda_h^{-3}\, I_{1/2}(\alpha_h)\,, \\ \mu_{\mathrm{HF}} &= -\frac{e^2}{4\pi\varepsilon}[\Lambda_e^{-1} I_{-1/2}(\alpha_e) + \Lambda_h^{-1} I_{-1/2}(\alpha_h)]\,, \\ f_{\mathrm{HF}} &= -\frac{e^2}{4\pi\varepsilon}[\Lambda_e^{-4}\, e^{2\alpha_e} H(\alpha_e) + \Lambda_h^{-4}\, e^{2\alpha_h} H(\alpha_h)]\,. \end{aligned} \tag{6.116}$$

The second order terms are explicitly known only in the two limiting situation of low or high degeneracy respectively.

In the case of a two component Fermi system with $\bar{n} \ll 1$ one can write the correlation contribution to the chemical potential, as follows:

$$\begin{aligned} -\mu^{(2)}/\mathrm{Ry} &= 2\bar{n}^{1/2}\left(\frac{4a^3 k_B T}{\varepsilon^6 \pi\, \mathrm{Ry}}\right)^{1/4} - \bar{n}\frac{a}{\varepsilon^2}(1+\gamma)^{1/2}\, k(T) + \dots\,, \\ k(T) &= \frac{1}{2} + \frac{1}{4}\left(\frac{2}{1+\gamma}\right)^{1/2}[1 + g_e^{-1}\ln 2 + \sqrt{\gamma} + \sqrt{\gamma}\, g_h^{-1}\ln 2]\,. \end{aligned} \tag{6.117}$$

where g_e and g_h denote the many-valley degeneracy of the electrons and holes respectively. At $\bar{n} \gg 1$ KRAEFT and FENNEL (1976) and ZIMMERMANN et al. (1984) gave a generalization of the Gellmann-Brückner formula which reads

$$\mu^{(2)} = c(\ln r_s - K_c) + \mu_{2X}\,, \tag{6.118}$$

$$C = \frac{2}{\pi^2} g_e \{\varrho(\gamma, \delta) + \delta^{-1}\varrho(\gamma^{-1}, \delta^{-1})\}\,,$$

$$K_c = 1.2250 - \ln\{\tfrac{1}{2}(g_e g_h)^{2/3}[1 + \delta^{1/3}\gamma^{2/3}]\,\delta^{-1/3}\gamma^{-2/3}\}\,,$$

$$\mu_{2X} = 0.04836\,(1+\gamma)^2\,\gamma^{-1}\,,$$
$$\gamma = m_e^*/m_h^*\,,\qquad a = m_e^*/m_e\,,\qquad \delta = g_e/g_h$$
$$\varrho(\gamma,\delta) = (1+\gamma)\,[1 - \ln 2 + \delta^{-1/3} - \gamma \ln (1 + \gamma^{-1}\delta^{-1/3})]\,,$$

m_c^* — effective masses, g_c — many-valley degeneracy.

From these expressions the temperature functions μ_0, $\tilde{\mu}_1$, μ_2 and $\tilde{\mu}_3$ which appear in the limiting expansions

$$\begin{aligned} -\mu^{(2)}/\mathrm{Ry} &= \mu_0 \bar{n}^{1/2} + \tilde{\mu}_1\,\bar{n} + \ldots\,,\\ -\mu^{(2)}/\mathrm{Ry} &= \mu_2 \ln \bar{n} + \tilde{\mu}_3 + \ldots\,,\\ \bar{n} &= (\bar{n}_e + \bar{n}_h)/2\,, \end{aligned} \tag{6.119}$$

may be calculated explicitly. We should note that the mass dependence of K_c and therefore also $\tilde{\mu}_3$ is not exactly known. The value used here was proposed by ZIMMERMANN et al. (1984).

6.4.2. Padé Approximations

If we restrict ourselves to a symmetrical e—h-plasma (sy), we can describe it with the knowledge of the electron gas values, it we pay attention to the scaling rules (EBELING et al., 1981). This means that we have to choose the bound energy of an excited pair E_{BO} as energy unit and the system Bohr radius a_{BO} as length unit.

$$E_{BO} = \frac{a_m}{\varepsilon_r^2}\,\mathrm{Ry}\,,\qquad a_{BO} = \frac{\varepsilon_r}{a_m}\,a_B\,. \tag{6.120}$$

The dimensioneless density parameter of the symmetrical system becomes than

$$r_s^{sy} = \frac{1}{4g^{4/3}}\,r_s\,. \tag{6.121}$$

As the temperature we use

$$T^{sy} = 8\,\frac{E_{BO}}{\mathrm{Ry}}\,g^2 T = 8\,\frac{\mu^*}{\varepsilon^2}\,g^2 T\,. \tag{6.122}$$

With these conventions we get the thermodynamic functions of the symmetrical plasma

$$\begin{aligned} \frac{f_{sy}^{(2)}(r_s^{sy}, T^{sy})}{n^{sy} E_{BO}} &= \frac{8g f_{eg}^{(2)}(r_s, T)}{n\,\mathrm{Ry}}\\ \frac{\mu_{sy}^{(2)}(r_s^{sy}, T^{sy})}{E_{BO}} &= \frac{8g \mu_{eg}^{(2)}(r_s, T)}{\mathrm{Ry}}\,. \end{aligned} \tag{6.123}$$

The Hartree-Fock contribution shows the same behaviour.

Fig. 6.11 shows the chemical potential for a symmetrical e—h-modell. As we can see a critical point at $T_c = 0.079 E_{BO}/k_B$ and $r_s^c = 3.4$. If we neglect the e^4-exchange term we get $T_c = 0.099 E_{BO}/k_B$ and $r_s^c = 4.4$ for the critical point (EBELING et al., 1981).

The framework of an asymmetrical e—h-model has to take into account the conduction and valence band structure. We take for m_a^* a simple version of the model due to Brinkmann and Rice (RICE et al., 1977), i. e. we assume $m_e = m_{de}$ (m_{de} — the density of states mass) $m_e^* = 0.22\, m_e$ (m_e — free electron mass) with a many-valley

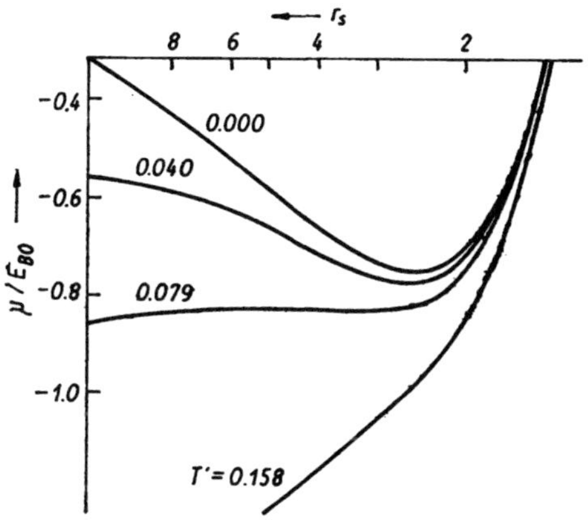

Fig. 6.11. The chemical potential of a symmetrical e—h-plasma at different temperatures ($T = T' E_{BO}/k_B$) as a function of the density parameter r_s (up to order e^4)

degeneration $g_e = 4$. Furtheron we neglect the light hole band so that $m_h^* = m_{nh} = 0.347 m_s$ and $g_h = 1$ with a dielectric constant $\varepsilon_r = 15.3$. In order to construct the Padé approximation we have to calculate the coefficients $\mu_0, \dots, \mu_3$. The explicite expressions of these coefficients which appear in eq. (6.119) are

$$\mu_0 = 2\left(\frac{4a_m^3 k_B T}{\varepsilon_r^6 \pi \,\mathrm{Ry}}\right)^{1/4} \cdot r^{-1/2}, \qquad \tilde{\mu}_1 = -\frac{a_m}{\varepsilon_r^2}(1+\gamma)^{1/2}\, k(T),$$

$$\mu_2 = \frac{2g_e a_m}{3\pi^2(1+\gamma)\,\varepsilon_r^2}\,[\varrho(\gamma) + \delta^{-1}\,\varrho(\gamma^{-1})],$$

$$\tilde{\mu}_3 = -\,3\mu_2\{\ln\,[3ra_{BO}/4\pi)^{1/3}] - K_c\} - \frac{a_m(1+\gamma)}{\varepsilon_r^2\gamma}\,0.04836\,.$$

The abbreviations are defined as above. These coefficients have to be introduced into the Padé approximation

$$-\frac{\mu^{(2)}}{\mathrm{Ry}} = \frac{\mu_0\bar{n}^{1/2} + 3\mu_2\bar{n}^{3/2}\ln\,[1 + \exp\,(7\tilde{\mu}_3/10\mu_2)\,\bar{n}^{7/10}]}{1 - \tilde{\mu}_1\mu_0^{-1}\,\bar{n}^{1/2} + 0.75\,\bar{n}^{3/2}}.$$

Taking into account the strong cancellation effects in the temperature dependence of the Hartree-Fock term and the Montroll-Ward term a unified treatment of both terms seems to be in order. ZIMMERMANN et al. (1984) propose the Padé approximation for zero temperature

$$\mu_{xc}(r_s, \tau = 0) = -\frac{k}{r_s} - e\,\ln\,(1 + f r_s^{-7/10}),$$

$$r_s = (3/4\pi n a_{BO}^3)^{1/3}, \qquad a_{BO} = a_B\varepsilon(1+\gamma)/a,$$

$$e = 10c/7, \qquad f = \exp\,(7K_c/10),$$

$$k = \frac{2}{\pi\alpha'}(g_e^{-1/3} + g_h^{-1/3}), \qquad \alpha' = 0.521,$$

$$r = \left(\frac{1}{2g_e} + \frac{\gamma^{3/2}}{2g_h}\right).$$

For finite temperatures the corresponding formulae are

$$\mu_{xc}(r_s, \tau) = \frac{\mu_{xc}(r_s, 0) - a\tau^2 r_s^3 - d\tau^{5/2} r_s^{9/2}}{1 + \{1 + 8 \ln(1 + 6\tau^{-1} r_s^{-3/2})\} u r_s^6 \tau^3}, \tag{6.124}$$

$$a = \frac{(1+\gamma)^2 g_e(1+\gamma\delta)}{24\pi(1+\gamma^{3/2}\delta)^2}, \qquad b = \frac{\sqrt{3\pi}}{32}\left\{1 + \frac{1 + \sqrt{\gamma} + (\ln 2)(g_e^{-1} + \sqrt{\gamma}\, g_h^{-1})}{\sqrt{2(1+\gamma)}}\right\},$$

$$d = \frac{3(1+\gamma)^3}{18\pi(g_e^{-1} + \gamma^{3/2} g_h^{-1})^2}, \qquad u = \frac{(1+\gamma)^3}{72\pi}(g_e^{-1} + \gamma^{3/2} g_h^{-1})^{-2}.$$

For the simple models of Ge and GaAs as well as for the limits of an electron gas and a symmetrical e—h-plasma Tab. 6.4 gives the values of the seven model parameters. Comparision with numerical calculations shows that the error of the Padé formulae is less than 0.1 meV for Ge and less than 0.2 ,meV for GaAs (ZIMMERMANN et al., 1984).

Tab. 6.4. Parameters of the Padé approximations for the exchange correlation contribution to the chemical potential to the order e^4 (ZIMMERMANN et al., 1984)

parameter	e-gas	symmetrical plasma	Ge	GaAs
h	3.3156 −3	2.6526 −2	7.1210 −2	1.6828 −2
b	3.2487 −1	2.5837 −1	2.1677 −1	2.4092 −1
d	2.7072 −3	6.1259 −2	4.4003 −1	3.8863 −2
e	0.8883 −1	7.1064 −1	2.0632	1.3988
f	3.6092	1.7955	0.8000	0.8838
k	1.2218	2.4435	1.7394	2.4435
u	1.1052 −3	8.8419 −3	6.3513 −2	5.6094 −3

The formulae given above may be used for the discussion of the droplet transition in semiconductors. The transition region corresponds to a nonmonotonic behaviour of the chemical potential versus the density. The density dependence of μ is demonstrated in Fig. 6.12. In the framework of this model we obtain for Ge the critical point $T_c = 8$ K and $n_c = 1.0 \cdot 10^{17}$ cm^{-3} (EBELING et al., 1981).

To get better results one should take a better account of the bandstructure and anisotropy. As a review of the electron-hole plasma theory we refer to the monograph of RICE et al. (1977, 1980). In conclusion, we may state that the Padé approximations for the thermodynamic functions derived here are a useful tool for the study of many thermoydnamic properties, phase equilibria etc. of e—h-plasmas in a wide range of density and temperature. Finally let us not that the theory developed here is restricted to the region of high ionization, i.e. $(1 - \alpha) \ll 1$. The therory of partially ionized exciton plasmas has been given elsewhere (KRAEFT et al., 1975; EBELING et al., 1976; RICHERT and EBELING, 1984). Methods for the study of partially ionized systems will be developed in the next Sections.

6.4.3. Ionization Equilibrium

Let us first consider the case of stable plasmas, i.e. the temperature is beyond the critical value and the chemical potential shows a monotonous behaviour. Following the considerations in Section 6.2. the condition of chemical equilibrium between the free charges and excitons may be expressed by the Saha equation for the exciton density n_0:

$$n_0 = (2s_e + 1)(2s_h + 1)\Lambda_0^{-3}\sigma(T, n)\exp(\beta\mu(n, T)) \tag{6.125}$$

where $\mu(n, T)$ is the chemical potential of the system of charges which in the following will be approximated by the contributions up to the order e^4 which have been calculated above,

$$\mu(n, T) = \mu_e^{id} + \mu_h^{id} + \mu_e^{HF} + \mu_h^{HF} + \mu^{(2)}(n, T),$$

and further

$$\sigma(T, n) = \sum_{s,l}(2l + 1)[\exp(-\beta E_{sl}) - 1 + \beta E_{sl}]$$

is the partition function calculated with the discrete energy levels which are obtained from the solution of the homogeneous Bethe-Salpeter equation. We mention that eq. (6.125) is in fact identical with the condition of chemical equilibrium between the charges and the excitons which reads

$$\mu(n, T) = \mu_e + \mu_h = \mu_0 = k_B T \ln\left[\frac{n_0\Lambda_0^3}{4\sigma(T, n)}\right].$$

Furthermore we note that the approximation of the chemical potential of the charges by the contributions up to the order e^4 is indeed justified for plasmas with nearly symmetrical masses since all the terms of order e^{2n} with n odd cancel in the case of charge- and mass-symmetry and the terms of order e^{2n} with $n \geqq 4$ (odd) are taken into account by the partition function as to be seen, e.g., from eq. (5.99). Finally let us underline that at any given density the partition function contains a finite number of terms corresponding to the energy levels which exist for that density. As we know from earlier considerations in Sections 3.2. and 5.4. at a critical density $n_M(T)$ which is called Mott density the last bound state disappears. Approximating the bound state levels by the unperturbed ones and the lowering of the continuum edge by the exchange-correlation contribution to the chemical potential, the Mott condition reads

$$E_{B0} = |E_{10}| = |\mu_{xc}(n, T)|.$$

At densities beyond the Mott density eq. (6.125) tells us, that the density of excitons is zero:

$$n = n_e, \quad n_0 = 0 \quad \text{if} \quad n > n_M(T).$$

At densities below the Mott density one part of the electrons and holes form neutral excitons which density is to be calculated from eq. (6.125) together with the balance for the total number of electrons (and holes). We note that the Mott condition may be approximated by the estimate $r_0 = a_B$ which is shown in Fig. 6.10. Further we find there an estimate for the densities where the degree of ionization is 0.5. Now let us consider the case of plasmas which are unstable with respect to the separation into two phases with the volumes V' and V'' where $V' + V'' = V$. The separation into two phases happens if the chemical potential shows wiggles as, e.g., in Figs. 6.11–6.12

for subcritical temperatures. Then thermodynamic equilibrium requires

$$\mu_e' = \mu_e'' , \quad \mu_h' = \mu_h'' , \quad \mu_0' = \mu_0'' ,$$
$$\mu_e' + \mu_h' = \mu_0' , \qquad \mu_e'' + \mu_h'' = \mu_0'' .$$

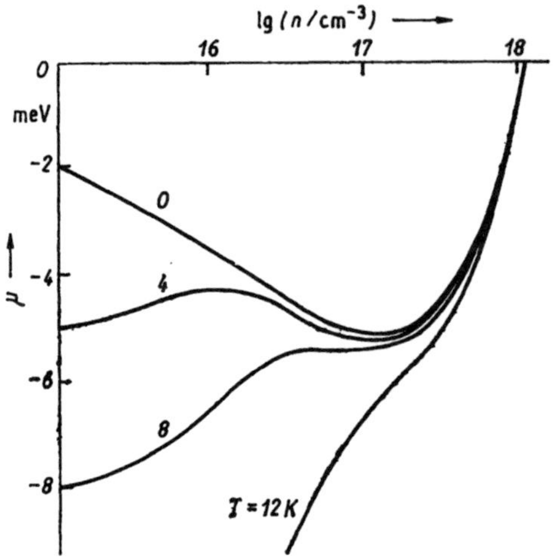

Fig. 6.12. The chemical potential of the unsymmetrical e—h-plasma up to order e^4 vs. the free electron density n at several temperatures

As we see the chemical potential of the plasma in both phases should be equal which leads us to a kind of Maxwell condition (EBELING et al., 1976, 1979). From the equilibrium conditions follows that the relation between the exciton densities in both phases is given by

$$\frac{n_0'}{n_0''} = \frac{\sigma(T, n')}{\sigma(T, n'')} .$$

Typically the upper part of the coexistence region (hatched part of the n—T-plane in Fig. 6.10) is beyond the Mott curve, i.e. one of the coexisting phases is fully ionized and the other one is only partially ionized. Roughly speaking fully ionized electronhole droplets are imbedded into a weakly ionized exciton gas.

6.5. Hydrogen Plasmas

6.5.1. The Two-Fluid Model

The contribution of the charged components to the thermodynamic functions of H-plasmas is analysed on the basis of the present theoretical knowledge. In the high-temperature low-density limit the Debye formula with quantum corrections is used. In the low-temperature high-density limit the Gellman-Brueckner formula for the electrons and a lattice-energy formula for the protons is used. Padé approximations are constructed which contain these limiting situations as special cases and describe also the region in between in a certain quality. As pointed out in earlier Sections the correct thermodynamic description of hydrogen plasmas is important for many applications connected, e.g., with the fusion problem and with astrophysical calculations; we remember that a large fraction of the matter in the universe is a dense hydrogen plasma. Considering the prevalence of hydrogen in the universe and its importance in fusion research much theoretical and experimental work has been devoted to the problem of finding the equation of state and the thermodynamic functions since the pioneering study of WIGNER and HUNTINGTON (1935) (see, e.g., HARRIS et al., 1960; ASHCROFT, 1968; KHALATNIKOV, 1971; BROWN and MARCH, 1972; BROVMAN et al., 1972; ROGERS and DE WITT, 1973; EBELING et al., 1976, 1979; MINOO et al., 1976; DHARMA-WARDANA and TAYLOR, 1981; DHARMA-WARDANA and PERROT, 1982; DHARMA WARDANA, PERROT and AERS, 1983; CEPERLEY, 1983). In a first approximation a hydrogen plasma be considered to be a mixture of two fluids: the fluid of light electrons and the fluid of the heavy protons. The nonideality of these fluids is given by the relations of the mean kinetic and potential energies respectively (Fig. 2.2).:

$$\Gamma_e = \frac{e^2}{4\pi\varepsilon_0 d\theta_e}, \qquad \Gamma_p = \frac{e^2}{4\pi\varepsilon_0 d\theta_p}$$

where

$$d = \left(\frac{3}{4\pi n}\right)^{1/3}, \qquad \theta_k = \frac{2I_{3/2}(\alpha_k)}{n\Lambda_k^3} k_B T . \tag{6.126}$$

Here n is the density of free electrons or free protons respectively, I_k are the Fermi-functions, Λ_k the thermal De Broglie wave length and α_k the ideal chemical potentials determined by the equations

$$2I_{1/2}(\alpha_k) = n\Lambda_k^3 , \qquad \Lambda_k = h[2\pi m_k k_B T]^{-1/2} .$$

For sufficently low densities the mean kinetic energy per degree of freedom is $k_B T/2$. In the latter case $\Gamma_e \approx \Gamma_p \approx \Gamma$ holds, with

$$\Gamma = (e^2/k_B T\, d\, 4\pi\varepsilon) .$$

In the opposite limit of strong degeneration there is another relation:

$$\theta_k = \tfrac{2}{5}\alpha_k\, k_B T \qquad (\alpha_k k_B T = \varepsilon_F \text{ — the Fermi energy}).$$

The nonideality parameter is now temperature independent:

$$\begin{aligned} \Gamma_e &= 5(\alpha')^2 r_s , \quad \alpha' = (4/9\pi)^{1/3} , \\ \Gamma_p &= 1836.15\Gamma_e ; \end{aligned} \tag{6.127}$$

r_s is the usual dimensionless density parameter

$$r_s = (d/a_B) , \qquad a_B = (4\pi\varepsilon\hbar^2/me^2) .$$

In the following we concentrate on the region in the temperature-density plane where the protons form a classical nonideal fluid (hatched in Fig. 2.2). Due to the large mass the protons may be treated classically here because

$$n\Lambda_p^3 \ll 1 .$$

The two limiting states for the proton fluid are a *bcc*-lattice state (if the temperatures are sufficiently low) and a Debye gas state (if the temperatures are sufficiently high). For the electron fluid we need expressions which cover the degenerate as well as the nondegenerate region. Here we may use expressions which were developed in Section 6.3.2. on the basis of Pade approximations for the thermodynamic functions of electron gases (EBELING et al., 1981; EBELING and RICHERT, 1982, 1985; EBELING 1984, 1985).

6.5.2. Basic Formulae for the Limiting Situations and Padé Approximations

Let us consider a hydrogen plasma consisting of n electrons, n_p protons and n_H atoms per cm³. A basic quantity for the description of such a plasma is the chemical potential of the charged components which is defined as the sum of the chemical potentials of the electrons and the protons

$$\mu(n, n_H, T) = \mu_e + \mu_p .$$

This quantity determines the chemical equilibrium between free charges and atoms

$$\mu(n, n_H, T) = \mu_H(n, n_H, T) .$$

Therefore μ is a central quantity. The plasma contributions to the pressure and to the free energy may be obtained by the following thermodynamic relations (Ebeling et al. 1979):

$$p = n\mu - f , \quad \mu = \partial f/\partial n$$

where f is the free energy density $f = F/V$.

The plasma chemical potential may be split into the ideal part, the Hartree-Fock part, and the correlation part:

$$\begin{aligned} &\mu = \mu_{id} + \mu_{int} , \ \mu_{int} = \mu_{HF} + \mu_{corr} , \\ &\mu_{id} = k_B T\alpha_e + k_B T\alpha_p , \\ &\mu_{HF} = -\frac{e^2}{\Lambda_e} I_{-1/2}(\alpha_e) - \frac{e^2}{\Lambda_p} I_{-1/2}(\alpha_p) . \end{aligned} \tag{6.128}$$

The correlation part is analytically known only in limiting cases. At very low densities and high temperatures we may write following Section 5.4. or 6.1.—6.2.:

$$\mu_{corr}/\text{Ry} = -2\left(\frac{4k_B T}{\pi\,\text{Ry}}\right)^{1/4} \bar{n}^{1/2} + (1+\gamma)^{1/2}\,(k(T) + \delta_1 k)\,\bar{n} + \dots \tag{6.129}$$

where

$$\bar{n} = n\Lambda_e^3 , \quad \text{Ry} = \frac{e^2}{8\pi\varepsilon a_B} , \quad a_B = \frac{4\pi\varepsilon\hbar^2}{me^2} , \quad \gamma = \frac{m_e}{m_p} ,$$

$$\begin{aligned} k(T) = &\frac{1}{2} + \frac{1}{4}\left(\frac{2}{1+\gamma}\right)^{1/2} (1 + \ln 2)\,(1 + \sqrt{\gamma}) - \frac{\xi}{3\sqrt{\pi}} \ln\left(\frac{2\sqrt{\gamma}}{1+\gamma}\right) \\ &- \frac{4}{\xi^2\sqrt{\pi}} \cdot Q_4(-\xi) - \frac{2}{\xi^2\sqrt{\pi}}\left(\frac{2}{1+\gamma}\right)^{3/2} \left\{Q_4\left(-\xi\sqrt{\frac{1+\gamma}{2}}\right)\right. \\ &\left. + \gamma^{3/2} Q_4\left(-\xi\sqrt{\frac{1+\gamma}{2\gamma}}\right) - \frac{1}{2} E_3\left(-\xi\sqrt{\frac{1+\gamma}{2}}\right) - \frac{\gamma^{3/2}}{2} E_3\left(-\xi\sqrt{\frac{1+\gamma}{2\gamma}}\right)\right\} . \end{aligned}$$

The virial functions Q_4 and E_3 are given by

$$Q_k(\xi) = \sqrt{\pi} \sum_{n=k}^{\infty} \frac{\zeta(n-2)}{\Gamma(\frac{1}{2}n + 1)} \left(\frac{1}{2}\xi\right)^n ,$$

$$E_k(\xi) = \sqrt{\pi} \sum_{n=k}^{\infty} \frac{(1-2)^{2-n}\,\zeta(n-1)}{\Gamma(\frac{1}{2}n + 1)} \left(\frac{1}{2}\xi\right)^n$$

where $\zeta(x)$ is the Riemann function, $\Gamma(x)$ the gamma function and ξ is the Born parameter for the electron-proton interaction:

$$\xi = (e^2/k_BT\,\lambda_{ep}\,4\pi\varepsilon)\,, \qquad \lambda_{ep} = \hbar\,[2m_{ep}k_BT]^{-1/2}\,.$$

Further

$$\delta_1 k = 8\xi^{-2}\cdot[\Sigma(T, n = 0) - \sigma(T)]$$

where $\sigma(T)$ is the Planck-Brillouin-Larkin partition function and $\Sigma(T, n)$ the actual partition function used in the calculations for which we have several possibilities of choice (EBELING and SÄNDIG, 1973). The correction disappears for the Planck-Brillouin Larkin choice. We shall come back to this important point again. We underline one again that eq. (6.128) describes the plasma contribution to the chemical potential, i.e. only the contribution of free charges. The idea which leads from the equations for the free energy given in Section 5.4. to the chemical potential eq. (6.129) is the following. The virial function Q in eq. (2.52) of Chapter 5. which describes free as well as bound pairs is split into two parts:

$$Q(\xi) = Q^b + Q^f \qquad (\xi > 0)\,,$$

where the bound state part is per definition

$$Q^b = 2\sqrt{\pi}\,\Sigma(T, n = 0)\,.$$

The corresponding contribution of free states is therefore

$$Q^f = Q(\xi) - Q^b(\xi), \qquad \xi > 0\,.$$

To the chemical potential of the plasma (free states) contributes only the part Q^f. In this way eq. (6.129) is obtained (see EBELING, 1969).

In the limit of high density the electrons and the protons behave quite in a different way. Therefore it seems to be useful to split the plasma potential in an electron gas contribution and a proton gas contribution:

$$\mu(n, n_H, T) = \mu_{eg} + \mu_{pg}\,. \tag{6.130}$$

Here μ_{eg} corresponds to the chemical potential of an electron gas in a uniform positive background and μ_{pg} includes all the other contributions. From the physical point of view μ_{pg} describes a proton gas with screened interactions. The screening effect is due to some increase of the electron density in the neighborhood of each proton. We should underline that due to their definition μ_{eg} and μ_{pg} are not strictly identical to the chemical potentials of the electrons and the protons in the real plasma. Following the derivations in Section 6.3.2. the chemical potential of the electron gas is in the e^4-approximation

$$\begin{aligned}\mu_{eg} &= k_BT\alpha_e - \mathrm{Ry}\,[\mu_0\bar{n}^{1/2} + \mu_5\bar{n} + \{9.7736r_s^{-1}\\ &+ 0.71064\ln(1 + 3.6094r_s^{-7/10})\}\,\bar{n}^2]\\ &\cdot[1 + 8\bar{n}^2 + 8\ln\{1 + 0.07483\,\mu_0^{-1}\bar{n}^{1/2}\}]^{-1}\,,\\ \mu_0 &= (k_BT/\pi\,\mathrm{Ry})^{1/4}\,, \qquad \mu = \tfrac{1}{2}\,(k_BT/\pi\,\mathrm{Ry})^{1/2}\,.\end{aligned} \tag{6.131}$$

The high-density behaviour of the proton gas is calculated in the adiabatic approximation (KRASNY and ONISHCHENKO, 1972; BAUS and HANSEN, 1980; EBELING et al.,

1983). We start from the formula for the interaction part of the free energy density (De Witt, 1976; Kalman and Carini, 1978)

$$f_{pg}^{int} = nk_BT\,[-0.89461\Gamma + 3.2660\,\Gamma^{1/4} - 0.50123\ln\Gamma - 2.809] + \delta f_{pg}\,. \tag{6.132}$$

De Witt (1976) derived the first term in eq. (6.132) by fitting the Monte-Carlo. data for proton gases obtained by Hansen (1973) for the fluid phase of dense proton systems. The influence of electron screening is described by the second term in eq. (6.132). The free energy contribution of the screening effects can be written as

$$\delta f_{pg} = \frac{1}{3\pi}\int_0^\infty dq\; q^2 S^{(0)}(q)\, w(q)\cdot(nk_BT) \tag{6.133}$$

with

$$w(q) = \frac{3\Gamma}{q^2}\left(\varepsilon^{-1}(q) - 1\right), \qquad q = kd$$

and $S^{(0)}(q)$ is the structure factor.

This gives in a certain approximation (see Kalman and Carini, 1978)

$$\delta f_{pg} = (nk_BT)\, r_s(-\,0.045\Gamma - 0.755\Gamma^{1/4} + 0.267)\,. \tag{6.134}$$

The screening contribution is a function of the nonideality parameter Γ as well as of the pure density parameter $r_s = d/a_B$.

In the representation with

$$\tilde{n} = nl^3 = n(e^2/4\pi\varepsilon k_BT)^3$$

as nonideality parameter we obtain finally for the interaction chemical potential

$$\mu_{pg}^{int}/k_BT = -1.9228\,\tilde{n}^{1/3} + 3.9868\,\tilde{n}^{1/12} - 0.16708\ln\tilde{n} - 3.215$$
$$-\frac{r_s}{1+r_s^2}\,\tilde{n}^{1/3}\,[0.0726 + 0.6383\tilde{n}^{-1/4} - 0.1779\tilde{n}^{-1/3}]\,.$$

Another writing is

$$\mu_{pg}^{int} = -\frac{e^2 n^{1/3}}{4\pi\varepsilon}\, B_\mu$$

where B_μ is so to speak the effective lattice constant for screened protons which is in our approximation (Ebeling and Richert, 1982, 1984)

$$B_\mu = 1.9228 - 3.9868\tilde{n}^{-1/4} + 3.215\tilde{n}^{-1/3}$$
$$+\,0.1671\tilde{n}^{-1/3}\ln\tilde{n} + \frac{r_s}{1+r_s^2}[0.0726 - 0.6383\,\tilde{n}^{-1/3} + 0.1779\,\tilde{n}^{-1/4}]\,.$$

Here as well as in eq. (6.135) we have introduced a cutting factor $(1 + r_s^2)$ in the denominator of the screening term since the expression (6.134) is valid only for $r_s \ll 1$ In order to improve this point we would need better expressions for the ground state energy of nondegenerate proton fluids which would replace

$$B_\mu = 1.9228 + 0.0726\,\frac{r_s}{1+r_s^2}, \qquad T = 0\,.$$

Now it remains to find a Padé approximation for the proton gas contribution. Again we know analytical expressions for the limits of low and high densities respectively which are given by eqs. (6.129) and (6.132). Assuming that the transition from the $n^{1/2}$-behaviour to the $n^{1/3}$-behaviour occurs in the region where $n = nl^3 \approx 1$, we propose for the chemical potential the following Padé approximation (EBELING and RICHERT, 1982, 1985):

$$\mu_{pg} = k_B T \left\{\alpha_p - \frac{p_0 \tilde{n}^{1/2} + B_\mu p_2 \tilde{n}^{3/2}}{1 - (p_1/p_0)\, \tilde{n}^{1/2} + p_2 \tilde{n}^{7/6}}\right\},$$

$$p_0 = 3.2408\,, \tag{6.136}$$

$$p_1 = -\pi^{3/2}(1+\gamma)^{1/2}\left[k + \delta_1 k - \frac{\gamma\pi^{-1/2}}{\xi(1+\gamma)} - \frac{1+\ln 2}{2^{3/2}(1+\gamma)^{1/2}}\right]\left(\frac{k_B T}{\mathrm{Ry}}\right)^{1/2}.$$

For the definition of the function k see eq. (6.129). Now all constants have been fixed except p_2. This constant regulates the finer details of the transition between the $n^{1/2}$- and the $n^{1/3}$-behaviour. We propose the choice $p_2 = 150$ which gives in the classical case with $p_1 = 0$, $r_s = 0$ a reasonable fit of the Monte Carlo data by HANSEN (1973). Another possible choice is $p_2 = -p_0 p_1$.

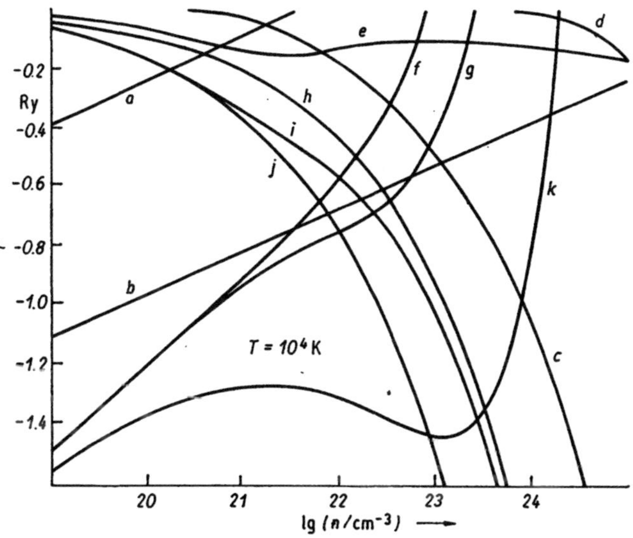

Fig. 6.13. The different contributions to the H-plasma potential as a function of the free electron density
a — μ_e^{id}, b — μ_p^{id}, c — μ_e^{HF}, d — μ_p^{HF}, e — μ_e^{corr}, f — μ_{id}, g — $\mu_{id} + \mu_{HF}$, h — μ_p^{corr}, i — μ_{corr}, j — $\mu_{xc} = \mu_{HF} + \mu_{corr}$, k — $\mu(n, 0, T) = \mu_{id} + \mu_{xc}$. Here the indices a and p correspond to the electron gas and the proton gas; id, HF and corr are the ideal, Hartree-Fock and correlation contribution, respectively

The different contributions to the plasma potential may be easily calculated now in a wide temperature-density range. For example, Fig. 6.13 shows various contributions at the relative low temperature $T = 10000$ K in dependence on the free electron density n. The density region covers (as can be seen from Fig. 2.2.) the region from $\Gamma \approx 1$ up to $\Gamma \approx 155$.

One important result is that the correlation part comes into the order of magnitude of the ideal part at densities $n \approx 10^{21}$ cm^{-3}. Further the Hartree-Fock contribution of the electrons and the correlation contribution of the protons are of certain importance. For densities less than 10^{24} cm^{-3} the protons can be considered as classical particles with a classical ideal part and a correlation part. The density of degeneracy for the electrons is about 10^{21} cm^{-3}. For this reason we have to take into account the Hartree-Fock contribution and quantum expressions for the other terms of the thermodynamical functions.

We want to emphasize also the smoothness of the sum of the exchange and correlation contributions (Fig. 6.13, j).

6.5.3. Ionization Equilibrium and Phase Diagram

Here we want to give an estimate of the degree of ionisation of the plasma. for the region where $\alpha > 1/3$, i.e. more than 50% of the particles are charged. Taking into account in this approximation only the formation of atoms we need an expression for the atomic chemical potential. Considering an atom in first approximation as a hard sphere with internal degrees of freedom we find (EBELING et al., 1976)

$$\mu_{\mathrm{H}} = k_{\mathrm{B}}T \ln \left[\frac{n_{\mathrm{H}} \Lambda_{\mathrm{H}}^3}{4\Sigma(T, n, n_{\mathrm{H}})} \right] + \mu_{\mathrm{H}}^{\mathrm{HC}}(n_{\mathrm{H}}) - 2n_{\mathrm{H}} A_{\mathrm{HH}} \tag{6.137}$$

where n_{H} is the density of atoms and $n_{\mathrm{p}} = n_{\mathrm{e}} = n + n_{\mathrm{H}}$ is the total density of the protons. Further A_{HH} is the van der Waals constant of the atoms. The chemical potential of hard spheres of density n_{H} is given by (CARNAHAN and STARLING, 1969)

$$\mu_{\mathrm{H}}^{\mathrm{HC}} = k_{\mathrm{B}}T\, \eta(8 - 9\eta + 3\eta^2)\,(1 - \eta)^{-3} \tag{6.138}$$

with $\eta = 4\pi n_{\mathrm{H}} R^3/3$ and R is an effective hard core radius of an H atom. The value of R we choose for hydrogen as

$$R = 1.54 a_{\mathrm{B}}\,.$$

$\Sigma(T, n, n_{\mathrm{H}})$ is the atomic partition function. The choice of the atomic partition function is not unique (EBELING and SÄNDIG, 1973), we can vary our choice (i.e. the distinction between bound and free states) within reason. In a complete theory this would not matter; what we remove from one page of the ledger (say the page of bound states) would be entered elsewhere (on the page of free states) with the same effect (after a comment made by Lars Onsager at the conference on electrochemistry in Montpellier, 1968). We propose here two alternative possibilities:

(1) Planck-Brillouin-Larkin convention, i.e. smooth border between bound and free states:

$$\Sigma_{\mathrm{P}} = \sigma(T, n, n_{\mathrm{H}}) = \sum_{s,l} (2l + 1) \left[\exp(-\beta E_{sl}) - 1 + (\beta E_{sl})\right] \tag{6.139}$$

where

$$E_{sl} = E_{sl}(n, n_{\mathrm{H}}, T)$$

are the eigen energies of the Bethe-Salpeter equation discussed earlier (Fig. 6.14). At zero density we get ($I = -E_0$)

$$\Sigma_P = \sigma(T) = \sum_{s=1}^{\infty} s^2[\exp(I/s^2) - 1 - (I/s^2)] . \tag{6.140}$$

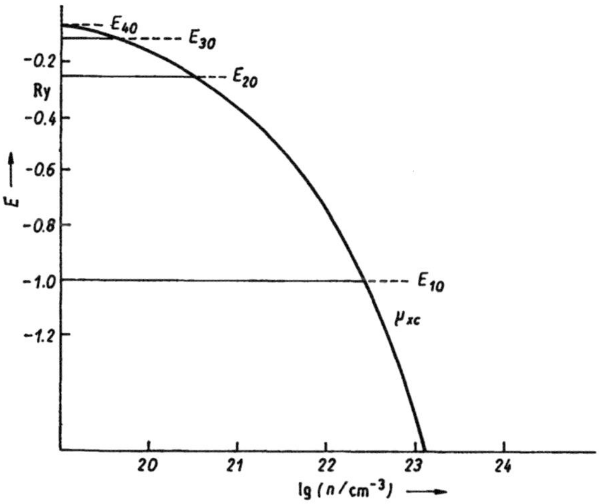

Fig. 6.14. Discrete two-particle energies and continuum edge as a function of the densities for hydrogen plasmas with $T = 10000$ K

(2) Riewe-Rompe renormalization, i.e. sharp border between bound and free states at the thermal energy:

$$\Sigma_R = \sum_{E_{sl} < \varepsilon^*} (2l + 1) \exp(-\beta E_{sl}) ,$$
$$\varepsilon^* = -\zeta k_B T . \tag{6.141}$$

The numerical constant ζ we put equal to one ($\zeta = 1$) in the following.

Both conventions are nearly identical in numerical calculation and both express the same physical idea: Discrete states with a binding energy less than $k_B T$ have to be considered (in thermodynamic calculations) as free states since their life time is limited by the time between two collisions. In other words the damping of those quasifree states with $|E_{sl}| < k_B T$ is so high that they are practically extincted. Since only little is known about the solutions of the Bethe-Salpeter equation at high density we have used in our calculations the confined atom model (Ebeling and Richert, 1982; Graboske et al. 1969; Fortov et al., 1976, 1979; Bushman and Fortov, 1983). In the framework of this model the atoms are assumed to be enclosed into spheres of radius r_0, which is the half of the mean distance between two protons in the system:

$$(4\pi/3)\, n_p \cdot r_0^3 = 1 .$$

The wave functions of bound electrons have to fulfil boundary conditions at the surface of the sphere. We may require, e.g., that the radial wave functions disappear, $R_{sl}(r_0) = 0$, or that the radial derivatives disappear, $R_{sl}'(r_0) = 0$. The first conditions corresponds to inpenetrability of the boundaries and the second one to the vanishing of the fluxes at the boundaries. The energy levels for free electrons enclosed in spherical boxes read

$$E_{sl} = \text{Ry}\; x_{sl}^2 (a_B^2/r_0^2) \tag{6.142}$$

where x_{sl} are the zeroes of the spherical Bessel functions or their radial derivatives respectively. In the first case we have, e.g., $x_{10} = 3.142$, $x_{11} = 4.493$, $x_{12} = 5.763$, $x_{20} = 6.283$. Assuming a simple addition rule for the enclosure energy and the Coulomb energy we may use the approximation

$$E_{sl} = \text{Ry}\left[x_{sl}^2 \frac{a_B^2}{r_0^2} - \frac{1}{s^2}\right] \quad \text{if} \quad s \leqq \frac{r_0}{a_B x_{sl}} \tag{6.143}$$

and $E_{sl} = 0$ otherwise (Ebeling and Richert, 1982). We note that with the present assumptions the energy levels depend only on the total proton densities n_p. Above a density of about 5×10^{22} cm^{-3} all the bound states disappear, i.e. a Mott transition is observed. The density dependence of the partition function is a very essential point of the theory which has important physical consequences. Taking, e.g., the approximation of density-independent energy levels we would find atoms at arbitrary high densities which clearly is in contradiction to the experimental observations. On the other hand approximations which take into account only the influence of free particles on the partition functions lead to serious unphysical instabilities of the theory. Therefore we have chosen the present variant which gives so to say an upper border for the disappearance of the energy levels.

Now the atomic partition function may be calculated. The degree of ionization is obtained by solving the following equations consistently:

$$\mu(n, n_H, T) = \mu(n, n_H, T)\,, \qquad n_p = n + n_H\,. \tag{6.144}$$

The degree of ionization is defined by $\alpha = (n/n_p)$ (see Fig. 6.15).

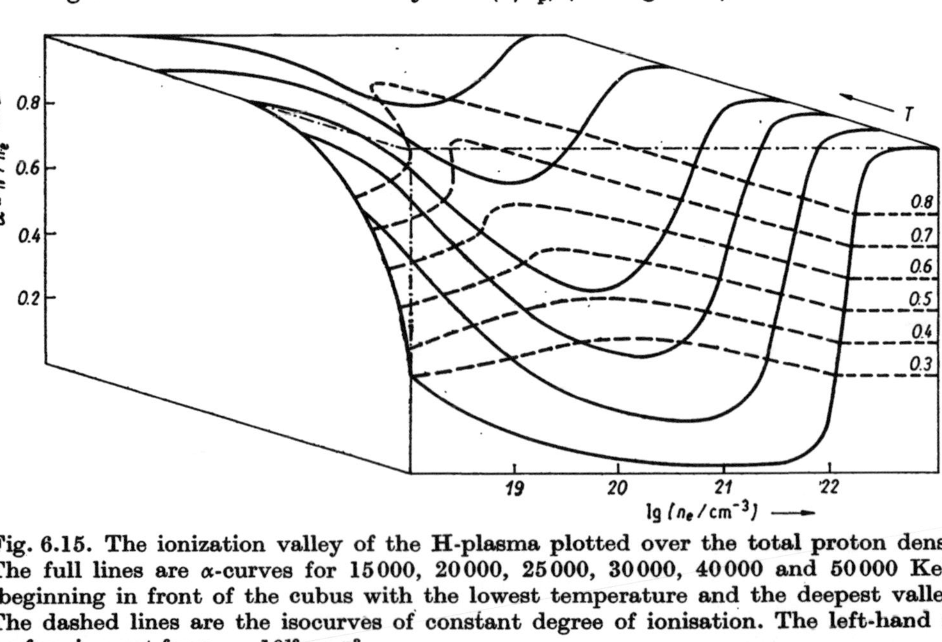

Fig. 6.15. The ionization valley of the H-plasma plotted over the total proton density. The full lines are α-curves for 15000, 20000, 25000, 30000, 40000 and 50000 Kelvin (beginning in front of the cubus with the lowest temperature and the deepest valley). The dashed lines are the isocurves of constant degree of ionisation. The left-hand side surface is a cut for $n_e = 10^{18}$ cm^{-3}

So far we have assumed that all the bound states together are to be considered as one species. There exist still another possibility (ZIMDAHL and EBELING, 1977; REDMER and RÖPKE, 1984; RICHERT et al., 1984). We may consider each bound state as a separate species. In this variant we have for each species characterized by the quantum numbers n, l, m a separate equilibrium relation

$$f(n, n_{slm}, T) = \min, \quad \mu_{slm} = \mu_e + \mu_p = \mu$$

and the total balance is

$$n + \sum n_{slm} = n_p \,.$$

Here the sum is to be extended over all states with $E_{sl} < \varepsilon^*$ were ε^* is the upper border of bound states by definition. The partition function for each state is

$$\Sigma_{slm} = \exp\left(-\beta E_{sl}\right) .$$

Further the chemical potential reads

$$\mu_{slm} = k_B T \ln \left[n_{slm} \Lambda_H^3\right] + E_{sl} + \mu_{s\,m}^{HC}(R_{slm}) - 2n_{slm} A_{slm} \,.$$

Here R_{slm} is the hard core radius and A_{slm} the van der Waals constant of the (slm)-state. We note that R_{slm} is increasing with s^2, i.e. the excited states have very big volumes and are thermodynamically unfavourable. Considering each (slm)-state as a new species it seems to be natural to consider the upper limit ε^* as an optimizable quantity. In other words

$$\varepsilon^* = \varepsilon^*(n_p, T)$$

is considered as a free function of the density and the temperature which is obtained by minimization of the free energy with respect to the variation of ε^*. The existence of this minimum is closely connected with the increase of the radius R_{slm} with increasing quantum numbers. Since the many-variable minimization with respect to n_{slm} and ε^* is a difficult procedure we restrict the following consideration to the simplest variant given by eqs. (6.139), (6.143) and (6.144). The more refined variant with (slm)-species is considered elsewhere (RICHERT et al., 1984).

The result of our calculation is shown in Fig. 6.15. One observes firstly a lowering of the degree of ionisation with increasing total proton density and then an increase again which is called sometimes pressure ionisation. The pressure ionisation is due to the decrease of the ionisation energy caused by the correlation contributions to the plasma potential and further by the influence of the repelling forces bewteen the atoms which make the atomic state unfavourable at high densities. The main contribution is due to lowering of the continuum edge. The 1s-state diseappears at the total proton density 5.2×10^{22} cm^{-3}.

We have to underline that the effects of hard spheres and the limitation of the wave functions are described here only approximately. The correct theory has to improve the calculation of the continuum edge and to handle the density dependence of the energy levels E_{sl} in a more consistent way, i.e. one has to solve the Bethe-Salpeter equation in a rigorous many-body theory (see Sections 5.4. and 6.1.). In the Fig. 6.16 we present the chemical plasma potential as a function of the total proton density. As critical parameters for a phase transition we have found $T_c \approx 13000$ K $n_p^c \approx 3\,1.'0^{22}$ cm^{-3}. Due to several approximations especially to the crude approximation for the interaction with neutrals the theory given here is restricted to the region where $\alpha > 1/3$. Therefore, the phase transition to a liquid-like state which is evidently determined by the neutral particles cannot be described in the framework of our theory of plasma effects.

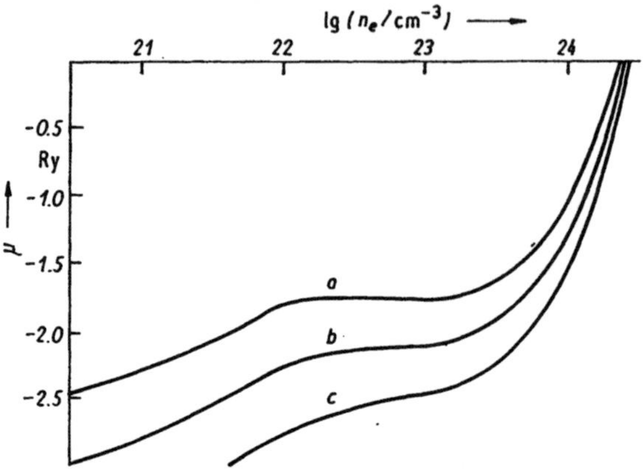

Fig. 6.16. The H-plasma potential vs. the total proton density for three temperatures a – 15000 K, b – 20000 K, c – 25000 K

The atomic and molecular phases of hydrogen is outside the scope of this book which is devoted to the effect of Coulombic interactions. Therefore, we restrict ourselves to a few remarks. It is well known that the phase diagram of hydrogen has considerable structure (KRUMHANSL, 1965; ASHCROFT, 1968; SCHNEIDER, 1969, 1971; GINZBURG, 1971, 1981; KAGAN et al., 1978; FRANCK, 1980, 1981; KRANENDONK, 1982; CEPERLEY, 1983; ROBNIK and KUNDT, 1983). At low pressure and low temperatures electrons and protons are bound in molecules.

In the molecular region there exists a critical point with the parameters

$$T_c = 33\,\mathrm{K}\,, \qquad n_p^c = 1.9 \cdot 10^{22}\,\mathrm{cm}^{-3}\,, \qquad p_c = 1.3\,\mathrm{MPa}\,.$$

This first critical point is not described by our theory (Fig. 6.17). At pressures above 100 GPa a phase transition to an unusual solid state occurs; it is believed that the protons are ordered in this molecular phase in an α—N_2-lattice.

At still higher pressure the electrons become delocalized and a (molecular) metal phase is formed. With increasing pressure the protons may change from a diatomic lattice to a monatomic lattice to lower the kinetic energy. At very high pressures of about 10^{19} Pa the proton lattice will melt to form a two-component Fermi liquid. Here again the system is strongly ionized and may be described by the methods outlined in this work (EBELING 1984, 1985).

At temperatures above 10^3 K the phase diagram of hydrogen shows less structure which is connected with the instability of molecules in this region. However, one expects the existence of a dielectric-metal transition (ZELDOVICH and LANDAU, 1944; NORMAN and STAROSTIN, 1968, 1970; FRANCK, 1981). For hydrogen this hypothetical transition is expected to occur at densities about 1 gcm^{-3} and at pressures about 10^7 bar. In our approximation the dielectric-metal transition has a critical point at about (EBELING and RICHERT, 1982, 1985)

$$T_c \approx 16500\,\mathrm{K}, \quad n_p^c \approx 0.8 \cdot 10^{23}\,\mathrm{cm}^{-3}, \quad p_c \approx 22.8\,\mathrm{GPa}; \quad \varrho_c \approx 0.13\,\mathrm{gcm}^{-3}$$

At temperatures above the critical value the plasma shows a continuous transition from the dielectric state with small degree of ionization to the metal-like state with complete ionization (Fig. 6.15). Below the critical temperature the system is separated into two phases, one with complete ionization (liquid metal like) and one with partial ionization. This is the so-called plasma phase transition studied first by NORMAN and STAROSTIN (1968, 1970). For temperatures about 10000 K our (approximative) theory predicts that the system is divided into a liquid metal-like phase consisting of free electrons and protons with densities about 10^{23} cm^{-3} and a weakly ionized gas consisting mainly of hydrogen atoms (EBELING and SÄNDIG, 1973). A more extended discussion of plasma phase transitions, which are hypothetical so far has been given in an earlier book (EBELING et al., 1976, 1979).

6.6. Alkali Plasmas and Noble Gas Plasmas

6.6.1. Pseudopotentials

For the behaviour of alkali plasmas and noble gas plasmas the short range forces between the particles are of great importance. Since the ions are extended particles we observe deviations from Coulomb's law at small distances between the charges which are mainly due to the influence of the core electrons. Similar as in solid state theory the effective interaction between electrons and ions may be described by pseudopotential which contain free parameters, which are fitted in such a way that the lower bound states and possibly also the scattering data agree with the behaviour observed experimentally. In earlier work (EBELING et al., 1976, 1979) ZIMDAHL and EBELING, 1977) it was proposed to describe the electron-ion interaction by a pseudopotential of Hellmann type

$$V_{\mathrm{ie}}(r) = -\frac{e^2}{4\varepsilon\pi r}[1 - A\exp(-\alpha r)]\,. \tag{6.145}$$

Furthermore it was proposed to describe the ion-ion interaction by the model of charged hard spheres with the crystallographic radii R_i. The electrical part of the ion-ion interaction will be described by a pseudopotential

$$V_{\mathrm{ii}}(r) = \begin{cases} e^2/4\pi\varepsilon r & \text{if} \quad r \geqq 2R_\mathrm{i}\,, \\ \infty & \text{if} \quad r < 2R_\mathrm{i}\,. \end{cases} \tag{6.146}$$

For the case of low temperature alkali systems similar models were treated by KRASNYI and ONISHENKO (1972) (see also RICHERT et al. 1984). In Tab. 6.5 numerical values for the potential parameters taken from HART and GOODFRIEND (1971) are given.

Tab. 6.5. Pseudopotential parameters for alkali plasmas

	Na	K	Rb	Cs
A	2.685	1.813	1.494	1.371
α^{-1}, nm	0.0366	0.0636	0.0801	0.0964
R_i, nm	0.0950	0.133	0.148	0.169

For alkali plasmas at higher densities the short range forces between charged and neutral particles and between atoms become important (ALEKSEEV and IAKUBOV, 1983). For simplicity we have chosen very simple model potentials:

$$V_{aa}(r) = \begin{cases} 0 & \text{if} \quad r \geqq 2R_a\,, \\ \infty & \text{if} \quad r < 2R_a\,, \end{cases}$$
$$V_{ea}(r) = \begin{cases} V_{pol} & \text{if} \quad r \geqq R_a\,, \\ \infty & \text{if} \quad r < R_a\,, \end{cases} \tag{6.147}$$

$$V_{ia}(r) = \begin{cases} V_{pol}(r), & r \geqq (R_a + R_i)\,, \\ \infty, & r < (R_a + R_i)\,. \end{cases} \tag{6.148}$$

Where the indices e, i, a denote the electrons, ions, and atoms respectively, R_a and R_i are the atomic and ionic radii and V_{pol} is the polarization interaction (ALEKSEEV and IAKUBOV, 1983; LAGARKOV and SARYCHEV, 1979). The polarization potential may be taken in the form

$$V_{pol}(r) = -\frac{\alpha^* e^2}{8\pi\varepsilon(r_a^2 + r^2)^2}\,. \tag{6.149}$$

An estimate for the potential parameters of Cs is $r_a = 9.4a_B$ and $\alpha^* = 400\,a_B^3$, for mercury one has $r_a = 4.6a_B$. A quantum statistical derivation of the polarization potential using a Green function approach is given by REDMER and RÖPKE (1985) (see also RÖPKE et al., 1982; HÖHNE et al., 1983).

6.6.2. The Chemical Potential of the Neutral Component

The chemical potential of the neutral atoms consists of the contribution of the bound states and that of the interactions. We shall use in the following (RICHERT et al., 1984)

$$\mu_a(n_a, n, T) = k_B T \ln\left[\frac{n_a \Lambda_a^3}{4\,\Sigma(T, n, n_a)}\right] + \mu_a^{HC}(n_a, T) + 2k_B T n_a B_{aa}^{att}(T)$$
$$+ 2k_B T n\,[B_{ea}(T) + B_{ia}(T)] - k_B T n_a \frac{\partial}{\partial n_a} \ln \Sigma\,(T, n\,, n_a)\,. \tag{6.150}$$

Here B_{aa}^{att} denotes the attracting part of the second virial coefficient for the atom-atom interaction

$$B_{aa}^{att}(T) = B_{aa}(T) - \frac{16\pi}{3} R_a^3 \approx -\frac{A_{aa}}{k_B T}$$

(A — van der Waals constant for the atom-atom interaction). Further μ_a^{HC} denotes the hard-core contribution from the atom-atom interaction which is taken into account in the approximation by CARNAHAN and STARLING (1969):

$$\mu_a^{HC}(n_a, T) = k_B T \eta_a (8 - 9\eta_a + 3\eta_a^2)\,(1 - \eta_a)^{-3}\,, \tag{6.151}$$

$$\eta_a = 4\pi n_a R_a^3/3\,. \tag{6.152}$$

The charged-neutral interaction however is taken into account in the approximation of the linearized screened second virial coefficient (Redmer and Röpke, 1984):

$$B_{ea} = \frac{2\pi}{3} R_a^3 + \frac{2\pi}{k_B T} \int\limits_{R_a}^{\infty} \tilde{V}_{pol}(r)\, r^2\, dr\,,$$
$$B_{ia} = \frac{2\pi}{3} (R_i + R_a)^3 + \frac{2\pi}{k_B T} \cdot \int\limits_{R_i+R_a}^{\infty} \tilde{V}_{pol}\, r^2\, dr\,, \tag{6.153}$$

$\tilde{V}_{pol}$ — screened polarization potential.

Let us discuss now the partition function. As in the case of hydrogen plasmas there exist two reasonable variants:

(i) Planck-Brillouin-Larkin renormalization procedure:

$$\Sigma_P = \sum_{sl} (2l + 1) [\exp(-\beta E_{sl}) - 1 + (\beta E_{sl})] \tag{6.154}$$

where E_{sl} are the eigen values of the Bethe-Salpeter equation (see Sections 4.4. and 6.2.).

(ii) Riewe-Rompe renormalization procedure:

$$\Sigma_R = \sum_{E_{sl} < \varepsilon^*} (2l + 1) \exp(-\beta E_{sl}) \tag{6.155}$$

where ε^* is the upper border of bound states.

Both variants nearly equivalent from the physical point of view since both express the same physical idea which is: In thermodynamic calculations all the discrete states with binding energies smaller than $k_B T$ have to be considered as quasifree states. These discrete states near to the continuum edge have a very small liefe time which actually is limited by the time between two collisions. In other words the discrete states near to the continuum are strongly damped and are to be treated on the same footing as the continuous states.

Due to our little knowledge about the solutions of the Bethe-Salpeter equation we have to use for concrete calculations simple approximations which reflect the main physical effects. In a recent paper (Richert et al., 1984) the continuum edge has been approximated by the very simple formula

$$\bar{\Delta} = \mu_{xc}(n, n_a, T)\,, \tag{6.156}$$

i.e. by the exchange-correlation part of the chemical potential. Further the energies E_{sl} have been identified with the unperturbed levels (Fig. 6.18). In the calculations a state E_{sl} has been omitted from the sum at that density where $E_{sl} = \bar{\Delta}$, i.e. where the unperturbed level touches the continuum. In order to avoid the step-like character a smoothing procedure based on quadratic interpolations was also used. The model assumption (6.156) predicts a disappearance of the ground state level at a free electron density of about 10^{21} cm^{-3}. Another possibility which may be used is the confined atom model developed in 6.5. which gives for the density dependence of the levels

$$E_{sl}(n_e) = E_{sl}(0) + x_{sl}^2 (a_B/r_0)^2 \,\mathrm{Ry}\,.$$

This model predicts for the disappearance of the ground state level a total density of about 10^{22} cm^{-3}. Evidently the confined atom model gives an upper border for the disappearance of the discrete levels. Due to the lack of sufficiently correct solutions of the Bethe-Salpeter equation in dense media the decision between the various models is not possible at the present ime.

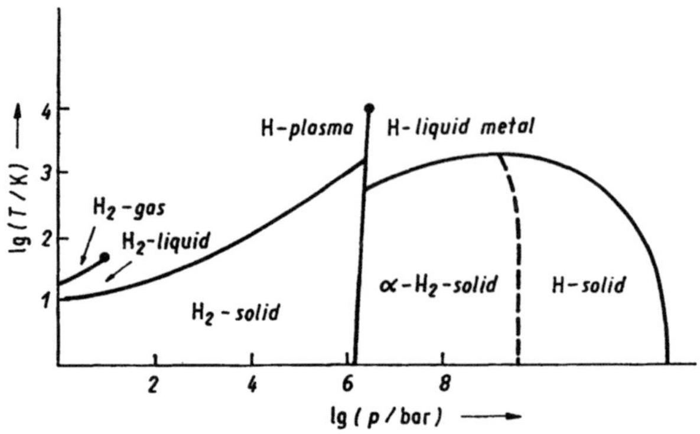

Fig. 6.17. The hypothetical phase diagram of hydrogen (Franck, 1981). The data of the critical point of the dielectric-metal transition calculated from our theory are $T_c = 16500$ K and $p_c = 0.23$ Mbar

6.6.3. The Chemical Potential of the Charged Component

As in the case of hydrogen plasmas the correlation part of the chemical potential may be described by a series expansion with respect to the degeneracy parameter

$$\bar{n} = n\Lambda_e^3$$

which starts with the terms

$$\mu_{\text{int}}/\text{Ry} = -2\left(\frac{4k_BT}{\pi\,\text{Ry}}\right)^{1/4}\bar{n}^{1/2} - \mu_5\bar{n} + \frac{32\pi}{3}\left(\frac{R_i}{\Lambda_e}\right)^3\left(\frac{k_BT}{\text{Ry}}\right)\bar{n}$$
$$+ (1+\gamma)^{1/2}\,[k(T) + \delta_1 k + \delta_2 k]\,\bar{n} + \dots, \tag{6.157}$$
$$\delta_1 k = 8\xi^{-2}\,[\Sigma(T,0,0) - \sigma(T)]\,.$$

All the notations used here are identical with those for hydrogen plasmas. The only difference to hydrogen consists in the hard-core term and the contribution $\delta_2 k$ to the virial function which is due to the additional short range forces between the electrons and the ions which are described by the Hellmann potential. Following Zimdahl and Ebeling (1977) as well as Richert et al. (1984) this correction is

$$\delta_2 k(T) = \left(\frac{4A}{x} - \frac{xA^2}{2}\right)\frac{1}{x}\left[1 - \exp\left(-\frac{x^2}{4}\right)\left(1 - \Phi\left(\frac{x}{2}\right)\right)\right]$$
$$+ \frac{1}{2}A^2 - \frac{1}{x}A^2\pi^{-1/2}, \qquad x = \alpha\lambda^{ie} \tag{6.158}$$

where $\Phi(x)$ is the error function. We note further that $\delta_1 k = 0$ for the Brillouin-Planck-Larkin choice of the partition function. In the opposite range of high density we may use the expansion

$$\mu_{\text{int}} = -\text{Ry}\,[7.222r_s^{-1} + 0.1167 - 0.0622 \ln r_s - k_B T\, \tilde{n}^{1/3} B_i\,,$$
$$B_i = B_p + \left(\frac{4\pi}{3}\right)^{1/3} \frac{6A}{\alpha^2 a_B^2 r_s^2}\,. \tag{6.159}$$

In comparison to the hydrogen case we have again a correction due to the short range electron-ion interactions which is proportional to the parameter A. In order to get the expression (6.159) we followed the work of BROVMAN and KAGAN (1967, 1969), BROVMAN et al. (1972) and RICHERT et al. (1984). All the other terms not proportional to A are identical with those for the hydrogen plasma. Since the differences between hydrogen plasmas and alkali plasmas consists only in the values of a few constants we may use the same Padé formula. In this way we get finally

$$\mu_{\text{int}} = -\text{Ry}\,\frac{\mu_0 \bar{n}^{1/2} + \mu_5 \bar{n} + C(r_s)\,\bar{n}^2}{1 + 8\bar{n}^2 + 8 \ln\,[1 + 0.07483\mu_0^{-1}\bar{n}^{1/2}]}$$
$$- k_B T\,\frac{p_0 \tilde{n}^{1/2} + B_i p_2 \tilde{n}^{3/2}}{1 - (p_1/p_0)\,\tilde{n}^{1/2} + p_2 \tilde{n}^{7/6}} + k_B T \eta_i \frac{(8 - 9\eta_i + 3\eta_i^2)}{(1 - \eta_i)^3}$$
$$+ 2k_B T n_a\,[B_{ea}(T) + B_{ia}(T)] - k_B T n_a \frac{\partial}{\partial n} \ln \Sigma\,(T, n, n_a)\,, \tag{6.160}$$

$$\eta_i = \frac{4\pi}{3} n R_i^3\,,$$

$$C(r_s) = 9.7736 r_s^{-1} + 0.71064 \ln\,(1 + 3.6094 r_s^{-7/10})\,. \tag{6.161}$$

Let us discuss now this quite long formula. The first two terms correspond to the Padé approximations for the electron gas and the ion gas respectively, the third term represents the contributoins from the hard-core part of the ion-ion interactions, the fourth part gives the contribution of the charge-neutral interaction and finally the last part comes from the density dependence of the partition function we have used in the chemical potential of the atoms. We note that the chemical potential should satisfy the consistency condition

$$\frac{\partial \mu_a}{\partial n} = \frac{\partial \mu}{\partial n_a}$$

which expresses the requirement that the second derivatives of the free energy are interchangeable. Fulfilling this condition the last terms in eq. (6.160) follow immediately.

We note that the derivatives of the partition function which appear here as well as in the chemical potential of the atoms in eq. (6.150) arise from the partition function term in the free energy. In order to get reasonable derivatives the partition function is to be smoothed in an appropriate way (RICHERT et al., 1984).

6.6.4. Saha Equation and Ionization Equilibrium

In thermodynamical equilibrium the chemical potentials of the plasma and the neutrals should be equal

$$\mu(n, n_a, T) = \mu_a(n, n_a, T)\,. \tag{6.162}$$

Using the balance equation

$$n_e = n_i = n + n_a \tag{6.163}$$

the equilibrium condition assumes the form

$$n_e = n + 4\Lambda_a^{-3} \exp\Big[\beta\mu(n, n_e - n, T) - \beta\mu_a^{HC} - 2n(B_{ea} + B_{ia}) - 2n_a B_a^{att} + (n_e - n)\frac{\partial}{\partial n_a} \ln \Sigma(T, n, n_e - n)\Big] \cdot \Sigma(T, n, n_e - n)\,. \tag{6.164}$$

In the case that Σ depends on n only another form of writing is obtained which is useful for the numerical treatment:

$$n_a \exp\left[\beta(\mu_a^{HC} - n_a A_{aa}) - 2n_a(B_{ea} + B_{ia}) + n_a \frac{\partial}{\partial n} \ln \Sigma\right]$$
$$= 4\Lambda_a^{-3} \exp\left[\beta(\mu_{id} + \mu_{xc}) - 2n(B_{ea} + B_{ia})\right] \Sigma(T, n)\,.$$

Let us still note that eq. (6.164) is equivalent to the familiar form of the Saha equation

$$n_a = n^2 K_{eff}(T, n, n_a) = \Lambda_{ie}^3 \exp[\beta I_{eff}]\,, \tag{6.165}$$
$$I_{eff} = (\ln \Sigma + \mu_{xc}) - (\mu_a^{HC} - n_a A_{aa}) + (\mu_{id} - \mu_{id}^{Boltz})$$
$$+ 2(n_a - n)(B_{ea} + B_{ia}) - n_a \frac{\partial}{\partial n} \ln \Sigma\,.$$

Here the first term gives the main contribution. In simplest approximation we may neglect the density dependence of Σ by putting

$$I_{eff} = [\ln \sigma(T) + \mu_{xc}] - [\mu_a^{HC} - n_a A_{aa}] + 2(n_a - n)(B_{ea} + B_{ia}) \quad \text{if} \quad n_a < n_M\,, \tag{6.166}$$
$$K_{eff} = 0 \quad \text{if} \quad n > n_M$$

where the Mott density is given by the condition of disappearance of the ground state

$$E_{10} = \mu_{xc}(n_M, T) \tag{6.167}$$

(see Fig. 6.18). The solution of the Saha equation is most easily found by iteration of eq. (6.165), calculating for a given temperature n_a as a function of n and starting with a guess for n_a, say $n_a = 0$ or $n_a = n$. We note that in our theory n_a is obtained as a step function of n, reaching the Mott density n_M the atoms disappear in the plasma. The knowledge of $n_a = F(n, T)$ gives us $n_e = n + F(n, T)$ and by inversion follows $n = i(n_e, T)$ and the degree of ionization

$$\alpha = n/n_e = i(n, T)/n_e\,. \tag{6.168}$$

The chemical potential of the plasma $\mu(n, n_a, T)$ is a key quantity of the plasma theory. By using the Saha equation we may obtain the plasma chemical potential as a function of the total electron density:

$$\mu(n_e, T) = \mu[i(n_e, T), n_e - i(n_e, T), T]\,. \tag{6.169}$$

This quantity is connected with the free energy density by

$$\mu(n_e, T) = \frac{\partial}{\partial n_e} f(n_e, T) \tag{6.170}$$

and with the pressure of the plasma by

$$p = -\frac{\partial}{\partial V}[Vf(n_e, T)] = n_e\mu(n_e, T) - f(n_e, T) . \tag{6.171}$$

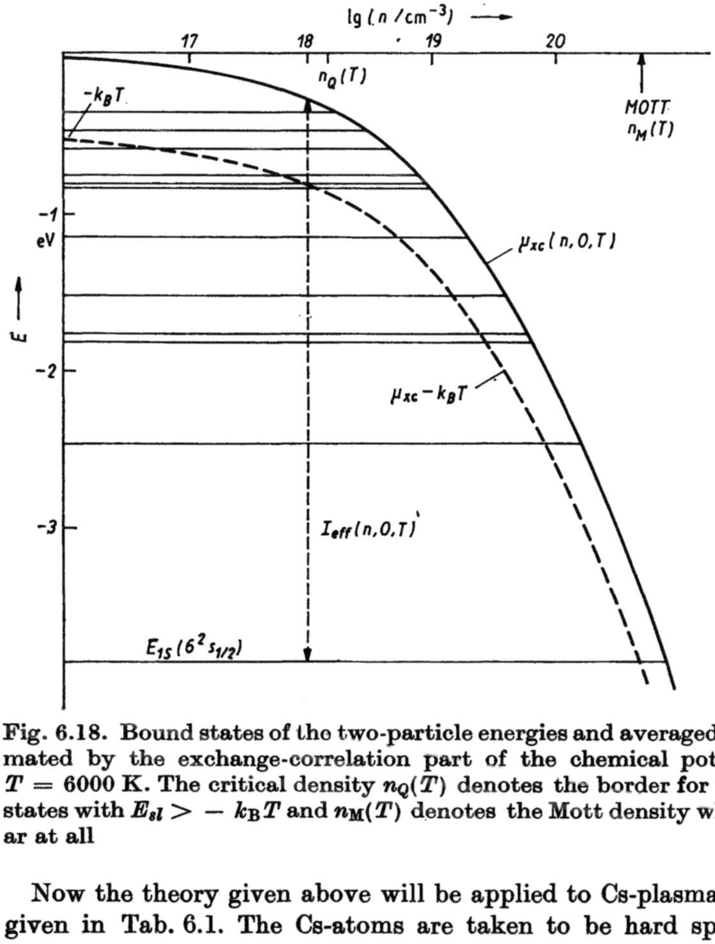

Fig. 6.18. Bound states of the two-particle energies and averaged continuum edge (approximated by the exchange-correlation part of the chemical potential) for Cs-plasmas at $T = 6000$ K. The critical density $n_Q(T)$ denotes the border for the existence of quasi-free states with $E_{sl} > -k_BT$ and $n_M(T)$ denotes the Mott density where bound states diasppear at all

Now the theory given above will be applied to Cs-plasma with the potential data given in Tab. 6.1. The Cs-atoms are taken to be hard spheres having the radius $R_i = 0.267$ nm. First we calculate $\mu = \mu(n, n_a, T)$ from eq. (6.160) and then $n = f(n_e)$ from the Saha equation (6.154) (note that $n = n_e = n_i$ is the density of free electrons and n_e the total electron density, free and bound). The chemical potential of the plasma as a function of the density of free electrons n is shown in Fig. 6.19 for different temperatures. We observe the appearance of a wiggle in the plasma potential below the critical temperature $T_c \approx 4600$ K. In Fig. 6.20 the density of free electrons is shown as a function of the total electron density n_e. The transformation between n_e and n is given by eq. (6.164); we see that wiggles of μ as a function of n transform into wiggles of μ as a function of n_e. See also DIENEMANN, CLEMENS and KRAEFT (1980).

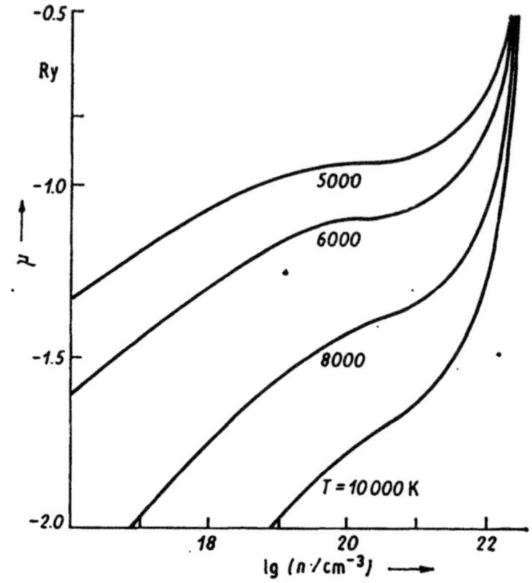

Fig. 6.19. The chemical potential of Cs-plasmas as a function of the free electron density for zero atom density $n_a = 0$ at several temperatures

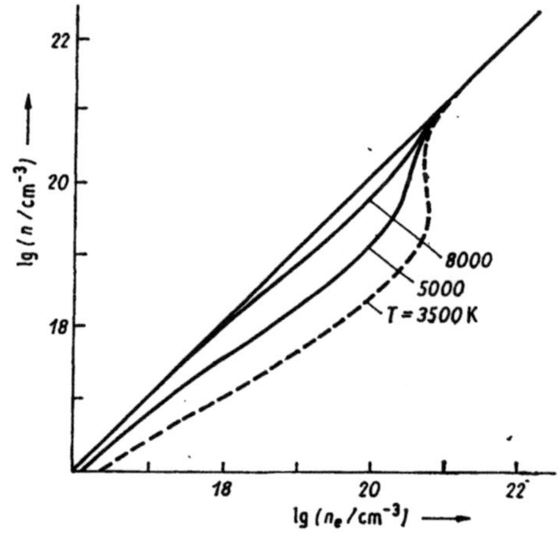

Fig. 6.20. The free electron density vs. the total electron density for Cs-plasmas

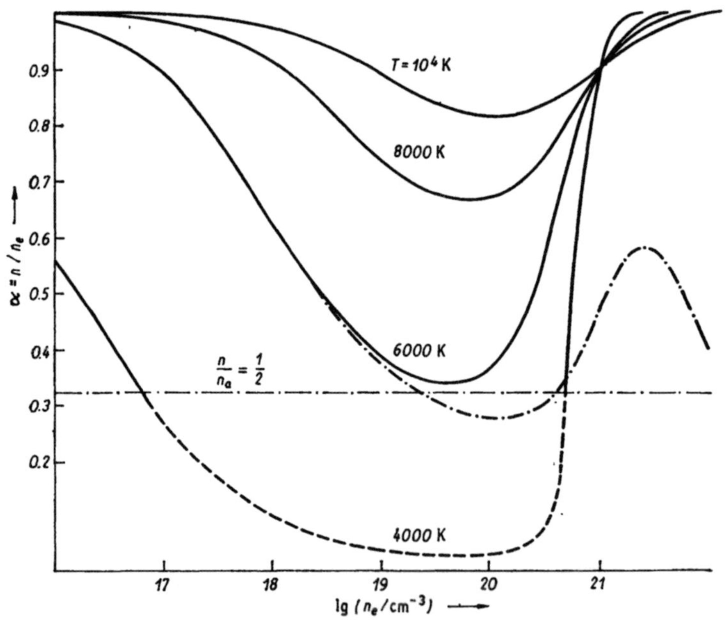

Fig. 6.21. The degree of ionization of Cs-plasmas as a function of the total electron density. The dash-dotted line for $T = 6000$ K is obtained if the lowering of the continuum edge is neglected

Several curves for the degree of ionization at temperatures above the critical value are shown in Fig. 6.21. A characteristic property of the α-isoterms is the appearance of a minimum (at densities $n \approx 10^{19} - 10^{20}\,\mathrm{cm}^{-3}$) and the turning back to high degree of ionization ($\alpha \approx 1$) at higher densities. This increase of the of ionization is caused by the nonideality of the plasma. (The theory would be improved by taking into account more exactly the density dependence of the sum of bound states).

In nonideal plasmas the interaction is of dominating importance. Due to the interaction of the charges the effective ionization energy

$$I_{\text{eff}} \approx -E_{10} - \frac{e^2\varkappa}{4\pi\varepsilon} = \frac{e^2}{4\pi\varepsilon}\left[\frac{1}{2a_{10}} - \frac{1}{r_0}\right] \tag{6.172}$$

decreases with increasing densities. The effective ionization energy approaches zero if $r_0 \lessapprox 2a_{10}$, where a_{10} is the ground state radius. Thus the α-increase is the expected transition to a fully ionized highly conducting state which appears if the screening radius is of the order of the Bohr radius. This transition is often called the Mott transition or pressure ionization. We mention that there exist different interpretations for the physical mechanism underlying the Mott transition, e.g., based on the consideration of the energy levels of screened potentials. In the termodynamic picture the condition for the Mott transition follows from the mass action law. Another picture was demonstrated in 5.4.

For alkali plasmas this condition means that a pressure ionization may be expected if the interaction part of the chemical potential consisting of the Hartree-Fock and the correlation contribution exceeds a few electron volt ($E_{10} = -3.9$ eV for Cs).

Quantitatively we define the transition density $\bar{n}_e(T)$ for a given temperature as the density where $\partial\alpha(n_e, T)/\partial n_e$ is positive and has a maximum, i.e. the second derivative is equal to zero:

$$\frac{\partial^2}{\partial n_e^2}\alpha(n_e, T) = 0 \quad \text{if} \quad n_e = \bar{n}_e(T)\,. \tag{6.173}$$

The transition to the highly ionized highly conducting state occurs in a rather narrow region of densities and may be considered as a diffuse phase transition above the critical point (Hensel, 1980, 1981).

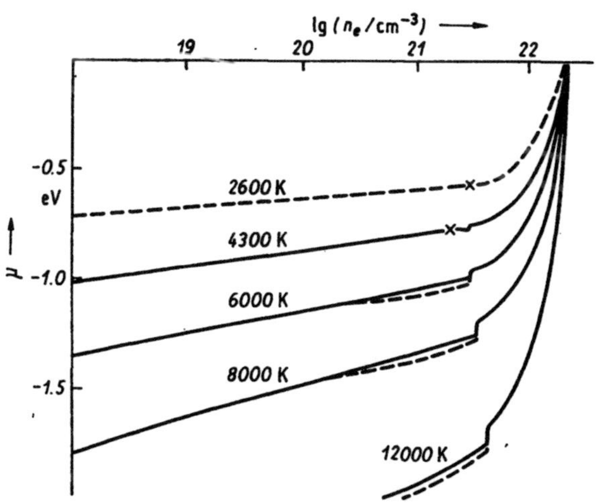

Fig. 6.22. Chemical potential of Cs-plasmas as a function of the total electron density (dashed line theory with polarization)

Using the solutions of the Saha equation the thermodynamic functions $\bar{\mu}(n_e, T)$, $p(n_e, T)$ etc. may be calculated. Some results are presented in Figs. 6.22–6.24.

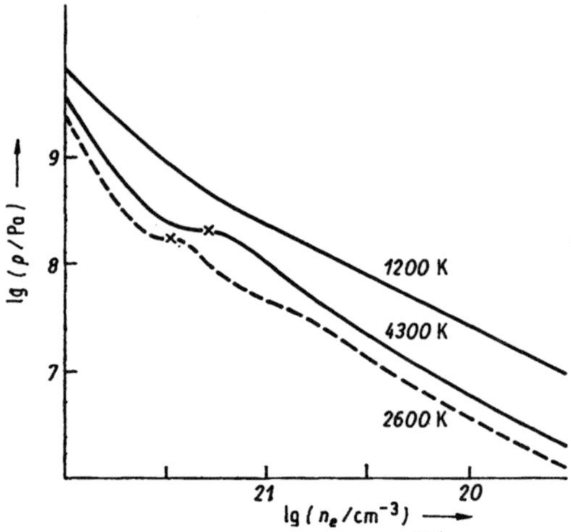

Fig. 6.23. Pressure of Cs-plasmas as a function of the total electron density (dashed line theory with polarization)

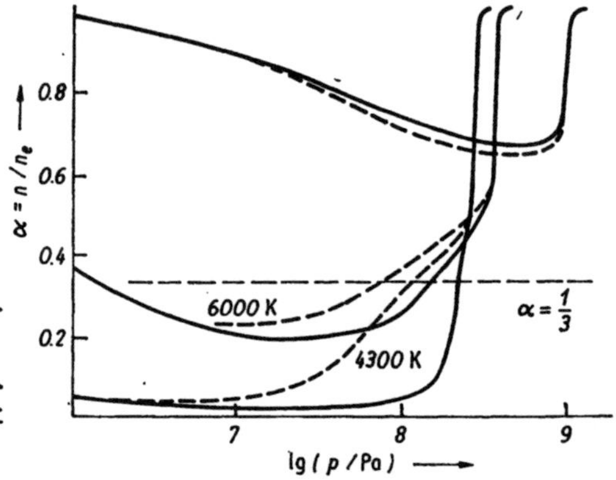

Fig. 6.24. Degree of ionization as a funktion of the pressure (dashed line theory with polarization). Upper curve 12000 K

In the case of lower temperatures we observe the formation of a wiggle of the isotherms. Similar as for the van der Waals isotherms, the wiggle of the isotherms indicate a thermodynamic instability of the system.

In a stable plasma $\bar{\mu}$ and p must be nondecreasing functions of the total electron density n_e. Clearly in our example this is the case for $T > 4300$ K but for $T < 4300$ K the curves $\bar{\mu}$ and p show a very flat minimum. Since the wiggles in our curves are very flat, a precise prediction of the critical point is impossible up to now. In our case of Cs-plasmas the appearance of a wiggle has the critical data (RICHERT et al., 1984):

without polarization: $T_c = 4300$ K, $n_c^c = 2 \cdot 10^{21}$ cm^{-3}, $p_c = 180$ MPa.

With polarization: $T_c = 2600$ K, $n_e^c = 3\ 10^{21}$cm^{-3}, $p_c = 180$ MPa.

We note that the critical data obtained from a theory depend very sensitively from the interactions assumed. The polarization between charges and neutrals tend to lower the critical temperature and to increase the critical density (compare KREMP, HARONSKA and SCHLANGES, 1984). In contrast to the situation for hydrogen plasmas, evidently only one critical point exists for alkali plasmas (RICHERT et al., 1984).

We note further that we obtained earlier for the model of pure Coulomb forces (EBELING and SÄNDIG, 1973)

$$T_c \approx 12600\ \text{K}\,, \qquad n_c^c \approx 4 \cdot 10^{21}\ \text{cm}^{-3}$$

and in a pseudopotential approach (EBELING et al., 1979)

$$T_c \approx 4600\ \text{K}\,, \qquad n_c^c \approx 4.7 \cdot 10^{20}\ \text{cm}^{-3}\,.$$

Evidently the short range forces between the electrons, ions and atoms stabilize the plasma and are responsible for a drastic lowering of the critical temperature. The pseudopotential approach given here is more complete than the approach presented earlier (ZIMDAHL and EBELING, 1977; EBELING et al., 1979), since we have taken into account here at least partially the effects of degeneracy of the electrons and the packing effects of the ions at higher densities. Both these effects are responsible for a steep increase of the plasma potential at densities larger than 10^{21} cm^{-3}.

Furtheron due to these effects the plasma potential is at all densities higher than the result of the simple Debye theory

$$\mu = \mu_{\text{id}} - \frac{e^2}{4\pi\varepsilon r_0}.$$

It was noted also in experimental work on the thermodynamic properties of Cs-plasmas that the Debye correction overestimates the interaction of free charges.

The experimental findings are that a phase transition of first order — the usual gas-liquid transition — is obtained at temperature $T < T_c = 2030$ K. The critical density is $\varrho^c = 0.4$ g cm^{-3}, $n^c = 1.8 \cdot 10^{21}$ cm^{-3} (Alekseev and Iakubov, 1983). Since the liquid metal-vapor phase transition is essentially influenced by the interaction of the charged particles with the neutrals which where taken into account only in a rough approximation we cannot give a correct description of this transition by means of our theory. For a survey of the theory of plasmas which are dominated by the charge-neutral interactions we refer to the work of Alekseev and Iakubov (1983). Let us further note that the details of the phase diagram of metals may be very complicated and are known only in part (Hensel, 1980, 1981). Therefore let us underline again that our theory is valid only for ionized plasmas ($\alpha > 1/3$) and cannot explain the whole phase diagram. In principle the theory developed in this Section may also be applied to noble gas plasmas. We underline however that in our work only the first ionization step is taken into account. The generalization of this theory to more complicated situations makes no difficulties. We note however that for complex chemical equilibria with more than three species the minimization of the free energy is more useful than the consideration of Saha equation (Patch, 1969; Richert and Ebeling, 1984).

Finally let us underline that the restriction to the plasma state in this Section eliminates from consideration systems such as liquid metals and semiconductors (Brovman et al., 1972; Ashcroft and Stroud, 1978; Krasny and Onishchenko, 1972; Shimoji, 1977; Ebina, 1983; Ebeling et al., 1984) or liquid metal-ammonia solutions (Thomson, 1976). In the liquid state of matter the structure is dominated by short range forces which indirectly are due to Coulombic forces and the Pauli exclusion principle (Matsubara, 1982; Montroll and Lebowitz, 1982).

7. Transport Properties

7.1. Linear Response Theory

7.1.1. Many-Body Effects and Transport Properties in Non-Ideal Plasmas

In contrast to thermal equilibrium where the state of the system is characterized by given mean values of the conserved quantities as energy and particle number, the non-equilibrium state of a system is characterized by additional mean values of macroobservables. In the hydrodynamic description of a non-equilibrium state, such additional observables are the currents of particles, energy and momentum, respectively, which are determined by the corresponding equations of continuity for the conserved quantities. In the kinetic description, the mean values of the occupation numbers, i.e. the distribution functions for the different species, are considered as relevant observables.

The generalized currents which characterize the non-equilibrium state can be considered as response of the system on external influences as, for instance, external fields, coupling to reservoirs with different temperatures, and different chemical potentials. If within a thermodynamic description, the production of entropy is known for the non-equilibrium state, generalized forces can be defined. It is assumed within the phenomenological theory of irreversible processes, see De Groot and Mazur (1962), that near the equilibrium these currents are linear functions of the generalized forces. The validity of this assumption up to macroscopic deviations from equilibrium is confirmed by experiments as, e.g., the validity of Ohm's law. We will restrict our treatment of Coulomb systems in non-equilibrium to this regime of linear response and anticipate the possibility of the expansion of the density matrix with respect to the generalized forces up to first order.

The linear response of the system can be treated by different methods, see Kubo (1957), Mori (1965), Zubarev (1974), Klimontovich (1975, 1980), Fujita (1966, 1969), which are equivalent in the sense that the Liouville-von-Neumann equation for the statistical operator is formally solved. Corresponding to the fluctuation-dissipation theorem, the transport coefficients are expressed by equilibrium correlation functions. In this way, it is possible to use the formalism derived in Chapters 3., 4. to formulate a theory of transport coefficients which is valid over a wide density range. All many-body effects discussed up to now can be included into a quantum statistical theory of transport properties in non-ideal plasmas, and in particular the approximations described in Chapter 4. which are of importance if bound states occur can be applied also to the theory of transport coefficients.

In detail we will restrict ourselves to the thermoelectric effects. The electrical current j_e and the heat current j_q are related to the corresponding generalized forces, the gradient of the temperature ∇T and the gradient of the electrochemical potential $\nabla\zeta = -\nabla(\varphi + \mu/e)$ ($\varphi(r)$ is the external potential, $\mu(r)$ the chemical potential), by the linear relations

$$\begin{aligned} j_e &= e^2 L_{11} \nabla\zeta - e L_{12} \nabla T/T , \\ j_q &= e L_{21} \nabla\zeta - L_{22} \nabla T/T . \end{aligned} \tag{7.1}$$

The electrical conductivity σ, the thermopower α and the thermal conductivity λ are defined by the Onsager coefficients $L_{ik} = L_{ki}$ according to

$$\begin{aligned} \sigma &= e^2 L_{11} , \\ \alpha &= \frac{1}{eT} L_{12}/L_{11} , \\ \lambda &= \frac{1}{T} (L_{22} - L_{12}L_{21}/L_{11}) . \end{aligned} \tag{7.2}$$

All quantities may be considered as complex frequency dependent scalar functions, the generalization to anisotropic systems, different components and the inclusion of an external magnetic field, for instance, is straightforward (see, e.g., KLIMONTOVICH, 1964, 1975; LANDAU and LIFSHITS, 1981, 1983).

By non-equilibrium statistical mechanics the transport coefficients are determined from the microscopic scattering process. Historically, this concept was first realized by the Boltzmann equation which is valid for systems with short-range interaction in the low density limit. We give some special results for the conductivity σ often used in the literature (LANDAU and LIFSHITS, 1981, 1983; CHAPMAN and COWLING, 1958; MITCHNER and KRUGER, 1973; SPITZER, 1962), these conductivity formulas are derived from a unified approach within the linear response theory below.

The Lorentz plasma is a system of non-interacting electrons which are scattered by ions at fixed positions. In the case of elastic, isotropic scattering, the Boltzmann equation is solved by the relaxation time ansatz. The electron-ion interaction is described by the statically screened Coulomb interaction, and in the first Born approximation we obtain the Brooks-Herring formula (BLATT, 1966)

$$\begin{aligned} \sigma_{BH} &= \frac{2^{5/2}}{\pi^{3/2}} \frac{(4\pi\varepsilon\varepsilon_0)^2 (k_B T)^{3/2}}{e^2 m_e^{*1/2}} \frac{1}{\Phi_{BH}(b)} , \\ \Phi_{BH}(b) &= \ln\sqrt{1+b} - \frac{1}{2}\frac{b}{1+b} , \qquad b = 12\,\varepsilon\varepsilon_0 \frac{m_e^*(k_B T)^2}{e^2\hbar^2 n} . \end{aligned} \tag{7.3}$$

This formula is used, for instance, to describe the ionized impurity scattering part of the electron mobility in semiconductors. The electron gas is considered to be non-degenerate; m_e^* is the the effective mass of the electron, ε the dielectric constant of the non-excited crystal, n the electron density.

For liquid metals, the Ziman formula for the electrical conductivity is generally applied (ZIMAN, 1961; FABER, 1972):

$$\sigma_{ZI}(T=0) = \frac{12\pi^3 e^2 n^2 \hbar^3}{m_e^2} \left\{ \int_0^{2k_F} dq\, q^3\, |V_{ei}(q)|^2 \right\}^{-1} . \tag{7.4}$$

In this formula, the electrons are considered to be strongly degenerate; $\hbar k_F$ is the Fermi momentum. The Born approximation has been extended to arbitrary order of the potential $V_{ei}(q)$ by introducing the T-matrix, see BUSCH and GÜNTHERODT (1974). In the Born approximation, $|V_{ei}(q)|^2$ is given by the product of the squared electron-ion pseudopotential $|V(q)|^2$ and the static ion structure factor $S(q, 0)$, see PINES and NOZIÈRES (1957).

A generalization of the Ziman formula (7.4) for finite temperatures has been considered by FABER (1972):

$$\sigma_0(T) = \frac{3}{4\sqrt{2\pi}} \frac{(4\pi\varepsilon_0)^2(k_B T)^{3/2}}{e^2 m_e^{1/2} \Phi_0},$$

$$\Phi_0 = \frac{2}{n}\left(\frac{m_e k_B T}{2\pi\hbar^2}\right)^{3/2} \int_0^\infty dE(k)\left(-\frac{df(E)}{dE}\right) \int_0^{2k_F} dq\, q^3\, |V_{ei}(q)\, \varepsilon_0/e^2|^2 . \tag{7.5}$$

In the non-degenerate limit where the Fermi distribution $f(E)$ may be replaced by the Maxwell distribution, the Coulomb logarithm Φ_0 coincides with Φ_{BH} if the statically screened interaction potential $V_{ei}(q) = e^2/(q^2 + \varkappa^2)\,\varepsilon_0$ is taken.

The electrical conductivity of the ideal, fully ionized plasma is well described by the Spitzer formula (SPITZER, 1962; EBELING et al., 1983)

$$\sigma_{SP} = 0.591 \frac{(4\pi\varepsilon_0)^2(k_B T)^{3/2}}{e^2 m_e^{1/2} \Phi_{sp}(\mu)}, \qquad \Phi_{SP}(\mu) = \ln\frac{3}{\mu}, \qquad \mu = \frac{e^2}{4\pi\varepsilon_0 k_B T}\left(\frac{ne^2}{\varepsilon_0 k_B T}\right)^{1/2} \tag{7.6}$$

is the plasma parameter. Eq. (7.6) is derived for the non-degenerate case taking into account the influence of electron-electron interaction on the conductivity. We remark that the Coulomb logarithm $\Phi_{Sp}(\mu)$ is derived from a cut-off procedure taking into account the exact trajectories of the Kepler problem instead of the Born approximation.

The different expressions for the conductivity given above should be considered as special limiting cases of a more general theory of transport properties of systems with Coulomb interaction. Such a theory is presented in this Chapter on the basis of linear response theory. The main idea of this Chapter is to show how the method elaborated in this book may be used to formulate the transport theory. Similar results can also be obtained from other approaches as kinetic theory, see 7.1.3.

We give a survey of many-particle effects to be treated in this Chapter. In the same way as in the theory of thermodynamic properties, the Coulomb interaction has to be screened in order to get convergent results for the conductivity. Therefore, a consistent theory of plasma conductivity can be formulated only on the basis of a many-particle theory including *dynamical screening*.

Furthermore, the appearence of *bound states* influences the plasma conductivity, and the Mott transition caused by the disappearence of bound states is at a first glance connected with the transition from the insulating to the metallic behaviour. Therefore, the problem of bound states occuring in a partially ionized plasma is of special interest in formulating the theory of transport coefficients in non-ideal plasmas.

A more involved many-particle effect is the *Debye-Onsager relaxation effect* as known from the theory of electrolytes. The conductivity is reduced because of the formation of a retarded screening cloud which reduces the effective electrical field within the plasma.

Similar effects are obtained from the treatment of *self-energy effects* in the quantum statistical theory of conductivity.

In dense plasmas, the effect of equilibrium correlations of the ions as described by the *structure factor* will be of importance as well known from the theory of liquid metals. A further effect is the *hopping conductivity* which is of importance for a system of weakly bound electrons. This transport mechanism which is well known from the theory of disordered semiconductors, may become important near the Mott transition. In ana ogy to the Boltzmann equation where the differential cross section can be presented by a Born series, the incorporation of higher order Born approximations in the linear response theory leads to a *T-matrix approach*. Furthermore, the treatment of *degeneration* and *exchange* of electrons is necessary in evaluating the theory of transport processes in dense Coulomb systems.

As a consequence, a consistent theory of transport processes can be given only within a quantum statistical approach. Beside the linear response theory, equivalent approaches can be formulated also within the kinetic theory, see KLIMONTOVICH (1975, 1980), or from the quantum statistical Boltzmann equation as developed by KADANOFF and BAYM (1962).

7.1.2. Transport Coefficients and Correlation Functions

In the framework of linear response theory, transport coefficients are expressed by equilibrium correlation functions of the many-particle system. A well known approach was given by KUBO (1957) where the linear response of an electrical field E yielding the electrical conductivity is considered (see also KRAEFT et. al, 1983), cf. also 4.2.1.

A more general approach which includes also non-mechanical perturbations was given by ZUBAREV (1974, 1980). We will give here an extension of the Zubarev method which relates the response quantities, such as the distribution functions, e.g., to the external perturbations (electrical field and temperature gradient) for an open system. Within this formulation, a direct relation between linear response theory and kinetic theory can be given (CHRISTOPH and RÖPKE, 1978; RÖPKE, 1983; EBELING et al., 1983).

We are interested in the formulation of linear response theory for a system containing free particles in states $|p\rangle$ (delocalized states; p describes species, spin and momentum of free particles) as well as in bound states $|nP\rangle$ (composites, localized states; n — internal quantum number and species, P — total momentum). Such systems are, for instance, the partially ionized plasma, excited semiconductors, heavily doped semiconductors, etc.

The set of occupation numbers for the relevant states $\{|\nu\rangle\} = \{|p\rangle, |nP\rangle\}$ is needed in order to formulate the principle of weakening of initial correlations which represents the initial value problem to the Liouville-von-Neumann equation. This principle is well known also from kinetic theory (BOGOLJUBOV, 1946; KLIMONTOVICH and KREMP, 1982). According to ZUBAREV (1971, 1974), the principle of weakening of initial correlations can be formulated by introducing a source term into the Liouville-von-Neumann equation (3.34) for the statistical operator

$$\frac{\partial}{\partial t} \varrho + \frac{i}{\hbar} [H, \varrho] = -\varepsilon(\varrho - \varrho_q) , \tag{7.7}$$

where the limit $\varepsilon \to +0$ is to be taken after the thermodynamic limit. In contrast to the Kubo theory which works in terms of the equilibrium statistical operator (3.62)

$$\varrho_0 = Z_0^{-1} \exp\{-\beta H_s + \sum_c \beta\mu_c N_c\} \tag{7.8}$$

(c denotes the species) rather than in terms of the quasiequilibrium statistical operator ϱ_q (generalized Gibbs state), we use (RÖPKE, 1983; HÖHNE et al., 1983)

$$\varrho_q = Z_q^{-1} \exp\{-\int \beta(r)\, h_s(r)\, \mathrm{d}\boldsymbol{r} - \sum_c \int \beta(r)\, e_c \zeta_c(r)\, n_c(r)\, \mathrm{d}\boldsymbol{r} + \sum_j \beta F_j P_j\}, \tag{7.9}$$

$$\beta(r) = \beta + \boldsymbol{r}\cdot\nabla\beta \quad \text{and} \quad e_c\zeta_c(r) = -\mu_c + (e_c\boldsymbol{E} - \nabla\mu_c)\,\boldsymbol{r}$$

are the given fields of temperature and of chemical potential, respectively; $n_c(r) = \psi_c^+(r)\psi_c(r)$ is the particle density of species c, and the density of the system Hamiltonian is given by

$$h_s(r) = \sum_c \psi_c^+(r)\left(-\frac{\hbar^2}{2m_c}\Delta\right)\psi_c(r) + \frac{1}{2}\sum_{cd}\int \mathrm{d}\boldsymbol{r}'\, \psi_c^+(r)\,\psi_d^+(r')$$
$$\cdot\, V_{cd}(\boldsymbol{r}-\boldsymbol{r}')\,\psi_d(r')\,\psi_c(r)\,. \tag{7.10}$$

Further observables P_j are introduced which characterize the non-equilibrium state. We will discuss two special choices of the set of observables $\{P_j\}$: the set of occupation numbers $\{n_\nu\}$ for the free-particle and two-particle bound states, and the set of momenta $\{P_n\} = \{\sum_k \hbar k_z(\beta E_k)^{n/2}\, n_k\}$ of the electron occupation numbers, with P_0 being the total electron momentum, so that in adiabatic approximation holds $j_e = e\langle P_0\rangle/\Omega m_e$, and $k_B T\langle P_2\rangle/\Omega m_e$ has the meaning of the ideal part of the electronic contribution to the energy current. Supposing that the mean values of these operators $\langle P_j\rangle_q$ in the quasiequilibrium state coincide with the actual mean values $\langle P_j\rangle$ in the non-equilibrium state (ZUBAREV, 1971, 1974),

$$\langle P_j\rangle \equiv \mathrm{Tr}\{\varrho P_j\} = \mathrm{Tr}\{\varrho_q P_j\} = \langle P_j\rangle_q\,, \tag{7.11}$$

which means that ϱ_q is the optimal generalized Gibbs state corresponding to ϱ having at given constraints (7.11) the extremal entropy, the relations (7.11) determine the response parameters F_j implicitly. From (7.7) and (7.11) follows that the operators P_j obey Hamiltonian dynamics

$$\frac{\mathrm{d}}{\mathrm{d}t}\langle P_j\rangle = \mathrm{Tr}\left\{\varrho\,\frac{i}{\hbar}[H, P_j]\right\}, \tag{7.12}$$

what is equal to zero in the stationary case considered here. Explicitly, the response parameters F_j are found by using the formal solution of (7.7)

$$\varrho(t) = \varrho_q(t) - \int_{-\infty}^{t} \mathrm{d}t'\, \mathrm{e}^{\varepsilon(t'-t)}\, \mathrm{e}^{\frac{i}{\hbar}H(t'-t)}\, \frac{i}{\hbar}[H, \varrho_q(t')]\, \mathrm{e}^{-\frac{i}{\hbar}H(t'-t)}\,. \tag{7.13}$$

Expanding this expression up to first order with respect to ΔT, $\Delta\zeta_c$ and F_j (linear response) we find (species c are omitted)

$$\varrho = \varrho_q - \varrho_0 \int_{-\infty}^{0} \mathrm{d}t \int_0^{\beta} \mathrm{d}\tau\, \mathrm{e}^{\varepsilon t}\left\{-e\nabla\zeta\, \dot{P}_0(t - i\hbar\tau) + \frac{\nabla T}{T}(k_B T\, \dot{P}_2(t - i\hbar\tau)\right.$$
$$\left. - \mu \dot{P}_0(t - i\hbar\tau)) + \sum_j F_j \dot{P}_j(t - i\hbar\tau)\right\}. \tag{7.14}$$

Inserting this expression in eq. (7.12), we get for the stationary case, $\mathrm{d}\langle P_j\rangle/\mathrm{d}t = 0$, the following equations of balance:

$$e\,\nabla\zeta N_{0m} - \frac{\nabla T}{T}(k_B T\, N_{2m} - \mu N_{0m}) = \sum_n F_n d_{nm} \tag{7.15}$$

with

$$\begin{aligned} N_{nm} &= \bar{N}_{nm} + \frac{1}{m_e}\langle P_n(\varepsilon); \dot{P}_m\rangle\,, \\ \bar{N}_{nm} &= \frac{1}{m_e}(P_n; \dot{P}_m)\,, \\ d_{nm} &= (\dot{P}_n; P_m) + \langle \dot{P}_n(\varepsilon); \dot{P}_m\rangle \end{aligned} \tag{7.16}$$

and the following notation for the correlation functions

$$\begin{aligned} (A; B) &= \int_0^\beta \mathrm{d}\tau\, \mathrm{Tr}\,\{\varrho_0\, A(-i\hbar\tau)\, B\}\,, \\ \langle A(\varepsilon); B\rangle &= \int_{-\infty}^0 \mathrm{d}t\, \mathrm{e}^{\varepsilon t}\, \big(A(t); B\big)\,, \\ A(t) &= \exp\,(iH_s t/\hbar)\, A\, \exp(-iH_s t/\hbar)\,. \end{aligned} \tag{7.17}$$

The linear response parameters F_j are found from (7.15) by use of Cramer's rule.

Equation (7.15) may be considered as a generalization of the linearized Boltzmann equation, where instead of binary collisions the collision integral d_{nm} is determined by correlation function of the whole plasma. Approximative expressions for F_j are obtained by using a variational principle (Kohler, 1949; Prigogine, 1955; Möbius, Goedsche and Vojta, 1979) which is valid because of the symmetry of the "collision integral" d_{nm} (Onsager symmetry relations) and its positive definitness (positive entropy production). Different sets of observables $\{P_j\}$ correspond to different choices of a trial function within the variational principle.

Having the statistical operator ϱ (7.14) to our disposal, the mean values of the electrical and the thermal currents can be evaluated:

$$\begin{aligned} j_e &= \frac{e}{\Omega}\sum_n F_n N_{0n}\,, \\ j_q &= \frac{1}{\Omega}\sum_n F_n(k_B T N_{2n} - \mu N_{0n})\,, \end{aligned} \tag{7.18}$$

where only the ideal part of the heat current has been considered. As can be shown, the non-ideal parts give no contribution to the transport coefficients in Born approximation. With the solution of (7.15) we obtain for the transport coefficients (Höhne et al. 1984)

$$L_{ik} = -\frac{1}{\Omega}(-\mu)^{i+k-2}\begin{vmatrix} 0 & \frac{k-1}{\beta\mu}N_2 - N_0 \\ \frac{i-1}{\beta\mu}\bar{N}_2 - \bar{N}_0 & d \end{vmatrix}|d|^{-1} \tag{7.19}$$

with

$$N_n = (N_{n0}, N_{n1}, \dots, N_{nL}) ,$$

$$\bar{N}_n = \begin{pmatrix} \bar{N}_{n0} \\ \bar{N}_{n1} \\ \vdots \\ \bar{N}_{nL} \end{pmatrix}, \qquad d = \begin{pmatrix} d_{00} & d_{01} & \dots & d_{0L} \\ d_{10} & d_{11} & \dots & d_{1L} \\ \vdots & & & \\ d_{L0} & d_{L1} & \dots & d_{LL} \end{pmatrix}.$$

In this way, the problem of the calculation of transport coefficients is reduced to the problem of evaluation of equilibrium correlation functions. This problem, in turn, is solved within the framework given in Chapters 3., 4.

We discuss in more detail the choice of the set $\{P_j\}$ of relevant operators in describing the non-equilibrium state. The most detailed description is obtained if we take the occupation numbers $\{n_\nu\} = \{n_p, n_{nP}\}$ of single particle states and bound states. Then, the equation of balance (7.15) corresponds to the linearized Boltzmann equation for the single particle and bound state distribution function

$$\langle n_\nu \rangle = f_\nu = f_\nu^0 + \delta f_\nu = \mathrm{Tr}\,\{\varrho_0 n_\nu\} + \sum_{\nu'} F_{\nu'}(n_{\nu'}; n_\nu) . \tag{7.20}$$

In order to solve the system of equations (7.15) we can apply the Kohler variational principle as mentioned above. Using different sets of orthonormal functions we obtain the Chapman-Enskog method (Sonine polynomials; CHAPMAN and COWLING, 1958) or the Grad method (Hermite polynomials; GRAD, 1957). In this way we see the equivalence of a generalized Boltzmann equation in a finite polynomial approximation and an ab initio calculation containing only these polynomials in ϱ_q (7.9) in determining the transport coefficients. As discussed below in Section 7.4. in detail, also in first Born approximation different results follow for the transport coefficients L_{ik} if finite sets of polynomials P_n are used to solve the generalized Boltzmann equation (7.15) in the sense of a variational approach.

Especially, working only with the total momentum of all electrons P_0, the force-force correlation function expression for the resistivity

$$R = \frac{1}{\sigma} = \frac{\Omega}{3N^2e^2} \frac{\langle \dot{P}_0(\varepsilon); \dot{P}_0 \rangle}{1 + (3Nm)^{-1} \langle \dot{P}_0(\varepsilon); \dot{P}_0 \rangle} \tag{7.21}$$

is obtained which is widely discussed in the literature, see HUBERMANN and CHESTER (1975), BALLENTINE and HEANEY (1976), RÖPKE and CHRISTOPH (1975). It has been pointed out there that the exact result for the resistivity in the weak scattering limit (first Born approximation) is obtained from formula (7.21) after performing partial summations in higher order expansions with respect to the scattering strength. This partial summation is avoided if we start with a more detailed description of the system by including more moments in the quasi-equilibrium statistical operator ϱ_q.

Similarly, in the Kubo theory where no additional operators are introduced in (7.9) so that the quasi-equilibrium statistical operator is replaced by the equilibrium statistical operator, partial summations have to be performed to obtain correct results for the conductivity. For instance, evaluating the Kubo expressions for the electrical conductivity by a series expansion with respect to the scattering strength would lead to divergent expressions for metallic systems. Also in the case of hopping conductivity it is well known (BÖTTGER and BRYKSIN, 1979) that no correct results are obtained from the Kubo approach without performing partial summations.

7.1.3. Further Approaches

In the framework of this chapter we present a powerful method of the determination of thermoelectric effects which consists of a generalized version of Zubarev's method using nonequilibrium statistical operators. The thermoelectric quantities are represented in terms of correlation functions, which, in turn, may be determined in the framework of Green's functions technique (see Chapters 3., 4.). In simple situations it is possible to take into account only linear deviations from the corresponding equilibrium situations; for this reason Zubarev's method is sometimes referred to as linear response theory.

However, we want to point out that also nonequilibrium processes far from equilibrium may be of high interest for Coulomb systems. In principle, the Zubarev approach (and the kinetic theory as well) is a variational approach which applies also to states far from equilibrium; however, we restrict ourselves here to the usual transport coefficients which are defined in the linear regime. In a more restricted sense, also the Kubo theory is a linear response theory. However, in most cases only the grand canonical equilibrium statistical operator is used which takes into account only the conservation of the energy and the number of particles. If one has to consider only electron-ion scattering, electron-impurity scattering or electron-atom scattering, one gets reasonable results (for e—i-scattering see, e. g., Blümlein, Kraeft and Meyer, 1980; Kraeft, Blümlein and Meyer, 1983). The neglection of the e—e-interaction corresponds to the "Lorentz plasma" model; here exact solutions are reproduced,

In the framework of the restricted Kubo theory the e—e-interaction may not be accounted for. But it may be observed that the e—e-interaction becomes less important in more dense systems. In the framework of the method of the nonequilibrium statistical operator one gets proper results for the included e—e-interaction, too, see the next Section 7.2.

An alternative approach for the determination of thermoelectric quantities is the method of kinetic equations, which is the oldest method from the historical point of view. Starting from Boltzmann's ideas, who firstly formulated a closed equation for the determination of a single particle distribution function ("kinetic equation") there was a vast growth in kinetic theory which leads, for different situations, to the invention of the Landau collision term, to the Vlasov equation, to the Lenard-Balescu collision integral and more complicated kinetic equations, which include in more detail the screening in higher collisions. For recent work see, e.g., Klimontovich (1975) and Ebeling et al. (1983, 1984).

Besides the standard results given by Spitzer (1962) there are many details given in textbooks, e.g., in those by Mitchner and Kruger (1973), Landau and Lifshits, Vol. X (1983), Biberman, Vorob'ev and Yakubov (1982), Suchy (1977). Moreover the reader may find a lot of hints in Ebeling et al. (1983, 1984), and Günter and Radtke (1984).

In kinetic theory, the main tasks are the determination of transport cross sections for the different scattering procedures taken into account and the determination of the distribution functions under the influence of external fields.

In the framework of the Lorentz plasma, the results of the kinetic theory agree with those of the Kubo theory as well as with those of the generalized Zubarev theory. However, also for the general case, the results of the kinetic theory (see, e.g., Ebeling et al., 1984; Kremp and Schlanges, 1983) are in agreement with the results of the generalized Zubarev theory; for this reason it was possible to decide to present in more detail only the latter method.

Both in the kinetic theory and in the generalized Zubarev method, the determination of the thermoelectric quantities is mapped to sets of linear equations, in which the coefficients are the relevant correlation functions.

The situation becomes more complicated, if (instead of a fully ionized plasma) a partially ionized plasma is considered. Now chemical reactions have to be considered and consequently a set of coupled (different) "Boltzmann" equations. Also such an inclusion of bound states is possible in the kinetic theory and in the linear response theory according to Zubarev as well. The calculations lead to the same results (see, e.g., Ebeling et al., 1983, 1984; Klimontovich and Kremp, 1981).

It should be remarked, that modern theories are equivalent and give the same results. Especially in more dense systems when dynamical screening and the "Mott transition" are typical effects, the modern many-particle theory (such as kinetic theory and linear response theory) is a powerful tool which allows, in principle, for the extension up to any order in some approximation scheme. Especially the Green's function method should be mentioned, which is equivalent to the quantum BBGKY hierarchy.

In conclusion of this Subsection we want to give some additional hints to the literature, in theory and experiment as well. First we want to draw the reader's attention to the publication of the biannual ICPIG. Furthermore see GÜNTHER and RADTKE (1984). Further recent publications are those by GRYAZNOV et al. (1980), GILL (1981), and ZHDANOV (1982).

7.2. Evaluation of Collision Integrals Using Green's Functions

7.2.1. Green's Functions, Diagrams and Correlation Functions

As shown in the previous Chapters, the transport coefficients are related to equilibrium correlation functions which, in turn, can be evaluated by using many-body theory. So, the transport coefficients are not solely determined by the binary collision process as in simple kinetic approach but the system reacts as a whole and may be described, e.g., by collective modes. A many-body approach is necessary especially in nonideal plasmas, and the Green's function technique outlined in Chapters 3. and 4. may be directly applied to the evaluation of nonequilibrium properties (RÖPKE, EBELING, and KRAEFT, 1980; KRAEFT, BLÜMLEIN, and MEYER, 1983).

Especially, we have

$$
\begin{aligned}
d_{nm} &= \langle \dot{P}_n(\varepsilon); \dot{P}_m \rangle = \hbar^2 \sum_{k_1 k_2} k_{1,z} k_{2,z} (\beta E_{k_1})^{n/2} (\beta E_{k_2})^{m/2} \langle \dot{n}_{k_1}(\varepsilon); \dot{n}_{k_2} \rangle \\
&= -\int_{-\infty}^{0} \mathrm{d}t \; \mathrm{e}^{\varepsilon t} \int_{0}^{\beta} \mathrm{d}\tau \cdot \frac{1}{\Omega^2} \sum_{kpqk'p'q'} e_k e_p V_q e_{k'} e_{p'} V_{q'} F(k, p, q, k', p', q', t - i\hbar\tau) \\
&\quad \cdot E^{(n)}(k, q) \, E^{(m)}(k', q') , \qquad (7.22) \\
E^{(n)}(k, q) &= \{ k_z (\beta E_k)^{n/2} - (k_z + q_z) (\beta E_{k+q})^{n/2} \} ,
\end{aligned}
$$

where the correlation functions $\langle \dot{n}_{k_1}(\varepsilon); \dot{n}_{k_2} \rangle$ are expressed by Green's functions according to

$$
\begin{aligned}
F(k, p, q, k', p', q', t) &= \mathrm{Tr} \{ \varrho_0 \, \mathrm{e}^{iH_s t/\hbar} \, a^+_{k+q} a^+_{p-q} a_p a_k \, \mathrm{e}^{-iH_s t/\hbar} \, a^+_{k'+q'}, a^+_{p'-q'}, a_{p'}, a_{k'} \} \\
&= -\int_{-\infty}^{\infty} \frac{\mathrm{d}\hbar\omega}{2\pi i} n_{\mathrm{B}}(\omega) \, \mathrm{e}^{i\omega t} [G_4(\omega + i0) - G_4(\omega - i0)]] \qquad (7.23)
\end{aligned}
$$

which contains the Green's function $G_4(z)$; $n_{\mathrm{B}}(\omega) \equiv n(\hbar\omega) = [\exp \hbar\beta\omega - 1]^{-1}$ is the Bose distribution function. In dependence of the term (k or $k + q$) in $E^{(n)}(k, q)$ which is taken in (7.22), the expression (7.18) consists of four summands which can be represented by diagrams of the following type

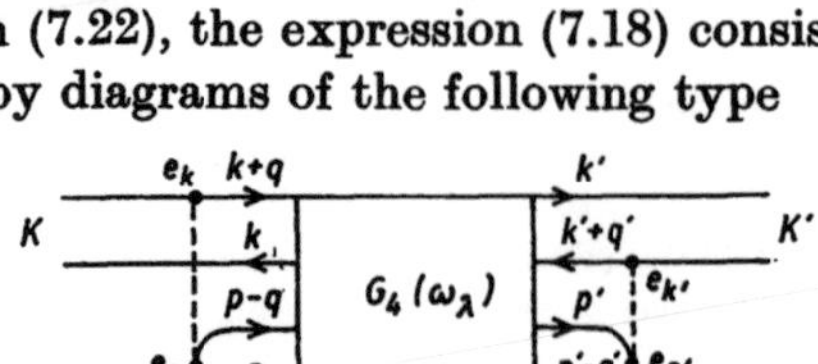

(7.24)

and three further diagrams having endpoints e_k, $e_{k'}$ which are situated on different lines.

Within the polarization approximation (first Born approximation with respect to the scattering system) we have

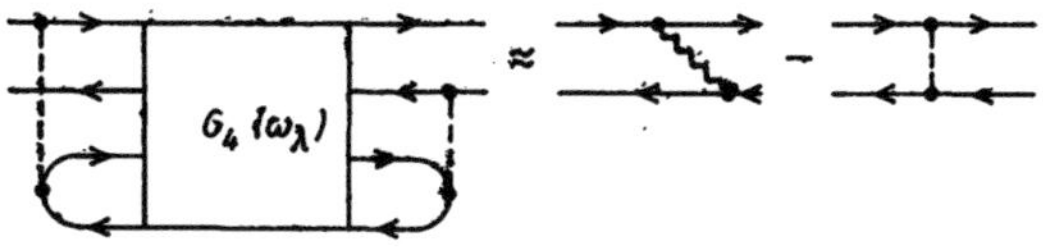

(7.25)

where the wavy line denotes the screened potential $V^s(q, z)$:

$$V^s(q, z) = V_q\left(1 - \int_{-\infty}^{\infty} \frac{d\omega}{\pi} \operatorname{Im} \varepsilon^{-1}(q, \omega + i0) \frac{1}{\omega - z}\right), \tag{7.26}$$

$\varepsilon(q, z)$ being the dielectric function, see also 3.3.1., 4.1.2., 4.2.3., and 4.4.3. The evaluation of the approximation (7.25) yields

$$d_{nm}^{(1)} = \frac{\beta\hbar^2\pi}{\Omega} \sum_{kq} e_k^2 V_q \int d\omega \operatorname{Im} \varepsilon^{-1}(q, \omega + i0) \left(1 + n_B(\omega)\right)$$
$$\cdot f_k(1 - f_{k+q})\, \delta(E_k - E_{k+q} - \hbar\omega)\, E^{(n)}(k, q)\, E^{(m)}(k + q, -q) . \tag{7.27}$$

In order to treat the partially ionized plasma, beyond the introduction of quasiparticle states $|p\rangle$, $|nP\rangle$ to characterize the quasi-equilibrium state which, in turn, yields the coupled Boltzmann equations for the free quasiparticle and bound state distribution functions, we can use the cluster decomposition of the dielectric function as described in Sections 4.1., 4.5. to treat free particles and bound states on the same footing. In the extended random phase approximation we have (4.52), (4.264)

$$\varepsilon(q, z) = 1 - V_q\tilde{\Pi}(q, z) ,$$
$$\Omega\tilde{\Pi}(q, z) = \sum_p e_p^2 \frac{f_p - f_{p+q}}{\varepsilon_p + \hbar z - \varepsilon_{p+q}} + \frac{1}{2} \sum_{nn'P} |\Gamma_{nPn'P+q}|^2 \frac{f_{nP} - f_{n'P+q}}{\varepsilon_{nP} + \hbar z - \varepsilon_{n'P+q}}$$
$$- \sum_{pp'} e_p^2 \frac{f_2(\varepsilon_p + \varepsilon_{p'}) - f_2(\varepsilon_{p+q} + \varepsilon_{p'})}{\varepsilon_p + \hbar z - \varepsilon_{p+q}}, \tag{7.28}$$

where

$$f_{nP} = f_2(\varepsilon_{nP}) = [\exp \beta(\varepsilon_{nP} - \mu_1 - \mu_2) - 1]^{-1} , \qquad V_q = \frac{1}{\varepsilon_0 q^2},$$
$$\Gamma_{nPn'P+q} = \int dk\, \psi_n(k) \left\{e_1\psi_{n'}\left(k + \frac{m_2}{m_1 + m_2} q\right) + e_2\psi_{n'}\left(k - \frac{m_1}{m_1 + m_2} q\right)\right\}; \tag{7.29}$$

ε_p and ε_{nP} are the single particle and two particle energies, respectively; $\psi_{nP}(k)$ is the two particle wave function, see Section 4.4. The collision term obtained from (7.28) will be discussed in Section 7.4. It does not represent the full first Born approximation because exchange contributions between the considered electron and the scattering system are not contained in eq. (7.27). An expression similar to (7.27) which relates the collision integral to the imaginary part of the dielectric function has also been given by KLIMONTOVICH (1978). More general expressions which relate the resistivity to generalized suszeptibilities are given by GOEDSCHE, VOJTA and MÖBIUS (1979).

Notice that the evaluation of the correlation functions and, in this way also that of the transport coefficients were performed within the grand canonical ensemble. It is more convenient to give expressions for the transport coefficients in terms of temperature and density or of pressure, respectively. To achieve this we have to replace the chemical potential by the density n or the pressure p which may be done using the equation of state as derived in Chapter 6., see 6.5.

7.2.2. Evaluation of Correlation Functions in First Born Approximation

The weak scattering limit is obtained by evaluating the expressions for the coefficients L_{ik} (7.19) in the first Born approximation with respect to the screened potential. To this order the Debye-Onsager relaxation term $\langle P_n(\varepsilon); \dot{P}_m\rangle$ vanishes, and we have

$$\bar{N}_{nm} = N_{nm} = N \frac{\Gamma\left(\frac{n+m+5}{2}\right)}{\Gamma(5/2)} \frac{I_{(n+m+1)/2}}{I_{1/2}} \tag{7.30}$$

N being the number of electrons, $I_\nu(z)$ denote the Fermi integrals.

Using the extended RPA expression (7.28) for the dielectric function, we find for the collision term (7.27) decomposition

$$d_{nm} = d_{nm}^{\mathrm{ei}} + d_{nm}^{\mathrm{ee}} + d_{nm}^{\mathrm{ea}}$$

with

$$d_{nm}^{\mathrm{es}} = \frac{2\beta\hbar^2\pi}{\Omega^2} \sum_{kpq} \int \mathrm{d}\omega \left|\frac{V_{\mathrm{es}}(q)}{\varepsilon(q,\omega)}\right|^2 f_k^{\mathrm{e}}(1 - f_{k+q}^{\mathrm{e}})\, f_p^{s}(1 - f_{p-q}^{s})$$
$$\cdot\, \delta(E_k - E_{k+q} - \hbar\omega)\, \delta(E_p - E_{p-q} + \hbar\omega)\, E^{(n)}(k, q)\, E^{(m)}(k+q, -q)\,, \tag{7.31}$$

$s = \mathrm{e}, \mathrm{i}$; and the electron-atom (e—a) contribution

$$d_{nm}^{\mathrm{ea}} = \frac{2\beta\hbar^2\pi}{\Omega^2} \sum_{kq} \int \mathrm{d}\omega\, |V(q,\omega)|^2 f_k(1 - f_{k+q})\, \delta(E_k - E_{k+q} - \omega)\, E^{(n)}(k, q)$$
$$\cdot\, E^{(m)}(k+q, -q) \Big\{ \sum_{nn'P}^{(b)} |\Gamma_{nP,n'P-q}|^2\, g_{nP}(1 + g_{n'P-q})\, \delta(E_{nP} - E_{n'P-q} + \hbar\omega)$$
$$+ \sum_{npP}^{(b)} |\Gamma_{nP,p,q}|^2\, g_{nP}\big(1 + f_2(\varepsilon_p + \varepsilon_{P-q-p})\big)\, \delta(\varepsilon_{nP} - \varepsilon_{P-q-p} - \varepsilon_p + \hbar\omega)$$
$$+ \sum_{N'pP}^{(b)} |\Gamma_{p,q,n'P}|^2 f_2(\varepsilon_p + \varepsilon_{P-p})\, (1 + g_{n'P-q})\, \delta(\varepsilon_p + \varepsilon_{P-p} - \varepsilon_{n'P-q} + \hbar\omega)\Big\} \tag{7.32}$$

where the superscript (b) means that the sum over n is restricted to the bound state part of the energy spectrum. The second term in (7.32) describes ionization processes, and the third one recombination processes. In the first Born approximation, the vertex function $\Gamma_{nP,p,q}$ is given by

$$|\Gamma_{nP,p,q}|^2 = e^2 \left|\psi_n\left(p - \frac{m_{\mathrm{e}}}{m_{\mathrm{e}} + m_{\mathrm{p}}} P\right) - \psi_{n\cdot}\left(p - q - \frac{m_{\mathrm{e}}}{m_{\mathrm{e}} + m_{\mathrm{p}}} P\right)\right|^2 \tag{7.33}$$

if the scattering states are replaced by free particle states $|p\rangle$.

Because of the long-range character of the Coulomb potential, the unscreened Born approximation for the collision term diverges, and we have to take into account screening effects to get convergent results. The use of the dynamically screened potential (7.31) yields the Lenard-Balescu expression for the transport coefficients as will be shown below in Section 7.2.5. In a more simple approximation, the dynamically screened potential $V^s(q, \omega)$ is replaced by the statically screened potential $V^s(q, 0)$, and we have (EBELING, RÖPKE, 1979)

$$V^s(q, 0) = \frac{V_q}{1 + \varkappa^2/q^2}, \qquad \varkappa^2 = \sum_{s=e,i} \varepsilon_0^{-1} n_s^* e_s^2 I_{-1/2}(\alpha_s)/I_{1/2}(\alpha_s) . \tag{7.34}$$

If we furthermore neglect the ion and atom motion (adiabatic approximation) we get for the e—i and e—a scattering processes

$$d_{nm}^{es} = \frac{4}{3} \frac{m_e}{\pi\beta^2\hbar^3} N_s \int_0^\infty dx\, x^{\frac{n+m+4}{2}} f(x)\,(1 - f(x))\, Q_{es}(x) \tag{7.35}$$

with the e—i transport cross section n ($x = \beta E_k$)

$$Q_{ei}(x) = \frac{1}{a_0^2 k^4} \left(\ln\left(1 + \frac{4k^2}{\varkappa^2}\right) - (1 + \varkappa^2/4k^2)^{-1} \right) \tag{7.36}$$

and the e — a transport cross section

$$Q_{ea}(x) = Q_{ea}^{el} + Q_{ea}^{in} ,$$

where Q_{ea}^{el} describes the elastic processes (non-degenerate case),

$$Q_{ea}^{el}(x) = \frac{e^4}{16\pi E_k^2} \int_0^{2k} dq\, q^3\, |V^s_{(q,0)}|^2 \sum_n \Big| \sum_p \psi_n(p + q)\, \psi_n(p) - 1\Big|^2 e^{-\beta E_n} , \tag{7.37}$$

Q_{ea}^{in} contains the inelastic processes as excitation

$$Q_{ea}^{ex}(x) = \frac{e^4}{16\pi E_k^2} \int_0^{2k} dq\, q^3\, |V^s(q, 0)|^2 \sum_{nn'} \Big| \sum_n \psi_n(p + q)\, \psi_{n'}(p) - 1\Big|^2 e^{-\beta E_n} , \tag{7.38}$$

ionization

$$Q_{ea}^{ion}(x) = \frac{e^4}{16\pi E_k^2} \int_0^{2k} dq\, q^3\, |V^s(q, 0)|^2 \sum_{np} |\psi_n(p) - \psi_n(p - q)|^2 e^{-\beta E_n} , \tag{7.39}$$

and recombination

$$Q_{ea}^{rec}(x) = \frac{e^4}{16\pi E_k^2} \int_0^{2k} dq\, q^3\, |V^s(q, 0)|^2 \sum_{n'p} |\psi_{n'}(p) - \psi_{n'}(p + q)|^2 \frac{N_i N_e}{N_a} . \tag{7.40}$$

The e—e correlation function d_{nm}^{ee} can also be related in the non-degenerate case to transport cross sections Q_{nm}^{ee}:

$$d_{nm}^{ee} = \tfrac{8}{3}\, n_e N_e (m_e/\beta)^{1/2} \int_0^\infty dx\, x^3 Q_{nm}^{ee}(x)\, e^{-x} \tag{7.41}$$

with

$$x = \beta\hbar^2 P^2/m_e .$$

Especially, we have

$$Q_{22}^{ee}(x) = \frac{\Omega^2\beta^2e^4}{16\pi x^2}\int_0^{2P} dq\, q^3\left(1-\frac{q^2}{4P^2}\right)V^s(q)\left(V^s(q)-\frac{1}{2}V^s(\sqrt{4P^2-q^2})\right) \quad (7.42)$$

which is obtained by introducing the centre of mass system, $2\hbar P/m_e$ is the relative velocity. The last term in (7.42) describes exchange processes.

7.2.3. Results for a Hydrogen Plasma

Before presenting the results for the transport coefficients in first Born approximation at arbitrary degeneracy we will discuss the limiting cases of non-degeneracy and strong degeneracy. In the non-degenerate limit the main contribution to the e—i correlation functions (7.35) arises from the maximum of the integrand near $x = (n + m)/2$. Then the e—i correlation function is given approximatively by (for a numerical evaluation see KRAEFT, BLÜMLEIN and MEYER, 1983)

$$d_{nm}^{ei} = \Gamma\left(\frac{n+m}{2}+1\right)\frac{4}{3}\sqrt{2\pi}\, N_i n_e m_e^{1/2}\beta^{3/2}\frac{e^4}{(4\pi\varepsilon_0)^2}\,\Phi \quad (7.43)$$

with

$$\Phi = \frac{1}{2}\left\{\ln\left(1+\frac{12m_e}{\beta\hbar^2\varkappa^2}\right)-\left(1+\frac{\beta\hbar^2\varkappa^2}{12m_e}\right)^{-1}\right\}.$$

The e—e correlation function can be approximated in the same manner. The contribution of the e—a scattering d_{nm}^{ea} can be neglected in the low density limit where the ionization degree is nearly 1. Then we find in first Born approximation for the transport coefficients σ (electrical conductivity, see eq. (7.3)), α (thermopower) , and λ_e (electronic part of the heat conductivity)

$$\sigma = f\,\frac{(k_BT)^{3/2}}{e^2m_e^{1/2}}\frac{4\pi\varepsilon_0}{\Phi} = \sigma^*\,\frac{(k_BT)^{3/2}}{e^2m_e^{1/2}}\,, \quad (7.44)$$

$$\alpha = ak_B/e\,, \quad (7.45)$$

$$\lambda_e = L(k_B/e)^2\,T\sigma\,. \quad (7.46)$$

The latter relation (7.46) represents the Wiedemann-Franz law. The prefactors f, a, and L are numbers which depend on the choice of the set of moments in (7.9), see RÖPKE and HÖHNE (1979), HÖHNE et al. (1983). There it is shown that the values for f, a, and L will converge if the set of momenta P_n is extended. The results for the set (P_0, P_1, ... , P_5) in the low density limit are given in Tab. 7.1.

In the Lorentz model where the electron-electron interaction is omitted, the exact results for the transport coefficients $f = 2^{5/2}/\pi^{3/2}$, $a = 1.5$, $L = 4$ are reproduced if the moments P_n with $n = 0, 1, 2, 3, 4, 5$ are used. The e—e scattering processes reduce the magnitude of the transport coefficients at low densities by about 35% (for the thermal conductivity the product fL has to be considered). If only the set $\{P_0, P_2\}$ is taken (particle and heat current), the results for the transport coeffi-

Tab. 7.1. Values of the quantities f, a, and L occuring in (7.44), (7.45), (7.46) for different sets of P_n; in addition the corresponding quantities for the relaxation time approximation are given.

	f		a		L	
Relaxation time result	1.0159		1.5		4.0	
Different sets P_n with n:	without/with e-e contribution		without/with e-e contribution		without/with e-e contribution	
0	0.2992	0.2992	0	0	0	0
0; 2	0.9724	0.5781	1.1538	0.8040	0.5917	0.6935
0; 2; 4	1.0145	0.5808	1.5207	0.9032	3.6781	2.7627
0; 2; 4; 6	1.0157	0.6184	1.5017	1.1316	4.0013	3.8858
0; 2; 4; 6; 8	1.0158	0.6208	1.5004	1.0934	3.9978	3.7628
0; 1	0.8322	0.5738	0.7792	0.5822	0.3409	0.3694
0; 1; 2	1.0077	0.5823	1.3818	0.7204	2.0027	1.6449
0; 1; 2; 3	1.0159	0.6109	1.5	0.9279	3.7055	2.4456
0; 1; 2; 3; 4	1.0159	0.6230	1.5	1.0584	3.9946	3.2592
0; 1; 2; 3; 4; 5	1.0159	0.6234	1.5	1.0781	4.0	3.7446

cients are identical with those obtained by solving the Boltzmann equation using the Grad thirteen moment method (KLIMONTOVICH and EBELING, 1962) or the Kohler variational method (APPEL, 1961).

In the opposite case of strong degeneration, the evaluation of the correlation functions leads to the Ziman formula (E_F, k_F — Fermi energy and momentum, respectively) for the electrical conductivity

$$\sigma = \frac{12\pi^3 e^2 n_e^2 \hbar^3}{m_e^2 \Phi_0}, \qquad \Phi_0 = \int_0^{2k_F} dq\, q^3\, |V_{ei}(q)|^2\,, \tag{7.47}$$

the Mott formula for the thermopower

$$\alpha = \frac{\pi^2 k_B^2}{3eE_F} T \left(3 - E_F \frac{\Phi_0'(E_F)}{\Phi_0(E_F)}\right), \tag{7.48}$$

and for the heat conductivity to the Wiedemann-Franz relation

$$\lambda_e = \frac{\pi^2 k_B^2}{3e^2} T\sigma \tag{7.49}$$

with the Lorenz number $L = \pi^2 k_B^2/3e^2$. Such expressions for the transport coefficients are common in metal physics (ZIMAN 1961).

Because of the sharp Fermi surface, the e—e scattering processes are quenched in the strong degeneration limit due to the Pauli exclusion principle. As pointed out by HUBERMAN and CHESTER (1975), the correct result for the electrical conductivity is obtained in the case of strong degeneration if only P_0 is taken into account. Generally spoken, the e—e interaction becomes more essential in the low density limit.

Evaluating the correlation functions in first Born approximation, the results for the transport coefficients for arbitrary degeneracy are shown in Figs. 7.1—7.3, if only the momenta P_0, P_2 are taken. The degree of ionization was determined by a simple mass action law which takes into account only the Debye shift of the free particles, see

also 6.5.3. and EBELING, KRAEFT, KREMP (1976). The thermal and electrical conductivities are reduced if the formation of bound states is taken into account. This effect is large in the region of small degree of ionization, i.e. for densities well below the Mott density and for low temperatures, and a minimum for σ and λ_e may occur. Similar results were obtained by KREMP and SCHLANGES (1982), KREMP, SCHLANGES and KILIMANN (1984), and SCHLANGES, KREMP and KRAEFT (1985).
Notice that the influence of bound state scattering on the thermopower yields a change of the sign for low temperatures.

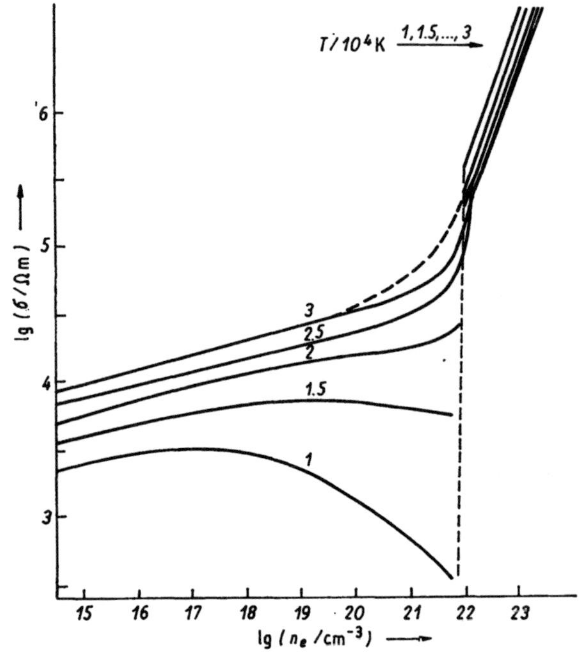

Fig. 7.1. Electrical conductivity σ in statically screened first Born approximation versus the total electron density n_e for different temperatures $T/10^4$ K. Broken lines for $T/10^4$ K $= 1$, 1.5, 2 denote the region of instability. The dotted line for $T/10^4$ K $= 3$ indicates the electrical conductivity for the model of the fully ionized plasma, neglecting the e—a scattering contribution

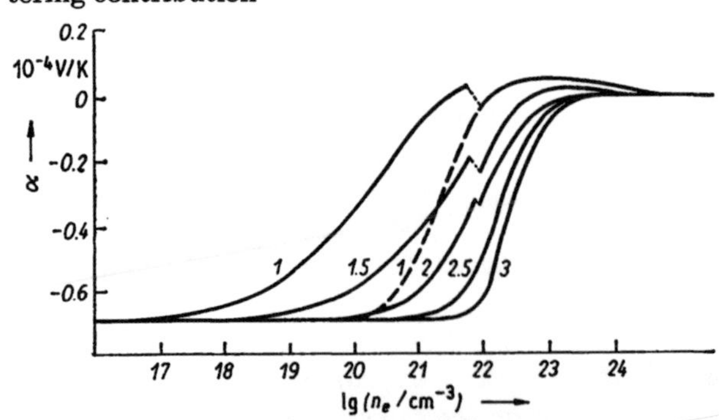

Fig. 7.2. Thermopower α in statically screened first Born approximation versus the total electron density n_e for different temperatures $T/10^4$ K. Broken lines denote the region of instability. For $T/10^4$ K $= 1$ is shown additionally the case of fully ionized plasma (broken line)

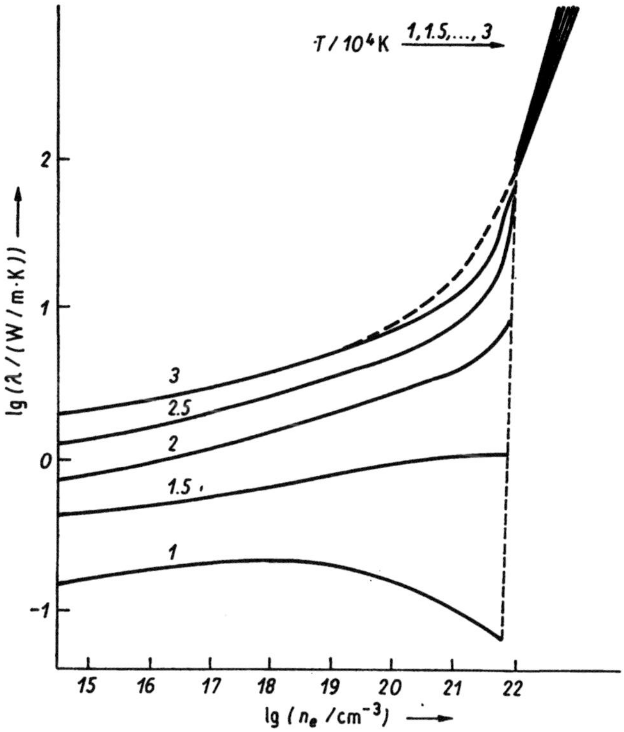

Fig. 7.3. Electronic part of the heat conductivity λ in statically screened first Born approximation versus the total electron density n_e for different temperatures $T/10^4$ K. Broken lines for $T/10^4$ K = 1, 1.5, 2 denote the region of instability. The broken line for $T/10^4$ K = 3 indicates the electronical part of the heat conductivity for the model of fully ionized plasma, neglecting the e—a scattering

In Figs. 7.1 — 7.3, only the scattering on the ground state was considered. The influence of the excited states, of recombination and ionization on the correlation function d_{00} is shown in Fig. 7.4. It can be seen that the influence of these scattering effects is negligible in comparison with the elastic scattering at the 1s ground state for the temperatures considered here.

The Mott effect (pressure ionization) leads in the high density region to the destruction of bound states (see also 4.4. and EBELING, KRAEFT, KREMP, 1976). Then, the e—i scattering mechanism will dominate, and the formulae for strong degeneracy (7.47) — (7.49) become valid. At low temperatures, a phase transition from the plasma to a liquid metal may occur. Of course, the evaluation of the correlation functions in first Born approximation as given here is not valid near the phase transition region and near the Mott transition. According to Section 4.4. we assumed that the bound states are not influenced very strongly by the surrounding plasma. However, near the Mott transition a more detailed description of bound states and the continuum of scattering states is necessary.

Let us repeat here once again what we are understanding in this book under the term "bound states". Taking into account the specificity of the Coulomb interaction which possesses an infinite number of discrete states we take as bound by definition all the discrete eigenstates of the electron-ion relative pair energy with a binding energy larger

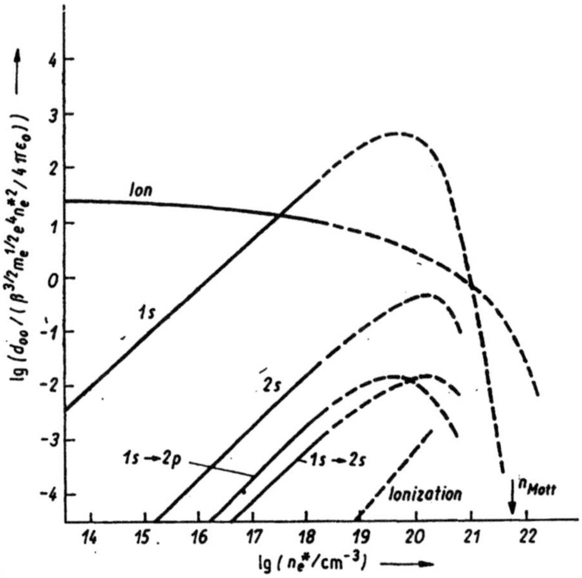

Fig. 7.4. Contributions to the correlation function d_{00} in first Born approximation versus the free electron density n_e for $T = 10^4$ K containing e—i scattering, elastic e—a scattering on the 1s- and 2s state, inelastic e—a scattering for $1s-2s$ and $1s-2p$ transitions and ionization. The region of instability below the Mott density is indicated by broken lines

a finite border, given by the thermal energy

$$E_n < \varepsilon^* = \zeta k_B T\,.$$

The free number ζ is arbitrary in certain limits, therefore we assume here simply $\zeta = 1$. The discrete states which are above this border level ε^* are considered as free due to the large extent of their wave functions and the instability of these states with respect to thermal collisions. This is an intrinsic difficulty connected with Coulombic systems. Fortunately the division of the discrete states of the relative pair energy into bound states with

$$-\infty < E_n < \varepsilon^*$$

and quasi-free states with

$$\varepsilon^* \leqq E_n < 0$$

which are to be treated together with the continuum (extended states) is necessary only in very dilute plasmas. With increasing density the number of quasi-free states is decreasing due to the lowering of the averaged continuum edge (see Sections 4.4., 6.5., 6.6.). In a first approximation the averaged continuum edge is given by the exchange correlation part of the chemical potential

$$\Delta(n, T) \approx \mu_{xc}(n, T)$$

which is an increasing function of the plasma density. Above a certain density $n_Q(T)$ (border density of quasi-free states) quasi-free states do not exist anymore. At a second critical density $n_M(T)$ all the bound states disappear due to the Mott effect. In this way at a given temperature T we have to differ between three regions of density (Figs. 6.14 and 6.18):

(i) small density region: $n < n_Q(T)$.
Here discrete bound states and discrete quasi-free states exist; the latters have to be treated together with the continuum.

(ii) intermediate density region: $n_Q(T) < n < n_M(T)$.
Here we find in the plasma only discrete bound states.
(iii) high density region: $n > n_M(T)$.
There exist only continuum states of the energy of internal motion.

7.2.4. Inclusion of the Ionic Structure Factor

In the previous Section, the first Born approximation with respect to the screened interaction potential was considered in the evaluation of the correlation functions which describe the collision integral. In dense systems like liquid metals, the equilibrium correlations within the system of scatterers are of importance for the transport coefficients so that the Born approximation is considered with respect to the screened interaction between the electrons and the whole subsystem of scatterers. For instance, in adiabatic approximation the equilibrium correlations within the ion subsystem are given by the static structure factor, whereas collective motions of the ion subsystem such as phonons are contained in the dynamic structure factor $S(q, \omega)$.

Structure factor effects are immediately obtained within the linear response theory by evaluating the equilibrium correlation functions. Denoting with $R_i(t)$ the ion position at time t, the electron-ion interaction is described by

$$V_{ei}(t) = \sum_n^{N_i} v_{ei}(\boldsymbol{r} - \boldsymbol{R}_n(t)) , \tag{7.50}$$

and the correlation functions can be written in the form

$$d_{nm}^{ei} = \frac{2\pi\hbar}{3} \left(\frac{\hbar^2}{mk_BT}\right)^{\frac{n+m}{2}} \frac{1}{\Omega^2} \sum_{kq} \int d(\hbar\omega) \left(-\frac{df}{dE_k}\right) \delta(E_{k+q} - E_k - \hbar\omega)\, k^{n+m}\, q^2 N \cdot S(q, \omega)\, |v_{ei}^s(q)|^2 \tag{7.51}$$

with the dynamical structure factor (see, e.g., PINES and NOZIÈRES, 1958)

$$S(q, \omega) = N_i^{-1} \int dt\, e^{-i\omega t} \langle \sum_{nn'} \exp\left(i\boldsymbol{q}\, (\boldsymbol{R}_n(0) - \boldsymbol{R}_{n'}(t)\right)\rangle ; \tag{7.52}$$

$v_{ei}(q)$ is the Fourier transform of the electron-ion interaction

$$v_{ei}(q) = \int \boldsymbol{dr}\, e^{i\boldsymbol{qr}}\, v_{ei}(r) .$$

Expressions for the transport coefficients in terms of the dynamical structure factor are given by BAYM (1964). In the static case where the ion positions are fixed we have $S(q, \omega) = S(q)\, \delta(\omega)$, and for the conductivity follows

$$\sigma(T = 0) = \frac{12\pi^3 e^2 n^2 \hbar^3}{m_e^2} \left\{ \int_0^{2k_F} dq\, q^3\, |v_{ei}^s(q)|^2\, S(q) \right\}^{-1} . \tag{7.53}$$

This expression was given by ZIMAN (1974) and is widely applied in the theory of liquid metals. The phonon dynamics can be evaluated from the dynamic structure factor.

The static structure factor $S(q)$ is related to the ion pair distribution function $g_{ii}(r)$ according to

$$S(q) = 1 + \int n_i\, g_{ii}(r)\, e^{i\boldsymbol{qr}}\, \boldsymbol{dr} . \tag{7.54}$$

The ion pair distribution function is determined by the ion-ion interaction. In the case of a pure Coulomb interaction between the ions we may use (for $\varkappa e^2/k_BT < 1$) the well known nonlinear Debye expression

$$g_{ii}(r) = \exp\left(-\frac{e^2}{4\pi\varepsilon_0 r k_B T} e^{-\varkappa r}\right). \tag{7.55}$$

In the case that there is additionally a hard core contribution to the ion-ion interaction (hard-core radius R_i), an expression for $g_{ii}(r)$ can be found in the mean spherical approximation or, neglecting the long-range interaction, by the Percus-Yevick solution for uncharged hard spheres. Results for the electrical conductivity of a hydrogen plasma taking into account the ion structure factor in Debye approximation and the conductivity of a cesium plasma with a structure factor in mean spherical approximation (MSA) are given by MEISTER and RÖPKE (1982), see Fig. 7.5. For cesium, the ion-ion interaction was described by the Coulomb interaction which was cut off at the crystallografic hard-core radius $R_i^{Cs} = 0.169$ nm. Structure factor effects change the conductivity by more than 20% for parameter values $\gamma > 2.3$ at $T = 10^4$ K, and for $\gamma > 25$ at $T = 303.15$ K, $\gamma = e^2/(4\pi\varepsilon_0 k_B T r_D)$.

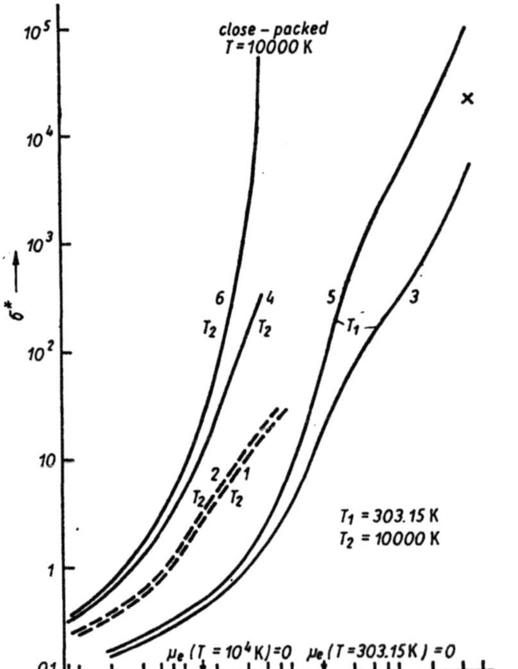

Fig. 7.5. Influence of the structure factor $S(q)$ on the reduced conductivity $\sigma^*(\gamma)$. Dashed lines — Coulomb potential: 1 — $S(q) \equiv 1$; 2 — $S(q)$ in Debye approximation; full lines — Hellmann potential: 3, 4 — $S(q) = 1$; 5, 6 — $S(q)$ in MSA (cf. MEISTER and RÖPKE, (1982); × — experimental value for Cs at the melting point (GINGRICH and ROY HEATON, 1961)

The incorporation of the dynamic structure factor into the theory of transport properties in dense Coulomb systems is widely used in the theory of liquid metals, amorphous and crystalline solids. The concept of dynamic structure factors was further evaluated in the theory of liquid metals, see, e.g., KOVALENKO and KRASNY in: EBELING et al. (1983, 1984).

7.2.5. Dynamically Screened Second Born Approximation

The dynamically screened second Born approximation is given by contributions to $G_4(\omega_\lambda)$, eq. (7.24), of the following type:

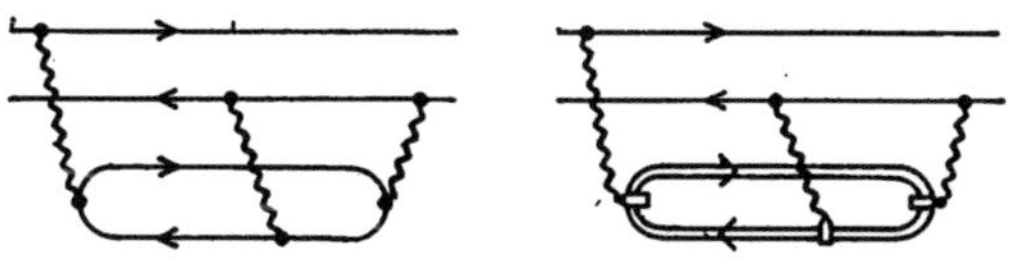

(7.56)

The first diagram describes the scattering by free particles, the second one the scattering by two-particle complexes. The evaluation of the collision term for the scattering by free particles yields in the second Born approximation (RÖPKE, 1983)

$$\langle \dot{n}_{k_1}(\varepsilon);\, \dot{n}_{k_2}\rangle^{(2)}_{\text{free}} = \frac{4\pi\beta}{\hbar\Omega^2} \sum_{kk'pp'} g(E_k + E_p)\, \delta(E_k + E_p - E_{k'} - E_{p'})$$

$$\cdot\, V^{(1)}(p, k, p', k')\, V^{(2)}(p', k', p, k)\, \delta_{k_1,k}(\delta_{k_2,k'} - \delta_{k_2,k_1})\, \delta_{p+k,p'+k'} \tag{7.57}$$

with

$$V^{(1)}(p, k, p', k') = e_p e_k V(\mathbf{k}' - \mathbf{k}) \left[1 + \int \frac{d\omega}{\pi} \operatorname{Im} \varepsilon^{-1}(k' - k, \omega^+) \frac{1}{\hbar\omega + E_{k'} - E_k}\right],$$

$$\Omega V^{(2)}(p, k, p', k') = e_p^2 e_k^2 \sum_{p''k''} \frac{\delta_{p+k,p''+k''}}{E_p + E_k - E_{p''} - E_{k''}} V(\mathbf{k} - \mathbf{k}'')\, V(\mathbf{k}'' - \mathbf{k}')$$

$$\cdot \left[1 - \int \frac{d\omega}{\pi} \operatorname{Im} \varepsilon^{-1}(k - k'', \omega^+) \left\{\frac{n_B(\omega)}{\hbar\omega + E_k - E_{k''}} + \frac{1 + n_B(\omega)}{E_a - \hbar\omega - E_{p''}}\right\}\right.$$

$$\cdot - \int \frac{d\omega}{\pi} \operatorname{Im} \varepsilon^{-1}(k'' - k', \omega^+) \left\{\frac{n_B(\omega)}{E_{k''} + \hbar\omega - E_{k'}} + \frac{1 + n_B(\omega)}{E_{p''} - \hbar\omega - E_{p'}}\right\}$$

$$+ \int \frac{d\omega}{\pi} \int \frac{d\omega'}{\pi} \operatorname{Im} \varepsilon^{-1}(k - k'', \omega^+) \operatorname{Im} (\varepsilon^{-1}k'' - k', \omega'^+) \frac{1}{\hbar\omega' + \hbar\omega - E_k + E_{k'}}$$

$$\left.\cdot \left\{\frac{1 + n_B(\omega)}{E_k - \hbar\omega - E_{k''}} + \frac{1 + n_B(\omega)}{E_p + \hbar\omega - E_{p''}} + \frac{n_B(\omega')}{E_{k'} + \hbar\omega' - E_{k''}} + \frac{n_B(\omega')}{E_{p'} - \hbar\omega' - E_{p''}}\right\}\right]. \tag{7.58}$$

In deriving this expression, only the lowest order terms with respect to the density were taken into account, i.e. $f(k) \ll 1$. Furthermore, contributions to the second Born approximation were dropped which contain, instead of $\delta(E_k + E_p - E_{k'} - E_{p'})$, an expression of the form $\delta(E_k - E_{k'} - \hbar\omega)$.

As shown in Section 4.4., the dynamically screened potential $V^s(q, z)$ may be replaced by an effective one-frequency potential with respect to the particle-particle channel (the kernel of the Bethe-Salpeter equation)

$$V(k, p, k', p', \Omega) = e_k e_p V(\boldsymbol{k}' - \boldsymbol{k}) \left[1 - \int_{-\infty}^{\infty} \frac{d\omega}{\pi} \operatorname{Im} \varepsilon^{-1}(k' - k, \omega^+)\right.$$

$$\left.\cdot \left\{\frac{n_B(\omega)}{\hbar\Omega + \hbar\omega - E_{p'} - E_k} + \frac{1 + n_B(\omega)}{\hbar\Omega - \hbar\omega - E_{k'} - E_p}\right\}\right] \delta_{k+p, k'+p'}, \tag{7.59}$$

which gives for $\hbar\Omega = E_p + E_k = E_{p'} + E_{k'}$ the effective dynamically screened potential $V^{(1)}(p, k, p', k')$. Within the Shindo approximation, this effective one-frequency potential $V^{(1)}$ may be used to construct higher Born approximations or a screened T-matrix approximation. However, the use of this potential does not give the exact dynamics of the multiple scattering processes. So, the second Born approximation $V^{(2)}(p, k, p', k')$ is not immediately obtained from $V^{(1)}$.

In the full second Born approximation, the collision term d_{nm}^{ei} reads

$$d_{nm}^{ei} = \frac{2\pi\beta\hbar}{\Omega^2} \sum_{kpq} f_2(E_k + E_p)\, \delta(E_k + E_p - E_{k+q} - E_{p-q})\, E^{(n)}(k, q)$$

$$\cdot E^{(m)}(k + q, -q)\, |V^{(1)}(p, k, \boldsymbol{p} - \boldsymbol{q}, \boldsymbol{k} + \boldsymbol{q}) + V^{(2)}(p, k, \boldsymbol{p} - \boldsymbol{q}, \boldsymbol{k} + \boldsymbol{q})|^2 . \tag{7.60}$$

Similarly, for the scattering by bound states we find in second Born approximation

$$\langle \dot{n}_{k_1}; \dot{n}_{k_2} \rangle_{ea}^{(2)} = \frac{4\pi\beta}{\hbar\Omega^2} \sum_{kk'nn'PP'} f_3(E_k + E_{nP})\, \delta(E_k + E_{nP} - E_{k'} - E_{n'P'})$$

$$\cdot V^{(1)}(k, n, P, k', n', P')\, V^{(2)}(k', n', P', k, n, P)\, \delta_{k_1,k}(\delta_{k_2,k'} - \delta_{k_2,k_1})\, \delta_{k+P,k'+P'},$$

with

$$V^{(1)}(k, n, P, k', n', P')$$

$$= e_k \Gamma_{nPn'P'} V(\boldsymbol{k}' - \boldsymbol{k}) \left[1 + \int \frac{d\omega}{\pi} \operatorname{Im} \varepsilon^{-1}(k' - k, \omega^+) \frac{1}{\hbar\omega + E_{k'} - E_k}\right] \tag{7.61}$$

and $V^{(2)}$ (k, n, P, k', n', P') has the same form as $V^{(2)}(k, p, k', p')$ but p replaced by n, P; p' by n', P'; p'' by n'', P'', and e_p^2 by $M_{nPn''P''} M_{n''P''n'P'}$ (Röpke, 1983). The evaluation of the third and further Born approximations will give rather involved expressions. It is a consequence of the neutrality of bound states in a Coulomb system that the main part to the interaction between a charged particle and a neutral atom is caused by the polarization of the atom. This effect is not contained in the first Born approximation, and it is necessary to treat the second Born approximation to incorporate the effect of the polarization potential.

The collision term d_{nm}^{ea} reads in full second Born approximation

$$d_{nm}^{ea} = 2\pi\beta \frac{\hbar}{\Omega} \sum_{nn'kPq} f_3(E_k + E_{nP})\, \delta(E_k + E_{nP} - E_{k+q} - E_{n'P-q})$$

$$\cdot E^{(n)}(k, q)\, E^{(m)}(k + q, -q)\, |V^{(1)}(k, n, P, \boldsymbol{k} + \boldsymbol{q}, n', \boldsymbol{P} - \boldsymbol{q})$$

$$+ V^{(2)}(k, n, P, \boldsymbol{k} + \boldsymbol{q}, n', \boldsymbol{P} - \boldsymbol{q})|^2 . \tag{7.62}$$

These expressions (7.61), (7.62) may be considered as generalization of the Lenard-Balescu collision term (Landau and Lifshits, 1981) which describes only the first order Born approximation.

The evaluation of the dynamically screened second Born approximation is rather involved, and we give some results in static approximation. Neglecting the energy transfer $E_k - E_{k+q}$ in the denominators of (7.59), (7.61), according to (4.77) the statically screened potential $V^s(q) = V(q)/(l + \varkappa^2/q^2)$ occurs,

$$V^{(1)}(p, k, p - q, k + q) = e_p e_k V^s(q) ,$$
$$\Omega V^{(2)}(p, k, p - q, k + q) = e_p^2 e_k^2 \sum_{q'} \frac{V^s(-q') \, V^s(q' - q)}{E_p + E_k - E_{k+q'} - E_{p-q'}} , \tag{7.63}$$

if only first order terms in the free particle density n_e are considered. In fact, this approximation which is often used is only justified if the energy transfer is small compared with the plasma frequency $\omega_{pl} = (ne^2/\varepsilon_0 m_e)^{1/2}$.

The resulting expressions for the correlation functions can be written in terms of the transport cross section. In the non-degenerate limit, we have according to NEWTON (1966) in adiabatic approximation

$$Q^{ei(2)}(x) = \frac{\beta^2 e^4}{32\pi\varepsilon_0^2 x^2} \left\{ \ln(1 + B) - \frac{B}{1 + B} - \frac{2}{\lambda(\frac{3}{4}B + 1)} \left[4\left(1 + \frac{1}{B}\right) \right. \right.$$
$$\left. \left. \cdot \ln\left(1 + \frac{B}{4(1 + 1/B)}\right) - \ln(1 + B) \right] + \frac{\beta m_e e^4}{8\pi^2 \varepsilon_0^2 \hbar^2} \frac{1}{Bx} \left[\ln\left(1 + \frac{B}{4(1 + 1/B)}\right) - \frac{1}{(1 + 2/B)^2} \right] \right\} \tag{7.64}$$

with the abbreviation $\lambda = \varkappa a_B$, $x = \beta E_k$, $B = 8m_e x/\beta\hbar^2\varkappa^2$. A similar expression can be found for the e—e contribution by introducing a centre of mass system.

In the electron-atom contributions d_{nm}^{ea} we also neglect the energy transfer $E_{k+q} - E_k$ in comparison with ω, so that

$$V^{(1)}(k, n, P, k', n', P - q) = e_k M_{nPn'P-q} \, V^s(q) . \tag{7.65}$$

In the second Born term $V^{(2)}(k, n, P, k', n', P')$, the n''-sum is divided into the term $n'' = n$ and the remaining sum over $n'' \neq n$. The energy differences $E_n - E_{n''}$, $n'' \neq n$, are considered to be large as compared with the relevant values ω_{pl} for ω and $k_B T$ for E_k. Then we arrive at

$$V^{(2)}(k, n, P, k', n', P - q) = \sum_{n'' \neq n, q'} \frac{e_k^2}{4\varepsilon_0 \Omega} \frac{M_{nPn''P-q'} \, M_{n''P-q'n'P-q}}{E_n - E_{n''}}$$
$$\cdot \{V_{-q'} + V^s(-q')\} \{V_{q'-q} + V^s(q' - q)\} = V_p^s(q) . \tag{7.66}$$

This approximation for dynamical screening effects can be interpreted as follows. If the energy differences of the virtual transitions in the atom are large compared with the plasma frequency, the plasma cannot follow these fluctuations and the corresponding contribution is unscreened. In contrast, the small angle scattering is connected with small energy transfer so that the plasma follows these fluctuations quasistatically.

The Fourier transform $V_p^s(r)$ of the polarization potential $V_p^s(q)$ (7.66) can be given in form of a Padé approximation

$$V_p^s(r) = -\frac{e^2 \alpha_p^{(1s)}}{(r^2 + r_0^2)^2} \frac{1}{32\pi\varepsilon_0} \{1 + e^{-\varkappa r}(1 + \varkappa r)\}^2 \tag{7.67}$$

which describes correctly the behaviour for $r \to \infty$ and $r = 0$. The ground state is considered having the polarizability

$$\alpha_{\mathrm{p}}^{(1s)} = -\frac{2e_{\mathrm{p}}^2}{4\pi\varepsilon_0} \sum_{n'' \neq 1s} \frac{|\langle 1s| z |n''\rangle|^2}{E_{1s} - E_{n''}}, \tag{7.68}$$

and the value $r_0 = a_{\mathrm{B}}$ is obtained from

$$V_{\mathrm{p}}^{\mathrm{s}}(r = 0) = \sum_{n'' \neq n} \frac{e_p^2 e_k^2}{E_n - E_{n''}} \left\{ \int_0^\infty \mathrm{d}r\, r\, R_{no}(r)\, R_{nn''}(r)\, \mathrm{e}^{-\varkappa r} \right\}^2 \tag{7.69}$$

using hydrogen wave functions.

Before presenting some explicit results for the transport coefficients we will discuss the influence of the higher Born approximations in the next Section. Because of the neutrality of the atom, not the first Born approximation but the second Born approximation gives the main contribution to the scattering of charged particles by neutral atoms. The minimum in the electrical as well as thermal conductivity and the maximum in the thermopower is more pronounced for a partially ionized plasma, see Figs. 7.1—7.3, if the second Born approximation is taken into account (Höhne et al., 1984; Ebeling et al., 1983, 1984).

7.2.6. Statically Sreened T-Matrix Approximation. Results

The correlation functions which represent the collision integral may be evaluated in Born approximation with respect to the dynamically screened potential as shown in the previous Section. Results beyond the second Born approximation are very involved. Instead of summing ring diagrams which leads to the dynamically screened potential, we can perform the summation of ladder diagrams so that T-matrices replace the interaction potential. Different possibilities in performing partial summations of diagrams to evaluate the correlation function $\langle \dot{P}_0(\varepsilon); P_0 \rangle$ can be considered to yield the dynamically screened potential (Lenard-Balescu collision Integral) or the T-matrix expressions (Boltzmann collision integral), and further improvements are discussed considering further diagram classes, see Röpke, Ebeling, and Kraeft (1980). The T-matrix for a Debye potential was also used by Meister and Röpke (1982) and that for a cut-off potential by Kraeft, Luft and Mihajlov (1983). Another approach to include multiple scattering effects is given by the method of Kubo-Greenwood, see, f. i., Goedsche, Richter and Vojta (1979), Röpke and Höhne (1981).

In order to evaluate electron-ion and electron-electron correlation functions including multiple scattering, we use different approximations which can be justified by the diagram technique. For the evaluation of the electron-ion correlation functions in higher order of scattering effects we consider the problem of scattering of electrons by a statically screened electron-ion potential $V_{\mathrm{ie}}^{\mathrm{s}}$. This corresponds to the replacement of H_{s} by H_{ei}:

$$H_{\mathrm{ei}} = \sum_k \frac{\hbar^2 k^2}{2m_{\mathrm{e}}} a_k^+ a_k + \sum_{kq} V_{\mathrm{ei}}^{\mathrm{s}}(q)\, a_{k+q}^+ a_k \tag{7.70}$$

(the electron-electron interaction $V_{ee}(q)$ is taken into account only by the statically screened electron-ion potential V^s_{ie}). We obtain

$$\langle \dot{P};\, \dot{P}\rangle = \frac{2\pi\beta\hbar}{3} \int_{-\infty}^{\infty} \mathrm{d}E\, f(E)\,(1 - f(E)) \cdot \mathrm{Tr}\,\{\delta(E - H_{ei})$$

$$\times\, \dot{P}^{(ei)}_{n'}\, \delta(E - H_{ei})\, P^{*(ei)}_{n}\}\,. \tag{7.71}$$

Introducing the electron-ion T-matrix

$$T^s_{ei} = V^s_{ei} + V^s_{ei}(E - H_{ei} + i\varepsilon)^{-1}\, V^s_{ei} \tag{7.72}$$

($\varepsilon \to +0$) and using the optical theorem it follows from (7.71)

$$\langle \dot{P}^{(ei)}_{n};\, \dot{P}^{(ei)}_{n'}\rangle = \frac{2\pi\beta\hbar}{3}\left(\frac{\hbar\beta}{m_e}\right)^{\frac{n+n'}{2}} \sum_{kk'}\left(-\frac{\mathrm{d}f(E)}{\mathrm{d}E}\right)\delta(E_k - E_{k'})$$

$$\cdot\, k^{n+n'}(\boldsymbol{k} - \boldsymbol{k}')^2\, |T^{ei}_s(k' - k')|^2\,. \tag{7.73}$$

For further calculation we express the many-particle electron-ion T-matrix $T^s_{ei}(q)$ in the single-site approximation by a two-particle one $t^s_{ei}(q)$ (Evans et al., 1973)

$$|T^s_{ei}(q)|^2 \approx NS(q)\, |t^s_{ei}(q)|^2\,, \tag{7.74}$$

where $S(q)$ is the structure factor of the ions defined by (7.51). This approximation is exact in the second order of the electron-ion potential and has been used, e.g., in the theory of conductivity of liquid metals and alloys, see Busch and Güntherodt (1974), Faber (1972), Mott and Davis (1971, 1979).

On the other hand, in order to take into account multiple scattering effects between two electrons in evaluating the electron-electron correlation functions, we include all orders of the statically screened electron-electron interaction $V^s_{ee}(q)$ and neglect the electron-ion interaction $V^s_{ei}(q)$. Replacing H_s by H_{ee},

$$H_{ee} = \sum_k \frac{\hbar^2 k^2}{2m_e} a^+_k a_k + \frac{1}{2}\sum_{kk'q} V^s_{ee}(q)\, a^+_{k+q} a^+_{k'-q} a_{k'} a_k\,, \tag{7.75}$$

we obtain, e.g.,

$$\langle \dot{P}^{(ee)}_2; \dot{P}^{(ee)}_2\rangle = \int_{-\infty}^{\infty} \mathrm{d}E \int_{-\infty}^{\infty} \mathrm{d}E' \int_{-\infty}^{0} \mathrm{d}t\, \mathrm{e}^{\varepsilon t} \int_0^{\beta} \mathrm{d}\lambda\, \frac{1}{z} \exp\,(-\beta E)$$

$$\cdot\, \exp\left\{\frac{i}{\hbar}(t - i\hbar\lambda)\,(E - E')\right\} \mathrm{Tr}\,\{\delta(E - H_{ee})\, \dot{P}^{(ee)}_2\, \delta(E' - H_{ee})\, \dot{P}^{(ee)}_2\}\,. \tag{7.76}$$

Introducing the electron-electron T-matrix

$$T^s_{ee} = V^s_{ee} + V^s_{ee}(E - H_{ee} + i\varepsilon)^{-1}\, V^s_{ee} \qquad (\varepsilon \to +0) \tag{7.77}$$

and using the optical theorem, we have ($H_0 = \sum_k E(k)\, a^+_k a_k$)

$$\langle \dot{P}^{(ee)}_2;\, \dot{P}^{(ee)}_2\rangle = -\frac{1}{\hbar^2}\int_{-\infty}^{0} \mathrm{d}t\, \mathrm{e}^{\varepsilon t} \int_0^{\beta} \mathrm{d}\lambda\, \mathrm{Tr}\,\{Z^{-1}\, \mathrm{e}^{-\beta H^0}\, \mathrm{e}^{i(t - i\hbar\lambda)H^0/\hbar}$$

$$\cdot\, [T^-_{ee}, \dot{P}^{(ee)}_2]\, \mathrm{e}^{-i(t - i\hbar\lambda)H_0/\hbar}\, [T^+_{ee}, \dot{P}^{(ee)}_2]\}\,. \tag{7.78}$$

Taking into account only binary electron-electron collisions (screened ladder T-matrix approximation)

$$T^s_{ee} = \sum_{kk'q} T^s_{ee}(k, k'; k+q, k'-q)\, a^+_{k+q}\, a^+_{k'-q}\, a_{k'}\, a_k \tag{7.79}$$

with

$$T^s_{ee}(k, k'; k+q, k'-q) = V^s_{ee}(q) + \sum_{q'} V^s_{ee}(q')\,(E - E(k+q') - E(k'-q') + i\varepsilon)^{-1}$$

$$\cdot\, T\, t^s_{ee}(k+q', k'-q'; k+q, k'-q) \tag{7.80}$$

we find

$$\langle \dot{P}^{(ee)}_2; \dot{P}^{(ee)}_2 \rangle = \frac{2\pi\beta^3\hbar}{3m_e^2} \sum_{kk'q} f(k+q)\, f(k'-q)\,(1 - f(k))\,(1 - f(k'))$$

$$\cdot\, \delta(E(k+q) + E(k'-q) - E(k) - E(k'))$$

$$\cdot\, \{|T^s_{ee}(k, k'; k+q, k'-q)|^2 - T^s_{ee}(k, k'; k+q, k'-q)\, T^s_{ee}(k; k'; k'-q, k+q)\}$$

$$\cdot\, \{(\boldsymbol{k}+\boldsymbol{q})\,(\boldsymbol{k}+\boldsymbol{q})^2 + (\boldsymbol{k}'-\boldsymbol{q})\,(\boldsymbol{k}'-\boldsymbol{q})^2 - \boldsymbol{k}\boldsymbol{k}'^2 - \boldsymbol{k}'\boldsymbol{k}'^2\}^2\,. \tag{7.81}$$

In the case of elastic scattering the T-matrices T^s_{ei}, T^s_{ee} are connected with the scattering phase shifts $\delta_l(E)$ by the relation

$$T^s_{ab}(k - k') = -\frac{2\pi\hbar^2}{\Omega m_{ab}} \sum_{l=0}^{\infty} (2l + 1) P_l(\cos\vartheta)\, \frac{1}{k}\, e^{i\delta l(E)} \sin \delta_l(E) \tag{7.82}$$

(see Newton, 1966). Here $E = \hbar^2k^2/2m_{ab}$ is the energy of relative motion corresponding to the reduced mass $m_{ab} = m_a m_b (m_a + m_b)^{-1}$, l is the angular momentum quantum number and ϑ the scattering angle.

Returning to the calculation of the electron-ion correlation functions we substitute (7.75), (7.82) into (7.81) and perform the integrals over the angular variables. For arbitrary degeneration and putting $S(q) = 1$, we obtain

$$\langle \dot{P}^{(ei)}_n; \dot{P}^{(ei)}_{n'} \rangle = \frac{4N\hbar}{3\pi} \left(\frac{\beta\hbar^2}{m_e}\right)^{\frac{n+n'+2}{2}} \int_0^{\infty} dk\, k^{3+n+n'} f(k)\,(1 - f(k))$$

$$\cdot \sum_{l=0}^{\infty} (l+1) \sin^2 \{\delta^{ei}_l(E(k)) - \delta^{ei}_{l+1}(E(k))\}\,. \tag{7.83}$$

A corresponding expression can be found for the electron-electron correlation function (7.78) in the non-degenerate limit if we introduce the centre of mass system. Performing the integrals over the angular variables we have

$$\langle \dot{P}^{(ee)}_2; \dot{P}^{(ee)}_2 \rangle = \frac{128\sqrt{\pi}\beta^{7/2}\hbar^8 n_e}{3m^{7/2}}\, \Omega \int_{\infty}^{0} dP\, P^5 \exp\left(-\beta\,\frac{\hbar^2P^2}{m_e}\right)$$

$$\cdot \sum_{l=0}^{\infty} \frac{(l+1)\,(l+2)}{(2l+3)} \sin^2\left\{\delta^{ee}_l\left(\frac{\hbar^2P^2}{m_e}\right) - \delta^{ee}_{l+2}\left(\frac{\hbar^2P^2}{m_e}\right)\right\}\left(1 - \frac{(-1)^l}{2}\right). \tag{7.854}$$

The further evaluation of (7.83), (7.84) needs the phase shifts $\delta_l(E)$ for the statically screened potential V^s_{ab}. Numerical integration of the radial Schrödinger equation has been performed by Meister and Röpke (1982) for different values of the screening length $\varkappa^{-1}$ and angular momentum 1. As a result, Fig. 7.6 shows a comparison of the reduced conductivity of the H-plasma at $T = 10^4$ K and $T = 10^5$ K in dependence on the plasma parameter μ calculated in different approximations: the Spitzer theory, the screened Born approximation and the screened T-matrix approximation. It is shown that in the high density region the screened T-matrix approximation approaches the Born approximation (for densities $n_e \gtrsim 10^{25}$ cm^{-3} at $T \sim 10^3$ K for a H-plasma). In the low density region the T-matrix approximation approaches the Spitzer theory. The screened T-matrix approximation, therefore, can be considered as an interpolation between the Spitzer theory and the Born approximation (Ziman theory); see eq. (7.4). A similar discussion has been given using the cut-off Coulomb potential instead of the statically screened Coulomb potential by KRAEFT, LUFT and MIHAJLOV (1983).

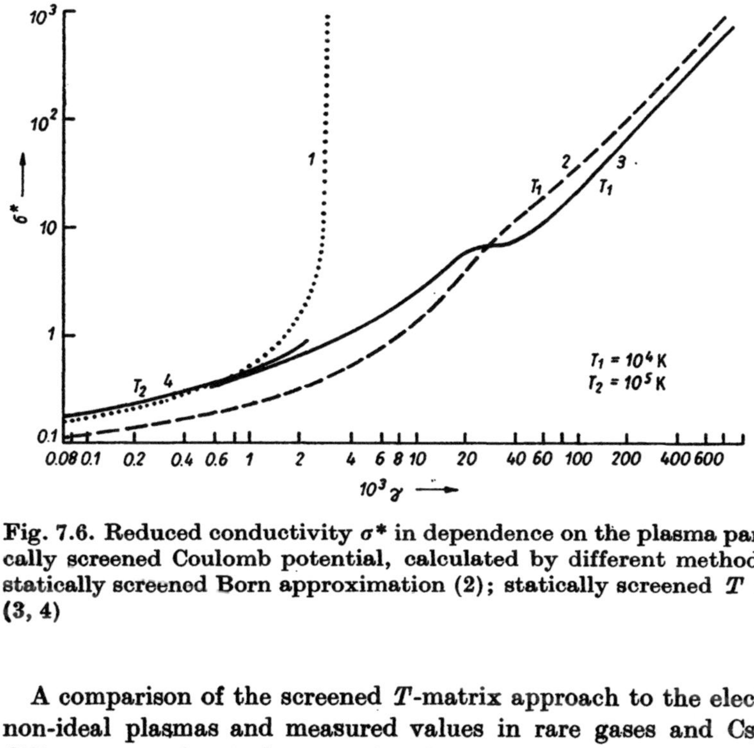

Fig. 7.6. Reduced conductivity σ^* in dependence on the plasma parameter γ for the statically screened Coulomb potential, calculated by different methods: Spitzer theory (1); statically screened Born approximation (2); statically screened T matrix-approximation (3, 4)

A comparison of the screened T-matrix approach to the electrical conductivity of non-ideal plasmas and measured values in rare gases and Cs-plasmas taken from different experimental groups has been performed by MEISTER and RÖPKE (1982), see Fig. 7.7. Although the experimental errors are rather large in the high density region, the measured conductivity values are significantly smaller than the Spitzer values. The interval between the statically screened T-matrix approximation and the Born approximation fits better to the experimental values. Degeneracy effects, ion structure factor and pseudopotential effects are small in this region.

An additional effect lowering the plasma conductivity is the Debye-Onsager relaxation effect which is briefly outlined in the next Section. Furthermore, it is necessary to investigate the influence of bound states and complexes on the conductivity. This may explain, e. g., the comparatively low values of the conductivity of dense Cs-plasmas.

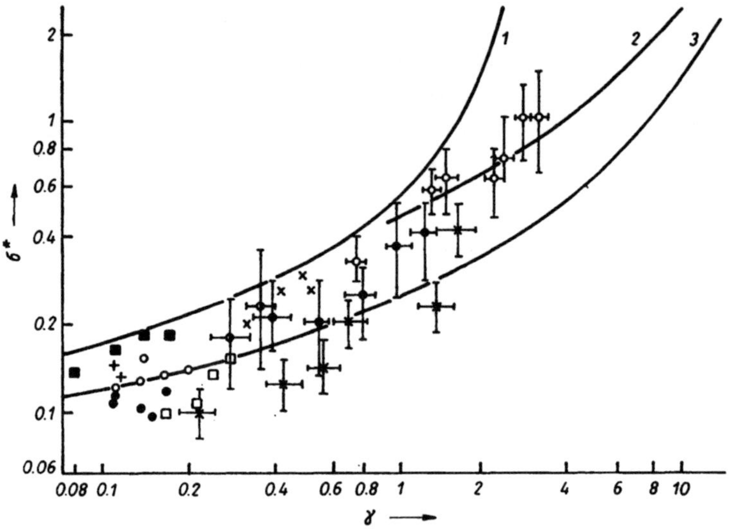

Fig. 7.7. Reduced conductivity σ^* in dependence on the plasma parameter γ. Comparison with experimental data. Theory: $T = 10^4$ K: 1 — Spitzer curve, 2 — statically screened t-matrix-approximation, 3 — statically screened Born approximation

Experiment: ■ Ar, 11750 K $\leqq T \leqq$ 15920 K [7], ⊢•⊣ Ar, ⊢○⊣ Xe, ⊢+⊣ Ne (with error bars), $T \sim$ 25000 K [5], • Ar, 12800 K $\leqq T \leqq$ 17400 K, □ Xe, 9000 K $\leqq T \leqq$ 13700 K [1], ⊢✱⊣ Cs (with error bars), 4000 K $\leqq T \leqq$ 25000 K [4], + H, 15400 K $\leqq T \leqq$ 21500 K [6], ○ air, 13500 K $\leqq T \leqq$ 18300 K [3], × C_2H_3Cl, 37000 K $\leqq T \leqq$ 39000 K [2]

[1] Bakeev and Rovinskii (1970).
[2] Ogurzova, Podmoshenskii and Smirnova (1974).
[3] Andreev and Gavrilova (1975).
[4] Sechenov, Son, and Shchekotov (1975).
[5] Ivanov, Mintsev, Fortov, and Dremin (1976).
[6] Radtke and Günther (1976); Günther, Popovic, Popovic and Radtke (1976).
[7] Günther, Lang and Radtke (1983)

7.3. Further Improvements of the Transport Theory

7.3.1. Self-Energy and Debye-Onsager Relaxation Effects

After having given a detailed discussion of the correlation functions occuring in the collision term of the generalized linear Boltzmann equation (7.15) we give a brief discussion of the correlation functions occuring in the drift term, i.e. being connected with the external fields. These terms have for instance the meaning of particle numbers which are modified by additional correlation functions and lead to corrections to the transport coefficients which are of higher order in the density of the plasma.

In detail let us consider the resistivity formula (7.21)

$$R = \frac{\Omega}{3N^2e^2} \frac{\langle \dot{P}_0(\varepsilon); \dot{P}_0 \rangle}{1 + (3Ne)^{-1} \sum_b \frac{e_b}{m_b} \langle \dot{P}_0(\varepsilon); P_0 \rangle} \tag{7.85}$$

which is obtained within the general approach (7.19) to the evaluation of transport coefficients if only the moment P_0 is taken into account. The relaxation function

$$Q = \sum_b \frac{e_b}{3eN_b m_b} \int_0^\infty \mathrm{d}t\, \mathrm{e}^{-\varepsilon t} \int_0^\beta \mathrm{d}\lambda\, \mathrm{Tr}\, \{\varrho_0\, \dot{P}_b(-t - i\hbar\lambda)\, P_0\} \tag{7.86}$$

can be evaluated in the non-degenerate limit as well as for arbitrary degeneracy in perturbation theory with respect to the interaction potential, see EBELING and RÖPKE (1979).

In the non-degenerate limit, the relaxation function is evaluated using classical trajectories, and for the statically screened potential we get

$$Q = \beta e^2 \varkappa / 6\,. \tag{7.87}$$

This result was first obtained by KADOMTSEV (1957) for a nonideal plasma, see also KLIMONTOVICH and EBELING (1972) where also quantum mechanical corrections were obtained:

$$Q = \frac{e^2\varkappa^2}{6\hbar\beta m_e (2\pi)^3} \int_0^\infty \mathrm{d}\tau\tau^2 \int_0^\infty \mathrm{d}k\; k^4 (V(k))^2$$
$$\cdot\, [1 + \varkappa^2 V(k)/4\pi]^{-1} \exp\,(-k^2\tau^2/2\beta m_e)\, \sin\,(\hbar k^2\tau/2m_e)\,. \tag{7.88}$$

Numerical values for the Debye-Onsager relaxation effect in the classical limit and including quantum mechanical corrections are given by EBELING and RÖPKE (1979). While quantum mechanical corrections are small in the region considered there, the resistivity increases by a factor 2, at 10^4 K and $\mu = \beta e^2 \varkappa \approx 2$.

The relaxation function Q must be also considered in the case that quasiparticles are introduced. For example, for band electrons in a periodic potential this term Q is of importance if Bloch electrons instead of free electrons are considered (CHRISTOPH, 1979). The connection between the relaxation function Q and the free particle self energy has also been discussed by KRAEFT, BLÜMLEIN and MEYER (1983). In the framework of Kubo theory, there is an "equilibrium" correcture to the conductivity of the order of μ (BLÜMLEIN and KRAEFT, 1984). There is also a relation between the relaxation function Q and the thermodynamic properties of the system. Especially the formation of bound states (localized states) is of importance in evaluating Q, see EBELING and RÖPKE (1985).

It should be noted that the relaxation function Q may be incorporated into the collision integral so that the correlation function of stochastic forces instead of real forces must be evaluated, see KALASHNIKOV (1978, 1982), RÖPKE (1983). A systematic approach to Q is not available yet.

7.3.2. Hopping Conductivity

As long as the contribution of carriers in extended states $|k\rangle$ (free particle states) is dominant in the evaluation of transport coefficients, the approach given up to now using moments of the free particle distribution function $P_n = \sum_k \hbar k (\beta E_k)^n n_k$ (7.8) may serve as a suitable choice of trial functions to construct the quasi-equilibrium statistical operator ϱ_q (7.9). However, in the region of the temperature-density plane where bound states are dominant, i.e. for low temperatures and densities below the Mott density, see Fig. 2.2, another choice of trial functions to construct the quasi-equilibrium statistical operator should be made describing the nonequilibrium state of the system by using the occupation numbers of the bound (localized) states, see RÖPKE and CHRISTOPH (1979). In detail we will employ the ansatz

$$\varrho_q = Z_q^{-1} \exp\,\{-\beta[H_s - \mu N - \sum_{in} B_{in} n_{in}\} \tag{789}$$

which is a specialization of the more general expression (7.8). Usually the system Hamiltonian H_s is taken in adiabatic approximation where the ion positions R_i are fixed, and in the representation with respect to a basis of linear combination of atomic orbitals (LCAO) $|in\rangle$ which are approximatively given by the free atom wave function for internal quantum number n and ion position R_i we have

$$H_s = \sum_{in} E_{in} c_{in}^+ c_{in} + \sum_{ii'n} t_{ii'n} c_{i'n}^+ c_{in} + \sum_{inn'} U_{inn'} n_{in} n_{in'} , \tag{7.90}$$

where the matrix elements E_{in}, $t_{ii'n}$ and $U_{inn'}$, are related to the operators of kinetic energy H^0 and interaction potential V according to

$$\begin{aligned} E_{in} &= \langle in| H^0 + V_{ei} |in\rangle + \sum_{i' \neq i,n'} U_{ii'nn'} n_{i'n'} , \\ t_{ii'n} &= \langle i'n| H^0 + V_{ei} | in\rangle , \\ U_{ii'nn'} &= \langle i'n', in| V_{ee} |in, i'n'\rangle , \\ U_{inn'} &= U_{ii'nn'} \delta_{ii'} . \end{aligned} \tag{7.91}$$

We have included in E_{in} a contribution due to the Coulombic interaction with occupied states in the surrounding to indicate that E_{in} fluctuates around the value $E_{in}^{(0)}$ of the isolated atom. The hopping matrix element $t_{ii'n}$ is also a fluctuating quantity depending on the distance $|\boldsymbol{R}_i - \boldsymbol{R}_{i'}|$. The Coulomb repulsion term $U_{inn'}$ excludes double occupation of one ion position if it is large in comparison with the kinetic energy and the temperature. Further matrix elements as, for instance, including three different electronic orbitals, are ommitted from eq. (7.90).

Systems described by a Hamiltonian (7.90) are widely discussed in the physics of crystalline and amorphous systems. They include systems to describe localization if U is neglected (Anderson, 1959) as well as the Hubbard model for ordered and disordered systems (Hubbard, 1957). These phenomena which may also be discussed in treating systems with Coulomb interaction will not be included in the present book because of the large extent to discuss the recent work in solving these model systems. We only point out the relation to the transport theory as formulated in the previous Sections.

A generalization of the model Hamiltonian (7.90) in adiabatic approximation may be given if phonons are included to describe the ion motion, see, for instance, Böttger and Bryksin (1976). In this case the quasi-equilibrium statistical operator ϱ_q (7.89) should also be generalized including the occupation numbers of phonon states.

As demonstrated in Section 7.2., the response parameters B_{in} are obtained from the stationarity conditions

$$\frac{\mathrm{d}}{\mathrm{d}t} \langle n_{in}\rangle = \frac{i}{\hbar} \langle [H, n_{in}]\rangle = 0 \tag{7.92}$$

which read

$$\begin{aligned} &\sum_{i'} \langle j_{i'i}\rangle + \langle j_{i,\mathrm{b}}\rangle = 0 ; \\ &j_{i'i} = \frac{i}{\hbar} (t_{i'in} c_{i'n}^+ c_{in} - t_{ii'n} c_{in}^+ c_{i'n}) , \\ &j_{i,\mathrm{b}} = \frac{i}{\hbar} [H_\mathrm{b}, n_{in}] , \end{aligned} \tag{7.93}$$

if $H = H_s + H_f + H_\mathrm{b}$, where $H_f = -e\boldsymbol{E} \sum_{in} n_{in} \boldsymbol{R}_i$, and H_b accounts for the input and output of the electrical current because of the interaction with the environment.

The mean values $\langle j_{ii'}\rangle$ are related to the current density in the system according to (Böttger and Bryksin, 1976; Böttger, 1983)

$$\boldsymbol{j} = \frac{e}{2\Omega} \sum_{ii'} (\boldsymbol{R}_{i'} - \boldsymbol{R}_i) \langle j_{i'i}\rangle , \tag{7.94}$$

and in linear response we have

$$\langle j_{i'i}\rangle = -(B_{in} - e\boldsymbol{E}\boldsymbol{R}_i - B_{i'n} + e\boldsymbol{E}\boldsymbol{R}_{i'}) \langle j_{i'i}; j_{ii'}\rangle . \tag{7.95}$$

The current from site i to i' splits into two parts, where the first part proportional to the parameters B_{in} is the diffusive part and the second part proportional to the electrical field E is the ohmic part. The parameters B_{in} must be choosen so that the balance equations (rate equations) (7.93) are fulfilled. In this way we are lead to the problem of random network of resistivities (Abraham-Miller network). The solution of rate equation (7.93) is in equivalence to the solution of the Boltzmann equation for extended states, whereas eq. (7.95) resembles the representation of the collision integral by correlation functions in kinetic theory. In order to derive the conductivity, explicit expressions for $\langle j_{i'i} \rangle$ must be found evaluating the correlation functions (7.95). For this the Green's function technique may be employed as given in Section 7.2.1.

A more general theory of transport properties of systems with Coulomb interaction would include both the contribution of free charge carriers and the hopping current. Such an approach is necessary for partially ionized plasmas with low ionization degree and, especially, near the Mott transition. A more general approach was developed recently (Röpke, 1982) taking the total electrical current as the sum of the free particle and the hopping current so that the conductivity consists on the free particle contribution σ_{free}, the hopping contribution σ_{hop} and a crossing term σ_{cr}, the latter containing correlation functions with free and bound state particles:

$$\sigma = \sigma_{\text{free}} + \sigma_{\text{hop}} + \sigma_{\text{cr}} \,. \tag{7.96}$$

The diagram technique to evaluate the correlation functions gives the possibility to consider different processes as, f.i., the plasmon induced hopping process.

Another example where the approach given here has been applied is the conductivity of small polarons (Röpke and Christoph, 1979; see also Böttger and Bryksin, 1976; Böttger, 1983). In conclusion of this Section we want to point out that in the last two decades there are large efforts in the investigation of disordered structures. In such systems the Coulomb interaction plays an important role. Effects which are determined by the Coulomb interaction are, e.g., the Coulomb gap and the correlated hopping transport, see, e.g. Böttger (1983).

7.3.3. Concluding Remarks

Transport properties are macroscopic phenomena of Coulomb systems which are determined by microscopic processes within the system which should be described by a many particle theory. In the last years there was a large success in applying consequently methods of many particle theory in the transport theory. One special approach, the linear response theory, was presented in this Chapter. Using the concepts developed in Chapters 3., 4., approximations for the transport coefficients may be obtained which are valid in a wider temperature and density range.

Especially, several many-body effects, such as dynamical screening, formation of bound states, local field effects and quasiparticle influences, were dealt with in contrast to the traditional versions of the older kinetic theory. In this way one is able to describe transport properties in nonideal systems as plasmas of arbitrary density and liquid metals from a unified point of view. It is possible to interpret experimental results for such systems, see, e.g., Fig. 7.7, and also Figs. 7.8, 7.9. A more detailed representation of the experimental situation is found, e.g., in Gryaznow et al. (1980), Ebeling et al. (1983, 1984), and Günther and Radtke (1984). There are many experimental papers which demonstrate that the behaviour of the conductance in nonideal metal vapours is quite different from that observed in ideal plasmas (Alekseev, 1970; Renkert et al., 1971; Likhalter, 1978; Issakov and Lomakin, 1979).

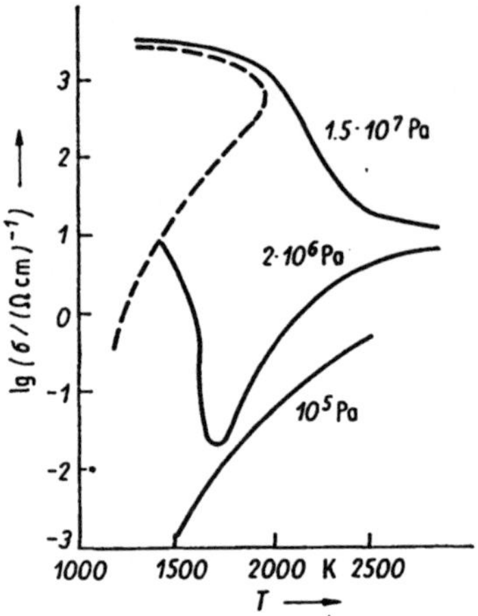

Fig. 7.8. Conductivity in dependence on the temperature — experimental curves

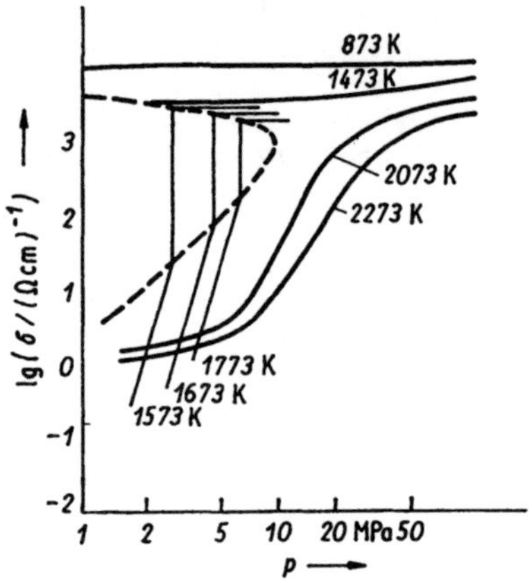

Fig. 7.9. Conductivity in dependence on the pressure — experimental curves

The qualitative behaviour of the conductance as a function of temperature and pressure is show in Figs. 7.8. and 7.9. The most remarkable observation is that the experimental curves for the conductance show a strong increase with increasing pressure. Quantitatively the conductance of a saturated vapour is by five or six orders of magnitude higher than the conductance obtained from Spitzers formula in combination with the Saha equation. This anomalously high conductance has even been the basis for a very promising idea of a new MHD-generator working with saturated Cs- or K-vapours which has been developed by Biberman, Likhalter and Iakubov(1982) based on an earlier idea to use nonideal plasmas developed by Iakubov and Vorobev (1974). The character of the temperature behaviour (Fig. 7.8.) depends on the given pressure. At small pressures ($p = 10^5$ Pa) the conductance increases, as usual, with increasing temperature. At very large (supercritical) pressures ($p = 1.5 \cdot 10^7$ Pa) in the opposite the conductance decreases with increasing pressure. In the mean region of pressures ($p = 2 \cdot 10^6$ Pa) the conductance goes through a very deep minimum (see also Section 7.2.3.). There is no final explanation for the observed facts since many effects come together. One possible explanation is based on the idea that in the plasma heavy charged clusters are formed (Lagarkov and Sarychev, 1979; Likhalter, 1981; Khrapak and Yakubov, 1981). The idea is in short the following. At temperatures below the critical the vapour has a very complicated composition. Most of the ions form clusters with several atoms and the electrons are also partially bound in negative clusters. However, the binding energy of the positive clusters is much higher than that of negative clusters what leads to an excess of free electrons. Denoting the density of positive clusters by n_k and neglecting negative clusters, neutrality requires

$$n_e \approx \sum_k n_k \gg n_i$$

(n_i — density of the free Cs-ions). From the Saha equation follows

$$\frac{n_0^*}{n_e^* n_i^*} \approx \Lambda_e^3 \exp\left[\frac{I}{k_B T}\right].$$

Using the neutrality condition we get

$$n_e^* \approx (n_e^*)^{1/2} \Lambda_e^{-3/2} \exp\left(-\frac{I}{2k_B T}\right) \sum_k (n_k/n_1)^{1/2}.$$

Without clusters we would find $n_e^* = n_i^*$ but now we have got a factor

$$(\sum_k n_k/n_1)$$

which leads to a strong increase of the electron density. This is a possible explanation for the anomalously high conductance of dense vapours.

8. Green's Function Approach to Optical Properties

8.1. General Formalism

8.1.1. Many-Body Theory of Absorption Spectra

One of the most interesting properties of Coulomb systems is the absorption spectrum. Particularly, spectral lines are important characteristics which are used in plasma diagnostics. A theoretical approach to the optical properties is necessary to relate the line shape to the microscopic interaction processes and, in consequence, to diagnose the density and temperature of the plasma from measured line profiles.

The influence of the natural and Doppler broadening on the shape of spectral lines is well known, see Section 8.1.3. In dense plasmas, however, pressure broadening is the dominant effect caused by the interaction with the surrounding particles.

Up to medium densities of some 10^{17} cm^{-3}, pressure broadening may be explained on the basis of traditional approaches such as the quasistatic microfield and the impact approximation. According to Lorentz, Lindholm, Foley, Weisskopf etc. (see Unsöld, 1955), in the impact approximation the radiating atoms are considered as classical oscillators the frequencies of which are disturbed because of the neighbouring particles' interaction. The Fourier transform of the phase perturbed radiation during binary collisions leads to a broadened and shifted Lorentz profile.

Within the quasistatic microfield approximation according to Holtsmark, Debye, Margenau, Verwey, Griem etc. (see Griem, 1974), the radiating atom is considered within the fixed configuration of the surrounding plasma particles. The resulting (static) microfield leads to the Stark shift of the atomic levels, and the distribution of the microfield yields the profile of the absorption line. Correlation effects were considered within the Debye screening theory by Ecker and Müller (1958), Mozer and Baranger (1960), Kelbg (1964, 1965). Improvements are given by Pfennig and Trefftz (1966) and Hooper (1968).

Both the traditional approaches may be considered as alternative approximations. Near the line center, $|\Delta\omega| < \omega_{W,e}(\omega_{W,i})$, the impact approximation can be used which takes into account especially the influence of the perturbing electrons, whereas in the line wings $|\Delta\omega| > \omega_{W,e}(\omega_{W,i})$, the line profile is given in good approximation by the ionic static microfield. $\omega_{W,a}$ denotes the Weisskopf frequencies ($\bar{v}_a$ —mean velocity of species a)

$$\omega_{W,a} = \bar{v}_a/r_{W,a}\,, \quad r_{W,a} \text{ — Weisskopf radius.} \tag{8.1}$$

For the validity of both approximations see also Traving (1968) and Seidel (1974). A detailed description of the impact approximation and of the microfield approximation is given in the literature, see f. i. GRIEM (1974), SOBELMAN, VAINSHTEIN and YUKOV (1981).

With regard to plasmas of higher densities a consequent many-body theoretical treatment of the pressure broadening is required. It is not a single perturbed atom which emits or absorbs radiation but the plasma as a whole which couples to the electromagnetic field. A many-body approach to the spectral line profile was given by BARANGER (1958, 1962) starting from the Fourier transform of the dipole-dipole autocorrelation function

$$L(\omega) = \frac{1}{2\pi} \operatorname{Re} \int_{-\infty}^{\infty} e^{i\omega t} C(t)\, dt\,, \qquad C(t) = \operatorname{Tr}\{\boldsymbol{d}U^{+}(t, 0)\, \boldsymbol{d}U(t, 0)\, \varrho\}\,, \tag{8.2}$$

which is directly related to the line profile. Different methods are used to evaluate this correlation function. A "unified theory" was given by VOSLAMBER (1969, 1972), VIDAL, COOPER and SMITH (1970, 1971, 1973). The relation to the microfield concept has been considered by BRISSAUD and FRISCH (1971, 1974), BRISSAUD et al. (1976), SEIDEL (1977, 1979), MAZURE and NOLLEZ (1978), PARGAMANIK and GINSBURG (1977, 1978, 1980).

The most adequate approach to the optical properties should be based on the dielectric function of the plasma which is related to the density-density correlation function. The appropriate language for discussing the problem within this framework is one which employs the Green's functions of the many-body system. Green's function approaches to the line shape have been given by KLEIN (1969), ROSS (1966), ZAIDI (1968), KUDRIN, (1974), DHARMA-WARDANA et al. (1980). An equivalent treatment may be given using the Liouville operator technique, see DUFTY and BOERCKER (1976), LEE (1973, 1979).

Recently, a consequent Green's function approach to the dielectric function of a partially ionized plasma was given by RÖPKE, SEIFERT and KILIMANN (1981), HITZSCHKE et al. (1985). This approach starts from the cluster decomposition of the dielectric function as given in Section 4.5. and will be presented in detail in this Chapter. Expressions for the shift as well as the broadening of spectral lines will be developed, and the evaluations are compared with experimental results. Using some approximations we show in Section 8.2.2. that our result may be identified with the shift and broadening expressions of the impact approach. Hence, a many-body generalization of the impact approach is obtained in this way. The inclusion of the quasistatic microfield into the Green's function concept, however, remains an open question, see for this also PARGAMANIK and GINSBURG (1980).

8.1.2. Dielectric Function and Spectral Line Shape of Plasmas

Our starting point of a many-body theoretical treatment of the optical properties of a plasma is its dielectric function. In a dense plasma, it is not a single atom but the plasma as a whole which emits or absorbs radiation, and a many-body treatment should be applied.

The dielectric function $\varepsilon(q, \omega)$ describes the plasma's reaction on an external perturbing field with wave number q and frequency ω within the framework of linear response. Actually, ε also depends on the thermodynamic variables of state, i.e. in particular on the density n_e and the temperature T of the plasma.

An approach to the optical properties is given by the long wave length limit ($q_{opt} \to 0$) of the imaginary part of ε (absorption coefficient). We have for the complex refraction index $n + i\varkappa$

$$(n + i\varkappa)^2 = \varepsilon_{tr}(q, \omega) , \tag{8.3}$$

and for $q \to 0$ we can replace the transversal dielectric function by the longitudinal one:

$$\varepsilon_{tr}(q, \omega) \approx \varepsilon(q, \omega) . \tag{8.4}$$

The dielectric function $\varepsilon(q, \omega)$ is related to the polarization function $\Pi(q, z)$:

$$\varepsilon(q, \omega) = 1 - \frac{e^2}{\varepsilon_0 q^2} \lim_{\delta \to 0} \Pi(q, \omega + i\delta) . \tag{8.5}$$

Now, there is a clear concept how to determine Π, see Chapters 3., 4.

Using diagrams, the polarization function Π may be represented as the sum of all the irreducible diagrams which connect an "inward" with an "outward" vertex point and do not separate if an inner interaction is cut. A certain approximation for Π includes a proper choice of diagrams giving the dominant contributions for the system under consideration.

Thus, for example, the most simple diagram consisting of a product of two free particle propagators G_0 yields the well known random phase approximation as discussed in Section 5.2.:

$\Pi_{RPA} =$ 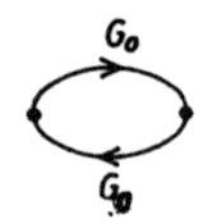

$$= \frac{1}{\Omega} \sum_{p,c} \frac{f_c(E_p^c) - f_c(E_{p-q}^c)}{E_p^c - E_{p-q}^c - \hbar(\omega + i\delta)} . \tag{8.6}$$

However, in this approximation no line spectra are obtained, and we have to improve the polarization function by including the contribution of bound states as demonstrated in Section 4.5.

An appropriate choice of contributions for $\Pi(q, z)$, may be given in the form, see Röpke and Der (1979), Röpke, Seifert and Kilimann (1981), Hitzschke et al. (1985),

$$\Pi(q, z) = \Pi_{RPA} + \Pi_{atom} \tag{8.7}$$

with

$\Pi_{atom} =$ 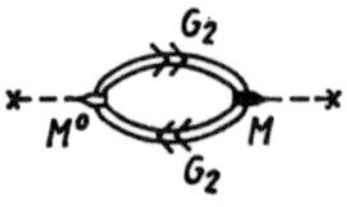 (8.8)

where G_2 denotes the two-particle propagator, and M is the full five-point vertex which couples the bound state to the Coulomb field.

In a simple approximation we obtain the extended RPA expression replacing G_2 by G_2^0 and the vertex $M_{nn'}(q)$ by $M_{nn'}^0(q)$ so that (cf. (4.265))

$$\Pi_{atom}(q, \omega + i\delta) = \sum_{nn'P} |M_{nn'}^0(q)|^2 \frac{g_{el}(E_{nP}) - g_{el}(E_{n'P-q})}{E_{nP} - E_{n'P-q} - \hbar(\omega + i\delta)} \tag{8.9}$$

with

$$M^0_{nn'}(q) = ie \sum_p \psi^*_n(p) \{\psi_{n'}(p) - \psi_{n'}(p + q)\} , \tag{8.10}$$

where terms of the order of the mass ratio $m_e/m_i \ll 1$ are neglected. This expression obviously describes sharp transitions with $\hbar\omega_{nn'} = E_n - E_{n'}$ if n, n' run over the discrete part of the energy spectrum. The dependence of $E_{n',P}$ on the centre of mass momentum describes the thermal motion of emitters which yields the Doppler broadening, see the next Section.

To take into account pressure broadening and shift of the spectral lines it is necessary to include the influence of the surrounding plasma on the radiating atoms. This may be done by using improved propagators G_2 which are to be derived from a Bethe-Salpeter equation. We write it in the form

$$G_2 = G_2^0 + G_2^0\Sigma_2 G_2 \tag{8.11}$$

where it is obvious that atoms are renormalized by a two-particle self energy Σ_2 due to plasma effects, see Section 5.4.

Further plasma corrections are obtained if an analogous equation is solved for the full vertex function

$$M = M^0 + K_4 G_2 G_2 M . \tag{8.12}$$

The kernel K_4 may be considered as the effective atom-atom interaction in the system, and is constructed from interactions between four particles including the interaction with the surrounding plasma.

As already discussed for the two-particle system, see Section 4.4., approximations have to be performed in the self energy and in the effective interaction K_4 on the same level. In this Chapter we will consider the spectral line shape in the framework of a V^s approximation, V^s being the dynamical RPA-screened Coulomb interaction. Within this approximation which takes into account the corrections proportional to the density of free electrons in Born approximation the atomic self energy reads

$$\Sigma_2 = M^0 V^s G_2^0 M^0 = \tag{8.13}$$

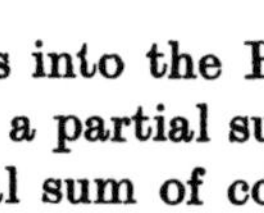

and the effective atom-atom interaction K_4 is determined by

$$K_4 = M^0 V^s M^0 = \tag{8.14}$$

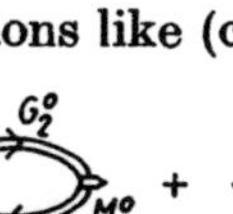

Inserting these expressions into the Bethe-Salpeter equation (8.11) or the vertex equation (8.12), respectively, a partial summation is performed, and the polarization function is given by a partial sum of contributions containing self-energy elements like (b) and vertex corrections like (c).

$$\Pi_{\text{atom}}(q, z) = \tag{8.15}$$

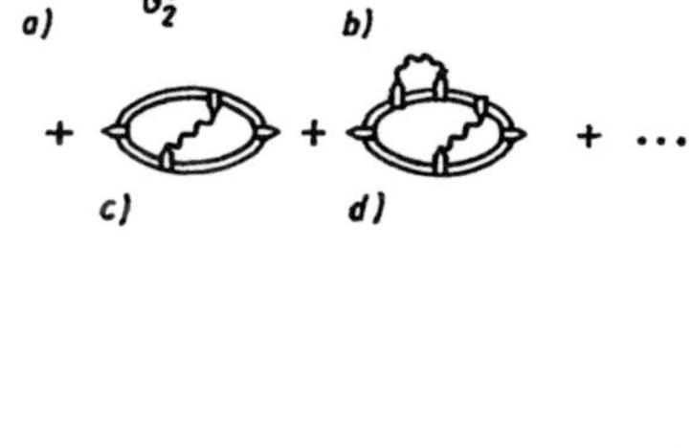

Of course, further improvements can be obtained by including more diagrams as shown, for instance, in Section 5.4.4. for the self energy.

8.1.3. Doppler Broadening

For illustration, we show how the expression (8.9) for the dielectric function gives the Doppler broadening in the case that no density effects are taken into account, which would lead to a further broadening of spectral lines. The expression (8.9) which is obtained in the approximation of isolated atoms leads to the following imaginary part:

$$\mathrm{Im}\, \varepsilon(q, \omega) = \frac{16\pi e^2}{\Omega} \sum_{Pnn'} |d_{nn'}|^2 \left[g_{\mathrm{ei}}(E_{nP}) - g_{\mathrm{ei}}(E_{n', P-q})\right]$$

$$\cdot\, \delta(\hbar\omega + E_{nP} - E_{n', P-q}) \tag{8.16}$$

with sharp lines corresponding to the transition frequencies

$$\omega_{n'n} = (E_{n', P-q} - E_{nP})/\hbar\,. \tag{8.17}$$

The natural line broadening because of the finite life time of the excited states is not described within the semiclassical radiation theory. However, the transition frequencies $\omega_{n'n}$ depend on the center of mass motion of the radiating atoms. With the Maxwell distribution

$$g_{\mathrm{ei}}(E_{nP}) \approx \tfrac{1}{4}\, n(E_n)\, \Lambda_{\mathrm{ei}}^3\, \mathrm{e}^{-\beta\hbar^2P^2/2M} \tag{8.18}$$

for the bound-bound part of (8.16) follows

$$\mathrm{Im}\, \varepsilon^{(b)}(q, \omega)\big|_{q=\frac{\omega}{c}} = \frac{2\epsilon^2}{\Omega} \sum_{nn'} |d|^2 \left(n(E_n) - n(E_{n'})\right)$$

$$\cdot\, \Lambda_{\mathrm{ei}}^3 \exp\left\{-\frac{\beta M c^2}{2\omega^2}(\omega - \omega_{nn'}^{(0)})^2\right\} \tag{8.19}$$

with

$$\omega_{nn'}^{(0)} = (E_{n'} - E_n)/\hbar\,,$$

what gives the characteristic Gauss profile of the Doppler broadened spectral line.

8.2. Evaluation of Line Shift and Broadening

8.2.1. Explicit Expressions for Shift and Broadening

As a starting point for the derivation of the line shape from the polarization function we use the approximation (8.15) which is written in compact form according to (8.13), (8.14)

$$\Pi_{\mathrm{atom}} = \Pi_{\mathrm{atom}}^{\mathrm{SE}}(q, \omega_\mu) + \Pi_{\mathrm{atom}}^{\mathrm{V}}(q, \omega_\mu) \tag{8.20}$$

or, in diagrams

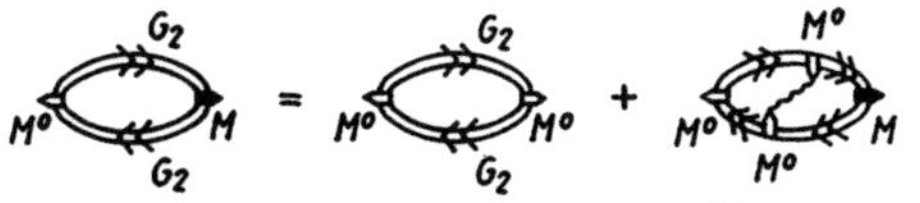

(8.21)

with

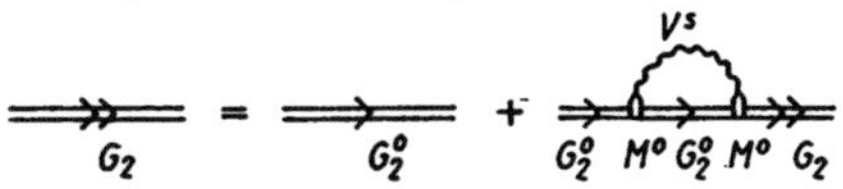

(8.22)

More explicitely this reads

$$\Pi_{\text{atom}}(q, \omega_\mu) = -\frac{1}{i\beta} \sum_{1,2} M^0_{12}(q)\, G_2(1)\, G_2(2)\, M(2, 1; -q, -\omega_\mu)$$

$$= -\frac{1}{i\beta} \sum_{1,2} M^0_{12}(q)\, G_2(1)\, G_2(2)\, M^0_{21}(-q)$$

$$+ \frac{1}{(-i\beta)^2} \sum_{1234,\bar{\mu}\,\bar{q}} M^0_{12}(q)\, M^0_{41}(\bar{q})\, i V^s(\bar{q}, \omega_{\bar{\mu}})\, M^0_{23}(\bar{q})$$

$$\cdot\, G_2(1)\, G_2(2)\, G_2(3)\, G_2(4)\, M(3, 4; -q - \omega_\mu)\,, \tag{8.23}$$

where abbreviations $1 = n_1,\ P,\ \Omega_\lambda;\ 2 = n_2,\ \boldsymbol{P} + \boldsymbol{q},\ \Omega_\lambda + \omega_\mu;\ 3 = n_3,\ \boldsymbol{P} + \boldsymbol{q} + \bar{\boldsymbol{q}},$ $\Omega_\lambda + \omega_\mu + \omega_{\bar{\mu}}$ and $4 = n_4, \boldsymbol{P} + \bar{\boldsymbol{q}}, \Omega_\lambda + \omega_{\bar{\mu}}$ are introduced.

The evaluation of the term $\Pi^{\text{SE}}_{\text{atom}}$ is straightforward, see Section 4.5.3. We use the solution of the Bethe-Salpeter equation including the self-energy of atoms in the plasma (8.13). The propagators G_2 are expressed by corresponding spectral functions A_1:

$$G_2(1, \Omega_\lambda) = \int\limits_{-\infty}^{\infty} \frac{d\omega}{2\pi} \frac{A_1(\omega)}{\Omega_\lambda - \omega}\,, \tag{8.24}$$

where it is assumed that G_2 and A_1 are diagonal with respect to the internal quantum numbers n_1 and the total momentum P_1 both denoted by 1.

The spectral function can be represented, for example, in the low density limit by a Lorentzian

$$A_1(\omega) = \frac{2\Gamma_1}{(\hbar\omega - E_1 - \Delta_1)^2 + \Gamma_1^2}\,; \tag{8.25}$$

Δ_1, Γ_1 are density and temperature dependent shift and broadening parameters of the two-particle state with unperturbed energy E_1, which may be determined by a perturbative treatment of the Bethe-Salpeter equation, and the resulting expression for $\Pi^{\text{SE}}_{\text{atom}}$ was given in Section 4.5.4.

A solution of the vertex equation like that of the Bethe-Salpeter equation is not possible. Therefore we have to evaluate the term $\Pi^{\text{V}}_{\text{atom}}$ in a more approximate way, see also for comparison Ross (1966).

A suitable approximation for the vertex may be derived from (8.23) with the assumption that $M(1, 2; -q, -\omega_\mu)$ is a sufficiently well-behaved function depending smoothly on its arguments. In particular M is assumed not to introduce new singularities such that we have to consider only the poles produced by G_2 and V^s. Accordingly, the Ω_λ and ω_μ^- summations in (8.23) may be performed with the result

$$\sum_{n_1n_2} M_{12}^0(q) \iint \frac{d\omega_1\, d\omega_2}{(2\pi)^2} \frac{A_1(\omega_1)\, A_2(\omega_2)}{\omega_1 - \omega_2 + \omega_\mu}$$

$$\cdot \{g(\hbar\omega_1)\, M_{21}(\omega_1 + \omega_\mu, \omega_1, -q, -\omega_\mu) - g(\hbar\omega_2)\, M_{21}(\omega_2, \omega_2 - \omega_\mu; -q, -\omega_\mu)\}$$

$$= \sum_{n_1n_2} |M_{12}^0(q)|^2 \iint \frac{d\omega_1\, d\omega_2}{(2\pi)^2} \frac{A_1(\omega_1)\, A_2(\omega_2)}{\omega_1 - \omega_2 + \omega_\mu} \{g(\hbar\omega_1) - g(\hbar\omega_2)\}$$

$$+ \sum_{n_1n_2n_3n_4} \sum_{\bar{q}} M_{12}^0(q)\, M_{23}^0(\bar{q})\, M_{41}^0(-\bar{q})\, V(\bar{q}) \int \frac{d\bar{\omega}}{\pi} (1 + n_B(\hbar\bar{\omega}))\, \mathrm{Im}\, \varepsilon_{\mathrm{RPA}}^{-1}(\bar{q}, \bar{\omega} + i0)$$

$$\cdot \frac{1}{(2\pi)^4} \int d\omega_1\, d\omega_2\, d\omega_3\, d\omega_4\, A_1(\omega_1)\, A_2(\omega_2)\, A_3(\omega_3)\, A_4(\omega_4)$$

$$\cdot \left\{ \frac{g(\hbar\omega_1 + \hbar\bar{\omega})\, M_{34}(\omega_1 + \omega_\mu + \bar{\omega}, \omega_1 + \bar{\omega}; -q, -\omega_\mu)}{(\omega_1 - \omega_2 + \omega_\mu)\, (\omega_1 - \omega_3 + \omega_\mu + \bar{\omega})\, (\omega_1 - \omega_4 + \bar{\omega})} \right.$$

$$- \frac{g(\hbar\omega_2 + \hbar\bar{\omega})\, M_{34}(\omega_2 + \bar{\omega}, \omega_2 - \omega_\mu + \bar{\omega}; -q, -\omega_\mu)}{(\omega_1 - \omega_2 + \omega_\mu)\, (\omega_2 - \omega_3 + \bar{\omega})\, (\omega_2 - \omega_4 - \omega_\mu + \bar{\omega})}$$

$$+ \frac{g(\hbar\omega_3)\, M_{34}(\omega_3, \omega_3 - \omega_\mu; -q, -\omega_\mu)}{(\omega_1 - \omega_3 + \omega_\mu + \bar{\omega})\, (\omega_2 - \omega_3 + \bar{\omega})\, (\omega_3 - \omega_4 - \bar{\omega}_\mu)}$$

$$\left. - \frac{g(\hbar\omega_4)\, M_{34}(\omega_4 + \omega_\mu, \omega_4; -q, -\omega_\mu)}{(\omega_1 - \omega_4 + \bar{\omega})\, (\omega_2 - \omega_4 - \omega_\mu + \bar{\omega})\, (\omega_3 - \omega_4 - \omega_\mu)} \right\}. \tag{8.26}$$

We have used here the relation

$$n(\hbar\bar{\omega})\, g(\hbar\omega) \approx (1 + n(\hbar\bar{\omega}))\, g(\hbar\omega + \hbar\bar{\omega})$$

which is correct up to the first order with respect to the density. There is no contribution of the pure Coulomb potential $V(\bar{q})$ contained in $V^s(\bar{q}, z)$ to the vertex term in (8.26) because it leads to terms of higher order of the density. Explicit momentum arguments are omitted referring to atoms at rest ($P = 0$, i.e. no Doppler broadening) and neglection of the change of the atoms' total momenta by collisions with electrons ($\pm\bar{q}$).

The expression (8.26) is essentially determined by the poles produced by the denominators which yield δ-functions if we proceed to imaginary parts according to $z = \omega + i0$. (Strictly speaking, the discrete Matsubara frequencies ω_μ are to be analytically continued to complex arguments z, and real and imaginary part of the polarization function follow then by $z = \omega + i0$.) Keeping this in mind, it turns out after a decomposition of the $\Pi_{\mathrm{atom}}^{\mathrm{V}}$ term of (8.26) into partial fractions that some contributions cancel if we take the g and M functions with the arguments of the corresponding poles. Thus we get

$$\sum_{n_1n_2} M_{12}^0(q) \iint \frac{d\omega_1\, d\omega_2}{(2\pi)^2} \frac{A_1(\omega_1)\, A_2(\omega_2)}{\omega_1 - \omega_2 + z} [g(\hbar\omega_1) - g(\hbar\omega_2)]\, M_{21}(\hbar\omega_2, \omega_1; -q, -z)$$

$$= \sum_{n_1n_2} |M_{12}^0(q)|^2 \iint \frac{d\omega_1\, d\omega_2}{(2\pi)^2} \frac{A_1(\omega_1)\, A_2(\omega_2)}{\omega_1 - \omega_2 + z} [g(\hbar\omega_1) - g(\hbar\omega_2)]$$

$$+ \sum_{n_1 n_2 n_3 n_4} \sum_{\bar{q}} M_{12}^0(q)\, M_{23}^0(\bar{q})\, M_{41}^0(-\bar{q})\, V(\bar{q}) \int \frac{d\bar{\omega}}{\pi} [1 + n(\hbar\bar{\omega})] \operatorname{Im} \varepsilon_{\mathrm{RPA}}^{-1} (\bar{q}, \bar{\omega} + i0)$$

$$\cdot \int d\omega_1\, d\omega_2\, d\omega_3\, d\omega_4 \frac{1}{(2\pi)^4} \frac{A_1(\omega_1)\, A_2(\omega_2)}{\omega_1 - \omega_2 - z} \frac{A_3(\omega_3)\, A_4(\omega_4)}{\omega_4 - \omega_3 + z}$$

$$\cdot \{(\bar{\omega} + \omega_1 - \omega_3 + z)^{-1} - (\bar{\omega} + \omega_2 - \omega_4 - z)^{-1}\}$$

$$\cdot [g(\hbar\omega_3) - g(\hbar\omega_4)]\, M_{34}(\omega_3, \omega_4; -q, -z)\,. \tag{8.27}$$

This represents an equation for the determination of the matrix elements of the vertex function, wherein all virtual transitions are taken into account. But there are different $M_{nn'}$ couples because of the summations over n_1 to n_4. In order to find a closed expression we restrict to the case that ω lies in the vincinity of an isolated transition frequency $\hbar\omega_{if} = E_f - E_i$, $E_{i,f}$ being the unperturbed energies of the initial and final states, respectively.

If we assume the spectral functions to be sharply peaked Lorentzians (8.25), we have

$$\iint \frac{d\omega_1\, d\omega_2}{(2\pi)^2} \frac{A_1(\omega_1)\, A_2(\omega_2)}{\omega_1 - \omega_2 + \omega + i0} = \frac{\hbar^{-1}}{\hbar\omega - \hbar\Omega_{12} + i(\Gamma_1 + \Gamma_2)}\,, \tag{8.28}$$

with

$$\hbar\Omega_{12} = E_2 + \Delta_2 - E_1 - \Delta_1\,. \tag{8.29}$$

The main contributions of (8.27) arise if $\omega_1 - \omega_2 = \omega_{if}$ and $\omega_4 - \omega_3 = \omega_{if}$. Since the ω_n ($n = 1, 2, 3, 4$) occur also as arguments of the spectral functions which may be considered as weighting factors, the sums reduce to the terms with $n_1 = n_4 = i$; $n_3 = n_2 = f$.

Next we have to discuss the $\bar{\omega}$ integration involving the terms

$$[1 + n(\hbar\bar{\omega})] \operatorname{Im} \varepsilon^{-1}(q, \bar{\omega} + i0)\, \{(\bar{\omega} + \omega_1 - \omega_3 + z)^{-1} - (\bar{\omega} + \omega_2 - \omega_4 - z)^{-1}\}\,. \tag{8.30}$$

Referring to $z \to \omega + i0$ we split the terms in the curly brackets into real and imaginary parts using Dirac's identity. The principal parts will be neglected henceforth. Because of the opposite signs they have the tendency to cancel, but they may be assumed to be small in general, as according to Ross (1966). Consequently, only the δ-functions of the imaginary parts play a role producing factors with $\operatorname{Im} \varepsilon^{-1}$ at arguments given by $\omega \approx \omega_{if}$ and the ω_n given above, if the $\bar{\omega}$ integration is performed.

As now regards the integrations over the ω_n the main contributions arise (as already discussed above) due to the spectral functions $A_n(\omega_n)$ at $\omega_n \approx E_n$. There we have a strong ω_n-dependence of the denominators $(\omega_n - \omega_{n'} \pm z)$ which become resonant at $z \to \omega_{if}$. In contrast to that, the ω_n-dependence of all the remaining terms may be considered to be smooth so we replace ω_n by E_n there.

In the framework of this approximation we may use (8.28) and solve (8.27) for the vertex function with the result

$$M_{fi}(-q, -\omega) \approx M_{fi}^0(-q)\, \{1 + i\Gamma_{if}^{\mathrm{V}} [\hbar\omega - \hbar\Omega_{if} + i(\Gamma_i + \Gamma_f)]^{-1}\}^{-1}\,, \tag{8.31}$$

with

$$i\Gamma^{\mathrm{V}}_{if} = \sum_{\bar{q}} M^0_{ii}(\bar{q})\, M^0_{ff}(-\bar{q})\, V(\bar{q}) \int_{-\infty}^{\infty} \frac{\mathrm{d}\bar{\omega}}{\pi} [1 + n(\hbar\bar{\omega})]$$

$$\cdot \operatorname{Im} \varepsilon^{-1}(\bar{q}, \bar{\omega} + i0) \left\{ \frac{1}{\bar{\omega} + i0} - \frac{1}{\bar{\omega} - i0} \right\}. \tag{8.32}$$

Now we may discuss the line shape function which results if (8.31) is inserted into the left-hand side of (8.23). We find with (8.28)

$$\Pi_{\mathrm{atom}}(q, \omega) = \sum_{\{i,f\}} [g(E_i) - g(E_f)] \frac{|M^0_{if}(q)|^2}{\hbar\omega - \hbar\Omega_{if} + i(\Gamma_i + \Gamma_f + \Gamma^{\mathrm{V}}_{if})}, \tag{8.33}$$

where $\{i, f\}$ are running over all contributing transitions. The imaginary part of this expression describes Lorentzian shaped spectral lines

$$I_{if}(\omega) \sim \frac{\Gamma_{if}}{(\hbar\omega - \hbar\omega_{if} - \Delta_{if})^2 + \Gamma^2_{if}}, \tag{8.34}$$

characterized by line shift and broadening parameters

$$\Delta_{if} = \Delta^{\mathrm{SE}}_i - \Delta^{\mathrm{SE}}_f, \qquad \Gamma_{if} = \Gamma^{\mathrm{SE}}_i + \Gamma^{\mathrm{SE}}_f + \Gamma^{\mathrm{V}}_{if}. \tag{8.35}$$

By inclusion of the improved vertex (8.31) we obtain an additional contribution Γ^{V}_{if} to the line width. Thus, besides the damping of the two-particle states also the influence of the plasma on transitions between them is taken into account. The line shift is directly related to the shifts of the two-particle energies.

The index SE denotes that these contributions are to be determined from the Bethe-Salpeter equation including the self-energy of the atoms in the plasma. According to Σ_2 (8.13) we obtain, see Röpke, Seifert and Kilimann (1981) for details,

$$\Delta^{\mathrm{SE}}_n + i\Gamma^{\mathrm{SE}}_n = -\sum_{\bar{q}} V(\bar{q}) \int_{-\infty}^{\infty} \frac{\mathrm{d}\hbar\bar{\omega}}{\pi} [1 + n(\hbar\bar{\omega})] \operatorname{Im} \varepsilon^{-1}(\bar{q}, \bar{\omega} + i0)\, D_n(\bar{q}, \bar{\omega}),$$

$$D_n(\bar{q}, \bar{\omega}) = \sum_{\alpha} |M^0_{n\alpha}(\bar{q})|^2 \{E_n - E_{\alpha, -\bar{q}} - \hbar\bar{\omega} - i0\}^{-1}. \tag{8.36}$$

Here n denotes the atomic level under consideration, whereas α runs over the whole two-particle spectrum. Correspondingly, D_n may be interpreted as a generalized atomic polarizability which is modified by the participation of collective plasma modes (plasmons) of energy $\hbar\bar{\omega}$ in bound-bound and bound-free transitions.

With (8.32) and (8.36) expressions are given for the shift and broadening parameters which determine a Lorentzian approximation of the spectral line shape. Using the RPA dielectric function these expressions are adequate for describing the perturbing effect of free electrons on radiating atoms in a plasma. We will illustrate that in the following Section by showing that our result contains the semi-classical impact approximation used by Griem (1974), and others, for the contribution of the electrons to the line shape, see also Hitzschke et al. (1985).

8.2.2. Relation to the Impact Approximation

The Green's function approach presented here may be considered as a many-body theoretical generalization of the well known semiclassical impact approximation.

We will show that by deriving identical expressions for the shift and broadening parameters from both approaches. For that purpose we first rewrite (8.32) and (8.36) using the approximate relation

$$\mathrm{Im}\ \varepsilon^{-1}(q,\omega) \approx -\mathrm{Im}\ \varepsilon(q,\omega) = 2\pi \frac{V(q)}{\Omega} \sum_p [f_e(p) - f_e(p-q)]\ \delta(E_p - E_{p-q} - \hbar\omega)\,. \tag{8.37}$$

In the low density limit the Fermi occupation of the electrons becomes a Boltzmann distribution

$$f_e(p) \approx \tfrac{1}{2}\, n_e (2\pi\hbar^2/m_e k_B T)^{3/2} \exp\left[-\hbar^2 p^2/2m_e k_B T\right], \tag{8.38}$$

n_e, m_e — electron density and mass, $E_p = \hbar^2 p^2/2m_e$.

Up to first order of density we may then replace

$$(1 + n(\hbar\bar{\omega}))\,[f_e(p) - f_e(\mathbf{p} - \mathbf{q})] \approx f_e(p) \tag{8.39}$$

and separate Δ and Γ according to Dirac's identity. Thus we get instead of (8.32) and (8.36), $n = i, f$,

$$\Delta_n^{\mathrm{SE}} = -\frac{2}{\Omega^2} \sum_{pq\alpha} f_e(p)\, V^2(q)\, \frac{|M^0_{n\alpha}(q)|^2}{E_n - E_\alpha + \hbar^2 \mathbf{pq}/m_e}\,, \tag{8.40}$$

$$\Gamma_n^{\mathrm{SE}} = \frac{2\pi}{\Omega^2} \sum_{pq\alpha} f_e(p)\, V^2(q)\, |M^0_{n\alpha}(q)|^2\, \delta(E_n - E_\alpha + \hbar^2 \mathbf{pq}/m_e)\,, \tag{8.41}$$

$$\Gamma_{if}^{\mathrm{V}} = -\frac{4\pi}{\Omega^2} \sum_{pq} f_e(p)\, V^2(q)\, M^0_{ii}(q)\, M^0_{ff}(-q)\, \delta(\hbar^2 \mathbf{pq}/m_e)\,. \tag{8.42}$$

(Here q^2-terms are neglected again referring to the optical limit $q \to 0$.) Now it will be shown how to derive (8.40)—(8.42) from the corresponding expressions of the impact approach. According to GRIEM (1974), SOBELMAN, VAINSHTEIN and YUKOV (1981), we have for the perturbed atomic levels

$$\Gamma_n - i\Delta_n = \int \mathrm{d}v\, P(v) \left\{ \frac{i}{\hbar} \int_{-\infty}^{\infty} \mathrm{d}t\, \langle V(t) \rangle_{nn} \right.$$
$$\left. + \frac{1}{\hbar^2} \int_{-\infty}^{\infty} \mathrm{d}t \int_{-\infty}^{\infty} \mathrm{d}t' \sum_\alpha \langle V(t) \rangle_{n\alpha} \langle V(t') \rangle_{\alpha n}\, \mathrm{e}^{i\omega_{n\alpha}(t-t')} \right\}, \tag{8.43}$$

and for the broadening and shift of spectral lines ($i \to f$), correspondingly,

$$\Gamma_{if} - i\Delta_{if} = -\frac{1}{\hbar^2} \int \mathrm{d}v\, P(v) \left\{ \iint_{-\infty}^{\infty} \mathrm{d}t\, \mathrm{d}t' \langle V(t) \rangle_{ii} \langle V(t') \rangle_{ff} \right.$$
$$\left. + \sum_{n=i,f} \int_{-\infty}^{\infty} \mathrm{d}t \int_{-\infty}^{t} \mathrm{d}t' \sum_\alpha \langle V(t) \rangle_{n\alpha} \langle V(t') \rangle_{\alpha n}\, \mathrm{e}^{i\omega_{n\alpha}(t-t')} \right\}. \tag{8.44}$$

These expressions are based on the perturbative treatment of a radiating atom colliding with classically moving electrons. Owing to the assumption of classical paths, he average over collision frequencies ν reads explicitly $\boldsymbol{v}$(, $\boldsymbol{\varrho}$ — velocity and impact parameter of the perturbers)

$$P(\nu)\,\mathrm{d}\nu = n_e v f_M(v)\; 2\pi\varrho\; d\varrho\; \mathrm{d}v\,, \tag{8.45}$$

where

$$f_M(v) = (2/\pi)^{1/2}\,(m/k_BT)^{3/2}\,v^2 \exp\left[-mv^2/2k_BT\right] \tag{8.46}$$

is the Maxwellian distribution of velocity. Besides that $\langle\ldots\rangle_{n\alpha}$ denote matrix elements, n and α the atomic states involved, and $\omega_{n\alpha}$ the corresponding transition frequencies. The α-summations run in principle over the whole spectrum, but in practice only the most important contributions depending on i, f are taken into account. Usually a multipole expansion of the atom-perturber interaction

$$V(t) = V(\boldsymbol{r} - \boldsymbol{R}(t)) - V(\boldsymbol{R}(t))\,, \qquad \boldsymbol{R}(t) = \boldsymbol{\varrho} + \boldsymbol{v}\cdot t\,, \tag{8.47}$$

is applied by which the line shape calculation reduces to the determination of dipole and quadrupole matrix elements, and some characteristic function, see GRIEM (1974).

However, the general expressions are considered here. (8.43) and (8.44) may be written in the form of (8.35), thus we have for the relation between the shift and broadening of atomic levels i, f, and the corresponding spectral lines

$$\begin{aligned} \Delta_{if} &= \Delta_f^{(2)} - \Delta_i^{(2)}\,, \\ \Gamma_{if} &= \Gamma_i^{(2)} + \Gamma_f^{(2)} + \Gamma_{if}^{\mathrm{V}}\,, \\ \Gamma_{if}^{\mathrm{V}} &= \int \mathrm{d}\nu\; P(\nu)\; \Delta_i^{(1)}(\nu)\; \Delta_f^{(1)}(\nu)\,. \end{aligned} \tag{8.48}$$

The upper indices (2) refer to the second order term of (8.43). There is no contribution of the first order term $\Delta_n^{(1)}(\nu)$ to the shift because its average vanishes for a neutral atom in an isotropic plasma. Γ_{if}^{V} is obviously given by the first term on the right-hand side of (8.44). In order to reproduce (8.42) from (8.48) we introduce the Fourier transforms

$$\langle v(t)\rangle_{n\alpha} = \frac{1}{\Omega}\sum_q V(q)\,\mathrm{e}^{-i\boldsymbol{q}\boldsymbol{R}(t)}\,\langle \mathrm{e}^{i\boldsymbol{q}\boldsymbol{r}} - 1\rangle_{n\alpha} \tag{8.49}$$

where $\langle \mathrm{e}^{i\boldsymbol{q}\boldsymbol{r}} - 1\rangle_{n\alpha} \equiv M_{n\alpha}^0(q)$, see (8.10).

Inserting this into (8.43) we obtain

$$\begin{aligned} \Gamma_n^{(2)} - i\Delta_n^{(2)} = \frac{1}{\hbar^2}\int \mathrm{d}\nu\; P(\nu) \int_{-\infty}^{\infty}\mathrm{d}t \int_{-\infty}^{t}\mathrm{d}t'\, \frac{1}{\Omega_d\Omega_{\tilde q}} \sum_{q\tilde q\alpha} |M_{n\alpha}^0(q)|^2 \\ \cdot\, V(q)\, V(\tilde q)\, \exp\{-i\boldsymbol{q}\boldsymbol{R}(t) + i\tilde{\boldsymbol{q}}\boldsymbol{R}(t') + i\omega_{n\alpha}(t - t')\}\,. \end{aligned} \tag{8.50}$$

Because of the geometry of collisions, we may rewrite

$$\exp\{-i\boldsymbol{q}\boldsymbol{R}(t) + i\boldsymbol{q}\boldsymbol{R}(t')\} = \exp\left(-i\tilde{\boldsymbol{R}}(\boldsymbol{q} - \tilde{\boldsymbol{q}})\right)\exp\left(-iv(qt - qt')\right), \tag{8.51}$$

with $\tilde{\boldsymbol{R}} = (v(t + t'),\, \varrho_2,\, \varrho_3)$.

Furthermore, we substitute the integrations using (8.38), (8.45), (8.46), and (8.51) as follows:

$$\frac{1}{\hbar^2}\int d\nu\, P(\nu) \int_{-\infty}^{\infty} dt \int_{-\infty}^{t} dt' \ldots = \frac{2}{\Omega} \sum_p f_e(p) \iint_{-\infty}^{\infty} d\varrho_2\, d\varrho_3 \int_{-\infty}^{\infty} d(v\tilde{t}) \int_0^{\infty} d\tau \ldots$$

$$= \frac{2}{\Omega} \sum_p f_e(p) \int d\tilde{\boldsymbol{R}} \int_0^{\infty} d\tau \ldots , \tag{8.52}$$

where $p = m_e v$, $\tilde{t} = t + t'$, and $\tau = t - t'$. Now the R- and $\tilde{q}$-integrations may be performed, and with (8.51) we finally obtain

$$\Gamma_n^{(2)} - i\Delta_n^{(2)} = \frac{2}{\Omega^2} \sum_{pq\alpha} f_e(p)\, V^2(q)\, |M_{n\alpha}^0(q)|^2 \int_0^{\infty} d\tau\, e^{i\tau(\omega_{n\alpha} - \hbar pq/m_e)} . \tag{8.53}$$

This expression exactly reproduces the self-energy contributions of (8.40), (8.41) because of $\hbar\omega_{n\alpha} = E_n - E_\alpha$ and

$$\int_0^{\infty} d\tau \exp\{i\tau(x + i0)\} = \frac{i}{x + i0} = \pi\delta(x) + iP\frac{1}{x} . \tag{8.54}$$

The same procedure is to be applied to identify the vertex term of (8.42) with Γ_{if}^{V} as defined by (8.48) and (8.44). Similar to (8.50) we start with

$$\Gamma_{if}^{V} = -\frac{1}{\hbar^2}\int d\nu\, P(\nu) \iint_{-\infty}^{\infty} dt\, dt' \frac{1}{\Omega^2} \sum_{\tilde{q}q} M_{ii}^0(q)\, M_{ff}^0(-\tilde{q})$$

$$\cdot\, V(q)\, V(\tilde{q}) \exp\{-i\boldsymbol{q}\boldsymbol{R}(t) + i\tilde{\boldsymbol{q}}\boldsymbol{R}(t')\} , \tag{8.55}$$

what becomes by appropriate use of (8.51) to (8.52)

$$\Gamma_{if}^{V} = -\frac{2}{\Omega^2} \sum_{pq} f_e(p)\, V^2(q)\, M_{ii}^0(q)\, M_{ff}^0(-q) \int_{-\infty}^{\infty} d\tau\, e^{-i\hbar\tau pq/m_e} . \tag{8.56}$$

Notice that an additional factor of two will be produced by applying (8.54), thus (8.42) follows from (8.56).

By the derivations given here we have shown that the semiclassical impact approximation is principially included within our approach to the line shape based on the approximation for Π_{atom} (8.15).

In particular, it becomes evident from the derivation of (8.40)–(8.42) which approximations have to be applied to come to the general impact shift and broadening expressions.

However, the essential value of our approach consists in the systematic concept how to generalize the impact approach on the basis of many-body theory.

8.2.3. Shift of Spectral Lines in Dense Hydrogen Plasmas

In order to show how explicit results can be obtained within our many-particle approach to the line profile in dense plasmas, we evaluate the line shift of hydrogen-like plasmas Δ_{if} (8.35) which is caused by the interaction with charged particles. It follows from theory that Δ_{if} which depends on the temperature, density, and plasma composition, starts with a linear term in the density of free carriers $n_{\text{free}} = n_e + n_i$ (a neutral plasma is considered, free electron density n_e equals the singly charged ion density n_i):

$$\lim_{n_{\text{free}} \to 0} \Delta_{if}(T, n_{\text{free}}) = -\delta_{if}(T)\, n_{\text{free}} \,. \tag{8.57}$$

Usually shifts are given in nanometers, which we obtain to be

$$\Delta\lambda_{if} = 2\pi\hbar c \cdot \hbar\Delta_{if}/(\hbar\omega_{if}^{(0)})^2 \,. \tag{8.58}$$

Following Röpke, Seifert and Kilimann (1981), the expressions for $\delta_{if}(T)$ will be evaluated numerically, where the plasmon pole approximation for the dynamically screened potential will be used. A comparison of our results with experimental values as well as with other theoretical approaches will also be given.

According to (8.35), the shift of spectral lines Δ_{if} is obtained from the self-energy shift of the two-particles states (8.36). The sum over the two-particle states α is decomposed into the bound part $D_n^{(b)}$ and the scattering part $D_n^{(sc)}$, respectively. The scattering states may be approximated by products of free particle wave functions. Then, because of the isotropy of the plasma, the integrals over angular variables are collected into the following expressions:

$$G_{nn'}(q) = \int \mathrm{d}\Omega_q \left| \int \frac{\mathrm{d}\boldsymbol{p}}{(2\pi)^3} \psi_n^*(p)\, \psi_{n'}(\boldsymbol{p} + \boldsymbol{q}) - \delta_{nn'} \right|^2 , \tag{8.59}$$

$$F_n(p, q) = \frac{1}{(2\pi)^3} \int \mathrm{d}\Omega_q\, \mathrm{d}\Omega_p\, |\psi_n(p) - \psi_n(\boldsymbol{p} - \boldsymbol{q})|^2. \tag{8.60}$$

These integrals have been performed analytically with hydrogen wave functions. We give some examples of the resulting expressions:

$$\begin{aligned}
G_{1s,1s}(q) &= 4\pi q^4(q^2 + 8)^2/(q^2 + 4)^4 \,, \\
G_{1s,2p}(q) &= 96\pi q^2/(q^2 + 2.25)^6 \,, \\
F_{1s}(p, q) &= 2^7\{[1/(1 + P)^2 - A]^2 + \tfrac{4}{3} A^3 B\} \,, \\
F_{2p}(p, q) &= \frac{2^{14}}{3} \{P/(1 - P)^6 - A[1 - A(P^2 - (Q + 1)^2)]\}/(1 + P)^3 \\
&\quad + A^4[C(C^2 + B) - A(C^4 + 2C^2B + 0.2B^2)]\} \,,
\end{aligned} \tag{8.61}$$

with $P = (np)^2$, $Q = (nq)^2$, n — main quantum number, and $C = P + Q + 1$, $B = 4PQ$, $A = 1/(C^2 - B)$.

The inverse dielectric function occuring in (8.36) can be derived from the RPA expression which has been described in Section 4.2. The $\overline{\omega}$-integral in (8.36) may be performed if a pole approximation is used for $\operatorname{Im} \varepsilon^{-1}$. Concerning the low density limit we take

$$\text{Im}\,\varepsilon^{-1}(q,\omega) = \frac{\pi n e^2}{2\varepsilon_0} \sum_{a=e,i} \frac{1}{m_a \omega_{q,a}} \{\delta(\omega + \omega_{q,a}) - \delta(\omega - \omega_{q,a})\} \tag{8.62}$$

where the dispersion is given by (plasma frequency $n_e e^2/\varepsilon_0 m_e \to 0$)

$$\omega_{q,a}^2 = \frac{c_a q^2}{m_a \beta} + \frac{\hbar^2 q^4}{4m_a^2}, \qquad c_e = 3, \qquad c_i = 0.6\,, \tag{8.63}$$

within the double plasmon pole approximation, see Section 5.2.5. Thus we obtain finally an expression for the shift Δ_n where the remaining integrations will have to be performed numerically. The shift

$$\Delta_n = n_e \frac{\hbar e^4}{\pi} \int \mathrm{d}q \int \mathrm{d}\Omega_q \sum_{a=e,i} \frac{1}{m_a \omega_{q,a}} \{(1 + n(\hbar\omega_{q,a}))\, \widetilde{D}_n(q, \omega_{q,a})$$
$$+\, n(\hbar\omega_{q,a})\, \widetilde{D}_n(q, -\omega_{q,a})\} = n_e \delta_n(T) \tag{8.64}$$

with

$$\int \mathrm{d}\Omega_q\, \widetilde{D}_n(q, \omega_q) = \sum_{n'}^{(b)} \frac{G_{nn'}(q)}{E_n - E_{n'} - \hbar\omega_q} + \int \mathrm{d}p \frac{p^2 F_n(p, q)}{E_n - \hbar^2 p^2/2m_a - \hbar\omega_q}$$

may be obviously divided into an ionic and an electronic contribution, respectively, according to the decomposition of $\text{Im}\,\varepsilon^{-1}$ (8.62). On the other hand the expressions for the shift consists of a bound-bound and a bound-free contribution.

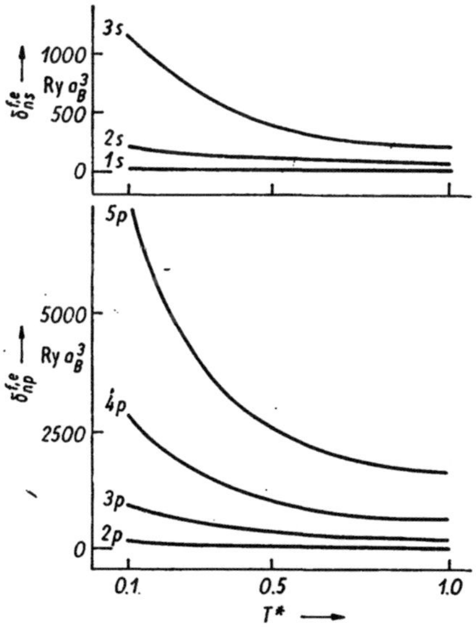

Fig. 8.1. The electronic contributions $\delta_n^{f,e}$ to the two-particle energy level shifts due to bound-free transitions within the temperature region $T^* = k_B T/\text{Ry} = 0.1 \ldots 1$

The electronic part of the bound-free contribution $\delta_n^{f,e}(T)$ is shown in Fig. 8.1. in dependence on the temperature T for $n = 1s, 2s, 3s$ and $2p, \ldots, 5p$. Notice the increase of $\delta_n^{f,e}$ with decreasing temperature. The corresponding ionic contributions yield some higher values of $\delta_n^{f,i}$ which depend only weakly on the temperature. These $\delta_n^{f,e}$ and $\delta_n^{f,i}$ lead to line shifts the behaviour of which can be compared qualitatively with the experimental findings (as regards, e.g., the order of magnitude or the increase of the shift with the quantum number n). However, these shift values are not in good quantitative agreement with measured one. Improved values may be achieved by introducing empirically a cut-off frequency $\hbar\omega_c = 0.03$ Ry into the integrations of (8.64) (Röpke, Seifert and Kilimann, 1981). Particularly, the large ionic contributions become small in this way. In Tab. 8.1, the shift values derived from the electronic contribution without cut-off and the total shifts according to a cut-off frequency are compared with experimental shift values. The empirically introduced cut-off should be considered as a hint that a closer agreement is achieved suppressing the low-frequency component of Im ε^{-1}, e.g., by improving the RPA dielectric function.

Tab. 8.1. Red shifts in nanometers for several Lyman and Balmer lines at density $n_e = 10^{23}$ m^{-3}
a) Shift values according to full $\delta_n^{f,e}$ at $T^* = 0.1$,
b) Total shifts $\delta_n^f = \delta_n^{f,e} + \delta_n^{f,i}$ evaluated with a frequency cut-off at 0.03 for $T^* = 0.1$,
c) Measured shift values normalized to 10^{23} m^{-3} assuming linear density dependence, temperatures between $T^* = 0.06 \ldots 0.085$.

sp. line	a)	b)	c)
L_α	0.0003	0.0004	0 ... −0.002[1]) 0.005[2])
L_β	0.0016	0.0014	±0.007[1]) 0.005[2])
L_γ	0.0044	0.0032	—
H_α	0.0508	0.0437	0.059 ± 0.009[3])
H_β	0.103	0.0703	0.072 ± 0.006[3]) 0.135 ± 0.017[4])
H_γ	0.220	0.0945	0.084 ± 0.018[3])

[1]) Fussmann (1974), [2]) Grützmacher and Wende (1977, 1978) [3]) Wiese, Kelleher and Paquette (1972), Wiese, Kelleher and Helbig (1975), [4]) Halenka and Musielok (1977)

Evaluating the bound-bound contributions it can be shown that only the near-resonant transitions prove to be of importance. It should be pointed out that the shift is over-estimated if the bound-bound and the bound-free contributions are summed up, see Röpke, Seifert and Kilimann (1981). It follows that the bound-bound contributions may be included in the calculation consistently only if the bound-free contributions are reduced by using the correct scattering states instead of free particle wave functions. The knowledge of the exact limiting values (polarizabilities of $D_n(q, \omega)$ permits the construction of a semi-empirical Padé approximation

$$\Delta_n^{\text{Padé}} = n_e \frac{\hbar e^2}{2} \sum_{a=e,i} \frac{1}{m_a} \int dq \frac{1 + n(\hbar\omega_{q,a})}{\omega_{q,a}} \frac{4\alpha_n q^2}{(1 + \hbar q^2 \sqrt{\alpha_n}/2e\sqrt{m_e})^2} \tag{8.65}$$

which reproduces the order of magnitude of the shift values shown in Fig. 8.1.

Using the approximant (8.65) to evaluate the shift of the alkali metal resonance lines, qualitative agreement with recently measured values (Purič et al., 1977) is obtained.

The approach presented here does not introduce the microfield concept and cannot be compared immediately with other theories based on it like in Griem (1974), Vidal, Cooper and Smith (1970), Brissaud et al. (1976).

8.2.4. Estimation of the Shift and Broadening of Spectral Lines for an Argon Plasma

In the case of hydrogen, described in the previous Section, the atomic wave functions are explicitely known, thus integrations involving the matrix elements $M^0_{nn'}(q)$ may be performed analytically. In order to illustrate the Green's function approach presented here and to give at least an estimation for more complex systems as inert gases in the framework of our approach, we may use a semiempirical approximation within which the shift and broadening parameters are related to spectroscopic oscillator strengths instead of explicit wave functions (matrix elements). For that purpose we introduce into (8.36) the plasmon pole approximation (8.62) replacing the inverse RPA dielectric function by

$$\mathrm{Im}\,\varepsilon^{-1}_{\mathrm{RPA}}(q,\omega) = 8\pi^2 n_e^* \frac{1}{\omega_q}\left(\delta(\omega+\omega_q) - \delta(\omega-\omega_q)\right), \tag{8.66}$$

where the pole obeys the dispersion relation

$$\omega_q = (16\pi n_e^* + 6T^* q^2 + q^4)^{1/2} \approx q(6T^* + q^2)^{1/2} . \tag{8.67}$$

Here only the electron contribution is taken into account, and Rydberg atomic units $\hbar = e^2/8\pi\varepsilon_0 = 2m_e = 1$ are used. Accordingly, density and temperature are given by $n_e^* = n_e a_B^3$, $T^* = k_B T\, 8\pi\varepsilon_0 a_B/e^2$, a_B — Bohr's radius.

Inserting (8.66) into (8.36) leads to

$$\Delta_n = \frac{8}{\pi} n_e^* P \int_0^\infty \frac{dq}{\omega_q} \int d\Omega_q \sum_{n'} |M^0_{nn'}(q)|^2 \left\{\frac{1+n(\omega_q)}{E_n - E_{n',-q} - \omega_q} - \frac{n(\omega_1)}{E_n - E_{n,'-q} + \omega_q}\right\},$$

$$\Gamma_n = 8n_e^* \int_0^\infty \frac{dq}{\omega_q} \int d\Omega_q \sum_{n'} |M^0_{nn'}(q)|^2 \left\{(1+n(\omega_q))\,\delta(E_n - E_{n',-q} - \omega_q) + n(\omega_q)\,\delta(E_n - E_{n',-q} + \omega_q)\right\} . \tag{8.68}$$

The integrands may be performed approximately using Padé approximants which are based on the limiting behaviour of the integrand functions. For $q \to 0$ the $M^0_{nn'}$ reduce to dipole matrix elements $d_{nn'}$. Therefore the expressions (8.68) are essentially determined by the oscillator strengths $f_{nn'}$, and by the static polarizability α_n, respectively, which are related to the dipole matrix elements

$$|d_{nn'}|^2 = f_{n'n}/(E_{n'} - E_n) , \qquad \alpha_n = 4 \sum_{n'} f_{n'n}/(E_{n'} - E_n)^2 . \tag{8.69}$$

On the other hand we find for $q \to \infty$ that the integrand of Δ_n goes with q^{-4}, whereas the Γ_n-integrand vanishes exponentially because of the Bose factors. According to that limiting behaviour the following Padé approximants may be used:

$$\Delta_n^{\text{Padé}} = 16 n_e^* \int_0^\infty dq \frac{1+n(\omega_q)}{\omega_q} \frac{\alpha_n q^2}{(1+\sqrt{\alpha_n/2}\,q^2)^2}, \tag{8.70}$$

$$\Gamma_n^{\text{Padé}} = 32\pi n_e^* \int_0^\infty dq \frac{q^2}{\omega_q} \sum_{n'} f_{n'n}/\omega_{n'n} \left\{(1+n(\omega_q))\,\delta(\omega_{n'n}+\omega_q) - n(\omega_q)\,\delta(\omega_{n'n}-\omega_q)\right\} . \tag{8.71}$$

Because of the δ-functions one of the terms of (8.71) is always vanishing for $\omega_{nn'} \gtrless 0$. Both terms may be put together using $1 + n(-\omega_q) = -n(\omega_q)$, thus we have after the q-integration

$$\Gamma_n^{\text{Padé}} = 16\pi n_e^* \sum_{n'} n(\omega_{n'n}) \frac{f_{n'n}}{\omega_{n'n}^2} \left\{ \frac{(\omega_{n'n}^2 + 9T^*)^{1/2} - 3T^*}{1 + (3T^*/\omega_{n'n})^2} \right\}^{1/2} . \tag{8.72}$$

We expect to obtain a rough estimation of the shift and broadening of atomic energy levels and the corresponding spectral lines if the approximants (8.70) and (8.72) are evaluated with spectroscopic f-values.

As a example we consider the $4s$ and $4p$ level of argon having configurations ($1s_4$), ($1s_5$), and ($2p_8$), ($2p_9$) in the Paschen notation. The summations over a set of n' in (8.70), (8.72) are performed with the $f_{nn'}$-values, and the energy terms E_n, $E_{n'}$ according to the tables of Wiese (1972). The remaining terms contributing to the level under consideration (especially the bound-free transitions) are summed up approximately using an effective transition frequency ($\omega_{nn'} \approx -E_n$) and the f-sum rule. The values we obtained for the polarizability α_n, the level shifts Δ_n, and the broadening parameters Γ_n are given in Tab. 8.2, see Hitzschke et al. (1985). The results for the

Tab. 8.2. Atomic polarizability α_n, level shift Δ_n, and broadening Γ_n for an Argon plasma with $n_e = 10^{17}$ cm^{-3} and $T = 15000$ K

level n		$\alpha_n(a_B^3)$	$\Delta_n(10^{-6}\, e^2/2a_B)$	$\Gamma_n(10^{-6}\, e^2/2a_B)$
$4s$	($1s_4$)	227	3.95	0.92
	($1s_5$)	336	5.4	1.6
$4p$	($2p_8$)	623.5	7.6	2.2
	($2p_9$)	800	8.9	2.8

Tab. 8.3. Argon line broadening Γ and shifts Δ corresponding to Tab. 8.2

transition	Hitzschke et al. (1985)		Griem (1974)		
	Δ, nm	Γ, nm	Δ, nm	Γ, nm	classf. number
$4p(2p_8) - 4s(1s_4)$	0.029	0.025	0.013	0.051	(33)
$4p(2p_9) - 4s(1s_4)$	0.040	0.030	—	—	—
$4p(2p_8) - 4s(1s_5)$	0.018	0.031	—	—	(26)
$4p(2p_9) - 4s(1s_5)$	0.028	0.036	0.032	0.060	(25)

shift and broadening of the corresponding argon lines are shown in Tab. 8.3. It should be noticed that the broadening parameters given here are to be considered as upper limits since we have neglected the vertex contribution (8.32) which is of opposite sign.

In two cases we may compare in Tab. 8.3 with the values given by Griem (1974) which are of the same order of magnitude. However, the reliability of our results cannot be estimated before discussing the whole line profile including also ionic contributions and comparing it with experimental findings, especially at densities higher than 10^{17} cm^{-3}. Especially for such high dense plasmas, where the conventional approaches to spectral lines become invalid, the Green's function approach given here may serve as a starting point to find relevant approximations.

8.3. Further Approaches and Concluding Remarks

The theory of the line profile of absorption spectra has been elaborated for a long time and describes the experimental findings in plasmas up to densities of the order 10^{17} cm^{-3} quite well. However, in very high dense plasmas a systematic many-body approach is needed because of the collective behaviour of the strongly correlated plasma. Thus, the Green's function approach to dense Coulombic systems presented in this book may serve also as an appropriate starting point to evaluate the spectroscopic properties.

The results given in this Chapter 8. are obtained by summing up special classes of diagrams for the polarization function (8.15) which are of relevance in the low density limit. At higher densities the class of diagrams for evaluating the polarization function has to be extended.

Furthermore, explicit results are obtained if the integrals are evaluated (8.36) with given wave functions and dielectric function. One aspect of further work would be the incorporation of ionic contributions to the line shape. Besides that, evaluations have to be performed for degenerate or closely lying levels instead of referring only to isolated, non-degenerate atomic energy levels as it was done here. It is to be expected that we will come this way to an adequate theoretical description of the optical properties of high dense plasmas for which the traditional approaches are no longer applicable.

We give two examples which need also a systematic many-body theoretical approach. Recently the question whether or not there exists a transparency window in high dense plasmas near the continuum edge has been disscueed in the literature. In contrast to arguments given by Kobzev, Kurilenko and Norman (1981), a more detailed investigation with correct oscillator strengths and energy levels of statically screened model potential given by Höhne and Zimmermann (1982) no transparency window was obtained. However, in the latter paper the two-particle problem was treated only on the base of an effective wave equation with a statically screened potential, and the broadening of spectral lines was introduced in a phenomenological manner. The behaviour of the optical spectra near the continuum edge should be obtained also from a consequent many-body theoretical approach.

Optical spectra are discussed for high dense plasmas recently also by Gryaznov et al. (1980), especially the behaviour of the absorption coefficient. A discussion on the base of a confined atom model was given in Ebeling et al. (1983), but the discussion of these high dense plasma results on the basis of a systematic many-body approach remains an open question.

9. References

ABÉ, R.: Progr. Theor. Phys. (Kyoto) **22**, 213 (1959).

ABRAHAMS, E., ANDERSON, P. W., LICCIARDELLO, D. C., RAMAKRISHNAN, T. V.: Phys. Rev. Lett. **42**, 673 (1979).

ABRIKOSOW, A. A.: Einführung in die Theorie normaler Metalle. Akademie-Verlag, Berlin 1976.

ABRIKOSOV, A. A., GOR'KOV, L., DZYALOSHINSKII, I. E.: Methods of Quantum Field Theory in Statistical Physics (in Russian). Moscow 1962.

AGARWAL, G. K., PATHAK, K. N.: J. Phys. **C15**, 5063, 5999 (1982); **C16**, 1887 (1983).

AHLBEHRENDT, N., BIGUN, G. I., JUCHNOVSKIJ, I. R., KELBG, G.: Ann. Physik (Leipzig) **24**, 188 (1970).

AIZENMAN, M., MARTIN, P. A.: Comm. Math. Phys. **78**, 99 (1980).

AKHIEZER, A. A., BERESTETSKII, V. B.: Quantum Electrodynamics (in Russian). Nauka, Moscow 1962.

ALASTUEY, A.: Phys. Lett. **A76**, 134 (1980); Physica **A110**, 293 (1982).

ALASTUEY, A., JANCOVICI, B.: Astrophys. J. **226**, 1034 (1978); Physica **A 97**, 349 (1979); **A102**, 327 (1980); J. Physique **42**, 1 (1981) .

ALBERTI, P., UHLMANN, A.: Dissipative Motion in State Spaces. BSG B.G. Teubner Verlagsgesellschaft, Leipzig 1981.

ALEKSEEV, V. A.: Teplofyz. Vys. Temp. **8**, 641 (1970).

ALEKSEEV, V. A., IAKUBOV, I. T.: Phys. Rep. **26**, 1 (1983).

ALEKSEEV, V. A., VEDENOV, A. A.: Usp. Fiz. Nauk **102**, 3 (1970).

ALEKSEEV, V. A.. et al.: The Effect of Saturation on the Thermo-emf of Cesium at High Temperatures and Pressures. Proc. 5th Conf. High Temp. High Press. Moscow 1975, p. 677.

ALFVÈN, H., ARRHENIUS, G.: Evolution of the Solar System. NASA, Washington 1967. Russ. Transl. Mir, Moscow 1979.

AMBEGAOKAR, V.: In: Astrophysics and Many Body Problem. Benjamin, New York 1963.

ANDERSON, H. C., CHANDLER, D.: J. Chem. Phys. **53**, 547 (1970); **54**, 26 (1971); **55**, 1497 (1971); **57**, 1918, 1930, 2626 (1972).

ANDERSON, P. W.: Phys. Rev. **109**, 1492 (1958).

ANDREEV, S. I., GAVRILOVA, T. V.: Teplofiz. Vys. Temp. **13**, 176 (1975).

AOKI, H., ANDO, T.: Solid State Commun. **38**, 1079 (1981).

APPEL, J.: Phys. Rev. **122**, 1760 (1961), **125**, 1815 (1962).

ARPONEN, J., PAJANNE, E.: Annals of Physics **91**, 450 (1975); J. Phys. **C12**, 3013 (1979); **15**, 2665 (1982).

ASHCROFT, N. W.: Phys. Rev. Lett. **21**, 1748 (1968); Dense Conducting Liquids. In: The Liquid State of Matter (E. W. MONTROLL, J. L. LEBOWITZ, eds.). North-Holland, Amsterdam 1982.

ASHCROFT, N. W., MERMIN, N.: Physics of Solid State (in Russian). Mir, Moscow 1979.

ASHCROFT, N. W., SCHAICH, W.: Phys. Rev. **B1**, 1370 (1970).

ASHCROFT, N. W., STROUT, D.: Solid State Physics **33**, 1 (1978).

BAKEEV, A. A., ROVINSKII, R. E.: Teplofiz. Vys. Temp. **8**, 1121 (1970).

BAKSHI, P., CALINOV, R., GOLDEN, K. I., KALMAN, G., MERLINI, D.: Phys. Rev. **A20**, 336 (1979); **A23**, 1915 (1981).
BALLENTINE, L. E., HEANEY, W. J.: J. Phys. **C7**, 1985 (1974).
BAMZAI, A. S., DEB, B. M.: Rev. Mod. Phys. **53**, 95 (1981).
BARKER, A. A.: Austr. J. Phys. **21**, 121 (1968); Phys. Rev. **171**, 186 (1968); **179**, 129 (1969); Chem. Phys. **55**, 1751 (1971).
BARRANGER, M.: Phys. Rev. **111**, 481, 494 (1958); **112**, 855 (1958); Spectral Line Broadering in Plasmas. In: Atomic and Molecular Processes (D. R. BATES, ed.). Acad. Press, New York 1962, p. 493.
BARROW, J., SILK, J.: Sci. Am. April, p. 98 (1980).
BARTH, U., HEDIN, L.: J. Phys. **C5**, 1629 (1972).
BARTSCH, G. P., EBELING, W.: Beitr. Plasmaphys. **11**, 393 (1971); **15**, 25 (1975).
BAUMGARTL, B. J.: Z. Physik **198**, 148 (1967)
BAUS, M.: J. Phys. **C13**, L41 (1980).
BAUS, M., HANSEN, J.-P.: Phys. Rep. **59**, 1 (1980); J. Stat. Phys. **31**, 409 (1983).
BAYM, G., KADANOFF, L. P.: Phys. Rev. **125**, 287 (1961).
BERG, L.: Asymptotische Darstellungen und Entwicklungen, Verlag der Wissenschaften, Berlin 1968.
BERNASCONI, J., SCHNEIDER, T. (eds.): Physics in One Dimension. Springer-Verlag, Berlin, Heidelberg, New York 1981.
BESSIS, N., BESSIS, G., DAKHEL, B., HADINGER, G.: J. Phys. **A11**, 467 (1978).
BIBERMAN, L. M., VOROB'EV, V. S., YAKUBOV, I. T.: Kinetika neravnovesnoi nizkotemperaturnoi plazmy. Nauka, Moscow 1982.
BINDER, K. (ed.): Monte Carlo Methods in Statistical Physics. Springer-Verlag, Berlin, Heidelberg, New York 1979. Russ. Transl. Mir, Moscow 1982.
BISHOP, R. F., LÜHRMANN, K.: Phys. Rev. **B17**, 3757 (1978).
BLATT, F. J., Solid State Phys. **4**, 199 (1957).
BLUM, L., GRUBER, C., LEBOWITZ, J.-L., MARTIN, P.: Phys. Rev. Lett. **48**, 1769 (1982).
BLUM, L., HENDERSON, D., LEBOWITZ, J. L., GRUBER, C., MARTIN, P. A.: J. Chem. Phys. **75**, 5974 (1981).
BLÜMLEIN, J.: On the Theory of the Electrical Conductivity of a Non-Ideal Quantum Plasma. Proc. 2nd Symp. Statistical Mechanics. Dubna 1981.
BLÜMLEIN, J., KRAEFT, W. D.: preprint 1984.
BLÜMLEIN, J., KRAEFT, W. D., MEYER, T.: Ann. Physik **37**, 379 (1980).
BOERCKER, D. B., DUFTY, J. W.: Annals of Physics **119**, 43 (1979).
BOGOLYUBOV, N. N.: Problems of the Dynamical Theory in Statistical Physics (in Russian). Gostekhizdat, Moscow 1946.
BOGOLYUBOV, N. N.: Selected Works, 1—3 (in Russian). Naukova Dumka, Kiev 1969, 1970, 1971.
BOHM, D., PINES, D.: Phys. Rev. **92**, 609 (1953).
BONCH-BRUEVICH, V. L.: The Electronic Structure of Heavily Doped Semiconductors, New York 1968.
BONCH-BRUEVICH, V. L., TYABLIKOV, S. V.: Method of Green Functions in Statistical Physics (in Russian). Nauka. Moscow 1961.
BONCH-BREUVICH, V. L., ZVYAGIN, I. P., KEIPER, R., MIRONOV, A. G., ENDERLEIN, R., ESSER, B.: Electronic Theory of Disordered Semiconductors (in Russian). Nauka, Moscow 1981.
BÖTTGER, H.: Principles of the Theory of Lattice Dynamics. Akademie-Verlag, Berlin 1983.
BÖTTGER, H., BRYKSIN, V. V.: Phys. stat. sol. (b) **68**, 285 (1975), **78**, **11**, **415** (1976), **89**, 9 (1976).
BRAUER, W.: Einführung in die Elektronentheorie der Metalle. Akad. Verlagsgesellsch. Geest & Portig, Leipzig 1972.
BRINKMAN, W. F., RICE, T. M., Phys. Rev. **B7**, 1508 (1973).
BRISSAND, A., FRISCH, U., J. Math. Phys. **15** (1974) 524.
BRISSAND, A., GOLDBACH, C., LEORAT, J., MAZURE, A., NOLLEZ, G., J. Phys. **B9**, 1129, 1147 (1976).

Brovman, E. G., Kagan, Yu.: Zh. Eksp. Teor. Fiz. **52**, 557 (1967), **57**, 10 (1969).
Brovman, E. G., Kagan, J., Kholas, A.: Zh. Eksp. Teor. Fiz. **62**, 1492 (1972).
Brown, R. C., March, N. H.: Phys. Earth Planet. Interiors **6**, 206 (1972).
Brush, S. G., Sahlin, H. L., Teller, E.: J. Chem. Phys. **45**, 2101 (1966).
Brydges, D., Federbusch, P.: Comm. Math. Phys. **78**, 197 (1980).
Bührig, W.: J. Math. Phys. **18**, 1121 (1977).
Busch, G., Güntherodt, H.-J.: Solid State Phys. **29**, 282 (1974).
Bushman, A. V., Lomakin, B. N., Secenov, V. A., Fortov, V. E., Scekotov, Sharindshanov, I. I.: Zh. Eksp. Teor. Fiz. **69**, 1624 (1975).
Bushman, A. V., et al.: Preprints Nrs. 532.593, 539.182, Chernogolovka 1983.
Callaway, J.: Energy Band Theory. Acad. Press. New York 1964. Russ. transl. Mir, Moscow 1969.
Callen, H., Swendson, R. H., Tahir Kheli, R. A.: Phys. Lett. **25A**, 505 (1967).
Cap, F.: Einführung in die Plasmaphysik. II. Wellen und Instabilitäten. Akademie-Verlag, Berlin 1972.
Carini, P., Kalman, G., Golden, K. I.: Phys. Lett. **A78**, 450 (1980).
Carnahan, N. F., Starling, K. E.: J. Chem. Phys. **51**, 632 (1969).
Carr, W. J., Maradudin, A. A.: Phys. Rev. **A133**, 371 (1964).
Carr, W. J., Coldwell-Horsefall, R. A., Feim, A. E.: Phys. Rev. **124**, 747 (1961).
Cauble, R., Boercker, D. B.: Phys. Rev. **A28**, 944 (1983).
Ceperley, D. M.: Phys. Rev. **B18**, 3126 (1978); Physica **B108**, 875 (1981); Proc. Conf. Monte Carlo Methods. Paris 1982. Preprint UCRL — 88642, Livermore 1983.
Ceperley, D. M., Alder, B. J.: Phys. Rev. Lett. **45**, 566 (1980); Physica **108B**, 875 (1981)
Ceperley, D., Kalos, M.: Quantum Many body problem. In: Monte Carlo Methods in Statistical Physics (K. Binder, ed.). Springer-Verlag, Berlin, Heidelberg, New York 1979.
Chakravarty, S., Dasgupta, C.: Phys. Rev. **B22**, 369 (1980).
Chapman, S., Cowling, T. G.: The Mathematical Theory of Non-uniform Gases. Univ. Press, Cambridge 1958.
Chihara, J.: Progr. Theor. Phys. **59**, 76 (1978).
Christoph, V., Röpke, G.: Phys. stat. sol. (b) **80**, K 117 (1977).
Coldwell-Horsfall, R. A., Maradudin, A. A.: J. Math. Phys. **1**, 395 (1960).
Cole, M.: Rev. Mod. Phys. **46**, 451 (1974).
Combescot, M., Nozières, P.: J. Phys. **C5**, 2369 (1972).
Czerwon, H. J.: Diplom Thesis, Wilhelm-Pieck-Universität Rostock 1972.
Dashen, R., Ma, S., Bernstein, H. J.: Phys. Rev. **187**, 345 (1969).
Dautcourt, G.: Relativistische Astrophysik. Akademie-Verlag, Berlin 1976.
Dawydow, A. S.: Quantenmechanik, Verlag der Wissenschaften, Berlin 1972.
De Boer, J.: Construction Operator Formalismen in Many Particle Systems. In: Studies in Statistical Mechanics (J. De Boer, G. E. Uhlenbeck, eds.) North-Holland, Amsterdam 1965.
De Groot, S. R., Mazur, P.: Nonequilibrium thermodynamics. North-Holland, Amsterdam 1962.
De Leeuw, S. W., Perram, J. W., Smith, E. R.: Physica **A119**, 441 (1983).
Del Rio, F., De Witt, H. E.: Phys. Lett. **A30**, 337 (1969); Phys. Fluids **12**, 791 (1969); Rev. Mex. Fis. (Mexico) **20**, 105 (1971).
Deutsch, C.: Physica Scripta **T2/1**, 192 (1982).
Deutsch, C., Lavaud, M.: Phys. Lett. **A39**, 253 (1972); **A43**, 193 (1973); **A46**, 349 (1974); Phys. Rev. Lett. **31**, 921 (1973); Phys. Rev. **A9**, 2598 (1974).
Deutsch, C., De Witt, H. E., Furutani, Y.: Phys. Rev. **A20**, 2631 (1979).
De Wette, F. W.: Phys. Rev. **135A**, 287 (1964).
De Witt, H. E.: J. Nucl. Energy **C2**, 27 (1961); J. Math. Phys. **3**, 1216 (1962); **7**, 616 (1966); in: S. Kumar (ed.), Low Luminosity Stars, New York 1968; Phys. Rev. **A14**, 816, 1290 (1976); Equilibrium Statistical Mechanics of Strongly Coupled Plasmas by Numerical Simulation. In: Strongly Coupled Plasmas (G. Kalman, P. Carini, eds.). Plenum Press, New York 1978.
De Witt, H. E., Hubbard, W. B.: Astron. J. **205**, 295 (1976).

De Witt, H. E., Graboske, H. C., Cooper, M. S.: Astrophys. J. **181**, 439 (1973).
Dharma-wardana, M. W. C., Perrot, F.: Phys. Rev. **A26**, 2096 (1982).
Dharma-wardana, W. M. C., Taylor, R.: J. Phys. **C14**, 629 (1981).
Dharma-wardana, M. W. C., Perrot, F., Aers, G. C.: Phys. Rev. **A28**, 344 (1983).
Dharma-wardana, M. W. C., Grimaldi, F., Lecourt, A., Pellisiert, J. L.: Phys. Rev. **A21**, 379 (1980).
Dienemann, H., Clemens, G., and Kraeft, W. D.: Ann. Physik (Leipzig) **37**, 444 (1980).
Dirac, P. A. M.: Principles of Quantum Mechanics. Clarendon Press, Oxford 1958 (fourth edition).
Dolgov, O. V., Maksimov, E. G.: Usp. Fiz. Nauk **138**, 95 (1982).
Dolgov, O. V., Kirzhnits, D. A., Maksimov, E. G.: Rev. Mod. Phys. **53**, 81 (1981).
Du Bois, D. F.: Ann. Phys. (N. Y.) **7**, 174 (1959); **8**, 24 (1959).
Du Bois, D. F., Kivelson, M. G.: Phys. Rev. **186**, 409 (1969).
Dufty, J. W., Boercker, D. B.: J. Quant. Spec. Rad. Trans. **16**, 1605 (1976).
Dunleavy, H. N., Jones, W.: J. Phys. **F8**, 1477 (1978).
Dyson, F. J.: J. Math. Phys. **8**, 1538 (1967).
Dyson, F. J., Lenard, A.: J. Math. Phys. **8** (1967) 423; **9**, 698 (1968).
Dyson, F. J., Montroll, E., Katz, M., Fisher, M.: Stability and Phase Transitions (in Russian). Mir, Moscow 1973.
Ebeling, W.: Ann. Physik **17**, 415 (1966); **19**, 104 (1967); **21**, 315 (1968); **22**, 33, 383, 392 (1969); Physica **38**, 378 (1968); **40**, 290 (1968); **43**, 293 (1969); **73**, 573 (1974); Strukturbildung bei irreversiblen Prozessen. BSB B. G. Teubner Verlagsgesellschaft, Leipzig 1976; Makroskopische Materie als Quantensystem von Punktladungen. In: 75 Jahre Quantentheorie. Festband zum 75. Jahrestag der Entdeckung der Planckschen Energiequanten. Akademie-Verlag, Berlin 1977; Kinetic Theory of Nonideal Plasmas. Invited Papers ICPIG XV. Minsk 1981; Effects of Nonideality in Quantum Kinetic Theory. Proc. RGD-13. Novosibirsk 1982; Correlations in Nonideal Plasmas. Proc. ICPIG XVI. Düsseldorf 1983; Dubna-Preprint E17-84-459 (1984); Physica A, in press.
Ebeling, W., Feistel, R.: Physik der Selbstorganisation und Evolution. Akademie-Verlag, Berlin 1982.
Ebeling, W., Grigo, M.: Ann. Physik **37**, 21 (1980); J. Solution Chem. **11**, 151 (1982); Z. physik. Chem. **265**, 1072 (1984).
Ebeling, W., Richert, W.: Ann. Physik **39**, 362 (1982); phys. stat. sol. (b) **128**, **167** (1985); Phys. Letters, in press.
Ebeling, W., Röpke, G.: Ann. Physik **36**, 429 (1979); preprint 1985;
Ebeling, W., Sändig, R.: Ann. Physik (Leipzig) **28**, 289 (1973).
Ebeling, W., Hoffmann, H. J., Kelbg, G.: Beitr. Plasmaphys. **7**, 233 (1967).
Ebeling, W., Kelbg, G., Rohde, K.: Ann. Physik **21**, 235 (1968); **22**, 1 (1968).
Ebeling, W., Kelbg, G., Schmitz, G.: Ann. Physik (Leipzig) **18**, 29 (1966).
Ebeling, W., Kelbg, G., Sändig, R.: Beitr. Plasmaphys. **10**, 507 (1970).
Ebeling, W., Kraeft, W.-D., Kremp, D.: Beitr. Plasmaphys. **10** (1970) 237.
Ebeling, W., Kraeft, W.-D., Kremp, D.: Theory of Bound States and Ionisation Equilibrium in Plasmas and Solids. Akademie-Verlag, Berlin 1976. Extended Russ. translation. Mir. Moscow 1979.
Ebeling, W., Kraeft, W. -D., Kremp, D.: Nonideal Plasmas. Invited Paper Proc. ICPIG-XIII. Berlin 1977.
Ebeling, W., Meister, C.-V., Sändig, R., Kraeft, W.-D.: Ann. Physik **36**, 321 (1979).
Ebeling, W., Richert, W., Kraeft, W.-D., Stolzmann, W.: Phys. stat. sol. (b) **104**, 193 (1981).
Ebeling, W., Kraeft, W.-D., Kremp, D., Kilimann, K.: Phys. stat. sol. (b) **78**, 241 (1976).
Ebeling, W., Kraeft, W.-D., Kremp, D., Röpke, G.: Astrophys. J. **290** 24 (1984).
Ebeling, W., et al.: Transport Properties of Dense Plasmas, Akademie-Verlag, Berlin 1983; Birkhäuser-Verlag, Basel, Boston, Stuttgart 1984.
Ebina, K.: Progr. Theor. Phys. **69**, 1686 (1983).
Ecker, G. H.: Theory of Fully Ionized Plasmas. New York, London 1972. Russ. transl. Moscow 1974.

ECKER, G., and MÜLLER, K. G.: Z. Physik **153** (1958) 317.
ECKER, G. H., WEIZEL, W.: Ann. Physik **17**, 126 (1956).
EFROS, A. L.: Usp. Fiz. Nauk **126**, 41 (1978).
EHRENREICH, H., COHEN, M.: Phys. Rev. **115**, 786 (1959).
EMERY, V. J.: Highly Conducting One-Dimension Solids (J. T. DEVREESE, R. P. EVRARD and V. E. DOREN, eds.). Plenum Press, New York 1979.
EWALD, P. P.: Ann. Physik **64**, 253 (1921).
FABER, T. E.: Introduction to the Theory of Liquid Metals. Univ. Press, Cambridge, 1972.
FALKENHAGEN, H., unter Mitwirkung von W. EBELING: Theorie der Elektrolyte. Hirzel Verlag, Leipzig 1971.
FALKENHAGEN, H., EBELING, W.: Equilibrium Properties of Dilute Electrolytes. In: Ionic Interactions, Vol. I (S. PETRUCCI, ed.) Acad. Press, New York, London 1971.
FEIX, M. R.: Computer Experiments in One-Dimensional Plasmas. In: KALMAN, G., CARINI, P. (eds.): Strongly Coupled Plasmas. Plenum Press, New York 1978.
FENNEL, W., WILFER, H. P.: Ann. Physik **32**, 265 (1975).
FENNEL, W., KRAEFT, W.-D., KREMP, D.: Ann. Physik **31**, 171 (1974).
FENNEL, W., KRAEFT, W.-D., KREMP, D., WILFER, H. P.: Wiss. Z. Univ. Rostock MNR **24**, 693 (1975).
FETTER, A. L., WALECKA, J. D.: Quantum Theory of Many Particle Systems. McGraw-Hill, New York 1971.
FICK, E.: Einführung in die Grundlagen der Quantentheorie. Akademische Verlagsgesellschaft Geest und Portig, K.-G., Leipzig 1968.
FILINOV, V. S.: Phys. Lett. **A54**, 259 (1975).
FINKELSTEIN, A. W.: Zh. Eksp. Teor. Fiz. **84**, 168 (1983).
FISCHBECK, H. J.: Eindimensionales Fermigas. Preprint 82-18. ZIE der AdW der DDR, Berlin 1982.
FISHER, D. S., HALPERIN, B. I., PLATZMAN, P. M.: Phys. Rev. Lett. **42**, 798 (1979).
FLECKINGER, R., GOMES, A., SOULET, Y.: Physica **A85**, 485 (1976); **A91**, 33 (1978).
FOMESTER, P. J., JANCOVICI, B., SMITH, E. R.: J. Stat. Phys. **31**, 129 (1983).
FORRESTER, P. J., JANCOVICI, B., SMITH, E. R.: J. Stat. Phys. **31**, 129 (1983).
FORTOV, V. E., et al.: Zh. Eksp. Teor. Fiz. **71**, 225 (1976). Preprint, Chernogolovka 1979.
FRANCK, S.: Ann. Physik **37**, 349 (1980); Adv. Space Res. **1**, 203 (1981).
FRANCK, E. U., HENSEL, F.: Phys. Rev. **141**, 1, 109 (1966); Ber. Bunsenges. **70**, 1154 (1966); Rev. Mod. Phys. **40**, 697 (1968).
FRIEDMAN, H.: Ionic Solution Theory. Interscience, New York, London 1962.
FRÖHLICH, J., PARK, Y. M.: Comm. Math. Phys. **59**, 235 (1978); J. Stat. Phys. **23**, 701 (1980).
FRÖHLICH, J., SPENCER, T.: Lecture presented at the Int. School of Mathematical Physics. Erice, Sicily 1980; Comm. Math. Phys. **81**, 527 (1981).
FUJITA, S.: Introduction to Non-Equilibrium Quantum Statistical Mechanics. W. B. Saunders, Philadelphia, London, 1966. Russ. transl. Mir, Moscow, 1969.
FUKUYAMA, H.: Solid State Commun. **19**, 551 (1976).
GALAM, S., HANSEN, J. P.: Phys. Rev. **A14**, 816 (1976).
GALITSKII, U. M., MIGDAL, A. R.: Zh. Eksp. Teor. Fiz. **34**, 139 (1958).
GALLET, F., DEVILLE, G., VALDÈS, A., WILLIAMS, F. I. B.: Phys. Rev. Lett. **49**, 212 (1982).
GANN, R. C., CHAKRAVARTY, S., CHESTER, G. V.: Phys. Rev. **B20**, 326 (1979).
GELL-MANN, M., BRUECKNER, K. A.: Phys. Rev. **106**, 364 (1957).
GILBERT, J. D.: J. Math. Phys. **18**, 791 (1977); **19**, 2605 (1978).
GILL, R. D. (ed.): Plasma Physics and Fusion Research. Acad. Press, London 1981.
GINGRICH, N. S., ROY HEATON, L.: J. Chem. Phys. **34**, 873 (1961).
GINZBURG, V. L.: Usp. Fiz. Nauk **103**, 87 (1971); **134**, 469 (1981).
GIRARDEAU, M. D.: J. Math. Phys. **16**, 1901 (1975).
GITTERMAN, M., STEINBERG, V.: Phys. Rev. Lett. **35**, 1588 (1975).
GLANSDORFF, P., PRIGOGINE, I.: Thermodynamics of Structure, Stability and Fluctuations. Wiley-Interscience, New York 1971.
GLASSER, M. L.: Phys. Lett. **A51**, 253 (1975).

GLICKSMAN, M.: Plasmas in Solids. Solid State Physics, Vol. 26. New York 1971.
GÖBEL, E. O., MAHLER, G.: Festkörperprobleme **19**, 105 (1979).
GOEDSCHE, F., RICHTER, R., VOJTA, G.: Phys. stat. sol. (b), 581 (1979).
GLUCK, P.: Nuovo Cimento **38**, 67 (1971).
GOLDBERGER, M. L., WATSON, K. M.: Collision Theory. Wiley, New York 1964.
GOLDEN, K. I., KALMAN, G.: Phys. Rev. **A14**, 1802 (1976); Annals of Physics **143**, 160 (1982).
GORECKI, J., POPIELAWSKI, J.: J. Phys. **F13**, 1197 (1983).
GORODSHENKO, V. D., MAKSIMOV, E. G.: Usp. Fiz. Nauk **130**, 65 (1980).
GRABOSKE, H. C.: Astrophys. J. **181**, 457 (1973).
GRABOSKE, H. C., HARWOOD, D. J., DE WITT, H. E.: Phys. Rev. **A3**, 1419 (1971).
GRABOSKE, H. C., HARWOOD, D. J., ROGERS, F. J.: Phys. Rev. **186**, 210 (1969).
GRABOSKE, H. C., DE WITT, H. E., GROSSMANN, A. S., COOPER, M. S., Astrophys. J. **181** (1973) 457.
GRAD, H.: Hdb. Phys., Bd. XII (S. FLÜGGE, ed.). Springer-Verlag, Berlin 1958.
GREEN, A. E. S., et al.: Phys. Rev. **A24**, 1 (1981); Phys. Rev. Lett. **48**, 638 (1982); Int. J. Quant. Chem. Symp. **16**, 331 (1982).
GRIEM, H.: Spectral line Broadening by Plasmas. Acad. Press, New York 1974.
GRIEM, H. R.: Phys. Rev. **A28**, 1596 (1983).
GRIMES, C. C., ADAMS, G.: Phys. Rev. Lett. **42**, 795 (1979).
GRUBER, CH., MARTIN, PH. A.: Phys. Rev. Lett. **45**, 853 (1980).
GRUBER, CH., LEBOWITZ, J. L., MARTIN, PH. A.: J. Chem. Phys. **75**, 944 (1981).
GRÜTZMACHER, K., WENDE, B.: Phys. Rev. **A16**, 243 (1977); **A18**, 2140 (1978).
GÜNDEL, H.: Beitr. Plasmaphys. **10**, 455 (1970).
GÜNTHER, K., RADTKE, R.: Ergebnisse der Plasmaphysik und der Gaselektronik, Bd. 7. Akademie-Verlag, Berlin 1984, Birkhäuser 1984.
GÜNTHER, K., LANG, S.; RADTKE, R.: J. Phys. **D16**, 1235 (1983).
GÜNTHER, K., POPOVIC, M. M., POPOVIC, S. S., and RADTKE, R.: J. Phys. **D9**, 1139 (1976).
GRYAZNOV, V. K., IOSILEVSKII, I. L.: Some Problems of Thermodynamical Calculations of Manycomponent Plasmas. In: Thermophysical Properties of Low-Temperature Plasmas (in Russian). Nauka, Moscow 1976.
GRYAZNOW, V. K., ZHERNOKLETOV, M. V., ZUBAREV, V. N., IOSILEVSKII, I. L., FORTOV, V. E.: Zh. Eksp. Teor. Fiz. **78**, 573 (1980).
GRYAZNOV, V. K., IOSILEVSKII, I. L., KRASNIKOV, YU. G., KUZNETSOVA, N. N., KUCHERENKO, V. I., LAPPO, G. B., LOMAKIN, B. N., PAVLOV, G. A., SON, E. E., Fortov, V. E.: Thermophysical Properties of the Working Medium of Gas-Phase Nuclear Reactors (i n Russian), Atomizdat, Moscow 1980.
HAKEN, H.: Laser Theory. Springer-Verlag, Berlin, Heidelberg, New York 1970; Synergetics. An Introduction. Springer-Verlag, Berlin, Heidelberg, New York 1978.
HAKEN, H. (ed.): Synergetics. Teubner, Stuttgart 1973.
HAKEN, H., WAGNER, M. (eds.): Cooperative Phenomena. Springer-Verlag, Berlin, Heidelberg, New York 1973.
HALENKA, J., MUSIELOK, J.: XIII. ICPIG Berlin (1977) Contr. Papers I, 135.
HANAMURA, E.: J. Phys. Soc. Japan **29**, 50 (1970).
HANSEN, J. P.: Phys. Rev. **A8**, 3096 (1973).
HANSEN, J. P., LEVESQUE, D., WEIS, J. J.: Phys. Rev. Lett. **43**, 979 (1979).
HANSEN, J. P., TORRIE, G. M., VIEILLEFOSSE, P.: Phys. Rev. **A16**, 2153 (1977).
HARRIS, G. M., ROBERTS, J. E., TRULIO, J. G.: Phys. Rev. **119**, 1832 (1960).
HART, G. A., GOODFRIEND, P. L.: J. Chem. Phys. **53**, 448 (1970).
HAUBOLD, H. J., JOHN, R. W.: Astron. Nachr. **303**, 161 (1982).
HAUG, H., TRAN THOAI, D. B.: Phys. stat. sol. (b) **85**, 561 (1978).
HAUGE, E. H., HEMMER, P. C.: Phys. Norvegica **5**, 269 (1971).
HEDIN, L., LUNDQUIST, B. I.: J. Phys. **C4**, 2064 (1971).
HELD, B., DEUTSCH, C., GOMBERT, M.-M.: Phys. Rev. **A25**, 585 (1982); J. Phys. **A15**, 3845 (1982); Phys. Lett. **A94**, 40 (1983).

HENSEL, F.: Thermophysical Properties of Metallic Fluids in the Sub- and Supercritical Region. In: Proc. 8. Symp. Thermophys. Prop., June 15—18, Gaithersburg, Md. (1981); Angew. Chem. Int. Ed. **19**, 593 (1980).
HERRMANN, R., PREPPERNAU, U.: Elektronen im Kristall, Akademie-Verlag, Berlin 1979.
HETZHEIM, H.: Ann. Physik **19**, 380 (1967).
HITZSCHKE, L., RÖPKE, G., SEIFERT, T., ZIMMERMANN, R.: J. Phys. B, submitted.
HOCKNEY, R. W., BROWN, T. R.: J. Phys. **C8**, 1813 (1975).
HOFFMANN, H. J., EBELING, W.: Beitr. Plasmaphys. **8**, 43 (1968); Physica **39**, 593 (1968).
HOFFMANN, H. J., KELBG, G.: Ann. Physik **17**, 356 (1966); **19**, 186 (1967).
HOHENBERG, P., KOHN, W.: Phys. Rev. **136**, 864B (1964).
HÖHNE, F. E., ZIMMERMANN, R.: J. Phys. **B15**, 2551 (1982).
HÖHNE, F. E., REDMER, R., RÖPKE, G., WEGNER, H.: Physica **128A**, 643 (1984).
HOOPER jr., C. F.: Phys. Rev. **165** (1968) 215.
HOOVER, W. G., ROSS, M., BENDER, C. F., ROGERS, F. J., OLENSS, R. J.: Phys. Earth Planet. Interiors **6**, 60 (1972).
HOROVITZ, B., THIEBERGER, R.: Physica **71**, 99 (1974).
HUANG, K.: Statistical Mechanics. New York 1963. Germ. transl. Bibl. Inst., Mannheim 1964. Russ. transl. Mir, Moscow 1966.
HUBBARD, J.: Proc. Roy. Soc. **A240**, 539 (1957); **243A**, 336 (1958); Phys. Lett. **A25**, 709 (1967).
HUBBARD, W. B., SLATTERY, W. L.: Austral. J. Phys. **168**, 131 (1977).
HUBERMAN, M., CHESTER, G. V.: Adv. Phys. **24** (1975) 489.
HUGON, R. L., GHAZALI, A.: Phys. Rev. **B14**, 602 (1976).
IAKUBOV, I. T., VOROBIEV, V. S.: Astronautical Acta **18**, 79 (1974).
ICHIMARU, S.: Rev. Mod. Phys. **54**, 1015 (1982); Basic Principles of Plasma Physics. Benjamin, Reading, Mass. 1973.
ICHIMARU, S., URSUMI, K.: Phys. Rev. **B24**, 7385 (1981).
IGLESIAS, C. A., HOOPER jr., C. F., DE WITT, H. E.: Phys. Rev. **A28**, 361 (1983).
IGLESIAS, C. A., LEBOWITZ, J. L., MACGOWAN, D.: Phys. Rev. **A28**, 1667 (1983).
INGARDEN, R. S.: Acta Physica Polonica **A43**, 1, 15 (1973); Information Theory and Thermodynamics. N. Copernicus Univ. Press, Torun 1974.
IOSILEVSKII, I. L.: Teplofyz. Vys. Temp. **18**, 447 (1980); **19**, 680 (7981).
IOSILEVSKII, I. L., GRYAZNOV, V. K.: Teplofyz. Vys. Temp. **19**, 1121 (1981).
ISIHARA, A.: Phys. Rev. **172**, 166 (1968); **178**, 412 (1969); Statistical Physics. New York 1971, Russ. transl. Moscow 1973; Phys. cond. Matter **15**, 225 (1972).
ISIHARA, A., IORATTI, C.: Physica **A103**, 621 (1980).
ISIHARA, A., KOJIMA, D. Y.: Phys. cond. Matter **18**, 249 (1974); Physica **77**, 469 (1974).
ISIHARA, A., MONTROLL, E. W.: Proc. Nat. Acad. Sci. USA **68**, 3111 (1971).
ISIHARA, A., WADATI, M.: Phys. Rev. **183**, 312 (1969); Physica **57**, 237 (1972).
ISSAKOV, I. M., LOMAKIN, B. N.: Teplofyz. Vys. Temp. **17**, 262 (1979).
IVANOV, C.: Physica **A94**, 571 (1978).
IVANOV, IU. V., MINTSEV, V. B., FORTOV, V. E., DREMIN, A. N.: Zh. Eksp. Teor. Fiz. **71**, 216 (1976) (Sov. Phys. — JETP **44**, 112 (1976)).
IYETOMI, H., ICHIMARU, S.: Phys. Rev. **A25**, 2434 (1982).
IYETOMI, H., UTSUMI, K., ICHIMARU, S.: J. Phys. Soc. Japan **50**, 3769, 3778 (1981).
JACKSON, J. K., KLEIN, L. S.: Phys. Fluids **7**, 232 (1963); Phys. Rev. **177**, 352 (1969).
JANCOVICI, B.: J. Stat. Phys. **17**, 357 (1977); **28**, 43 (1982); **29**, 263 (1982); Molec. Phys. **32**, 1177 (1976); Phys. Rev Letters **46**, 386 (1981).
JASTROW, R.: Phys. Rev. **98**, 1478 (1955).
JAYNES, E. T.: Phys. Rev. **106**, 620 (1957), **108**, 171 (1957).
KADANOFF, L. P., BAYM, G.: Quantum Statistical Mechanics. New York 1962. Russ. transl. Moscow 1964.
KADOMTSEV, B. B.: Zh. Eksp. eor. Fiz. **33**, 151 (1957); Collective Effects in Plasma (in Russian). Nauka, Moscow 1976.
KAGAN, YU., PUSHKAREV, V., KHOLAS, A.: Zh. Eksp. Teor. Fiz. **73**, 967 (1977).
KALASHNIKOV, W. P.: Teor. Mat. Fiz. (USSR) **34**, 412 (1978); **35**, 127 (1978).
KALIA, R. K., VASHISTA, P., DE LEEUW, S. W.: Phys. Rev. **B23**, 4794 (1981).

KALIA, R. K., VASHISHTA, R., MAHANTI, S. D., QUINN, J. J.: J. Phys. **C16**, L491 (1983).
KALMAN, G.: Recent Progress in the Understanding of Strongly Coupled Coulomb Systems. Proc. II. Int. Conf. Many Body Physics. Oactepec 1981.
KALMAN, G., CARINI, P. (Eds.): Strongly coupled Plasma. Plenum Press, New York 1978.
KELBG, G.: Ann. Physik **12**, 219 (1963); **12**, 354 (1964), **14**, 394 (1964); Wiss. Z. Univ. Rostock **14**, 251 (1965); Ergebnisse der Plasmaphysik und Gaselektronik **3** (1972).
KELDYSH, L. V.: Usp. Fiz. Nauk **100**, 514 (1970).
KHALATNIKOV, I. M.: Priroda (Fizika) **10**, 8 (1971).
KHRAPAK, A. G., YAKUBOV, I. T.: Electrons in Dense Gases and Plasmas (in Russian). Nauka, Moscow 1981.
KIKOIN, I. K., SENCHENKOV, A. A.: Phys. Met. Metallorg. USSR **24**, 843 (1967).
KILIMANN, M. K.: Doctor thesis II, Wilhelm-Pieck-Universität Rostock, DDR, 1978.
KILIMANN, M. K., KRAEFT, W.-D.: AdW. der DDR, ZfI.-Mitt. **13**, 70 (1978).
KILIMANN, M. K., KRAEFT, W.-D., KREMP, D.: Phys. Lett. **A61**, 393 (1977).
KILIMANN, K., KREMP, D., RÖPKE, G.: Teor. Mat. Fiz. **55**, 448 (1983).
KIRZHNITS, D. A., LOZOVIK, YU. E., SHPATAKOVSKAYA, G. V.: Usp. Fiz. Nauk **117**, 3 (1975).
KITTEL, CH.: Einführung in die Festkörperphysik. Leipzig 1973.
KLEIN, L.: J. Quant. Spec. Rad. Trans. **9**, 199 (1969).
KLIMONTOVICH, YU. L.: Statistical Theory of Nonequilibrium Processes (in Russian). Izdat MGU, Moscow 1964; Kinetic Theory of Nonideal Gases and Nonideal Plasmas (in Russian). Nauka, Moscow 1975. Engl. transl. Academic Press, New York 1982; Kinetic Theory of Electrodynamic Processes (in Russian). Nauka, Moscow 1980, Engl. transl. Springer-Verlag, Berlin, Heidelberg, New York 1982; Statistical Physics (in Russian). Nauka, Moscow 1982.
KLIMONTOVICH, YU. L., EBELING, W.: Zh. Eksp. Teor. Fiz. **43**, 146 (1962).
KLIMONTOVICH, YU. L., EBELING, W.: Zh. Eksp. Teor. Fiz. **63**, 905 (1972).
KLIMONTOVICH, YU L., KRAEFT, W. C.: Teplofiz. Vys. Temp. **12**, 239 (1974).
KLIMONTOVICH, YU. L., KREMP, D.: Kinetische Gleichungen für Systeme mit Bindungszuständen. Preprint, Rostock 1980; Physica **109A**, 517 (1981).
KLIMONTOVICH, YU. L., SILIN, V. P.: Usp. Fiz. Nauk **70**, 247 (1960).
KOBZEV, G. A., KURILENKOV, YU. K.: Teplofiz. Vys. Temp. **16**, 458 (1978).
KOBZEV, G. A., KURILENKOV, YU. K., NORMAN, G. E.: Teplofiz. Vys. Temp. **15**, 153 (1977).
KOHLER, M.: Z. Physik **124**, 772 (1948); **125**, 679 (1948).
KOHN, W., SHAM, L. J.: Phys. Rev. **140**, 1133A (1965).
KOLESNICHENKO, E. G.: Teor. Mat. Fyz. (USSR) **30**, 114, 282 (1977).
KOPYSHEV, V. P.: Zh. Eks. Teor. Fiz. **55**, 1304 (1968).
KOSTERLITZ, J. M.: J. Phys. **C7**, 1046 (1974).
KOSTERLITZ, J. M., THOULESS, D. J.: J. Phys. **C6**, 1181 (1973).
KRAEFT, W.-D.: Nonideal Plasmas. In: R. K. JANEV (ed.): The Physics of Ionized Gases. Institute of Physics, Beograd 1978; Lecture at the International Workshop on the Statistical Mechanics of Ionic Matter. Les Houches 1982.
KRAEFT, W.-D., FENNEL, W.: Phys. stat. sol. (b) **73**, 487 (1976).
KRAEFT, W.-D., JAKUBOWSKI, P.: Ann. Physik **35**, 294 (1978).
KRAEFT, W.-D., KREMP, D.: Z. Physik **208**, 475 (1968).
KRAEFT, W.-D., STOLZMANN, W.: Phys. Lett. **56A**, 41 (1976); Physica **A79**, 306 (1979); Lecture at the Conference "Physik der flüssigen Phase", Rostock 1982; J. Phys. **C17**, 3561 (1984).
KRAEFT, W.-D., BLÜMLEIN, J. MEYER, T.: Contr. Plasma Phys. **23**, 9 (1983).
KRAEFT, W.-D., EBELING, W., KREMP, D.: Phys. Lett. **A29**, 466 (1969).
KRAEFT, W.-D., KILIMANN, M. K., KREMP, D.: Phys. stat. sol. (b) **72**, 461 (1975).
KRAEFT, W.-D., KREMP, D., KILIMANN, K.: Ann. Physik **29**, 177 (1973).
KRAEFT, W.-D., KREMP, D., RÖPKE, G.: Progress in the Theory of Nonideal Plasmas, Proc. XVI. ICPIG. Invited Papers. Düsseldorf 1983.
KRAEFT, W.-D., LUFT, M., MIHAJLOV, A. A.: Physica **A120**, 263 (1983).
KRAEFT, W.-D., EBELING, W., RICHERT, W., STOLZMANN, W.: Phys. Lett., in press (1985).
KRAEFT, W.-D., STOLZMANN, W., WIPPER, M., KREMP, D.: Ann. Physik **32**, 1 (1975).

KRASNY, YU. P., KOVALENKO, N. P.: Zh. Eksp. Teor. Fiz. **62**, 829 (1972).
KRASNY, YU. P., ONISHCHENKO, V. P.: Ukrain. Fiz. Zh. **17**, 1704 (1972).
KREMP, D., KRAEFT, W.-D.: Ann. Physik (Leipzig) **20**, 340 (1968); Phys. Lett. 38A, 167 (1972).
KREMP, D., SCHLANGES, M.: Second Int. Workshop Nonídeal Plasmas, Wustrow, DDR, p. 163 (1982)).Eds.: KRAEFT, W.- D., ROTHER, T.
KREMP, D., SCHLANGES, M.: In: W. EBELING et al. (R. ROMPE and M. STEENBECK, eds.): Ergebnisse der Plasmaphysik und der Gaselektronik, Vol. VI. Akademie-Verlag, Berlin 1983; Birkhäuser-Verlag Basel, Boston, Stuttgart 1984 (in English).
KREMP, D., SCHMITZ, G.: Z. Naturforsch. **22a**, 1366 (1967).
KREMP, D., HARONSKA, P., SCHLANGES, M.: Wiss. Z. Univ. Rostock MNR (1985).
KREMP, D., KRAEFT, W.-D., EBELING, W.: Physica **51**, 146 (1971).
KREMP, D., KRAEFT, W.-D., FENNEL, W.: Physica **62**, 461 (1972).
KREMP, D., KRAEFT, W.-D., KILIMANN, M. K.: AdW der DDR, ZfI.-Mitt. **13**, 66 (1978).
KREMP, D., KRAEFT, W.-D., LAMBERT, A. J. M. D.: Physica **127A**, 72 (1984).
KREMP, D., SCHLANGES, M., KILIMANN, M. K.: Phys. Lett. **110A**, 149 (1984).
KREMP, D., KILIMANN, M.K., KRAEFT, W.-D., STOLZ, H., ZIMMERMANN, R.: Physica **127A**, 646 (1984a).
KRIENKE, H., STIPS, A.: Z. Physik. Chem. (im Druck).
KRIENKE, H., EBELING, W., CZERWON, H. J.: Wiss. Z. Univ. Rostock MNR **24**, 5 (1975).
KRUMHANSL, J. A.: Physics of Solids at High Pressures. New York 1965.
KUBO, R.: J. Phys. Soc. Japan **12**, 570 (1957).
KUDRIN, L. P.: Statistical Physics of Plasmas (in Russian). Nauka, Moscow 1974.
KUNI, F. M.: Statistical Physics and Thermodynamics (in Russian). Nauka, Moscow 1981.
KWOK, P. C., SCHULTZ, T. D.: J. Phys. **C2**, 1196 (1969).
LAGARKOV, A. N., SARYCHEV, A. K.: Teplofyz. Vys. Temp. **17**, 466 (1979).
LAM, C. S., VARSHNI, Y. P.: Phys. Rev. **A4**, 1875 (1971).
LANDAU, L. D., LIFSCHITZ, E. M.: Lehrbuch der Theoretischen Physik, III. Quantenmechanik. Akademie-Verlag, Berlin 1977.
LANDAU, L. D., LIFSHITS, E. M.: Teoreticheskaya, Vol. X. Fizicheskaya Kinetika (LIFSHITS, E. M., PITAEVSKII, L. P.). Nauka, Moscow 1979. German transl. Akademie-Verlag, Berlin 1983.
LEBOWITZ, J. L.: Free Energy and Correlation functions of Coulomb Systems. Lecture notes. Int. School of Math. Phys., Erice 1980; Lectures presented at the Int. School of Mathematical Physics. Erice, Sicily 1980; Phys. Rev. **A27**, 1491 (1983).
LEBOWITZ, J. L., LIEB, E. H.: Phys. Rev. Lett. **13**, 631 (1969); **22**; Adv. Math. **9**, 317 (1972).
LEBOWITZ, J. L., PENA, R. E.: J. Chem. Phys. **59**, 1362 (1973).
LEE, R. W.: J. Phys. **B6**, 1044, 1060 (1973), **B12**, 1129, 1145 (1979).
LENARD, A.: Statistical Mechanics and Mathematical Problems. Lecture Notes in Physics, Vol. 20. Springer-Verlag, Berlin, Heidelberg, New York 1973; J. Math. Phys. **2**, 682 (1961).
LENARD, A., DYSON, F. J.: J. Math. Phys. **9**, 698 (1968).
LEUTHEUSER, E.: Phys. Rev. **A28**, 1762 (1983).
LEVEN, R.: Beitr. Plasmaphys. **9**, 29, 165, 293 (1969); **10**, 27, 347 (1970).
LIEB, E. H.: Rev. Mod. Phys. **48**, 553 (1976).
LIEB, E. H., LEBOWITZ, J. L.: Adv. Math. **9**, 316 (1972); Springer Lecture Notes in Physics **20** 1(973).
LIEB, E. H., MATTIS, D. C.: Mathematical Physics in One Dimension. Acad. Press. New York 1966.
LIEB, E. H., NARNHOFER, H.: J. Stat. Phys. **14**, 465 (1976).
LIEB, E. H., SIMON, B.: J. Chem. Phys. **61**, 735 (1977); Phys. Rev. Lett. **33**, 681 (1977).
LIEB, E. H., THIRRING, W.: Phys. Rev. Lett. **31**, 111 (1975).
LIFSCHITZ, E. M., PITAJEWSKI, L. P.: Statistische Physik, Teil 1 (5. Aufl.). Akademie-Verlag, Berlin 1979; Statistische Physik, Teil 2. Akademie-Verlag, Berlin 1980.
LIFSCHITZ, I. M., ASBEL, J. M., KAGANOW, M. I.: Elektronentheorie der Metalle. Akademie-Verlag, Berlin 1975.

LIFSHITS, I. M., GREDESKUL, S. A., PASTUR, L. A.: Introduction to the Theory of Disordered Systems (in Russian). Nauka, Moscow 1982.
LIKHALTER, A. A.: Teplofiz. Vys. Temp. **16**, 1219 (1978); **19**, 746 (1981).
LINDHARD, J.: Kgl. Danske Videnskab. Selskab. Mat. Fys. Medd. **28**, 8 (1954).
LUDWIG, G.: Einführung in die Grundfragen der theoretischen Physik, Bd. 3—4. Vieweg, Braunschweig 1980.
LUKYANOV, S. Y.: Hot Plasma and Controlled Fusion (in Russian). Nauka, Moscow 1975.
LUNDQUIST, B. J.: Phys. cond. Mater **6**, 206 (1967).
MAHLER, G., BIRMAN, J. L.: Phys. Rev. **B16**, 1552 (1977).
MARTIN, P. C.: Phys. Rev. **161**, 143 (1967).
MARTIN, P., SCHWINGER, J.: Phys. Rev. **115**, 1342 (1959).
MATSUBARA, T. (ed.): The Structure and Properties of Matter. Springer-Verlag, Berlin, Heidelberg, New York 1982.
MATTUCK, R. D.: A Guide to Feynman Diagrams in the Many Body Problem. McGraw Hill, London 1967.
MAZURE, A., NOLLEZ, G.: Z. Naturforsch. **A33**, 1575 (1978).
MEHROTRA, R., GUENIN, B. M., A. J.: Phys. Rev. Lett. **48**, 641 (1982).
MEISTER, C. V., RÖPKE, G.: Ann. Physik (Leipzig) **39**, 133 (1982).
MESSIAH, A.: Quantum Mechanics. North-Holland, Amsterdam 1961.
METROPOLIS, N., et al.: J. Chem. Phys. **21**, 1087 (1953).
MEYER, J. R., BARTOLI, F. J.: J. Phys. **C15**, 1987 (1982).
MIGDAL, A. B.: Theory of Finite Fermi Systems (in Russian), Nauka, Moscow 1965.
MIKHAILOVSKII, A. B.: Theory of Plasma Instabilities (in Russian). Atomizdat, Moscow 1977.
MINOO, H., DEUTSCH, C., HANSEN, J. P.: Phys. Rev. **14A**, 840 (1976).
MITCHNER, M., KRUGER, C. H.: Partially Ionized Gases. Wiley, New York 1973. Russ. transl. Mir, Moscow 1976.
MIYAGI, H.: Progr. Theor. Phys. **65**, 66 (1981).
MÖBIUS, A., GOEDSCHE, F., VOJTA, G.: Physica **A95**, 294 (1979).
MONTROLL, E. W., LEBOWITZ, J. L.: The Liquid State of Matter. North-Holland, Amsterdam 1982.
MONTROLL, E. W., WARD, J. C.: Phys. Fluids **1**, 55 (1958).
MOORE, C. E., MINNEART, M. G. F., HOUTGAST, F.: The Solar Spectrum 2935 Å to 8770 Å. NBS Monograph No 61.
MORITA, T.: Progr. Theor. Phys. (Japan) **22**, 757 (1959).
MOTT, N. F.: Metal Insulator Transition. Taylor & Francis, London 1974; Russ. transl. Mir, Moscow 1976.
MOTT, N. F., DAVIS, E. A.: Electronic Processes in Non-Crystalline Materials. Clarendon Press, Oxford 1971; Russ. transl. Mir, Moscow 1974.
MOZER, B., BARANGER, M., Phys. Rev. **118** (1960) 626.
NAKAYAMA, T., DE WITT, H. E.: J. Quant. Spectr. Rad. Transfer **4**, 623 (1964).
NEWTON, G.: Scattering Theory of Waves and Particles. McGraw-Hill, New York 1966.
NICOLIS, G., PRIGOGINE, I.: Self-Organization in Non-Equilibrium Systems. Wiley-Interscience, New York 1977.
NORMAN, G. E.: Zh. exp. teor. Fiz. **60**, 1686 (1971).
NORMAN, G. E., STAROSTIN, A. N.: Teplofiz. Vys. Temp. **6**, 410 (1968); 8, 413 (1979) (High Temp. Phys. (USSR) 8, 381).
NOVIKOV, I. D.: Evolution of the Universe (in Russian). Nauka, Moscow 1983.
NOZIERES, P.: Theory of Interacting Fermion Systems. New York 1964.
NOZIERES, P., PINES, D.: Phys. Rev. **111**, 442 (1958); **113**, 1254 (1959).
OGURZOVA, M. N., PODMOSHENSKII, I. V., SMIRNOVA, V. JA:. Teplofiz. Vys. Temp. **12**, 650 (1974).
ONISHCHENKO, V. P., KRASNY, YU. P.: Ukrain. Fiz. Zh. **18**, 1194 (1973).
ONSAGER, L., MITTAG, L., STEPHEN, M.: Ann. Physik **18**, 71 (1966).
PAJANNE, E.: J. Phys. C. Solid State Physics **15**, 5629 (1982).
PARGAMANIK, L. E., GINSBURG, M. D., Ukrain. Fiz. Zh. **22**, 938, 1611; **23**, 898 (1978).
PARGAMANIK, L. E., GINSBURG, M. D.: Ukrain. Fiz. Zh. **22**, 564, 1185 (1980).

Patashinskii, A. Z., Pokrovskii, V. L.: Fluctuation Theory of Phase Transitions. Pergamon Press, Oxford 1979.
Patch, R. W.: J. Quant. Spec. Rad. Trans. **9**, 63 (1969).
Pathak, K. N., Vashishta, P.: Phys. Rev. **B7**, 3649 (1973).
Peebles, P. E. J.: Physical Cosmology. Univ. Press, Cambridge 1971.
Peletminskii, S. W.: Teor. Mat. Fiz. **6**, 123 (1971).
Perrot, F.: Phys. Rev. **A25**, 498 (1982); **A76**, 1035 (1982).
Petrucci, S. (ed.): Ionic Interactions, Vol. 1. Acad. Press. New York, London 1971.
Pfennig, H., Trefftz, E.: Z. Naturforsch. **A21**, 697 (1966).
Pines, D.: Nuovo Cimento **7**, 329 (1958); The Many Body Problem. New York 1962. Russ. transl. Mir, Moscow 1963.
Pines, D., Nozieres, F.: Theory of Quantum Liquids. Benjamin, New York 1966. Russ. transl. Mir, Moscow 1967.
Platzmann, P. M., Wolf, P. A.: Waves and Interactions in Plasmas. New York 1973.
Pollock, E. L., Hansen, J. P.: Phys. Rev. **A8**, 3110 (1973).
Prigogine, I.: Introduction to Thermodynamics of Irreversible Processes (3rd Ed.). Interscience, New York 1967.
Prigogine, I., Defay, R.: Chemische Thermodynamik. Verlag für Grundstoffindustrie, Leipzig 1962.
Primas, H.: Chemistry, Quantum Mechanics and Reductionism. Perspectives in heoretical Chemistry. Springer-Verlag. Berlin, Heidelberg, New York 1981.
Puff, H.: AdW der DDR, Berlin, ZIE, preprint Nr. 76-11, 1976, Nr. 79-1, 1979.
Purič, J., Labat, J., Cirkovic, L., Lakicevic, I., Djenize, S.: J. Phys. **B10**, 2375 (1977).
Quinn, J. J., Ferrel, R. A.: Phys. Rev. **112**, 812 (1958).
Radtke, R., and Günther, K.: J. Phys. **D9**, 1131 (1976).
Ramos, J. G., Gomes, A. A.: Nuovo Cimento **3**, 441 (1971).
Redmer, R., Röpke, G.: Ann. Physik (1985), in press.
Renkert, H., Hensel, F., Franck, E. U.; Ber. Bunsenges. Phys. Chem. **75**, 507 (1971).
Rice, T. M., Hensel, J. C., Phillips, T. C., Thomas, G. A.: The Electron-Hole Liquid in Semiconductors. Solid State Physics, Vol. 32. Acad. Press, New York 1977 Russ. transl. Mir, Moscow 1980.
Richert, W.: Approximationen für die thermodynamischen Funktionen von Quantenplasmen. Diplom Thesis. W.-Pieck-Universität Rostock, 1979; Zur statistischen Thermodynamik von Vielteilchensystemen mit Coulombwechselwirkung. Dissertation A. Humboldt-Universität Berlin, 1982.
Richert, W., Ebeling, W.: Phys. stat. sol. (b) **121**, 633 (1984).
Richert, W., Insepov, S. A., Ebeling, W.: Ann. Physik **41**, 139 (1984)
Riewe, H. K., Rompe, R.: Ann. Physik **111**, 79 (Y938).
Riste, T. (ed.): Fluctuations, Instabilities and Phase Transitions. Plenum Press, New York 1975.
Robnik, M., Kundt, W.: Astron. Astrophys. **120**, 227 (1983).
Rogers, F. J.: Phys. Rev. **A4**, 1145 (1971); **A10**, 2441 (1974).
Rogers, F. J., De Witt, H. E.: Phys. Rev. **A1**, 1061 (1973).
Rogers, F. J., Graboske, H. C., De Witt, H. E.: Phys. Rev. **A1**, 1577 (1970); Phys. Lett. **A34**, 127 (1971).
Rogers, F. J., Graboske, H. C., Harwood, D. J.: Phys. Rev. **A1**, 1577 (1970).
Rohde, K., Kelbg, G., Ebeling, W.: Ann. Physik **22**, 1 (1968).
Rolov, B. N.: Diffuse Phase Transitions (in Russian), Zinatja, Riga 1973.
Rompe, R., Steenbeck, M.: Ergebnisse der exakten Naturwissenschaften **18**, 275 (1939).
Rompe, R., Treder, H. J.: Über Physik. Studien zu ihrer Stellung in Wissenschaft und Gesellschaft. Akademie-Verlag, Berlin 1979.
Röpke, G.: Physica **A86**, 147 (1977); Teor. Mat. Fiz. **46**, 279 (1981); Ann. Physik **39**, 35 (1982); Physica **A121**, 92 (1983).
Röpke, G., Christoph, V.: J. Phys. **C8**, 3615 (1975); Phys. stat. sol. (b) **95**, K15 (1979).
Röpke, G., Der, R.: Phys. stat, sol. (b) **92**, 501 (1979).
Röpke, G., Höhne, F. E.: Phys. stat. sol. (b) **107**, 603 (1981).
Röpke, G., Ebeling, W., Kraeft, W.-D.: Physica **A101**, 243 (1980).

Röpke, G., Münchow, L., Schulz, H.: Nucl. Phys. **A379**, 536 (1982); Phys. Lett. **B110**, 21 (1982).
Röpke, G., Schmidt, M., Redmer, R.: Wiss. Z. Univ. Rostock MNR **31**, 55 (1982).
Röpke, G., Seifert, T., Kilimann, K.: Ann. Physik **38**, 381 (1981).
Röpke, G., Kilimann, K., Kraeft, W.-D., Kremp, D.: Phys. Lett. **A68**, 329 (1978).
Röpke, G., Meister, C. V., Kollmorgen, K., Kraeft, W.-D.: Ann. Physik **36**, 377 (1979).
Röpke, G., Schmidt, M., Münchow, L., Schulz, H.: Nucl. Phys. **A399**, 587 (1983).
Röpke, G., Seifert, T., Stolz, H., Zimmermann, R.: Phys. stat. sol. (b) **100**, 215 (1980).
Röpke, G., Kilimann, K., Kremp, D., Kraeft, W.-D., Zimmermann, R.: Phys. stat. sol. (b) **88**, K59 (1978).
Ross, W. D.: Annals of Physics **36**, 458 (1966).
Rouse, C.: Phys. Rev. **159**, 41 (1967); **163**, 62 (1968); **188**, 525 (1969).
Rouse, C. A.: Astrophys. J. **272**, 377 (1983).
Ruelle, D.: Statistical Mechanisсs. Rigorous Results. New York 1969. Russ. transl. Moscow 1971.
Sadovskii, M. V.: Usp. Fiz. Nauk **133**, 743 (1981).
Sändig, R., Meister, C. V.: Wiss. Z. Univ. Rostock MNR **28**, 247 (1979).
Schlanges, M., Kremp, D.: Ann. Physik **39**, 69 (1982).
Schlanges, M., Kremp, D., Kraeft, W.- D.: Contr. Plasmaphys. (1985), in press.
Schmidt, M., Röpke, G., Haronska, P.: Wiss. Z. Univ. Rostock MNR **30** (1981) 59.
Schmitz, G.: Phys. Lett. **21**, 174 (1966); Ann. Physik **21**, 31 (1968).
Schmitz, G., Kremp, D.: Z. Naturforsch. **23a**, 1392 (1968).
Schneider, T.: Helvetia Phys. Acta **42**, 957 (1969); Physica **52**, 481 (1971).
Schrieffer, J. R.: Theory of Superconductivity (in Russian). Nauka, Moscow 1970.
Schweber, S. S.: An Introduction to Relativistic Quantum Field Theory. Row, Peterson & Co., Evanston, Ill., 1961.
Seidel, I.: Z. Naturforsch. **A32**, 1207 (1977); **A34**, 1385 (1979).
Seifert, T.: Ann. Physik **37**, 368 (1980).
Seshenov, V. A., Son, E. E., Stchekotov, O. E.: Pis'ma v. ZTF **1**, 891 (1975).
Shaner, J. W., Gathers, G. R., Minichino, C.: High Temp. — High Press. **8**, 125 (1976).
Shimoji, M.: Liquid Metals. Acad. Press, London 1977.
Shindo, K.: J. Phys. Soc. Japan **29**, 278 (1970).
Silin, V. P.: Introduction to the Kinetic Theory of Gases (in Russian). Nauka, Moscow 1971.
Sinai, Ya. G.: Theory of Phase Transitions. Rigorous Results. Pergamon Press, Oxford 1982.
Singwi, K. S., Sjölander, A., Tosi, M. P., Land, R. H.: Phys. Rev. **B1**, 1044 (1970).
Singwi, K. S., Tosi, M. P., Land, R. H., Sjölander, A.: Phys. Rev. **176**, 589 (1968).
Sjögren, L.: J. Phys. **C13**, L841 (1980).
Sjögren, L., Sjölander, A.: Ann. Physik (N.Y.) **110**, 122 (1978).
Slattery, W. L., Doolem, G. D., De Witt, H. E.: Phys. Rev. **A21**, 2087 (1980).
Sobelman, I. I., Vainshtein, L. A., Yukov, E. A.: Excitation of Atoms and Broadening of Spectral lines. Springer-Verlag, New York 1981.
Spitzer, R.: Physics of Fully Ionized Gases. Wiley, New York 1962.
Stevens, F. A., Pokrant, M. A.: Phys. Rev. **A8**, 990 (1973).
Stevenson, D. J., Ashcroft, N. W.: Phys. Rev. **A9**, 782 (1974).
Stevenson, D. J., Salpeter, E. E.: Astrophys. J. Suppl. **35**, 321 (1977).
Stillinger, F., Lovett, F.: J. Chem. Phys. **49**, 1991 (1968).
Stolz, H.: Einführung in die Vielelektronentheorie der Kristalle. Berlin 1974; AdW der DDR, Berlin, ZIE, preprint Nr. 75-6, 1975, Nr. 76-11, 1976; Supraleitung. Akademie-Verlag, Berlin 1979.
Stolz, H., Zimmermann, R.: Phys. stat. sol. (b) **94**, 135 (1979).
Stolz, H., Zimmermann, R., Röpke, G.: Phys. stat. sol. (b) **105**, 585 (1981)
Stolzmann, W., Kraeft, W.-D.: Ann. Physik **36**, 338 (1979); Greifswalder Phys. Hefte **5**, 91 (1980).
Storer, R. G.: J. Math. Phys. **9**, 964 (1968); Phys. Rev. **176**, 326 (1968).

STUDART, N., HIPOLITO, O.: Phys. Rev. **A19**, 1970 (1979); **A22**, 2860 (1980).
STURM, K.: Adv. Phys. **31**, 1 (1982).
SUBAREW, D. N.: Statistische Thermodynamik des Nichtgleichgewichts. Akademie-Verlag, Berlin 1976.
SUCHY, K.: Transport Coefficients and Collision Frequencies for Aeronomic Plasmas. Springer-Verlag, Berlin, Heidelberg, New York 1977.
TAKESHIMA, M.: Phys. Rev. **B17**, 3996 (1978).
TELLER, E.: Rev. Mod. Phys. **34**, 627 (1962).
THIRRING, W.: Quantenmechanik großer Systeme. Springer-Verlag, Wien 1980.
THOMSON, J. C.: Electrons in Liquid Ammonia. Univ. Press, Oxford 1976.
THOULESS, D. J.: Phys. Rep. **13**, 93 (1974).
TINKHAM, M.: Introduction to Superconductivity. McGraw-Hill, London 1975.
TOTSUJI, H., ICHIMARU, S.: Prog. Theor. Phys. **52**, 42 (1974).
TRAVING, G.: Interpretation of Line Broadening and Line Shift. In: W. LOCHTE-HOLTGREVEN (ed.): North-Holland, Amsterdam 1968.
TREDER, H. J.: Elementare Kosmologie. Akademie-Verlag, Berlin 1977.
TRUBNIKOV, B. A., ELESIN, V. F.: Zh. Eksp. Fiz. **47**, 1279 (1964).
UNSÖLD, A.: Physik der Sternatmosphären. Springer-Verlag, Berlin 1955.
USUI, T.: Progr. Phys. **23**, 787 (1960).
UTSUMI, K., ICHIMARU, S.: Phys. Rev. **B22**, 5203 (1980).
VALDESH, A.: Contribution a l'etude du solide electronique a deux dimensions. Thesis Université Paris Sud Orsay, 1982.
VAN HORN, H. M.: Phys. Rev. **157**, 342 (1967); Astrophys. J. **151**, 227 (1968); Phys. Lett. **A28**, 706 (1969).
VAN KRANINDONK, J.: Solid Hydrogen. Plenum Press, New York (1982).
VASHISHTA, P., KALIA, R. K.: Phys. Rev. **B25**, 6492 (1982).
VASHISHTA, P., SINGWI, K. S.: Phys. Rev. **B6**, 875 (1972).
VASHISHTA, P., BHATTACHARYA, P., SINGWI, K. S.: Phys. Rev. **B10**, 5108 (1974).
VASHUKOV, S. I., MARUSIN, V. V.: Teplofiz. Vys. Temp. **20**, 38 (1892).
VAVRUKH, M. V.: Ukrain. Fiz. Zh. **13**, 733 (1968).
VEDENOV, A. A.: Usp. Fiz. Nauk **84**, 833 (1964).
VEDENOV, A. A., LARKIN, A. I.: Zh. Eksp. Teor. Fiz. **36**, 1133 (1959).
VELO, G., WIGHTMAN, A. S. (eds.): Rigorous Atomic and Molecular Physics. Plenum Press, New York, London 1981.
VIDAL, C. R., COOPER, J., SMITH, E. W.: J. Quant. Spec. Rad. Trans. **10**, 1011 (1970); **11**, 263 (1971); Astrophys. J. Suppl. **25**, 37 (1973).
VILLAIN, J.: J. Physique **36**, 581 (1975).
VOROB'EV, V. S., KHOMKIN, A. L.: Teor. Mat. Phys. (USSR) 8, 101 (1971); **26**, 364 (1976); Teplofiz. Vys. Temp. **10**, 939 (1972); **12**, 1137 (1974); **13**, 245 (1975); **14**, 204 (1976).
VOROL'EV, V. S., NORMAN, G. E., FILINOV, V. S.: Zh. Eksp. Teor. Fiz. **30**, 459 (1980).
VORONTSOV-VELYAMINOV, P. N., ELYASHVICH, A. M., MORGENSTERN, V. P., CHASSOVSKICH, V. P.: Teplofiz. Vys. Temp. **14**, 199 (1976).
VOSKO, S. H., WILK, L.: Phys. Rev. **B22**, 3812, (1980).
VOSKO, S. H., WILK, L., NUSAIR, M.: Can. J. Phys. **58**, 1200 (1980).
VOSLAMBER, D.: Z. Naturforsch. **A24**, 1458 (1969); **A26**, 1558 (1971); **A27**, 1783 (1972); Phys. Lett. **A40**, 266 (1972).
VOSLAMBER, D., CAPES, H.: Phys. Rev. **A5**, 2528 (1972).
WAISMAN, E., LEBOWITZ, J. L.: J. Chem. Phys. **56**, 3086 (1972).
WASSERMANN, A., BUCKHOLTZ, T. J., DE WITT, H. E.: J. Math. Phys. **11**, 477 (1970).
WEINBERG, S.: Die ersten drei Minuten. Piper-Verlag, München 1976.
WIESE, W. L., KELLEHER, D. E., HELBIG, V.: Phys. Rev. **A11**, 1854 (1975).
WIESE, W. L., KELLEHER, D. E., PAQUETTE, D. R.: Phys. Rev. **A6**, 1132 (1972).
WIGNER, E. P.: Phys. Rev. **40**, 749 (1932); **46**, 1002 (1934); Trans. Far. Soc. **34**, 678 (1938); Phys. Rev. **94**, 77 (1954).
WIGNER, E., HUNTINGTON, H. B.: J. Chem. Phys. **3**, 764 (1935).
WILLIAMS, F. I. B.: Surface Science **113**, 371 (1982).

WILSON, B. A., ALLEN, S. J., TSUI, D. C.: Phys. Rev. Lett. **44**, 479 (1980); Phys. Rev. **B24**, 5887 (1981).
WISER, N., COHEN, M. H.: J. Phys. **C2**, 193 (1969).
WOOD, D. M., ASHCROFT, N. W.: Phys. Rev. **B25**, 2532 (1982).
WÜNSCHE, H. J.: Phys. stat. sol. (b) **92**, 379 (1979); Dichtefunktionaltheorie gebundener Mehrexzitonenkomplexe in Halbleitern. Dissertation B, Humboldt-Universität Berlin, 1983.
YANG, C. N.: Rev. Mod. Phys. **34**, 694 (1962)
YUKHNOVSKII, I. R.: Eksp. Teor. Fiz. **27**, 690 (1954); **34**, 179 (1958); Ukrain. Fiz. Zh. **4**, 167 (1959); **7**, 267 (1962); **14**, 705 (1969).
YUKHNOVSKII, I. R., BLAZHIEVSKII, L. F., Ukrain, Fiz. Zh. **11** (1966) 936.
YUKHNOVSKII, I. R., GOLOVKO, M. F.: Ukrain. Fiz. Zh. **14**, 1119 (1969); **15**, 1996 (1970); **16**, 1517 (1971); **17**, 756 (1972).
YUKHNOVSKII, I. R., HETZHEIM, H.: Ukrain. Fiz. Zh. **13**, 881 (1968).
ZAIDI, H. R.: Phys. Rev. **173**, 123, 132 (1968).
ZAMALIN, V. M., NORMAN, G. E., FILINOV, V. S.: Monte Carlo Method in Statistical Thermodynamics (in Russian). Nauka, Moscow 1977.
ZELDOVICH, I. B., NOVIKOV, I. D.: Structure and Evolution of the Cosmos (in Russian). Nauka, Moscow 1975.
ZELDOVICH, I. B., NOVIKOV, I. D.: Relativistic Astrophysics (in Russian). Nauka, Moscow 1967. Structure and Evolution of the Universe (in Russian). Nauka, Moscow 1976.
ZELENER, B. V., NORMAN, G. E., FILINOV, V. S., Teplofyz. Vys. Temp. **10**, 1160 (1972); **11**, 922 (1973); **12**, 267 (1974); **13**, 712, 913 (1975).
ZHDANOV, V. M.: Phenomena of Transport in Multicomponent Plasmas (in Russian). Energoizdat, Moscow 1982.
ZIMAN, J. M.: Phil. Mag. **6**, 1013 (1961).
ZIMAN, J. M.: Models of Disorder. Univ. Press, Cambridge 1979. Russ. transl. Mir, Moscow 1982; Prinzipien der Festkörpertheorie. Akademie-Verlag, Berlin 1974. (Engl. original Cambridge 1964. Russ. transl. Moscow 1966.)
ZIMDAHL, W., EBELING, W.: Ann. Physik **34**, 9 (1977).
ZIMMERMANN, R.: Phys. stat. sol. (b) **48**, 603 (1971); **76**, 191 (1976); Wiss. Z. Pädagogische Hochschule Güstrow, DDR, **21**, 7 (1983).
ZIMMERMANN, R., RÖSLER, M.: Solid State Commun. **25**, 651 (1978); Private communication 1982.
ZIMMERMANN, R., RÖSLER, M., RICHERT, W.: Phys. stat. sol. (1984).
ZIMMERMANN, R., KILIMANN, K., KRAEFT, W.-D., KREMP, D., RÖPKE, G.: Phys. stat. sol. (b) **90**, 175 (1978).
ZUBAREV, D. N.: Usp. Fiz. Nauk **71**, 71 (1960); Fortschritte der Physik **18**, 125 (1970). Nonequilibrium Statistical Thermodynamics (in Russian). Nauka, Moscow 1971. German transl. Akademie-Verlag, Berlin 1975. English transl. Plenum Press, New York London 1974; In: Itogi Nauki i Tekhniki (USSR), vol. 15, p. 131. VINITI, Moscow 1980.
ZUBAREV, D. N., NOVIKOV, M. YU.: Fortschritte der Physik **21**, 703 (1973).

10. Subject Index

Bei Fragen zur Produktsicherheit wenden Sie sich bitte an:
If you have any questions regarding product safety,
please contact:

Walter de Gruyter GmbH
Genthiner Straße 13
10785 Berlin
productsafety@degruyterbrill.com